Recursive Identification and Parameter Estimation

Recursive Identification and Parameter Estimation

Han-Fu Chen
Wenxiao Zhao

CRC Press
Taylor & Francis Group
Boca Raton London New York

CRC Press is an imprint of the
Taylor & Francis Group, an **informa** business

CRC Press
Taylor & Francis Group
6000 Broken Sound Parkway NW, Suite 300
Boca Raton, FL 33487-2742

Printed on acid-free paper
Version Date: 20140418

International Standard Book Number-13: 978-1-4665-6884-6 (Hardback)

Library of Congress Cataloging-in-Publication Data

Chen, Hanfu.
Recursive identification and parameter estimation / authors, Han-Fu Chen, Wenxiao Zhao.
pages cm
Includes bibliographical references and index.
ISBN 978-1-4665-6884-6 (hardback)
1. Systems engineering--Mathematics. 2. Parameter estimation. 3. Recursive functions. I. Zhao, Wenxiao. II. Title.

TA168.C476 2014
620'.004201519536--dc23 2014013190

Visit the Taylor & Francis Web site at
http://www.taylorandfrancis.com

and the CRC Press Web site at
http://www.crcpress.com

Contents

Symbols

Ω	basic space
ω	element, or sample
M^T	transpose of matrix M
A, B	sets in Ω
$\emptyset$	empty set
$\sigma(\xi)$	σ-algebra generated by random variable ξ
$\mathbb{P}$	probability measure
$A \bigcup B$	union of sets A and B
$A \bigcap B$	intersection of sets A and B
$A \Delta B$	symmetric difference of sets A and B
I_A	indicator function of set A
$E\xi$	mathematical expectation of random variable ξ
$\lambda_{\max}(X)$	maximal eigenvalue of matrix X
$\lambda_{\min}(X)$	minimal eigenvalue of matrix X
$\|X\|$	norm of matrix X defined as $(\lambda_{\max}(X^T X))^{\frac{1}{2}}$
$\xrightarrow{\text{a.s.}}$	almost sure convergence
$\xrightarrow{\mathbb{P}}$	convergence in probability
$\xrightarrow{w}$	weak convergence
$\mathbb{R}$	real line including $+\infty$ and $-\infty$
$\mathscr{B}$	1-dimensional Borel σ-algebra
$m(\xi)$	median of random variable ξ
$\|\nu\|_{\text{var}}$	total variation norm of signed measure ν
$a_N \sim b_N$	$c_1 b_N \leq a_N \leq c_2 b_N$ $\forall\, N \geq 1$ for some positive constants c_1 and c_2
$\otimes$	Kronecker product
$\det A$	determinant of matrix A
$\text{Adj}\, A$	adjoint of matrix A
$n!$	factorial of n
C_n^k	combinatorial number of k from n: $C_n^k = \frac{n!}{k!(n-k)!}$
$\text{Re}\{a\}$	real part of complex number a
$\text{Im}\{a\}$	imaginary part of complex number a
M^+	pseudo-inverse of matrix M
$[a]$	integer part of real number a

Abbreviations

AR	autoregression
ARMA	autoregressive and moving average
ARMAX	autoregressive and moving average with exogenous input
ARX	autoregression with exogenous input
DRPA	distributed randomized PageRank algorithm
EIV	errors-in-variables
ELS	extended least squares
GCT	general convergence theorem
LS	least squares
MA	moving average
MFD	matrix fraction description
MIMO	multi-input multi-output
NARX	nonlinear autoregression with exogenous input
PCA	principal component analysis
PE	persistent excitation
RM	Robbins–Monro
SA	stochastic approximation
SAAWET	stochastic approximation algorithm with expanding truncations
SISO	single-input single-output
SPR	strictly positive realness
a.s.	almost surely
iff	if and only if
iid	independent and identically distributed
i.o.	infinitely often
mds	martingale difference sequence

Preface

To build a mathematical model based on the observed data is a common task for systems in diverse areas including not only engineering systems, but also physical systems, social systems, biological systems and others. It may happen that there is no a priori knowledge concerning the system under consideration; one then faces the "black box" problem. In such a situation the "black box" is usually approximated by a linear or nonlinear system, which is selected by minimizing a performance index depending on the approximation error. However, in many cases, from physical or mechanical thinking or from human experiences one may have some a priori knowledge about the system. For example, it may be known that the data are statically linearly related, or they are related by a linear dynamic system but with unknown coefficients and orders, or they can be fit into a certain type of nonlinear systems, etc. Then, the problem of building a mathematical model is reduced to fixing the uncertainties contained in the a priori knowledge by using the observed data, e.g., estimating coefficients and orders of a linear system or identifying the nonlinear system on the basis of the data. So, from a practical application point of view, when building a mathematical model, one first has to fix the model class the system belongs to on the basis of available information. After this, one may apply an appropriate identification method proposed by theoreticians to perform the task.

Therefore, for control theorists the topic of system identification consists of doing the following things: 1) Assume the model class is known. It is required to design an appropriate algorithm to identify the system from the given class by using the available data. For example, if the class of linear stochastic systems is assumed, then one has to propose an identification algorithm to estimate the unknown coefficients and orders of the system on the basis of the input–output data of the system. 2) The control theorists then have to justify that the proposed algorithm works well in the sense that the estimates converge to the true ones as the data size increases, if the applied data are really generated by a system belonging to the assumed class. 3) It has also to be clarified what will happen if the data do not completely match the assumed model class, either because the data are corrupted by errors or because the true system is not exactly covered by the assumed class.

When the model class is parameterized, then the task of system identification consists of estimating parameters characterizing the system generating the data and also clarifying the properties of the derived estimates. If the data are generated by a system belonging to the class of linear stochastic systems, then the identification algorithm to be proposed should estimate the coefficients and orders of the system and also the covariance matrix of the system noise. Meanwhile, properties such as strong consistency, convergence rate and others of the estimates should be investigated. Even if the model class is not completely parameterized, for example, the class of Hammerstein systems, the class of Wiener systems, and the class of nonlinear ARX systems where each system in the class contains nonlinear functions, and the purpose of system identification includes identifying the nonlinear function $f(\cdot)$ concerned, the identification task can still be transformed to a parameter estimation problem. The obvious way is to parameterize $f(\cdot)$ by approximating it with a linear combination of basis functions with unknown coefficients, which are to be estimated. However, it may also be carried out in a nonparametric way. In fact, the value of $f(x)$ at any fixed x can be treated as a parameter to estimate, and then one can interpolate the obtained estimates for $f(x)$ at different x, and the resulting interpolating function may serve as the estimate of $f(\cdot)$.

As will be shown in the book, not only system identification but also many problems from diverse areas such as adaptive filtering and other problems from signal processing and communication, adaptive regulation and iterative learning control, principal component analysis, some problems connected with network systems such as consensus control of multi-agent systems, PageRank of web, and many others can be transformed to parameter estimation problems.

In mathematical statistics there are various types of parameter estimates whose behaviors basically depend on the statistical assumptions made on the data. In the present book, with the possible exception of Sections 3.1 and 3.2 in Chapter 3, the estimated parameter denoted by x^0 is treated as a root of a function $g(\cdot)$, called the regression function. It is clear that an infinite number of functions may serve as such a $g(\cdot)$, e.g., $g(x) = A(x - x^0)$, $g(x) = \sin \|x - x^0\|$, etc. Therefore, the original problem may be treated as root-seeking for a regression function $g(\cdot)$. Moreover, it is desired that root-seeking can be carried out in a recursive way in the sense that the $(k+1)$th estimate x_{k+1} for x^0 can easily be obtained from the previous estimate x_k by using the data O_{k+1} available at time $k+1$. It is important to note that any data O_{k+1} at time $k+1$ may be viewed as an observation on $g(x_k)$, because we can always write $O_{k+1} = g(x_k) + \varepsilon_{k+1}$, where $\varepsilon_{k+1} \triangleq O_{k+1} - g(x_k)$ is treated as the observation noise. It is understandable that the properties of $\{\varepsilon_k\}$ depend upon not only the uncertainties contained in $\{O_k\}$ but also the selection of $g(\cdot)$. So, it is hard to expect that $\{\varepsilon_k\}$ can satisfy any condition required by the convergence theorems for the classical root-seeking algorithms, say, for the Robbins–Monro (RM) algorithm. This is why a modified version of the RM algorithm is introduced, which, in fact, is a stochastic approximation algorithm with expanding truncations (SAAWET). It turns out that SAAWET works very well to deal with the parameter estimation problems transformed from various areas.

In Chapter 1, the basic concept of probability theory and some information from martingales, martingale difference sequences, Markov chains, mixing processes, and stationary processes are introduced in such a way that they are readable without referring to other sources and they are kept to the minimum needed for understanding the coming chapters. Except material easily available from textbooks, here most of the results are proved for those who are interested in mathematical derivatives. The proof is placed in Appendix A at the end of the book, but without it the rest of the book is still readable. In Chapter 2, root-seeking for functions is discussed, starting from the classical RM algorithm. Then, SAAWET is introduced, and its convergence and convergence rate are addressed in detail. This chapter provides the main tool used for system identification and parameter estimation problems to be presented in the subsequent chapters.

In Chapter 3, the ARMAX (autoregressive and moving average with exogenous input) system is recursively identified. Since ARMAX is linear with respect to its input, output, and driven noise, the conventional least squares (LS) or the extended LS (ELS) methods work well and the estimates are derived in a recursive way. However, for convergence of the estimates given by ELS a restrictive SPR (strictly positive realness) condition is required. After analyzing the estimation errors produced by the recursive LS and ELS algorithms in the first two sections, we then turn to the root-seeking approach to identifying ARMAX systems without requiring the SPR condition. Since the coefficients of the AR part and the correlation functions/the impulse responses of the system are connected by a linear algebraic equation via a Hankel matrix, it is of crucial importance to have the row-full-rank of the Hankel matrix appearing in the linear equation. This concerns the identifiability of the AR part and is discussed here in detail. Then the coefficients of ARMAX systems are recursively estimated by SAAWET, and the strong consistency of the estimates is established. As concerns the order estimation for ARMA, almost all existing methods are based on minimizing some information criteria, so they are nonrecursive. In this chapter a recursive order estimation method for ARMAX systems is presented and is proved to converge to the true orders as data size increases. The recursive and strongly consistent estimates are also derived for the case where both the input and output of the ARMAX systems are observed with additive errors, i.e., the errors-in-variables (EIV) case.

Chapter 4 discusses identification of nonlinear systems: Unlike other identification methods, all estimation algorithms given in this chapter are recursive. The following types of nonlinear systems are considered: 1) the Hammerstein system composed of a linear subsystem cascading with a static nonlinearity, which is located at the input side of the system; 2) the Wiener system being also a cascading system composed of a linear subsystem and a static nonlinearity but with the nonlinearity following the linear part; 3) the Wiener–Hammerstein system being a cascading system with the static nonlinearity sandwiched by two linear subsystems; 4) the EIV Hammerstein and EIV Wiener systems; 5) the nonlinear ARX (NARX) system defining the system output in such a way that the output nonlinearly depends on the finite number of system past inputs and outputs. For the linear subsystems of these systems, SAAWET is applied to estimate their coefficients and the strong consistency of estimates is proved. The nonlinearities in these systems including the nonlinear function defining the NARX system are estimated by SAAWET incorporated with kernel functions, and the strong consistency of the estimates is established as well.

Chapter 5 addresses the problems arising from different areas that are solved by SAAWET. We limit ourselves to present the most recent results including principal component analysis, consensus control of the multi-agent system, adaptive regulation for Hammerstein and Wiener systems, and PageRank of webs. As a matter of fact, the proposed approach has successfully solved many other problems such as adaptive filtering, blind identification, iterative learning control, adaptive stabilization and adaptive control, adaptive pole assignment, etc. We decided not to include all of them, because either they are not the newest results or some of them have been presented elsewhere.

Some information concerning the nonnegative matrices is provided in Appendix B, which is essentially used in Sections 5.2 and 5.4.

The book is written for students, researchers, and engineers working in systems and control, signal processing, communication, and mathematical statistics. The target the book aims at is not only to show the results themselves on system identification and parameter estimation presented in Chapters 3–5, but more importantly to demonstrate how to apply the proposed approach to solve problems from different areas.

Han-Fu Chen and Wenxiao Zhao

Acknowledgments

The support of the National Science Foundation of China, the National Center for Mathematics and Interdisciplinary Sciences, and the Key Laboratory of Systems and Control, Chinese Academy of Sciences are gratefully acknowledged. The authors would like to express their gratitude to Professor Haitao Fang and Dr. Biqiang Mu for their helpful discussions.

About the Authors

Having graduated from the Leningrad (St. Petersburg) State University, Han-Fu Chen joined the Institute of Mathematics, Chinese Academy of Sciences (CAS). Since 1979, he has been with the Institute of Systems Science, now a part of the Academy of Mathematics and Systems Science, CAS. He is a professor at the Key Laboratory of Systems and Control of CAS. His research interests are mainly in stochastic systems, including system identification, adaptive control, and stochastic approximation and its applications to systems, control, and signal processing. He has authored and coauthored more than 200 journal papers and 7 books.

Professor Chen served as an IFAC Council member (2002–2005), president of the Chinese Association of Automation (1993–2002), and a permanent member of the Council of the Chinese Mathematics Society (1991–1999).

He is an IEEE fellow, IFAC fellow, a member of TWAS, and a member of CAS.

Wenxiao Zhao earned his BSc degree from the Department of Mathematics, Shandong University, China in 2003 and a PhD degree from the Institute of Systems Science, AMSS, the Chinese Academy of Sciences (CAS) in 2008. After this he was a postdoctoral student at the Department of Automation, Tsinghua University. During this period he visited the University of Western Sydney, Australia, for nine months. Dr. Zhao then joined the Institute of Systems of Sciences, CAS in 2010. He now is with the Key Laboratory of Systems and Control, CAS as an associate professor. His research interests are in system identification, adaptive control, and system biology. He serves as the general secretary of the IEEE Control Systems Beijing Chapter and an associate editor of the *Journal of Systems Science and Mathematical Sciences*.

Chapter 1

Dependent Random Vectors

CONTENTS

In the convergence analysis of recursive estimation algorithms the noise properties play a crucial role. Depending on the problems under consideration, the noise may be with various properties such as mutually independent random vectors, martingales, martingale difference sequences, Markov chains, mixing sequences, stationary processes, etc. The noise may also be composed of a combination of such kind of processes. In order to understand the convergence analysis to be presented in the coming chapters, the properties of above-mentioned random sequences are described here. Without any attempt to give a complete theory, we restrict ourselves to present the theory at the level that is necessary for reading the book.

Prior to describing random sequences, we first introduce some basic concepts of probability theory. This mainly is to provide a unified framework of notations and language.

1.1 Some Concepts of Probability Theory

Denote by Ω the basic space. A point $\omega \in \Omega$ is called the *element*, or *sample*. Denote by A, B, or C, etc. the sets in Ω and by $\emptyset$ the empty set. The *complementary set* of A in Ω is denoted by $A^c \triangleq \{\omega \in \Omega : \omega \notin A\}$ and the *symmetric difference* of sets A and B by $A\Delta B \triangleq AB^c \cup BA^c$.

Definition 1.1.1 A set-class $\mathscr{F}$ *is called a σ-algebra or σ-field if the following conditions are satisfied:*

(i) $\Omega \in \mathscr{F}$;

(ii) *If* $A \in \mathscr{F}$, *then* $A^c \in \mathscr{F}$;

(iii) *If* $A_n \in \mathscr{F}, n = 1,2,3,\cdots$, *then* $\cup_{n=1}^{\infty} A_n \in \mathscr{F}$.

Let $\mathscr{F}_1$ be a σ-algebra and $\mathscr{F}_1 \subset \mathscr{F}$. Then $\mathscr{F}_1$ is called a sub-σ-algebra of $\mathscr{F}$.

A conclusion immediately follows from Definition 1.1.1 that the σ-algebra $\mathscr{F}$ is closed under countable intersection of sets, i.e., $\cap_{n=1}^{\infty} A_n \in \mathscr{F}$ if $A_n \in \mathscr{F}, n = 1,2,3,\cdots$. A set A belonging to $\mathscr{F}$ is also named as A being *measurable with respect to* $\mathscr{F}$ or $\mathscr{F}$*-measurable* and the pair $(\Omega, \mathscr{F})$ is called a *measurable space.*

Definition 1.1.2 *For a sequence* $\{A_n\}_{n\geq 1}$ *of sets, define*

$$\liminf_{n\to\infty} A_n \triangleq \bigcup_{n=1}^{\infty}\bigcap_{k=n}^{\infty} A_k, \tag{1.1.1}$$

$$\limsup_{n\to\infty} A_n \triangleq \bigcap_{n=1}^{\infty}\bigcup_{k=n}^{\infty} A_k. \tag{1.1.2}$$

By Definition 1.1.2, it is clear that

$$\liminf_{n\to\infty} A_n = \{\omega \in \Omega : \omega \in A_n \text{ for all but a finite number of indices } n\},$$

$$\limsup_{n\to\infty} A_n = \{\omega \in \Omega : \omega \in A_n, \text{ i.o.}\},$$

and $\liminf_{n\to\infty} A_n \subset \limsup_{n\to\infty} A_n$, where the abbreviation "i.o." is to designate "infinitely often."

If $\liminf_{n\to\infty} A_n = \limsup_{n\to\infty} A_n = A$, then A is called the *limit* of the sequence $\{A_n\}_{n\geq 1}$.

Definition 1.1.3 A set function ϕ on $\mathscr{F}$ is called σ*-additive* or *countably additive* if $\phi(\cup_{n=1}^{\infty} A_n) = \sum_{n=1}^{\infty} \phi(A_n)$ for any sequence of disjoint sets $\{A_n\}_{n\geq 1}$, i.e., $A_n \in \mathscr{F}, n \geq 1$ and $A_n \cap A_m = \emptyset$ for any $n \neq m$. A nonnegative σ-additive set function ϕ is called a *measure* if $\phi(A) \geq 0$ for any $A \in \mathscr{F}$ and $\phi(\emptyset) = 0$. If ϕ is a measure, then the triple $(\Omega, \mathscr{F}, \phi)$ is called a *measure space. A measure ϕ is said to be σ-finite if there is $\{\Omega_n\}_{n\geq 1} \subset \mathscr{F}$ such that $\Omega = \cup_{n=1}^{\infty} \Omega_n$ and $\phi(\Omega_n) < \infty$, $n \geq 1$. For two measures ϕ_1 and ϕ_2, ϕ_1 is said to be* absolutely continuous *with respect to ϕ_2 if $\phi_1(A) = 0$ whenever $\phi_2(A) = 0$, $A \in \mathscr{F}$. This is denoted by $\phi_1 \ll \phi_2$.*

By Definition 1.1.3, the σ-additive set function ϕ is allowed to take values $+\infty$ or $-\infty$. For the σ-additive function ϕ and any $A \in \mathscr{F}$, define

$$\phi_+(A) \triangleq \sup_{B\in\mathscr{F}, B\subset A} \phi(B), \quad \phi_-(A) \triangleq -\inf_{B\in\mathscr{F}, B\subset A} \phi(B). \tag{1.1.3}$$

Theorem 1.1.1 *(Jordan–Hahn Decomposition) For any σ-additive set function ϕ on $\mathscr{F}$, there exists a set $D \in \mathscr{F}$ such that*

$$\phi_+(A) = \phi(AD), \quad \phi_-(A) = -\phi(AD^c), \tag{1.1.4}$$

where both ϕ_+ and ϕ_- are measures on $\mathscr{F}$ and $\phi = \phi_+ - \phi_-$.

The measures ϕ_+, ϕ_-, and $\overline{\phi} \triangleq \phi_+ + \phi_-$ are called the *upper*, *lower*, and *total variation measures* of ϕ.

Definition 1.1.4 A measure $\mathbb{P}$ defined on $(\Omega, \mathscr{F})$ *is called a probability measure if* $\mathbb{P}(\Omega) = 1$. *The triple* $(\Omega, \mathscr{F}, \mathbb{P})$ *is called a basic probability space and* $\mathbb{P}(A)$ *is called the probability of set* A, $A \in \mathscr{F}$, *respectively.*

For a measure space $(\Omega, \mathscr{F}, \phi)$, the subsets of an $\mathscr{F}$-measurable set A with $\phi(A) = 0$ may not belong to $\mathscr{F}$. It is natural to add all subsets of sets with zero measure to the σ-algebra $\mathscr{F}$ and define their measures to equal zero. Mathematically, this is expressed as

$$\mathscr{F}^* \triangleq \{A\Delta M : A \in \mathscr{F}, M \subset N \in \mathscr{F}, \phi(N) = 0\}, \tag{1.1.5}$$

$$\phi^*(A\Delta M) \triangleq \phi(A). \tag{1.1.6}$$

It can be proved that $\mathscr{F}^*$ is a σ-algebra and ϕ^* is a measure on $(\Omega, \mathscr{F}^*)$. The triple $(\Omega, \mathscr{F}^*, \phi^*)$ is called the *completion* of $(\Omega, \mathscr{F}, \phi)$. In the sequel, we always assume that the measure space and the probability space are completed.

Example 1.1.1 Denote the real line by $\mathbb{R} \triangleq [-\infty, \infty]$. The σ-algebra $\mathscr{B}$ generated by the class of infinite intervals of the form $[-\infty, x)$ with $-\infty < x < +\infty$ is called the *Borel σ-algebra*, the sets in $\mathscr{B}$ are called the *Borel sets*, and the measurable space $(\mathbb{R}, \mathscr{B})$ is known as the 1-dimensional *Borel space*.

Theorem 1.1.2 *Any nondecreasing finite function $m(\cdot)$ on $(-\infty, \infty)$ determines a complete measure space $(\mathbb{R}, \mathscr{M}_m, \nu_m)$ with $\mathscr{M}_m$ being the completed σ-algebra generated by the set class $\{[t : m(t) \in [a,b)], \ a, b \in \mathbb{R}\}$, where*

$$\nu_m([a,b)) = m(b-) - m(a-), \quad -\infty < a \le b < +\infty,$$
$$\nu_m(\{\infty\}) = \nu_m(\{-\infty\}) = 0,$$

where $m(x-)$ denotes the left limit of $m(\cdot)$ at x.

Definition 1.1.5 The measure ν_m defined by Theorem 1.1.2 is called the *Lebesgue Stieljies measure* determined by m and the complete measure space $(\mathbb{R}, \mathscr{M}_m, \nu_m)$ is named as the *Lebesgue–Stieljies measure space* determined by m. If $m(x) = x$ $\forall x \in (-\infty, \infty)$, then $\nu = \nu_m$ is referred to the *Lebesgue measure* and the sets in $\mathscr{M} \triangleq \mathscr{M}_m$ are called the *Lebesgue measurable sets*.

Definition 1.1.6 *For measurable spaces $(\Omega_i, \mathscr{F}_i)$, $i = 1, \cdots, n$, define*

$$\prod_{i=1}^{n} A_i \triangleq \{(\omega_1, \cdots, \omega_n) : \omega_i \in A_i,\ i = 1, \cdots, n\},\ A_i \in \mathscr{F}_i,$$

$$\prod_{i=1}^{n} \mathscr{F}_i \triangleq \sigma\left(\left\{\prod_{i=1}^{n} A_i : A_i \in \mathscr{F}_i,\ i = 1, \cdots, n\right\}\right),$$

$$\prod_{i=1}^{n} (\Omega_i, \mathscr{F}_i) \triangleq \left(\prod_{i=1}^{n} \Omega_i, \prod_{i=1}^{n} \mathscr{F}_i\right),$$

where $\prod_{i=1}^{n} \Omega_i$ is called the product space, $\prod_{i=1}^{n} (\Omega_i, \mathscr{F}_i)$ the n-dimensional product measurable space, and $\prod_{i=1}^{n} \mathscr{F}_i$ is the product σ-algebra.

Theorem 1.1.3 *Let $(\Omega_i, \mathscr{F}_i, \phi_i)$, $i = 1, 2$ be two σ-finite measure spaces and let $\Omega = \prod_{i=1}^{2} \Omega_i$, $\mathscr{F} = \prod_{i=1}^{2} \mathscr{F}_i$ be the product measurable spaces. Then there exists a σ-finite measure ϕ on $(\Omega, \mathscr{F})$ for which*

$$\phi(A_1 \times A_2) = \phi(A_1) \cdot \phi(A_2),\ \forall A_i \in \mathscr{F}_i,\ i = 1, 2.$$

The σ-finite measure ϕ on $(\Omega, \mathscr{F})$ is called the product measure of ϕ_1 and ϕ_2 and is often denoted by $\phi = \phi_1 \times \phi_2$.

Definition 1.1.7 *A real function $\xi = \xi(\omega)$ defined on $(\Omega, \mathscr{F})$ is called a random variable if it is $\mathscr{F}$-measurable, i.e., $\{\omega \in \Omega : \xi(\omega) \in A\} \in \mathscr{F}$ for any $A \in \mathscr{B}$ and if it takes finite value almost surely, i.e., $\mathbb{P}(|\xi(\omega)| < \infty) = 1$. The distribution function $F_\xi(\cdot)$ of a random variable ξ is defined as*

$$F_\xi(x) \triangleq \mathbb{P}(\omega : \xi(\omega) < x)\ \ \forall x \in \mathbb{R}. \tag{1.1.7}$$

A real number $m(\xi)$ is called the median of the random variable ξ if $\mathbb{P}(\xi \le m(\xi)) \ge \frac{1}{2} \le \mathbb{P}(\xi \ge m(\xi))$.

From the above definition we see that the random variables are in fact the measurable functions from $(\Omega, \mathscr{F})$ to $(\mathbb{R}, \mathscr{B})$. A measurable function f from $(\mathbb{R}, \mathscr{B})$ to $(\mathbb{R}, \mathscr{B})$ is usually called the *Borel measurable function*.

It can be shown that the distribution function is nondecreasing and left-continuous. If F_ξ is differentiable, then its derivative $f_\xi(x) \triangleq \mathrm{d}F_\xi(x)/\mathrm{d}x$ is called the *probability density function* of ξ, or, simply, density function.

The n-dimensional vector $\xi = [\xi_1 \cdots \xi_n]^T$ is called a *random vector* if ξ_i is a random variable for each $i = 1, \cdots, n$. The n-dimensional distribution function and density function are, respectively, defined by

$$F_\xi(x_1, \cdots, x_n) \triangleq \mathbb{P}(\xi_1 < x_1, \cdots, \xi_n < x_n), \tag{1.1.8}$$

and

$$F_\xi(x_1, \cdots, x_n) \triangleq \int_{-\infty}^{x_1} \cdots \int_{-\infty}^{x_n} f_\xi(t_1, \cdots, t_n) \mathrm{d}t_1 \cdots \mathrm{d}t_n. \tag{1.1.9}$$

Example 1.1.2 The well-known Gaussian density function is given by

$$\frac{1}{\sqrt{2\pi}\sigma}\exp\left\{-\frac{1}{2}\left(\frac{x-\mu}{\sigma}\right)^2\right\},\ x\in\mathbb{R} \tag{1.1.10}$$

for fixed scalars μ and σ with $\sigma > 0$. In the n-dimensional case the Gaussian density function is defined as

$$\frac{1}{(2\pi)^{\frac{n}{2}}(\det\Sigma)^{\frac{1}{2}}}\exp\left\{-\frac{1}{2}(x-\mu)^T\Sigma^{-1}(x-\mu)\right\},\ x\in\mathbb{R}^n \tag{1.1.11}$$

with fixed $\mu \in \mathbb{R}^n$ and positive definite $\Sigma \in \mathbb{R}^{n\times n}$.

We now introduce the concept of *mathematical expectation* of a random variable ξ, denoted by $E\xi \triangleq \int_\Omega \xi \mathrm{d}\mathbb{P}$, where $\int_\Omega \xi \mathrm{d}\mathbb{P}$ denotes the integral of ξ with respect to probability measure $\mathbb{P}$. We first consider the nonnegative ξ.

For each $n \geq 1$, consider ω-sets $A_{ni} = \left\{\omega \in \Omega : \frac{i}{2^n} < \xi \leq \frac{i+1}{2^n}\right\}$, $i = 0, \cdots, n2^n - 1$ and the sequence

$$S_n = \sum_{i=0}^{n2^n-1} \frac{i}{2^n}\mathbb{P}(A_{ni}) + n\mathbb{P}(\xi > n). \tag{1.1.12}$$

It can be shown that S_n converges as $n \to \infty$ and the limit is defined as the mathematical expectation of ξ, i.e., $E\xi = \int_\Omega \xi \mathrm{d}\mathbb{P} \triangleq \lim_{n\to\infty} S_n$.

In the following, for $A \in \mathscr{F}$, by $\int_A \xi \mathrm{d}\mathbb{P}$ we mean $\int_\Omega \xi I_A \mathrm{d}\mathbb{P}$ where I_A is the *indicator* of A:

$$I_A(\omega) \triangleq \begin{cases} 1, \text{ if } \omega \in A, \\ 0, \text{ otherwise.} \end{cases}$$

For a random variable ξ, define the nonnegative random variables

$$\xi_+ \triangleq \max\{\xi, 0\},\quad \xi_- \triangleq \max\{-\xi, 0\}. \tag{1.1.13}$$

It is clear that $\xi = \xi_+ - \xi_-$. The mathematical expectation of ξ is defined by $E\xi \triangleq E\xi_+ - E\xi_-$ if at least one of $E\xi_+$ and $E\xi_-$ is finite. If both $E\xi_+$ and $E\xi_-$ are finite, then ξ is called *integrable*. If both $E\xi_+$ and $E\xi_-$ are infinite, then we say that the mathematical expectation does not exist for ξ.

Theorem 1.1.4 *(Fubini) If $(\Omega, \mathscr{F}, \mathbb{P})$ is the product space of two probability spaces $(\Omega_i, \mathscr{F}_i, \mathbb{P}_i)$, $i = 1,2$ and $X = X(\omega_1, \omega_2)$ is a random variable on $(\Omega, \mathscr{F})$ for which the mathematical expectation exists, then*

$$\int_\Omega X\mathrm{d}\mathbb{P} = \int_{\Omega_1}\mathrm{d}\mathbb{P}_1\int_{\Omega_2}X(\omega_1,\omega_2)\mathrm{d}\mathbb{P}_2 = \int_{\Omega_2}\mathrm{d}\mathbb{P}_2\int_{\Omega_1}X(\omega_1,\omega_2)\mathrm{d}\mathbb{P}_1. \tag{1.1.14}$$

We now list some basic inequalities related to the mathematical expectation.

Chebyshev Inequality. For any $\varepsilon > 0$,

$$P(|\xi| > \varepsilon) \le \frac{1}{\varepsilon} E|\xi|. \tag{1.1.15}$$

Jensen Inequality. Let the Borel measurable function g be convex, i.e., $g(\theta_1 x_1 + \theta_2 x_2) \le \theta_1 g(x_1) + \theta_2 g(x_2)$ for any $x_1, x_2 \in \mathbb{R}$ and any $\theta_1 \ge 0, \theta_2 \ge 0$ with $\theta_1 + \theta_2 = 1$. If ξ is integrable, then

$$g(E\xi) \le Eg(\xi). \tag{1.1.16}$$

Lyapunov Inequality. For any $0 < s < t$,

$$(E|\xi|^s)^{\frac{1}{s}} \le (E|\xi|^t)^{\frac{1}{t}}. \tag{1.1.17}$$

Hölder Inequality. Let $1 < p < \infty$, $1 < q < \infty$, and $\frac{1}{p} + \frac{1}{q} = 1$. For random variables ξ and η with $E|\xi|^p < \infty$ and $E|\eta|^q < \infty$, it holds that

$$E|\xi\eta| \le (E|\xi|^p)^{\frac{1}{p}} (E|\eta|^q)^{\frac{1}{q}}. \tag{1.1.18}$$

In the case $p = q = 2$, the Hölder inequality is also named as the Schwarz inequality.

Minkowski Inequality. If $E|\xi|^p < \infty$ and $E|\eta|^p < \infty$ for some $p \ge 1$, then

$$(E|\xi + \eta|^p)^{\frac{1}{p}} \le (E|\xi|^p)^{\frac{1}{p}} + (E|\eta|^p)^{\frac{1}{p}}. \tag{1.1.19}$$

$\mathbf{C_r}$-Inequality.

$$\left(\sum_{i=1}^{n} |\xi_i|\right)^r \le C_r \sum_{i=1}^{n} |\xi_i|^r, \tag{1.1.20}$$

where $C_r = \begin{cases} 1, & \text{if } r < 1, \\ n^{r-1}, & \text{if } r \ge 1. \end{cases}$

We now introduce the concepts of convergence of random variables.

Definition 1.1.8 *Let ξ and $\{\xi_n\}_{n\ge1}$ be random variables. The sequence $\{\xi_n\}_{n\ge1}$ is said to converge to ξ with probability one or almost surely, denoted by $\xi_n \xrightarrow[n\to\infty]{\text{a.s.}} \xi$, if $\mathbb{P}\left(\omega : \xi_n(\omega) \underset{n\to\infty}{\longrightarrow} \xi(\omega)\right) = 1$. $\{\xi_n\}_{n\ge1}$ is said to converge to ξ in probability, denoted by $\xi_n \xrightarrow[n\to\infty]{\mathbb{P}} \xi$, if $\mathbb{P}(|\xi_n - \xi| > \varepsilon) = o(1)$ for any $\varepsilon > 0$. $\{\xi_n\}_{n\ge1}$ is said to weakly converge or to converge in distribution to ξ, denoted by $\xi_n \xrightarrow[n\to\infty]{w} \xi$, if $F_{\xi_n}(x) \underset{n\to\infty}{\longrightarrow} F_\xi(x)$ at any x where $F_\xi(x)$ is continuous. $\{\xi_n\}_{n\ge1}$ is said to converge to ξ in the mean square sense if $E|\xi_n - \xi|^2 \underset{n\to\infty}{\longrightarrow} 0$.*

In what follows iff is the abbreviation of "if and only if." The relationship between various types of convergence is demonstrated by the following theorem.

Theorem 1.1.5 *The following results on convergence of random variables take place:*

(i) *If* $\xi_n \xrightarrow[n\to\infty]{\text{a.s.}} \xi$, *then* $\xi_n \xrightarrow[n\to\infty]{\mathbb{P}} \xi$.

(ii) *If* $\xi_n \xrightarrow[n\to\infty]{\mathbb{P}} \xi$, *then* $\xi_n \xrightarrow[n\to\infty]{w} \xi$.

(iii) *If* $E|\xi_n - \xi|^2 \xrightarrow[n\to\infty]{} 0$, *then* $\xi_n \xrightarrow[n\to\infty]{\mathbb{P}} \xi$.

(iv) $\xi_n \xrightarrow[n\to\infty]{\text{a.s.}} \xi$ *iff* $\sup_{j\geq n}|\xi_j - \xi| \xrightarrow[n\to\infty]{\mathbb{P}} 0$ *iff* $\sup_{m>n}|\xi_m - \xi_n| \xrightarrow[n\to\infty]{\mathbb{P}} 0$.

(v) $\xi_n \xrightarrow[n\to\infty]{\mathbb{P}} \xi$ *iff* $\sup_{m>n}\mathbb{P}(|\xi_m - \xi_n| > \varepsilon) = o(1)\ \forall\, \varepsilon > 0$.

The following theorems concern the interchangeability of taking expectation and taking limit.

Theorem 1.1.6 *(Monotone Convergence Theorem) If* $\{\xi_n\}_{n\geq 1}$ *nondecreasingly converges to* ξ *with probability one and* $\xi_n \geq \eta$ *a.s. for some random variable* η *with* $E\eta_- < \infty$, *then* $\lim_{n\to\infty} E\xi_n = E\xi$.

Theorem 1.1.7 *(Fatou Lemma) If* $\xi_n \geq \eta$, $n = 1,2,3,\cdots$ *for some random variable* η *with* $E\eta_- < \infty$, *then* $E\liminf_{n\to\infty}\xi_n \leq \liminf_{n\to\infty} E\xi_n$.

Theorem 1.1.8 *(Dominated Convergence Theorem) If* $|\xi_n| \leq \eta$, $n = 1,2,3,\cdots$, $E\eta < \infty$, *and* $\xi_n \xrightarrow[n\to\infty]{\mathbb{P}} \xi$, *then* $E|\xi| < \infty$, $E|\xi_n - \xi| \xrightarrow[n\to\infty]{} 0$, *and* $E\xi_n \xrightarrow[n\to\infty]{} E\xi$.

We now introduce an important concept in probability theory, *the conditional expectation.*

Let A and B be $\mathscr{F}$-measurable sets. The conditional probability of set A given B is defined by $\mathbb{P}(A|B) = \frac{\mathbb{P}(AB)}{\mathbb{P}(B)}$ whenever $\mathbb{P}(B) \neq 0$. Intuitively, $\mathbb{P}(A|B)$ describes the probability of set A with *a priori* information of set B. We now generalize this elementary concept from conditioned on a set to conditioned on a σ-algebra.

Theorem 1.1.9 *(Radon–Nikodym Theorem) Let* $\mathscr{F}_1$ *be a sub-*σ*-algebra of* $\mathscr{F}$. *For any random variable* ξ *with* $E\xi$ *well defined, there is a unique (up to sets of probability zero)* $\mathscr{F}_1$*-measurable random variable* η *such that*

$$\int_A \eta \, d\mathbb{P} = \int_A \xi \, d\mathbb{P} \quad \forall\, A \in \mathscr{F}_1. \tag{1.1.21}$$

Definition 1.1.9 *The* $\mathscr{F}_1$*-measurable function* η *defined by (1.1.21) is called the conditional expectation of* ξ *given* $\mathscr{F}_1$, *denoted by* $E(\xi|\mathscr{F}_1)$. *In particular, if* $\xi = I_B$, *then* η *is the conditional probability of* B *given* $\mathscr{F}_1$ *and we write it as* $\mathbb{P}(B|\mathscr{F}_1)$.

For random variables ξ and ζ, the conditional expectation of ξ given ζ is defined by $E(\xi|\zeta) \triangleq E(\xi|\sigma(\zeta))$, where $\sigma(\zeta)$ is the σ-algebra generated by ζ, i.e., $\sigma(\zeta)$ being the smallest σ-algebra containing all sets of the form $\{\omega : \zeta(\omega) \in B\}$, $B \in \mathscr{B}$.

It is worth noting that if $\xi = I_B$, $\zeta = I_D$, then $\sigma_\zeta = \{\emptyset, \Omega, D, D^c\}$ and $\mathbb{P}(B|\zeta) = \mathbb{P}(B|D)I_D + \mathbb{P}(B|D^c)I_{D^c}$, where $\mathbb{P}(B|D)$ and $\mathbb{P}(B|D^c)$ are defined for the elementary case.

For the conditional expectation, the following properties take place.

(i) $E[a\xi + b\eta|\mathscr{F}_1] = aE[\xi|\mathscr{F}_1] + bE[\eta|\mathscr{F}_1]$ for any constants a and b.

(ii) If $\xi \leq \eta$, then $E[\xi|\mathscr{F}_1] \leq E[\eta|\mathscr{F}_1]$.

(iii) There exists a Borel-measurable function f such that $E[\xi|\eta] = f(\eta)$ a.s.

(iv) $E[E(\xi|\eta)] = E\xi$.

(v) $E[\eta\xi|\mathscr{F}_1] = \eta E[\xi|\mathscr{F}_1]$ a.s. for any $\mathscr{F}_1$-measurable η.

(vi) If σ-algebras $\mathscr{F}_1 \subset \mathscr{F}_2 \subset \mathscr{F}$, then $E[E(\xi|\mathscr{F}_2)|\mathscr{F}_1] = E[\xi|\mathscr{F}_1]$.

(vii) If $\mathscr{F}_1 = \{\Omega, \emptyset\}$, then $E[\xi|\mathscr{F}_1] = E\xi$ a.s.

For a sub-σ-algebra $\mathscr{F}_1$ of $\mathscr{F}$, the Chebyshev inequality, Jensen inequality, Lyapunov inequality, Hölder inequality, Minkowski inequality, and C_r-inequality also hold if the expectation operator $E(\cdot)$ is replaced by $E(\cdot|\mathscr{F}_1)$. The monotone convergence theorem, Fatou lemma, and the dominated convergence theorem also remain valid by replacing $E(\cdot)$ with $E(\cdot|\mathscr{F}_1)$. We just need to note that in the case of conditional expectation these inequalities and convergence theorems hold a.s.

1.2 Independent Random Variables, Martingales, and Martingale Difference Sequences

Throughout the book the basic probability space is always denoted by $(\Omega, \mathscr{F}, \mathbb{P})$.

Definition 1.2.1 *The events $A_i \in \mathscr{F}$, $i = 1, \cdots, n$ are said to be mutually independent if $\mathbb{P}\left(\cap_{j=1}^m A_{i_j}\right) = \prod_{j=1}^m \mathbb{P}(A_{i_j})$ for any subset $[i_1 < \cdots < i_m] \subset [1, \cdots, n]$. The σ-algebras $\mathscr{F}_i \subset \mathscr{F}$, $i = 1, \cdots, n$ are said to be mutually independent if $\mathbb{P}\left(\cap_{j=1}^m A_{i_j}\right) = \prod_{j=1}^m \mathbb{P}(A_{i_j})$ for any $A_{i_j} \in \mathscr{F}_{i_j}$, $j = 1, \cdots, m$ with $[i_1 < \cdots < i_m]$ being any subset of $[1, \cdots, n]$. The random variables $\{\xi_1, \cdots, \xi_n\}$ are called mutually independent if the σ-algebras $\sigma(\xi_i)$ generated by ξ_i, $i = 1, \cdots, n$ are mutually independent. Let $\{\xi_i\}_{i \geq 1}$ be a sequence of random variables. $\{\xi_i\}_{i \geq 1}$ is called mutually independent if for any $n \geq 1$ and any set of indices $\{i_1, \cdots, i_n\}$, the random variables $\{\xi_{i_k}\}_{k=1}^n$ are mutually independent.*

A sequence of random variables $\{\xi_k\}_{k \geq 1}$ is called *independent and identically distributed* (iid) if $\{\xi_k,\ k \geq 1\}$ are mutually independent with the same distribution function.

Definition 1.2.2 *The* tail σ-algebra *of a sequence* $\{\xi_k\}_{k\geq 1}$ *is* $\bigcap_{k=1}^{\infty}\sigma\{\xi_j, j\geq k\}$. *The sets of the tail σ-algebra are called* tail events *and the random variables measurable with respect to the tail σ-algebra are called* tail variables.

Theorem 1.2.1 *(Kolmogorov Zero–One Law) Tail events of an iid sequence* $\{\xi_k\}_{k\geq 1}$ *have probabilities either zero or one.*

Proof. See Appendix A. □

Theorem 1.2.2 *Let $f(x,y)$ be a measurable function defined on $(\mathbb{R}^l \times \mathbb{R}^m, \mathscr{B}^l \times \mathscr{B}^m)$. If the l-dimensional random vector ξ is independent of the m-dimensional random vector η and $Ef(\xi,\eta)$ exists, then*

$$E[f(\xi,\eta)|\sigma(\xi)] = g(\xi) \text{ a.s.} \tag{1.2.1}$$

where

$$g(x) = \begin{cases} Ef(x,\eta), & \text{if } Ef(x,\eta) \text{ exists,} \\ 0, & \text{otherwise.} \end{cases}$$

Proof. See Appendix A. □

Theorem 1.2.3 *(Kolmogorov Three Series Theorem) Assume the random variables $\{\xi_k\}_{k\geq 1}$ are mutually independent. The sum $\sum_{k=1}^{\infty}\xi_k$ converges almost surely iff the following three series converge:*

$$\sum_{k=1}^{\infty}\mathbb{P}(|\xi_k|>1)<\infty, \tag{1.2.2}$$

$$\sum_{k=1}^{\infty}E\xi_k'<\infty, \tag{1.2.3}$$

$$\sum_{k=1}^{\infty}E(\xi_k'-E\xi_k')^2<\infty, \tag{1.2.4}$$

where $\xi_k' \triangleq \xi_k I_{[|\xi_k|\leq 1]}$.

Theorem 1.2.4 *(Marcinkiewicz–Zygmund) Assume $\{\xi_k\}_{k\geq 1}$ are iid. Then*

$$\frac{\sum_{k=1}^{n}\xi_k - cn}{n^{\frac{1}{p}}} \xrightarrow[n\to\infty]{} 0 \text{ a.s. } p\in(0,2) \tag{1.2.5}$$

if and only if $E|\xi_k|^p<\infty$, where the constant $c=E\xi_k$ if $p\in[1,2)$, while c is arbitrary if $p\in(0,1)$.

As will be seen in the later chapters, the convergence analysis of many identification algorithms relies on the almost sure convergence of a series of random vectors, which may not satisfy the independent assumption and thus Theorem 1.2.3 is not directly applicable. Therefore, we need results on a.s. convergence for the sum of dependent random variables, which are summarized in what follows.

We now introduce the concept of *martingale*, which is a generalization of the sum of zero-mean mutually independent random variables, and is widely applied in diverse research areas.

Definition 1.2.3 *Let $\{\xi_k\}_{k\geq 1}$ be a sequence of random variables and $\{\mathscr{F}_k\}_{k\geq 1}$ be a sequence of nondecreasing σ-algebras. If ξ_k is $\mathscr{F}_k$-measurable for each $k \geq 1$, then we call $\{\xi_k, \mathscr{F}_k\}_{k\geq 1}$ an adapted process. An adapted process $\{\xi_k, \mathscr{F}_k\}_{k\geq 1}$ with $E|\xi_k| < \infty$ $\forall k \geq 1$ is called a submartingale if $E[\xi_n|\mathscr{F}_m] \geq \xi_m$ a.s. $\forall n \geq m$, a supermartingale if $E[\xi_n|\mathscr{F}_m] \leq \xi_m$ a.s. $\forall n \geq m$, and a martingale if it is both a supermartingale and a submartingale, i.e., $E[\xi_n|\mathscr{F}_m] = \xi_m$ a.s. $\forall n \geq m$. An adapted process $\{\xi_k, \mathscr{F}_k\}_{k\geq 1}$ is named as a martingale difference sequence (mds) if $E[\xi_{k+1}|\mathscr{F}_k] = 0$ a.s. $\forall k \geq 1$.*

Here we give a simple example to illustrate the definition introduced above.

Let $\{\eta_k\}_{k\geq 1}$ be a sequence of zero-mean mutually independent random variables. Define $\mathscr{F}_k = \sigma\{\eta_1, \cdots, \eta_k\}$ and $\zeta_k = \sum_{i=1}^{k} \eta_i$. Then $\{\eta_k, \mathscr{F}_k\}_{k\geq 1}$ is an mds and $\{\zeta_k, \mathscr{F}_k\}_{k\geq 1}$ is a martingale.

Theorem 1.2.5 *(Doob maximal inequality) Assume $\{\xi_k\}_{k\geq 1}$ is a nonnegative submartingale. Then for any $\lambda > 0$,*

$$\mathbb{P}\left\{\max_{1\leq j\leq n} \xi_j \geq \lambda\right\} \leq \frac{1}{\lambda} \int\limits_{\left[\max_{1\leq j\leq n} \xi_j \geq \lambda\right]} \xi_n d\mathbb{P}. \tag{1.2.6}$$

Further,

$$\left[E \max_{1\leq j\leq n} \xi_j^p\right]^{\frac{1}{p}} \leq \frac{p}{p-1} \left[E\xi_n^p\right]^{\frac{1}{p}} \tag{1.2.7}$$

if $E\xi_j^p < \infty$, $j = 1, \cdots, n$, where $1 < p < \infty$.

Proof. See Appendix A. □

Definition 1.2.4 *Let $\{\mathscr{F}_k\}_{k\geq 1}$ be a sequence of nondecreasing σ-algebras. A measurable function T taking values in $\{1, 2, 3, \cdots, \infty\}$ is called a stopping time with respect to $\{\mathscr{F}_k\}_{k\geq 1}$ if*

$$\{\omega : T(\omega) = k\} \in \mathscr{F}_k \ \ \forall k \geq 1. \tag{1.2.8}$$

In addition, if $\mathbb{P}(T = \infty) = 0$ then the stopping time T is said to be finite. A finite stopping time is also called a stopping rule or stopping variable.

Lemma 1.2.1 *Let $\{\xi_k, \mathscr{F}_k\}_{k\geq 1}$ be adapted, T a stopping time, and B a Borel set. Let T_B be the first time at which the process $\{\xi_k\}_{k\geq 1}$ hits the set B after time T, i.e.,*

$$T_B \triangleq \begin{cases} \inf\{k : k > T, \xi_k \in B\} \\ \infty, \ \text{if } \xi_k \notin B \text{ for all } k > T. \end{cases} \tag{1.2.9}$$

Then T_B is a stopping time.

Proof. The conclusion follows from the following expression:

$$[T_B = k] = \bigcup_{i=0}^{k-1} \{[T = i] \cap [\xi_{i+1} \notin B, \cdots, \xi_{k-1} \notin B, \xi_k \in B]\} \in \mathscr{F}_k \ \forall\, k \geq 1.$$

□

Let $\{\xi_k, \mathscr{F}_k\}$, $k = 1, \cdots, N$ be a submartingale. For a nonempty interval (a, b), define

$$T_0 \triangleq 0,$$

$$T_1 \triangleq \begin{cases} \min\{1 \leq k \leq N : \xi_k \leq a\}, \\ N+1, \ \text{if } \xi_k > a, \ k = 1, \cdots, N, \end{cases}$$

$$T_2 \triangleq \begin{cases} \min\{T_1 < k \leq N : \xi_k \geq b\}, \\ N+1, \ \text{if } \xi_k < b \ \forall k : T_1 < k \leq N, \text{ or } T_1 = N+1, \end{cases}$$

$$\vdots$$

$$T_{2m-1} \triangleq \begin{cases} \min\{T_{2m-2} < k \leq N : \xi_k \leq a\}, \\ N+1, \ \text{if } \xi_k > a \ \forall k : T_{2m-2} < k \leq N, \text{ or } T_{2m-2} = N+1, \end{cases}$$

$$T_{2m} \triangleq \begin{cases} \min\{T_{2m-1} < k \leq N : \xi_k \geq b\}, \\ N+1, \ \text{if } \xi_k < b \ \forall k : T_{2m-1} < k \leq N, \text{ or } T_{2m-1} = N+1. \end{cases}$$

The largest m for which $\xi_{2m} \geq b$ is called the number of *up-crossings* of the interval (a, b) by the submartingale $\{\xi_k, \mathscr{F}_k\}_{k=1}^N$ and is denoted by $\beta(a, b)$.

Theorem 1.2.6 *(Doob) For the submartingale $\{\xi_k, \mathscr{F}_k\}_{k=1}^N$ the following inequalities hold*

$$E\beta(a, b) \leq \frac{E(\xi_N - a)_+}{b - a} \leq \frac{E(\xi_N)_+ + |a|}{b - a} \tag{1.2.10}$$

where $(\xi_N)_+$ is defined by (1.1.13).

Proof. See Appendix A. □

Theorem 1.2.7 *(Doob) Let $\{\xi_k, \mathscr{F}_k\}_{k\geq 1}$ be a submartingale with $\sup_k E(\xi_k)_+ < \infty$. Then there is a random variable ξ with $E|\xi| < \infty$ such that*

$$\lim_{k\to\infty} \xi_k = \xi \ \text{ a.s.} \tag{1.2.11}$$

Proof. See Appendix A. □

Corollary 1.2.1 *If either (i) or (ii) are satisfied, where*

(i) $\{\xi_k, \mathscr{F}_k\}_{k\geq 1}$ *is a nonnegative supermartingale or nonpositive submartingale,*

(ii) $\{\xi_k, \mathscr{F}_k\}_{k\geq 1}$ *is a martingale with* $\sup_k E|\xi_k| < \infty$,

then

$$\lim_{k\to\infty} \xi_k = \xi \text{ a.s. and } E|\xi| < \infty.$$

We have presented some results on the a.s. convergence of some random series and sub- or super-martingales. However, a martingale or an mds may converge not on the whole space Ω but on its subset. In the following we present the set where a martingale or an mds converges.

Let $\{\xi_k, \mathscr{F}_k\}_{k\geq 0}$ with $\xi_k \in \mathbb{R}^m$ be an adapted sequence, and let G be a Borel set in $\mathscr{B}^m$. Then the first exit time T of $\{\xi_k\}_{k\geq 0}$ from G defined by

$$T = \begin{cases} \min\{k : \xi_k \notin G\} \\ \infty, \text{ if } \xi_k \in G, \ \forall\, k \geq 0 \end{cases}$$

is a stopping time. This is because $\{T = k\} = \{\xi_0 \in G, \xi_1 \in G, \cdots, \xi_{k-1} \in G, \xi_k \notin G\} \in \mathscr{F}_k$.

Lemma 1.2.2 *Let* $\{\xi_k, \mathscr{F}_k\}_{k\geq 0}$ *be a martingale (supermartingale, submartingale) and* T *be a stopping time. Then the process* $\{\xi_{T\wedge k}, \mathscr{F}_k\}_{k\geq 0}$ *is again a martingale (supermartingale, submartingale), where* $T \wedge k \triangleq \min(T, k)$.

Proof. See Appendix A. □

Theorem 1.2.8 *Let* $\{\xi_k, \mathscr{F}_k\}_{k\geq 0}$ *be a one-dimensional mds. Then as* $k \to \infty$, *the sequence* $\left\{\eta_k = \sum_{i=0}^k \xi_i\right\}_{k\geq 0}$ *converges on*

$$A \triangleq \left\{\omega : \sum_{k=1}^{\infty} E\left(\xi_k^2 | \mathscr{F}_{k-1}\right) < \infty\right\}. \tag{1.2.12}$$

Proof. See Appendix A. □

Theorem 1.2.9 *Let* $\{\xi_k, \mathscr{F}_k\}_{k\geq 0}$ *be an mds and let* $\eta_k = \sum_{i=0}^k \xi_i$, $k \geq 0$.

(i) If $E(\sup_k \xi_k)_+ < \infty$, *then* η_k *converges* a.s. *on* $A_1 \triangleq \{\omega : \sup_k \eta_k < \infty\}$.

(ii) If $E(\inf_k \xi_k)_- < \infty$, *then* η_k *converges* a.s. *on* $A_2 \triangleq \{\omega : \inf_k \eta_k > -\infty\}$.

Proof. It suffices to prove (i) since (ii) is reduced to (i) if ξ_k is replaced by $-\xi_k$. The detailed proof is given in Appendix A. □

Theorem 1.2.10 *(Borel–Cantelli–Lévy) Let $\{B_k\}_{k\geq 1}$ be a sequence of events, $B_k \in \mathscr{F}_k$. Then*

$$\sum_{k=1}^{\infty} I_{B_k} < \infty \tag{1.2.13}$$

iff

$$\sum_{k=1}^{\infty} \mathbb{P}(B_k|\mathscr{F}_{k-1}) < \infty, \tag{1.2.14}$$

or equivalently

$$\bigcap_{k=1}^{\infty}\bigcup_{i=k}^{\infty} B_i = \left\{\omega : \sum_{k=1}^{\infty} \mathbb{P}(B_k|\mathscr{F}_{k-1}) = \infty\right\}. \tag{1.2.15}$$

Proof. See Appendix A. □

Theorem 1.2.11 *(Borel–Cantelli) Let $\{B_k\}_{k\geq 1}$ be a sequence of events.*

(i) If $\sum_{k=1}^{\infty} \mathbb{P}(B_k) < \infty$, then the probability that B_k, $k \geq 1$ occur infinitely often is zero.

(ii) If B_k, $k \geq 1$ are mutually independent and $\sum_{k=1}^{\infty} \mathbb{P}(B_k) = \infty$, then $\mathbb{P}(B_k \text{ i.o.}) = 1$.

Proof. See Appendix A. □

Lemma 1.2.3 *Let $\{y_k, \mathscr{F}_k\}$ be an adapted process and $\{b_k\}$ a sequence of positive numbers. Then*

$$\left\{\omega : \sum_{k=1}^{\infty} y_k \text{ converges}\right\}\bigcap A = \left\{\omega : \sum_{k=1}^{\infty} y_k I_{[|y_k|\leq b_k]} \text{ converges}\right\}\bigcap A \tag{1.2.16}$$

where

$$A = \left\{\omega : \sum_{k=1}^{\infty} \mathbb{P}\big(|y_k| > b_k|\mathscr{F}_{k-1}\big) < \infty\right\}. \tag{1.2.17}$$

Proof. See Appendix A. □

The following result is a generalization of Theorem 1.2.3.

Theorem 1.2.12 *Denote by S the ω–set where the following three series converge:*

$$\sum_{k=1}^{\infty} \mathbb{P}(|y_k| > c|\mathscr{F}_{k-1}) < \infty, \tag{1.2.18}$$

$$\sum_{k=1}^{\infty} E\big(y_k I_{[|y_k|\leq c]}|\mathscr{F}_{k-1}\big) < \infty, \tag{1.2.19}$$

$$\sum_{k=1}^{\infty} \left\{E\big(y_k^2 I_{[|y_k|\leq c]}|\mathscr{F}_{k-1}\big) - \big(E\big(y_k I_{[|y_k|\leq c]}|\mathscr{F}_{k-1}\big)\big)^2\right\} < \infty, \tag{1.2.20}$$

where c is a positive constant. Then $\eta_k = \sum_{i=1}^{k} y_i$ *converges on S as* $k \to \infty$.

Proof. See Appendix A. □

Theorem 1.2.13 *Let* $\{\xi_k, \mathscr{F}_k\}$ *be an mds. Then* $\eta_k \triangleq \sum_{i=1}^{k} \xi_i$ *converges on*

$$A \triangleq \left\{ \omega : \sum_{k=1}^{\infty} E\left(|\xi_k|^p | \mathscr{F}_{k-1}\right) < \infty \right\} \text{ for } 0 < p \leq 2. \tag{1.2.21}$$

Theorem 1.2.13 generalizes Theorem 1.2.8. For its proof we refer to Appendix A.

For analyzing the asymptotical properties of stochastic systems we often need to know the behavior of partial sums of an mds with weights. In the sequel we introduce such a result to be frequently used in later chapters.

For a sequence of matrices $\{M_k\}_{k\geq 1}$ and a sequence of nondecreasing positive numbers $\{b_k\}_{k\geq 1}$, by $M_k = O(b_k)$ we mean

$$\limsup_{k\to\infty} \|M_k\|/b_k < \infty$$

and by $M_k = o(b_k)$,

$$\lim_{k\to\infty} \|M_k\|/b_k = 0.$$

We introduce a technical lemma, known as the *Kronecker lemma.*

Lemma 1.2.4 *(Kronecker lemma) If* $\{b_k\}_{k\geq 1}$ *is a sequence of positive numbers nondecreasingly diverging to infinity and if for a sequence of matrices* $\{M_k\}_{k\geq 1}$,

$$\sum_{k=1}^{\infty} \frac{1}{b_k} M_k < \infty, \tag{1.2.22}$$

then $\sum_{i=1}^{k} M_i = o(b_k)$.

Proof. See Appendix A. □

Based on Lemma 1.2.4, the following estimate for the weighted sum of an mds takes place.

Theorem 1.2.14 *Let* $\{\xi_k, \mathscr{F}_k\}$ *be an l-dimensional mds and* $\{M_k, \mathscr{F}_k\}$ *a matrix adapted process. If*

$$\sup_k E\left(\|\xi_{k+1}\|^\alpha | \mathscr{F}_k\right) \triangleq \sigma < \infty \text{ a.s.}$$

for some $\alpha \in (0,2]$, *then as* $k \to \infty$

$$\sum_{i=0}^{k} M_i \xi_{i+1} = O\left(s_k(\alpha)\left(\log(s_k^\alpha(\alpha) + e)\right)^{\frac{1}{\alpha}+\eta} \right) \text{ a.s. } \forall\, \eta > 0, \tag{1.2.23}$$

where $s_k(\alpha) = \left(\sum_{i=0}^{k} \|M_i\|^\alpha\right)^{\frac{1}{\alpha}}$.

Proof. See Appendix A. □

Example 1.2.1 In the one dimensional case if $\{\xi_k\}_{k\geq 1}$ is iid with $E\xi_k = 0$ and $E\xi_k^2 < \infty$, then from Theorem 1.2.14 we have $\sum_{i=1}^{k} \xi_i = O\big(k^{\frac{1}{2}}(\log k)^{\frac{1}{2}+\eta}\big)$ a.s. $\forall\ \eta > 0$. Thus the estimate given by Theorem 1.2.14 is not as sharp as those given by the law of the iterative logarithm but the conditions required here are much more general.

1.3 Markov Chains with State Space $(\mathbb{R}^m, \mathscr{B}^m)$

In systems and control, many dynamic systems are modeled as the discrete-time stochastic systems, which are closely connected with Markov chains. To see this, let us consider the following example.

Example 1.3.1 *The ARX system is given by*

$$y_{k+1} = a_1 y_k + \cdots + a_p y_{k+1-p} + b_1 u_k + \cdots + b_q u_{k+1-q} + \varepsilon_{k+1}, \tag{1.3.1}$$

where u_k and y_k are the system input and output, respectively, and $\{\varepsilon_k\}$ is a sequence of mutually independent zero-mean random variables. Define

$$\varphi_k = \begin{bmatrix} y_k \\ \vdots \\ y_{k+1-p} \\ u_k \\ \vdots \\ u_{k+1-q} \end{bmatrix}, A = \begin{bmatrix} a_1 & \cdots & \cdots & \cdots & a_p & b_1 & \cdots & \cdots & b_q \\ 1 & 0 & \cdots & & 0 & 0 & & & 0 \\ 0 & 1 & & & 0 & 0 & & & 0 \\ \vdots & \ddots & \ddots & \ddots & \vdots & \vdots & & & \vdots \\ 0 & & \ddots & 1 & 0 & 0 & \cdots & \cdots & 0 \\ 0 & & & 0 & 0 & 0 & \cdots & \cdots & 0 \\ \vdots & & \cdots & & 0 & 1 & \ddots & & \vdots \\ \vdots & & \cdots & & & \ddots & \ddots & \ddots & \vdots \\ 0 & \cdots & \cdots & \cdots & \cdots & \cdots & 0 & 1 & 0 \end{bmatrix}, \xi_k = \begin{bmatrix} \varepsilon_k \\ 0 \\ \vdots \\ 0 \\ u_k \\ 0 \\ \vdots \\ 0 \end{bmatrix}. \tag{1.3.2}$$

Then (1.3.1) can be rewritten as

$$\varphi_{k+1} = A\varphi_k + \xi_{k+1}, \tag{1.3.3}$$

and hence the regressor sequence $\{\varphi_k\}_{k\geq 0}$ is a Markov chain valued in $(\mathbb{R}^{p+q}, \mathscr{B}^{p+q})$ provided $\{\xi_k\}_{k\geq 0}$ is a sequence of mutually independent random vectors. Thus, in a certain sense, the analysis of the system (1.3.1) can resort to investigating the properties of the chain $\{\varphi_k\}_{k\geq 0}$.

In this section we introduce some results of vector-valued Markov chains. These results, extending the corresponding results of chains valued in a countable state space, will frequently be used in the later chapters to establish the a.s. convergence of recursive algorithms.

Assume $\{x_k\}_{k\geq 0}$ is a sequence of random vectors valued in $\mathbb{R}^m$. If

$$\mathbb{P}\{x_{k+1} \in A | x_k, \cdots, x_0\} = \mathbb{P}\{x_{k+1} \in A | x_k\} \tag{1.3.4}$$

for any $A \in \mathscr{B}^m$, then the sequence $\{x_k\}_{k\geq 0}$ is said to be a *Markov chain* with the state space $(\mathbb{R}^m, \mathscr{B}^m)$. Further, if the right-hand side of (1.3.4) does not depend on the time index k, i.e.,

$$\mathbb{P}\{x_{k+1} \in A | x_k\} = \mathbb{P}\{x_1 \in A | x_0\}, \tag{1.3.5}$$

then the chain $\{x_k\}_{k\geq 0}$ is said to be *homogenous*.

In the rest of the section, all chains are assumed to be homogenous if no special statements are claimed.

For the chain $\{x_k\}_{k\geq 0}$, denote the *one-step transition probability* and the *k-step transition probability* by $P(x,A) = \mathbb{P}\{x_1 \in A | x_0 = x\}$ and $P_k(x,A) = \mathbb{P}\{x_k \in A | x_0 = x\}$, respectively, where $x \in \mathbb{R}^m$ and $A \in \mathscr{B}^m$. It holds that

$$P_k(x,A) = \int_{\mathbb{R}^m} P_{k-1}(y,A) P(x,\mathrm{d}y). \tag{1.3.6}$$

Denote by $P_k(\cdot)$ the probability measure induced by x_k: $P_k(A) = \mathbb{P}\{x_k \in A\}$, $A \in \mathscr{B}^m$.

Assume

$$\int_{\mathbb{R}^m} P(x,A) P_0(\mathrm{d}x) = P_0(A) \ \ \forall A \in \mathscr{B}^m \tag{1.3.7}$$

for some initial probability measure $P_0(\cdot)$ of x_0. Then by (1.3.6) and Theorem 1.1.4, it can inductively be proved that for any $A \in \mathscr{B}^m$,

$$\begin{aligned}
P_1(A) &= \int_{\mathbb{R}^m} P(x,A) P_0(\mathrm{d}x) = P_0(A), \\
P_2(A) &= \int_{\mathbb{R}^m} P_2(x,A) P_0(\mathrm{d}x) = \int_{\mathbb{R}^m} \Big(\int_{\mathbb{R}^m} P(y,A) P(x,\mathrm{d}y) \Big) P_0(\mathrm{d}x) \\
&= \int_{\mathbb{R}^m} P(y,A) \Big(\int_{\mathbb{R}^m} P(x,\mathrm{d}y) P_0(\mathrm{d}x) \Big) = \int_{\mathbb{R}^m} P(y,A) P_1(\mathrm{d}y) \\
&= \int_{\mathbb{R}^m} P(y,A) P_0(\mathrm{d}y) = P_1(A) = P_0(A),
\end{aligned}$$

and further,

$$P_k(A) = P_0(A), \ k \geq 1.$$

The initial probability measure $P_0(\cdot)$ of x_0 satisfying (1.3.7) is called the *invariant probability measure* of the chain $\{x_k\}_{k\geq 0}$.

It should be noted that for a chain $\{x_k\}_{k\geq 0}$, its invariant probability measure does not always exist, and if exists, it may not be unique.

Denote the *total variation norm* of a signed measure $\nu(\cdot)$ on $(\mathbb{R}^m, \mathscr{B}^m)$ by $\|\nu\|_{\text{var}}$, i.e.,

$$\|\nu\|_{\text{var}} = \int_{\mathbb{R}^m} \nu_+(\mathrm{d}x) + \int_{\mathbb{R}^m} \nu_-(\mathrm{d}x),$$

where $\nu = \nu_+ - \nu_-$ is the Jordan–Hahn decomposition of ν (see Theorem 1.1.1).

Definition 1.3.1 *The chain* $\{x_k\}_{k\geq 0}$ *is called* ergodic *if there exists a probability measure* $P_{\text{IV}}(\cdot)$ *on* $(\mathbb{R}^m, \mathscr{B}^m)$ *such that*

$$\|P_k(x,\cdot) - P_{\text{IV}}(\cdot)\|_{\text{var}} \underset{k\to\infty}{\longrightarrow} 0 \tag{1.3.8}$$

for any $x \in \mathbb{R}^m$. *Further, if there exist constants* $0 < \rho < 1$ *and* $M > 0$ *possibly depending on* x, *i.e.,* $M = M(x)$ *such that*

$$\|P_k(x,\cdot) - P_{\text{IV}}(\cdot)\|_{\text{var}} \leq M(x)\rho^k, \tag{1.3.9}$$

then the chain $\{x_k\}_{k\geq 0}$ *is called* geometrically ergodic.

The probability measure $P_{\text{IV}}(\cdot)$ is, in fact, the invariant probability measure of $\{x_k\}_{k\geq 0}$. It is clear that if the chain $\{x_k\}_{k\geq 0}$ is ergodic, then its invariant probability measure is unique. In what follows we introduce criteria for ergodicity and geometric ergodicity of the chain $\{x_k\}_{k\geq 0}$ valued in $(\mathbb{R}^m, \mathscr{B}^m)$. For this, we first introduce some definitions and related results, which the ergodicity of Markov chains is essentially based on.

Definition 1.3.2 *The chain* $\{x_k\}_{k\geq 0}$ *valued in* $(\mathbb{R}^m, \mathscr{B}^m)$ *is called* μ-irreducible *if there exists a measure* $\mu(\cdot)$ *on* $(\mathbb{R}^m, \mathscr{B}^m)$ *such that*

$$\sum_{k=1}^{\infty} P_k(x,A) > 0 \tag{1.3.10}$$

for any $x \in \mathbb{R}^m$ *and any* $A \in \mathscr{B}^m$ *with* $\mu(A) > 0$. *The measure* $\mu(\cdot)$ *is called the* maximal irreducibility measure *of* $\{x_k\}_{k\geq 0}$ *if*

(i) $\{x_k\}_{k\geq 0}$ *is* μ*-irreducible;*

(ii) *for any other measure* $\mu'(\cdot)$ *on* $(\mathbb{R}^m, \mathscr{B}^m)$, $\{x_k\}_{k\geq 0}$ *is* μ'*-irreducible if and only if* $\mu'(\cdot)$ *is absolutely continuous with respect to* $\mu(\cdot)$;

(iii) $\mu\left(\left\{x : \sum_{k=1}^{\infty} P_k(x,A) > 0\right\}\right) = 0$ *whenever* $\mu(A) = 0$.

Formula (1.3.10) indicates that for the μ-irreducible chain $\{x_k\}_{k\geq 0}$, starting from any initial state $x_0 = x \in \mathbb{R}^m$, the probability that in a finite number of steps the sequence $\{x_k\}_{k\geq 0}$ enters any set A with positive μ-measure is always positive. In the following, when we say that the chain $\{x_k\}_{k\geq 0}$ is μ-irreducible, we implicitly assume that $\mu(\cdot)$ is the maximal irreducibility measure of $\{x_k\}_{k\geq 0}$.

Definition 1.3.3 *Suppose $A_1, \cdots, A_d$ are disjoint sets in $\mathscr{B}^m$. For the chain $\{x_k\}_{k\geq 0}$, if*

(i) $P(x, A_{i+1}) = 1 \;\; \forall\, x \in A_i, \; i = 1, \cdots, d-1,$

and

(ii) $P(x, A_1) = 1 \;\; \forall\, x \in A_d,$

*then $\{A_1, \cdots, A_d\}$ is called a d-*cycle *of $\{x_k\}_{k\geq 0}$.*

The d-cycle is called maximal *if*

(iii) *there exists a measure $\nu(\cdot)$ on $(\mathbb{R}^m, \mathscr{B}^m)$ such that*

$$\nu(A_i) > 0, \; i = 1, \cdots, d \;\text{ and }\; \nu\Big(\mathbb{R}^m / \bigcup_{i=1}^{d} A_i\Big) = 0, \tag{1.3.11}$$

and

(iv) *for any sets $\{A'_1, \cdots, A'_{d'}\}$ satisfying (i) and (ii) with d replaced by d', d' must divide d.*

The integer d is called the period *of $\{x_k\}_{k\geq 0}$ if the d-cycle of $\{x_k\}_{k\geq 0}$ is maximal. When the period equals 1, the chain $\{x_k\}_{k\geq 0}$ is called* aperiodic.

The small set is another concept related to ergodicity of Markov chains valued in $(\mathbb{R}^m, \mathscr{B}^m)$. Let us first recall the ergodic criterion for Markov chains valued in a countable state space.

Suppose that the chain $\{\varphi_k\}_{k\geq 0}$ takes values in $\{1, 2, 3, \cdots\}$ and its transition probability is denoted by $p_{ij} = \mathbb{P}\{\varphi_{k+1} = j | \varphi_k = i\}$, $i, j = 1, 2, 3, \cdots$. It is known that if $\{\varphi_k\}_{k\geq 0}$ is irreducible, aperiodic, and there exist a finite set $C \subset \{1, 2, 3, \cdots\}$, a nonnegative function $g(\cdot)$, and constants $K > 0$ and $\delta > 0$ such that

$$E[g(\varphi_{k+1}) | \varphi_k = j] < g(j) + K \;\; \forall\, j \in C, \tag{1.3.12}$$

$$E[g(\varphi_{k+1}) | \varphi_k = j] < g(j) - \delta \;\; \forall\, j \overline{\in} C, \tag{1.3.13}$$

then $\{\varphi_k\}_{k\geq 0}$ is ergodic, i.e.,

$$\lim_{k\to\infty} \mathbb{P}\{\varphi_k = j | \varphi_0 = i\} = \pi_j, \;\; j = 1, 2, \cdots \tag{1.3.14}$$

with $\pi_j \geq 0$ and $\sum_{j=1}^{\infty} \pi_j = 1$. The concept of a small set can be regarded as an extension of the above finite subset C.

In the sequel, we adopt the following notations,

$$E(s(x_n)|x_0 = x) \triangleq \int_{\mathbb{R}^m} s(y)P_n(x, \mathrm{d}y), \quad E_\nu(s) \triangleq \int_{\mathbb{R}^m} s(x)\nu(\mathrm{d}x) \tag{1.3.15}$$

where $P_n(x,\cdot)$ is the n-step transition probability of the chain $\{x_k\}_{k\geq 0}$, $s(x)$ is a measurable function on $(\mathbb{R}^m, \mathscr{B}^m)$, and $\nu(\cdot)$ is a measure on $(\mathbb{R}^m, \mathscr{B}^m)$.

Definition 1.3.4 *Assume $\{x_k\}_{k\geq 0}$ is a μ-irreducible chain. We say that $\{x_k\}_{k\geq 0}$ satisfies the* minorization condition $M(m_0, \beta, s, \nu)$, *where $m_0 \geq 1$ is an integer, $\beta > 0$ a constant, $s(x)$ a nonnegative measurable function on $(\mathbb{R}^m, \mathscr{B}^m)$ with $E_\mu(s) > 0$, and $\nu(\cdot)$ a probability measure on $(\mathbb{R}^m, \mathscr{B}^m)$, if*

$$P_{m_0}(x,A) \geq \beta s(x)\nu(A) \;\; \forall\, x \in \mathbb{R}^m \;\; \forall\, A \in \mathscr{B}^m. \tag{1.3.16}$$

The function $s(x)$ and the probability measure $\nu(\cdot)$ are called the small function *and* small measure, *respectively. If $s(x)$ equals some indicator function, i.e., $s(x) = I_C(x)$ for some $C \in \mathscr{B}^m$ with $\mu(C) > 0$ and*

$$P_{m_0}(x,A) \geq \beta\nu(A) \;\; \forall\, x \in C \;\; \forall\, A \in \mathscr{B}^m, \tag{1.3.17}$$

then C is called a small set.

Lemma 1.3.1 *Suppose that the μ-irreducible chain $\{x_k\}_{k\geq 0}$ satisfies the minorization condition $M(m_0, \beta, s, \nu)$. Then*

(i) the small measure $\nu(\cdot)$ is also an irreducibility measure for $\{x_k\}_{k\geq 0}$, and

(ii) the set $C \triangleq \{x : s(x) \geq \gamma\}$ for any constant $\gamma > 0$ is small, whenever it is μ-positive.

Proof. See Appendix A. □

Lemma 1.3.2 *Suppose that the chain $\{x_k\}_{k\geq 0}$ is μ-irreducible. Then,*

(i) for any set B with $\mu(B) > 0$, there exists a small set $C \subset B$;

(ii) if $s(x)$ is small, so is $E(s(x_n)|x_0 = x)$ $\forall\, n \geq 1$; and

(iii) if both $s(\cdot)$ and $s'(\cdot)$ are small functions, so is $s(\cdot) + s'(\cdot)$.

Proof. See Appendix A. □

The following results are useful in justifying whether a Markov chain is aperiodic or a set is small.

Theorem 1.3.1 *Suppose that the chain $\{x_k\}_{k\geq 0}$ is μ-irreducible. If either*

(i) there exists a small set $C \in \mathscr{B}^m$ with $\mu(C) > 0$ and an integer n, possibly depending on C, such that

$$P_n(x,C) > 0, \;\; P_{n+1}(x,C) > 0 \;\; \forall\, x \in C, \tag{1.3.18}$$

or

(ii) *there exists a set* $A \in \mathscr{B}^m$ *with* $\mu(A) > 0$ *such that for any* $B \subset A$, $B \in \mathscr{B}^m$ *with* $\mu(B) > 0$ *and for some positive integer* n *possibly depending on* B,

$$P_n(x,B) > 0, \ P_{n+1}(x,B) > 0 \ \forall\, x \in B, \tag{1.3.19}$$

then $\{x_k\}_{k\geq 0}$ *is aperiodic.*

Proof. See Appendix A. □

Theorem 1.3.2 *Suppose that* $\{x_k\}_{k\geq 0}$ *is a* μ*-irreducible, aperiodic Markov chain valued in* $(\mathbb{R}^m, \mathscr{B}^m)$.

(i) *Let* $s(x)$ *be a small function. Then, any set* C *with* $\mu(C) > 0$ *satisfying*

$$\inf_{x\in C} \sum_{k=0}^{l} E(s(x_k)|x_0 = x) > 0 \tag{1.3.20}$$

for some integer $l \geq 0$ *is a small set.*

(ii) *Any set* C *with* $\mu(C) > 0$ *satisfying the following condition is a small set: there exists some* $A \in \mathscr{B}^m$ *with* $\mu(A) > 0$ *such that for any* $B \subset A$ *with* $\mu(B) > 0$,

$$\inf_{x\in C} \sum_{k=0}^{l} P_k(x,B) > 0, \tag{1.3.21}$$

where the integer $l \geq 0$ *may depend on* B.

Proof. See Appendix A. □

Theorem 1.3.3 *Assume that the chain* $\{x_k\}_{k\geq 0}$ *is irreducible and aperiodic. If there exist a nonnegative measurable function* $g(\cdot)$, *a small set* S, *and constants* $\rho \in (0,1)$, $c_1 > 0$, *and* $c_2 > 0$ *such that*

$$E[g(x_{k+1})|x_k = x] \leq \rho g(x) - c_1 \ \ \forall\, x \overline{\in} S, \tag{1.3.22}$$

$$E[g(x_{k+1})|x_k = x] \leq c_2 \ \ \forall\, x \in S, \tag{1.3.23}$$

then there exist a probability measure $P_{\text{IV}}(\cdot)$ *and a nonnegative measurable function* $M(x)$ *such that*

$$\|P_k(x,\cdot) - P_{\text{IV}}(\cdot)\|_{\text{var}} \leq M(x)\rho^k. \tag{1.3.24}$$

Further, the nonnegative function $M(x)$ *in (1.3.24) can be selected such that* $M(x) = a + bg(x)$, *where* $a \geq 0$, $b \geq 0$ *are constants and* $\int_{\mathbb{R}^m} g(x)P_{\text{IV}}(\mathrm{d}x) < \infty$.

Under assumptions different from those required in Theorem 1.3.3 we have different kinds of ergodicity. To this end, we introduce the following definition.

Definition 1.3.5 *For the chain $\{x_k\}_{k\geq 0}$, the following property is called the* Doeblin condition*: There exist a probability measure $\nu(\cdot)$ and some constants $0<\varepsilon<1$, $0<\delta<1$ such that $P_{k_0}(x,A)\geq\delta\ \forall\, x\in\mathbb{R}^m$ for an integer k_0 whenever $\nu(A)>\varepsilon$.*

Theorem 1.3.4 *Suppose that the chain $\{x_k\}_{k\geq 0}$ is irreducible and aperiodic. If $\{x_k\}_{k\geq 0}$ satisfies the Doeblin condition, then there exist a probability measure $P_{\rm IV}(\cdot)$ and constants $M>0$, $0<\rho<1$ such that*

$$\|P_k(x,\cdot)-P_{\rm IV}(\cdot)\|_{\rm var}\leq M\rho^k. \tag{1.3.25}$$

Theorem 1.3.5 *Assume that the chain $\{x_k\}_{k\geq 0}$ is irreducible and aperiodic. If there exist a nonnegative measurable function $g(\cdot)$, a small set S, and constants $c_1>0$ and $c_2>0$ such that*

$$E[g(x_{k+1})|x_k=x]\leq g(x)-c_1\ \ \forall\, x\overline{\in} S, \tag{1.3.26}$$
$$E[g(x_{k+1})|x_k=x]\leq c_2\ \ \forall\, x\in S, \tag{1.3.27}$$

then there exists a probability measure $P_{\rm IV}(\cdot)$ such that

$$\|P_k(x,\cdot)-P_{\rm IV}(\cdot)\|_{\rm var}\underset{k\to\infty}{\longrightarrow}0. \tag{1.3.28}$$

Theorems 1.3.3, 1.3.4, and 1.3.5 are usually called the geometrically ergodic criterion, the uniformly ergodic criterion, and the ergodic criterion, respectively. It is clear that the geometrical ergodicity is stronger than the ergodicity, but weaker than the uniform ergodicity. Next, we show that for a large class of stochastic dynamic systems the geometrical ergodicity takes place if a certain stability condition holds.

By $\mu_n(\cdot)$ we denote the Lebesgue measure on $(\mathbb{R}^n,\mathscr{B}^n)$.

Let us consider the ergodicity of the single-input single-output (SISO) nonlinear ARX (NARX) system

$$y_{k+1}=f(y_k,\cdots,y_{k+1-p},u_k,\cdots,u_{k+1-q})+\varepsilon_{k+1}, \tag{1.3.29}$$

where u_k and y_k are the system input and output, respectively, ε_k is the noise, (p_0,q_0) are the known system orders, and $f(\cdot)$ is a nonlinear function.

The NARX system (1.3.29) is a straightforward generalization of the linear ARX system and covers a large class of dynamic phenomena. This point will be made clear in the later chapters.

By denoting

$$x_k\triangleq[y_k,\cdots,y_{k+1-p},u_k,\cdots,u_{k+1-q}]^T,$$
$$\varphi_1(x_k)\triangleq[f(y_k,\cdots,y_{k+1-p},u_k,\cdots,u_{k+1-q}),\ y_k,\cdots,y_{k+2-p}]^T,$$
$$\varphi_2(x_k)\triangleq[0,\ u_k,\cdots,u_{k+2-q}]^T,$$
$$\varphi(x_k)\triangleq[\varphi_1(x_k)^T\ \varphi_2(x_k)^T]^T,$$

and

$$\xi_{k+1} \triangleq [\underbrace{\varepsilon_{k+1}, 0, \cdots, 0}_{p}, \underbrace{u_{k+1}, 0, \cdots, 0}_{q}]^T,$$

the NARX system (1.3.29) is transformed to the following state space model

$$x_{k+1} = \varphi(x_k) + \xi_{k+1}. \tag{1.3.30}$$

Thus $\{x_k\}_{k\geq 0}$ is a Markov chain if $\{\xi_k\}_{k\geq 0}$ satisfies certain probability conditions, e.g., if $\{\xi_k\}_{k\geq 0}$ is a sequence of mutually independent random variables. Ergodicity of $\{x_k\}_{k\geq 0}$ can be investigated by the results given in the preceding sections. To better understand the essence of the approach, let us consider the first order (i.e., $p = q = 1$) NARX system:

$$y_{k+1} = f(y_k, u_k) + \varepsilon_{k+1}. \tag{1.3.31}$$

We need the following conditions.

A1.3.1 *Let the input $\{u_k\}_{k\geq 0}$ be a sequence of iid random variables with $Eu_k = 0$, $Eu_k^2 < \infty$, and with a probability density function denoted by $f_u(\cdot)$, which is positive and continuous on $\mathbb{R}$.*

A1.3.2 *$\{\varepsilon_k\}_{k\geq 0}$ is a sequence of iid random variables with $E\varepsilon_k = 0$, $E\varepsilon_k^2 < \infty$, and with a density function $f_\varepsilon(\cdot)$, which is assumed to be positive and uniformly continuous on $\mathbb{R}$;*

A1.3.3 *$\{\varepsilon_k\}_{k\geq 0}$ and $\{u_k\}_{k\geq 0}$ are mutually independent;*

A1.3.4 *$f(\cdot,\cdot)$ is continuous on $\mathbb{R}^2$ and there exist constants $0 < \lambda < 1$, $c_1 > 0$, $c_2 > 0$, and $l > 0$ such that $|f(\xi_1, \xi_2)| \leq \lambda|\xi_1| + c_1|\xi_2|^l + c_2 \ \ \forall\, \xi = [\xi_1\ \xi_2]^T \in \mathbb{R}^2$, where λ, c_1, c_2, and l may be unknown;*

A1.3.5 *$E|u_k|^l < \infty$ and the initial value y_0 satisfies $E|y_0| < \infty$.*

By denoting $x_k \triangleq [y_k\ u_k]^T$, $\varphi(x_k) \triangleq [f(y_k, u_k)\ 0]^T$, and $\xi_k \triangleq [\varepsilon_k\ u_k]^T$, the NARX system (1.3.31) is rewritten as follows:

$$x_{k+1} = \varphi(x_k) + \xi_{k+1}. \tag{1.3.32}$$

Under the conditions A1.3.1–A1.3.3, it is clear that the state vector sequence $\{x_k\}_{k\geq 0}$ defined by (1.3.32) is a time-homogeneous Markov chain valued in $(\mathbb{R}^2, \mathscr{B}^2)$. As to be seen in what follows, Assumption A1.3.4 is a kind of stability condition to guarantee ergodicity of $\{x_k\}_{k\geq 0}$.

Lemma 1.3.3 *If A1.3.1–A1.3.3 hold, then the chain $\{x_k\}_{k\geq 0}$ defined by (1.3.32) is μ_2-irreducible and aperiodic, and μ_2 is the maximal irreducibility measure of $\{x_k\}_{k\geq 0}$. Further, any bounded set $A \in \mathscr{B}^2$ with $\mu_2(A) > 0$ is a small set.*

Proof. See Appendix A. □

Theorem 1.3.6 *Assume A1.3.1–A1.3.5 hold. Then*

(i) there exist a probability measure $P_{\text{IV}}(\cdot)$ on $(\mathbb{R}^2, \mathscr{B}^2)$, a nonnegative measurable function $M(x)$, and a constant $\rho \in (0,1)$ such that

$$\|P_n(x,\cdot) - P_{\text{IV}}(\cdot)\|_{\text{var}} \le M(x)\rho^n \quad \forall\, x = [\xi_1\ \xi_2]^T \in \mathbb{R}^2; \qquad (1.3.33)$$

(ii) $\sup_n \int_{\mathbb{R}^2} M(x)P_n(\mathrm{d}x) < \infty$ and $\|P_n(\cdot) - P_{\text{IV}}(\cdot)\|_{\text{var}} \le c\rho^n$ for some constants $c > 0$ and $\rho \in (0,1)$;

(iii) $P_{\text{IV}}(\cdot)$ is with probability density $f_{\text{IV}}(\cdot,\cdot)$ which is positive on $\mathbb{R}^2$, and

$$f_{\text{IV}}(s_1,s_2) = \iint\limits_{\mathbb{R}^2} f_\varepsilon(s_1 - f(\xi_1,\xi_2))P_{\text{IV}}(\mathrm{d}x) f_u(s_2). \qquad (1.3.34)$$

Proof. We first prove (i). Define the Lyapunov function $g(x) \triangleq |\xi_1| + \beta|\xi_2|^l$, where $x = [\xi_1\ \xi_2]^T \in \mathbb{R}^2$ and $\beta > 0$ is a constant to be determined.

By A1.3.1–A1.3.5, we have

$$\begin{aligned}
&E[g(x_{k+1})|x_k = x] = E[|y_{k+1}| + \beta|u_{k+1}|^l|x_k = x] \\
&= E[|f(y_k,u_k) + \varepsilon_{k+1}||x_k = x] + \beta E|u_1|^l \le |f(\xi_1,\xi_2)| + E|\varepsilon_1| + \beta E|u_1|^l \\
&\le \lambda|\xi_1| + c_1|\xi_2|^l + c_2 + E|\varepsilon_1| + \beta E|u_1|^l.
\end{aligned}$$

Let $\beta = \frac{c_1}{\lambda}$. Then from the above inequalities it follows that

$$\begin{aligned}
E[g(x_{k+1})|x_k = x] &\le \lambda|\xi_1| + \lambda\beta|\xi_2|^l + c_2 + E|\varepsilon_1| + \beta E|u_1|^l \\
&\le \lambda g(x) + c_3, \qquad (1.3.35)\\
&= \lambda' g(x) - \big((\lambda' - \lambda)g(x) - c_3\big), \qquad (1.3.36)
\end{aligned}$$

where $c_3 \triangleq c_2 + E|\varepsilon_1| + \beta E|u_1|^l$ and $0 < \lambda < \lambda' < 1$.

Choose $K > 0$ large enough such that $(\lambda' - \lambda)K - c_3 > 0$, and define $S = \{x \in \mathbb{R}^2 | g(x) \le K\}$. Since S is a bounded set, by Lemma 1.3.3 S is a small set.

From (1.3.36) we have

$$E[g(x_{k+1})|x_k = x] \le \lambda' g(x) - \big((\lambda' - \lambda)K - c_3\big) \triangleq \lambda' g(x) - c_4 \quad \forall\, x \overline{\in} S, \qquad (1.3.37)$$

and from (1.3.35)

$$E[g(x_{k+1})|x_k = x] \le \lambda K + c_3 \triangleq c_5 \quad \forall\, x \in S. \qquad (1.3.38)$$

Noticing (1.3.37) and (1.3.38) and applying Theorem 1.3.3, we see that (1.3.33) holds.

We now prove (ii). By Theorem 1.3.3, the measurable function $M(x)$ actually can be taken as $a+bg(x)$, where a and b are positive constants and $g(x)$ is the Lyapunov function defined above. To prove (ii) we first verify that

$$\sup_{k\geq 0}\iint_{\mathbb{R}^2} g(x)P_k(\mathrm{d}x) < \infty. \tag{1.3.39}$$

Noticing A1.3.4, A1.3.5, and $g(x) = |\xi_1| + \beta|\xi_2|^l$, we have that

$$\begin{aligned}
\iint_{\mathbb{R}^2} g(x)P_k(\mathrm{d}x) &= Eg(x_k) = E|y_k| + \beta E|u_k|^l \\
&= E|f(y_{k-1}, u_{k-1}) + \varepsilon_k| + \beta E|u_1|^l \\
&\leq \lambda E|y_{k-1}| + c_1 E|u_{k-1}|^l + c_2 + E|\varepsilon_1| + \beta E|u_1|^l \\
&= \lambda E|f(y_{k-2}, u_{k-2}) + \varepsilon_{k-1}| + \left(c_1 E|u_1|^l + c_2 + E|\varepsilon_1|\right) + \beta E|u_1|^l \\
&\leq \lambda^2 E|y_{k-2}| + \lambda\left(c_1 E|u_1|^l + c_2 + E|\varepsilon_1|\right) + \left(c_1 E|u_1|^l + c_2 + E|\varepsilon_1|\right) + \beta E|u_1|^l \\
&\leq \cdots \\
&\leq \lambda^k E|y_0| + (\lambda^{k-1} + \lambda^{k-2} + \cdots + 1)\left(c_1 E|u_1|^l + c_2 + E|\varepsilon_1|\right) + \beta E|u_1|^l,
\end{aligned}$$

which imply (1.3.39) by noticing $0 < \lambda < 1$.

Then by Lemma 1.3.3 and (1.3.39) and by noticing the basic property of the total variation norm, for any $A \in \mathscr{B}^2$ we have

$$\begin{aligned}
|P_n(A) - P_{\mathrm{IV}}(A)| &\leq \iint_{\mathbb{R}^2} |P_n(x,A) - P_{\mathrm{IV}}(A)| P_0(\mathrm{d}x) \\
&\leq \iint_{\mathbb{R}^2} \|P_n(x,\cdot) - P_{\mathrm{IV}}(\cdot)\|_{\mathrm{var}} P_0(\mathrm{d}x) \leq \rho^n \iint_{\mathbb{R}^2} M(x)P_0(\mathrm{d}x) \leq M\rho^n,
\end{aligned}$$

and

$$\begin{aligned}
\|P_n(\cdot) - P_{\mathrm{IV}}(\cdot)\|_{\mathrm{var}} &= \sup_{A\in\mathscr{B}^2}(P_n(A) - P_{\mathrm{IV}}(A)) - \inf_{A\in\mathscr{B}^2}(P_n(A) - P_{\mathrm{IV}}(A)) \\
&\leq 2\sup_{A\in\mathscr{B}^2}|P_n(A) - P_{\mathrm{IV}}(A)| \leq 2M\rho^n.
\end{aligned}$$

Hence (ii) holds.

Finally we prove (iii). Noticing that both $\{u_k\}$ and $\{\varepsilon_k\}$ are sequences of iid random variables with densities $f_u(\cdot)$ and $f_\varepsilon(\cdot)$, respectively, and

$$P_{\mathrm{IV}}(A) = \iint_{\mathbb{R}^2} P_n(x,A)P_{\mathrm{IV}}(\mathrm{d}x) \quad \forall\, A \in \mathscr{B}^2 \quad \forall\, n \geq 1, \tag{1.3.40}$$

by (A.56) we have

$$P_{\rm IV}(A) = \iint_{\mathbb{R}^2} P(x,A)P_{\rm IV}(dx)$$
$$= \iint_A \iint_{\mathbb{R}^2} f_\varepsilon(s_1 - f(\xi_1,\xi_2))f_u(s_2)P_{\rm IV}(dx)ds_1 ds_2. \quad (1.3.41)$$

Hence, $P_{\rm IV}(A)$ is with density function

$$f_{\rm IV}(s_1,s_2) = \iint_{\mathbb{R}^2} f_\varepsilon(s_1 - f(\xi_1,\xi_2))P_{\rm IV}(dx)f_u(s_2). \quad (1.3.42)$$

According to A1.3.4, we have $\sup_{\|x\|\le K}|f(x_1,x_2)| < \infty$ for any fixed $K > 0$. As both $f_u(\cdot)$ and $f_\varepsilon(\cdot)$ are positive, for a large enough $K > 0$ it follows that

$$f_{\rm IV}(s_1,s_2) = \iint_{\mathbb{R}^2} f_\varepsilon(s_1 - f(\xi_1,\xi_2))P_{\rm IV}(dx)f_u(s_2)$$
$$\ge \iint_{\|x\|\le K} f_\varepsilon(s_1 - f(\xi_1,\xi_2))P_{\rm IV}(dx)f_u(s_2)$$
$$\ge \inf_{\|x\|\le K}\{f_\varepsilon(s_1 - f(\xi_1,\xi_2))\}f_u(s_2)P_{\rm IV}\{\|x\|\le K\} > 0.$$

This proves (iii). □

We now consider the NARX system (1.3.29) and (1.3.30) with $p_0 > 1$, $q_0 > 1$.

For (1.3.29), the assumptions A1.3.1, A1.3.2, and A1.3.3 remain unchanged, while A1.3.4 and A1.3.5 correspondingly change to the following A1.3.4' and A1.3.5'.

A1.3.4' *$f(\cdot)$ is continuous on $\mathbb{R}^{p+q}$ and there exist a vector norm $\|\cdot\|_v$ on $\mathbb{R}^p$ and constants $0 < \lambda < 1$, $c_1 > 0$, $c_2 > 0$, and $l > 0$ such that*

$$\|\varphi_1(x)\|_v \le \lambda\|s\|_v + c_1\sum_{i=1}^{q}|t_i|^l + c_2\ \forall\, x \in \mathbb{R}^{p+q}, \quad (1.3.43)$$

where $s \triangleq [s_1 \cdots s_p]^T \in \mathbb{R}^p$, $t \triangleq [t_1 \cdots t_q]^T \in \mathbb{R}^q$, and $x \triangleq [s^T\ t^T]^T \in \mathbb{R}^{p+q}$.

A1.3.5' *$E|u_k|^l < \infty$ and $E\|Y_0\| < \infty$, where $Y_0 \triangleq [y_0, y_{-1}, \cdots, y_{1-p}]^T$ is the initial value.*

The probabilistic properties of $\{x_k\}_{k\ge0}$ such as irreducibility, aperiodicity, and ergodicity for the case $p > 1$, $q > 1$ can be established as those for the first order system. In fact, we have the following theorem.

Theorem 1.3.7 *If A1.3.1–A1.3.3, A1.3.4', and A1.3.5' hold, then the chain $\{x_k\}_{k\ge0}$ defined by (1.3.30) is μ_{p+q}-irreducible, aperiodic, and* *(i)* *there exist a probability measure* $P_{\text{IV}}(\cdot)$ *on* $(\mathbb{R}^{p+q}, \mathscr{B}^{p+q})$*, a nonnegative measurable function* $M(x)$*, and a constant* $0 < \rho < 1$ *such that* $\|P_n(x,\cdot) - P_{\text{IV}}(\cdot)\|_{\text{var}} \leq M(x)\rho^n \; \forall\, x \in \mathbb{R}^{p+q}$;

(ii) $\sup_n \int_{\mathbb{R}^{p+q}} M(x)P_n(\mathrm{d}x) < \infty$ *and* $\|P_n(\cdot) - P_{\text{IV}}(\cdot)\|_{\text{var}} \leq c\rho^n$ *for some constants* $c > 0$ *and* $0 < \rho < 1$.

Further, $P_{\text{IV}}(\cdot)$ *is with probability density, which is positive on* $\mathbb{R}^{p+q}$.

Theorem 1.3.7 can be proved similarly to Lemma 1.3.3 and Theorem 1.3.6. Here we only give some remarks.

Remark 1.3.1 *Set* $n_0 \triangleq p \vee q = \max\{p,q\}$. *To establish irreducibility and aperiodicity in the case* $p = q = 1$*, the one-step transition probability* $P(x,A)$*,* $x \in \mathbb{R}^2$*,* $A \in \mathscr{B}^2$ *is considered, while for the case* $n_0 > 1$*, the* n_0*-step transition probability* $P_{n_0}(x,A)$*,* $x \in \mathbb{R}^{p+q}$*,* $A \in \mathscr{B}^{p+q}$ *should be investigated. To establish the geometrical ergodicity of* $\{x_k\}_{k\geq 0}$*, the Lyapunov function may be chosen as* $g(x) = \|s\|_v + \sum_{i=1}^{q} \beta_i |t_i|^l$*, where* $\beta_1 = \frac{qc_1}{\lambda^q}$ *and* $\beta_{i+1} = \lambda\beta_i - c_1$*,* $i = 1,\cdots,q-1$.

Remark 1.3.2 *We note that (1.3.34) gives the expression of the invariant probability density of the first order NARX system* $\big((p,q) = (1,1)\big)$. *For the general case* $p > 1$ *and* $q > 1$*, the invariant probability density and its properties can similarly be obtained from the* n_0*-step transition probability* $P_{n_0}(x,\cdot)$ *with* $n_0 = \max(p,q)$. *For example, for the case* $(p,q) = (2,1)$ *by investigating the two-step transition probability, we find that the invariant probability density is expressed as follows:*

$$
\begin{aligned}
f_{\text{IV}}(s_1,s_2,s_3) = \int_{\mathbb{R}^3} \Big(\int_{-\infty}^{\infty} & f_\varepsilon\big(s_1 - f(s_2,x_1,t)\big) f_u(t)\mathrm{d}t \Big) \\
& \cdot f_\varepsilon\big(s_2 - f(x_1,x_2,x_3)\big) P_{\text{IV}}(\mathrm{d}x) f_u(s_3),
\end{aligned}
$$

while for the case $(p,q) = (3,2)$ *considering the three-step transition probability leads to the invariant probability density*

$$
\begin{aligned}
& f_{\text{IV}}(s_1,s_2,s_3,s_4,s_5) \\
& = \int_{\mathbb{R}^5} \Big(\int_{-\infty}^{\infty} f_\varepsilon\big(s_1 - f(s_2,s_3,x_1,s_5,t)\big) f_\varepsilon\big(s_2 - f(s_3,x_1,x_2,t,x_4)\big) f_u(t)\mathrm{d}t \Big) \\
& \quad \cdot f_\varepsilon\big(s_3 - f(x_1,x_2,x_3,x_4,x_5)\big) P_{\text{IV}}(\mathrm{d}x) f_u(s_4) f_u(s_5).
\end{aligned}
$$

The properties of $f_{\text{IV}}(s_1,s_2)$*,* $f_{\text{IV}}(s_1,s_2,s_3)$*, and* $f_{\text{IV}}(s_1,s_2,s_3,s_4,s_5)$ *are derived from the above formulas by using the assumptions made in Theorem 1.3.7.*

Remark 1.3.3 *In A1.3.4', a vector norm rather than the Euclidean norm is adopted. This is because such a norm is more general than the Euclidean norm and* λ *in (1.3.43) for many NARX systems in such a norm may be taken smaller than* 1. *The fact that* $\lambda \in (0,1)$ *is of crucial importance for establishing stability and ergodicity* *of the NARX system (see the proof of Theorem 1.3.6). It is natural to ask what will happen if $\lambda \geq 1$. Let us consider the following example:*

$$y_{k+1} = y_k + \varepsilon_{k+1},$$

where $\{\varepsilon_k\}$ is iid. It is clear that $y_{k+1} = \sum_{i=1}^{k+1} \varepsilon_i$ if the initial value $y_0 = 0$. It is seen that for the above system, the constant λ equals 1 and $\{y_k\}_{k\geq 1}$ is not ergodic. So, in a certain sense, the condition $\lambda \in (0,1)$ is necessary for ergodicity of the NARX system.

For ergodicity of nonlinear systems, we assume that both $\{u_k\}_{k\geq 0}$ and $\{\varepsilon_k\}_{k\geq 0}$ are with positive probability density functions. In fact, these assumptions are sufficient but not necessary for ergodicity of stochastic systems. Let us consider the following linear process:

$$x_{k+1} = Fx_k + G\varepsilon_{k+1} \quad k \geq 0, \tag{1.3.44}$$

where $x_k \in \mathbb{R}^m$, $\varepsilon_k \in \mathbb{R}^r$, $F \in \mathbb{R}^{m\times m}$, and $G \in \mathbb{R}^{m\times r}$.

We make the following assumptions.

A1.3.6 *All eigenvalues of F are strictly inside the unit cycle;*

A1.3.7 *(F,G) is controllable, i.e.,* $\operatorname{rank}\begin{bmatrix} G & FG & \cdots & F^{m-1}G \end{bmatrix} = m$;

A1.3.8 *$\{\varepsilon_k\}_{k\geq 0}$ is iid with density which is positive and continuous on a set $U \in \mathscr{B}^r$ satisfying $\mu_r(U) > 0$, where $\mu_r(\cdot)$ is the Lebesgue measure on $(\mathbb{R}^r, \mathscr{B}^r)$.*

Theorem 1.3.8 *Assume that A1.3.6–A1.3.8 hold. Then the chain $\{x_k\}_{k\geq 0}$ defined by (1.3.44) is geometrically ergodic.*

To prove Theorem 1.3.8, we need an auxiliary lemma.

Lemma 1.3.4 *Given a matrix $A \in \mathbb{R}^{n\times n}$ and any $\varepsilon > 0$, there exists a vector norm $\|\cdot\|_v$ such that*

$$\|Ax\|_v \leq (\rho(A) + \varepsilon)\|x\|_v \quad \forall\, x \in \mathbb{R}^n, \tag{1.3.45}$$

where $\rho(A) \triangleq \max\{|\lambda_i|, i = 1, \cdots, n\}$ and $\{\lambda_i, i = 1, \cdots, n\}$ are the eigenvalues of A.

Proof. First, for the matrix A there exists a unitary matrix U such that

$$U^{-1}AU = \begin{bmatrix} \lambda_1 & t_{12} & t_{13} & \cdots & t_{1n} \\ 0 & \lambda_1 & t_{23} & \cdots & t_{2n} \\ \vdots & \ddots & \ddots & & \vdots \\ \vdots & & \ddots & \ddots & \vdots \\ 0 & \cdots & \cdots & 0 & \lambda_n \end{bmatrix}. \tag{1.3.46}$$

For any fixed $\delta > 0$, define

$$D_\delta \triangleq \text{diag}\{1, \delta, \cdots, \delta^{n-1}\}. \tag{1.3.47}$$

Then it follows that

$$(UD_\delta)^{-1}A(UD_\delta) = \begin{bmatrix} \lambda_1 & \delta t_{12} & \delta^2 t_{13} & \cdots & \delta^{n-1} t_{1n} \\ 0 & \lambda_1 & \delta t_{23} & \cdots & \delta^{n-2} t_{2n} \\ \vdots & \ddots & \ddots & & \vdots \\ \vdots & & \ddots & \ddots & \delta t_{n-1,n} \\ 0 & \cdots & \cdots & 0 & \lambda_n \end{bmatrix}. \tag{1.3.48}$$

For the given $\varepsilon > 0$, we can choose $\delta > 0$ small enough such that

$$\sum_{j=i+1}^{n} |t_{ij}|\delta^{j-i} < \varepsilon, \;\; i = 1, \cdots, n-1. \tag{1.3.49}$$

For any $x \in \mathbb{R}^n$, define the vector norm by

$$\|x\|_v \triangleq \|(UD_\delta)^{-1}x\|_\infty, \tag{1.3.50}$$

where $\|v\|_\infty \triangleq \max_{1\le i\le n}\{|v_i|\}$, $v = [v_1 \cdots v_n]^T \in \mathbb{R}^n$.

From (1.3.49) and (1.3.50), we have

$$\begin{aligned} \|Ax\|_v =& \|(UD_\delta)^{-1}Ax\|_\infty \\ =& \max_{1\le i\le n} \left|\left((UD_\delta)^{-1}Ax\right)_i\right| \\ =& \max_{1\le i\le n} \left|\left((UD_\delta)^{-1}A(UD_\delta)\cdot(UD_\delta)^{-1}x\right)_i\right| \\ \le& \max_{1\le i\le n}\left(\sum_{j=i+1}^{n}|t_{ij}|\delta^{j-i} + |\lambda_i|\right)\cdot \max_{1\le i\le n}\left|\left((UD_\delta)^{-1}Ax\right)_i\right| \\ \le& (\rho(A)+\varepsilon)\|x\|_v \;\; \forall\, x \in \mathbb{R}^n. \end{aligned} \tag{1.3.51}$$

This finishes the proof. □

Proof of Theorem 1.3.8. We only sketch the proof, since it can be done in similar fashion to Theorems 1.3.6 and 1.3.7.

From (1.3.44), it follows that

$$x_{k+1} = F^{k+1}x_0 + F^kG\varepsilon_1 + F^{k-1}G\varepsilon_2 + \cdots + G\varepsilon_{k+1}. \tag{1.3.52}$$

Denote by $\mathbb{E}$ the vector space spanned by vectors $\{F^{k+1}x_0(\omega) + F^kG\varepsilon_1(\omega) + F^{k-1}G\varepsilon_2(\omega) + \cdots + G\varepsilon_{k+1}(\omega),\ k \ge 1,\ \omega \in \Omega\}$. By A1.3.7 and A1.3.8, $\mathbb{E}$ is μ_m-positive. Denote by $\mathscr{C}^m$ the sub-σ-algebra of $\mathscr{B}^m$ restricted on $\mathbb{E}$. In the following we consider the measurable space $(\mathbb{E}, \mathscr{C}^m)$ and the Lebesgue measure on it. For simplicity of notations, the Lebesgue measure on $(\mathbb{E}, \mathscr{C}^m)$ is still denoted by $\mu_m(\cdot)$.

It can be shown that $\{x_k\}_{k\geq 0}$ defined by (1.3.44) is a Markov chain valued in $(\mathbb{E},\mathscr{C}^m)$. Carrying out a discussion similar to that for Theorems 1.3.6 and 1.3.7 and noticing that the distribution of ε_k is absolutely continuous with respect to $\mu_m(\cdot)$, we can show that $\{x_k\}_{k\geq 0}$ is $\mu_m(\cdot)$-irreducible, aperiodic, and any bounded set in $\mathscr{C}^m$ with a positive μ_m-measure is a small set.

By Lemma 1.3.4 for the matrix F there exist a vector norm $\|\cdot\|_v$ and $0<\lambda<1$ such that $\|Fx\|_v \leq \lambda\|x\|_v \ \forall\ x\in\mathbb{R}^n$. Then by choosing the Lyapunov function $g(\cdot)=\|\cdot\|_v$ and by applying Theorem 1.3.3, it is shown that $\{x_k\}_{k\geq 0}$ is geometrically ergodic. □

1.4 Mixing Random Processes

Consider the following linear systems

$$y_{1,k+1} = b_1 u_k + \varepsilon_{k+1}, \tag{1.4.1}$$

$$y_{2,k+1} = b_1 u_k + \cdots + b_q u_{k+1-q} + \varepsilon_{k+1}, \tag{1.4.2}$$

$$y_{3,k+1} = a_1 y_{3,k} + \cdots + a_p y_{3,k+1-p} + b_1 u_k + \cdots + b_q u_{k+1-q} + \varepsilon_{k+1}. \tag{1.4.3}$$

Suppose that $\{u_k\}_{k\geq 0}$ and $\{\varepsilon_k\}_{k\geq 0}$ are mutually independent and each of them is a sequence of iid random variables. Further, assume $A(z)=1-a_1z-\cdots-a_pz^p$ is stable, i.e., all roots of $A(z)$ lie strictly outside the unit disk. It is clear that $\{y_{1,k}\}_{k\geq 0}$ and $\{y_{2,qk+l}\}_{k\geq 0}$ are iid sequences for each $l=0,1,\cdots,q-1$. But, this does not hold for $\{y_{3,k}\}_{k\geq 0}$, since for each k, $y_{3,k}$ depends on the past inputs $\{u_i\}_{i=0}^{k-1}$ and noises $\{\varepsilon_i\}_{i=0}^{k}$. However, since $A(z)$ is stable, we can show that as l tends to infinity, $y_{3,k}$ and $y_{3,k+l}$ are asymptotically independent in a certain sense. In probability theory, this is called the *mixing*. In this section, we first introduce definitions of different types of mixing random processes and the related covariance inequalities, then give results on the almost sure convergence of mixing random series, and finally present the connection between the mixing random processes and the geometrically ergodic Markov chains. It is worth noting that all definitions and results given here can be applied to random vectors.

Let $\{\varphi_k\}_{k\geq 0}$ be a random sequence and let $\mathscr{F}_0^n \triangleq \sigma\{\varphi_k,\ 0\leq k\leq n\}$ and $\mathscr{F}_n^\infty \triangleq \sigma\{\varphi_k,\ k\geq n\}$ be the σ-algebras generated by $\{\varphi_k,\ 0\leq k\leq n\}$ and $\{\varphi_k,\ k\geq n\}$, respectively.

Definition 1.4.1 *The process* $\{\varphi_k\}_{k\geq 0}$ *is called an* α-mixing *or* strong mixing *if*

$$\alpha(k) \triangleq \sup_n \sup_{A\in\mathscr{F}_0^n, B\in\mathscr{F}_{n+k}^\infty} |\mathbb{P}(AB)-\mathbb{P}(A)\mathbb{P}(B)| \underset{k\to\infty}{\longrightarrow} 0, \tag{1.4.4}$$

a β-mixing *or* completely regular *if*

$$\beta(k) \triangleq \sup_n E\Big[\sup_{B\in\mathscr{F}_{n+k}^\infty} |\mathbb{P}(B|\mathscr{F}_0^n)-\mathbb{P}(B)|\Big] \underset{k\to\infty}{\longrightarrow} 0, \tag{1.4.5}$$

and a ϕ-mixing *or* uniformly strong mixing *if*

$$\phi(k) \triangleq \sup_n \sup_{A\in\mathscr{F}_0^n, \mathbb{P}(A)>0, B\in\mathscr{F}_{n+k}^\infty} \frac{|\mathbb{P}(AB)-\mathbb{P}(A)\mathbb{P}(B)|}{\mathbb{P}(A)} \underset{k\to\infty}{\longrightarrow} 0. \tag{1.4.6}$$

The sequences $\{\alpha(k)\}_{k\geq 0}$, $\{\beta(k)\}_{k\geq 0}$, and $\{\phi(k)\}_{k\geq 0}$ are called the *mixing coefficients*. It can be shown that

$$\alpha(k) \leq \beta(k) \leq \phi(k). \tag{1.4.7}$$

Lemma 1.4.1 *(i) Assume* $\{\varphi_k\}_{k\geq 0}$ *is an* α*-mixing. For* $\xi \in \mathscr{F}_0^k$ *and* $\eta \in \mathscr{F}_{n+k}^\infty$*, if* $E[|\xi|^p + |\eta|^q] < \infty$ *for some* $p > 1$, $q > 1$, *and* $\frac{1}{p}+\frac{1}{q} < 1$, *then*

$$|E\xi\eta - E\xi E\eta| \leq 10(\alpha(n))^{1-\frac{1}{p}-\frac{1}{q}}(E|\xi|^p)^{\frac{1}{p}}(E|\eta|^q)^{\frac{1}{q}}. \tag{1.4.8}$$

(ii) Assume $\{\varphi_k\}_{k\geq 0}$ *is an* ϕ*-mixing. For* $\xi \in \mathscr{F}_0^k$ *and* $\eta \in \mathscr{F}_{n+k}^\infty$*, if* $E[|\xi|^p + |\eta|^q] < \infty$ *for some* $p > 1$, $q > 1$, *and* $\frac{1}{p}+\frac{1}{q} = 1$, *then*

$$|E\xi\eta - E\xi E\eta| \leq 2(\phi(n))^{\frac{1}{p}}(E|\xi|^p)^{\frac{1}{p}}(E|\eta|^q)^{\frac{1}{q}}. \tag{1.4.9}$$

Proof. See Appendix A. □

The concept *mixingale* is generated from the mixing property and is defined as follows.

Definition 1.4.2 *Let* $\{\mathscr{F}_k\}_{k\geq 0}$ *be a sequence of nondecreasing* σ*-algebras. The sequence* $\{\varphi_k, \mathscr{F}_k\}_{k\geq 0}$ *is called a* simple mixingale *if* φ_k *is* $\mathscr{F}_k$*-measurable and if for two sequences of nonnegative constants* $\{c_k\}_{k\geq 0}$ *and* $\{\psi_m\}_{m\geq 0}$ *with* $\psi_m \to 0$ *as* $m\to\infty$*, the following conditions are satisfied:*

(i) $(E|E(\varphi_k|\mathscr{F}_{k-m})|^2)^{\frac{1}{2}} \leq \psi_m c_k \ \forall\, k\geq 0$ and $\forall\, m \geq 0$,

(ii) $E\varphi_k = 0$,

where $\mathscr{F}_k \triangleq \{\emptyset, \Omega\}$ *if* $k \leq 0$.

From the definition, we see that $\{c_k\}_{k\geq 0}$ and $\{\psi_m\}_{m\geq 0}$ reflect the moment and mixing coefficients of $\{\varphi_k\}_{k\geq 0}$, which are important for the almost sure convergence of $\sum_{k=0}^\infty \varphi_k$. In fact, we have the following result.

Theorem 1.4.1 *Let* $\{\varphi_k, \mathscr{F}_k\}_{k\geq 0}$ *be a simple mixingale such that*

$$\sum_{k=1}^\infty c_k^2 < \infty \tag{1.4.10}$$

and

$$\sum_{k=1}^\infty (\log k)(\log\log k)^{1+\gamma}\psi_k^2 \sum_{j=k}^\infty c_j^2 < \infty \ \text{ for some } \gamma > 0. \tag{1.4.11}$$

Then

$$\sum_{k=1}^{\infty} \varphi_k < \infty \quad \text{a.s.} \tag{1.4.12}$$

Proof. See Appendix A. □

Theorem 1.4.2 *Assume that* $\{\varphi_k\}_{k\geq 0}$ *is an* α*-mixing with mixing coefficients denoted by* $\{\alpha(k)\}_{k\geq 0}$*. Let* $\{\Phi_k(\cdot)\}_{k\geq 0}$ *be a sequence of functions* $\Phi_k(\cdot):\mathbb{R}\to\mathbb{R}$ *and* $E\Phi_k(\varphi_k)=0$*. If there exist constants* $\varepsilon>0$ *and* $\gamma>0$ *such that*

$$\sum_{k=1}^{\infty} \left(E|\Phi_k(\varphi_k)|^{2+\varepsilon}\right)^{\frac{2}{2+\varepsilon}} < \infty \tag{1.4.13}$$

and

$$\sum_{k=1}^{\infty} \log k (\log\log k)^{1+\gamma} (\alpha(k))^{\frac{\varepsilon}{2+\varepsilon}} < \infty, \tag{1.4.14}$$

then

$$\sum_{k=1}^{\infty} \Phi_k(\varphi_k) < \infty \text{ a.s.} \tag{1.4.15}$$

Proof. See Appendix A. □

By Theorem 1.4.2, to establish the almost sure convergence of series of mixing random variables, it suffices to verify the convergence of two deterministic series (1.4.13) and (1.4.14). The first series concerns the $(2+\varepsilon)$th moment of the variables, while the second one is related to the mixing coefficients.

If $\{\varphi_k\}_{k\geq 0}$ is a sequence of mutually independent random variables satisfying the assumptions required in Theorem 1.4.2, then the mixing coefficients $\alpha(k)=0,\ k\geq 1$. By the Lyapunov inequality we have $\left(E|\Phi_k(\varphi_k)|^2\right)^{\frac{1}{2}} \leq \left(E|\Phi_k(\varphi_k)|^{2+\varepsilon}\right)^{\frac{1}{2+\varepsilon}}$. Hence (1.4.13) implies

$$\sum_{k=1}^{\infty} E|\Phi_k(\varphi_k)|^2 < \infty. \tag{1.4.16}$$

Applying Theorem 1.2.8 to the sum of independent random variables, we obtain

$$\sum_{k=1}^{\infty} \Phi_k(\phi_k) < \infty \text{ a.s.}$$

So, Theorem 1.4.2 can be regarded as a generalization of the corresponding results for the sum of independent random variables.

The following theorem connects the mixing process with the transition probability of Markov chains.

Theorem 1.4.3 *Assume $\{x_k\}_{k\geq 0}$ is a Markov chain with state space $(\mathbb{R}^m, \mathscr{B}^m)$. The β-mixing coefficients of $\{x_k\}_{k\geq 0}$ are estimated as follows:*

$$\beta(k) \leq \sup_n \int_{\mathbb{R}^m} \|P_k(x,\cdot) - P_{k+n}(\cdot)\|_{\text{var}} P_n(\mathrm{d}x). \tag{1.4.17}$$

Proof. See Appendix A. □

For the NARX system (1.3.29) and the linear stochastic system (1.3.44), the following theorem takes place.

Theorem 1.4.4 *If A1.3.1–A1.3.3 and A1.3.4'–A1.3.5' hold, then*

(i) $\{x_k\}_{k\geq 0}$ *defined by (1.3.30) is an α-mixing with mixing coefficient, denoted by $\alpha(k)$, satisfying $\alpha(k) \leq c\rho^k$ $\forall$ $k \geq 1$ for some constants $c > 0$ and $\rho \in (0,1)$.*

(ii) *Similar results also hold for the process $\{F(x_k)\}_{k\geq 0}$, where $F(\cdot)$ is any measurable function defined on $(\mathscr{B}^{p+q}, \mathbb{R}^{p+q})$.*

If assumptions A1.3.6–A1.3.8 hold, then $\{x_k\}_{k\geq 0}$ defined by (1.3.44) is also an α-mixing with mixing coefficient geometrically tending to zero.

Proof. We first consider the NARX system (1.3.29).

By Theorem 1.4.3, the β-mixing coefficient $\beta(k)$ of the chain $\{x_k\}_{k\geq 0}$ defined by (1.3.30) can be estimated by the transition probability as follows:

$$\begin{aligned}\beta(k) \leq & \sup_n \int_{\mathbb{R}^{p_0+q_0}} \|P_k(x,\cdot) - P_{n+k}(\cdot)\|_{\text{var}} P_n(\mathrm{d}x) \\ \leq & \sup_n \int_{\mathbb{R}^{p_0+q_0}} \|P_k(x,\cdot) - P_{\text{IV}}(\cdot)\|_{\text{var}} P_n(\mathrm{d}x) \\ & + \sup_n \int_{\mathbb{R}^{p_0+q_0}} \|P_{n+k}(\cdot) - P_{\text{IV}}(\cdot)\|_{\text{var}} P_n(\mathrm{d}x).\end{aligned} \tag{1.4.18}$$

Further, by Theorem 1.3.7 we have

$$\begin{aligned}&\sup_n \int_{\mathbb{R}^{p_0+q_0}} \|P_k(x,\cdot) - P_{\text{IV}}(\cdot)\|_{\text{var}} P_n(\mathrm{d}x) \\ &\quad \leq \rho^k \sup_n \int_{\mathbb{R}^{p_0+q_0}} M(x) P_n(\mathrm{d}x) \leq c_1 \rho^k,\end{aligned} \tag{1.4.19}$$

and

$$\sup_n \int_{\mathbb{R}^{p_0+q_0}} \|P_{n+k}(\cdot) - P_{\text{IV}}(\cdot)\|_{\text{var}} P_n(\mathrm{d}x) \leq c_2 \rho^k, \tag{1.4.20}$$

where c_1, c_2, and $\rho \in (0,1)$ are constants.

Combining (1.4.18)–(1.4.20) and noticing (1.4.7), we know that $\{x_k\}_{k\geq 0}$ defined by (1.3.30) is an α-mixing with mixing coefficient $\alpha(k)$ satisfying $\alpha(k) \leq c\rho^k$, $k \geq 1$ for some constants $c > 0$ and $\rho \in (0,1)$. So the assertion (i) takes place.

For the process $\{F(x_k)\}_{k\geq 0}$, it is clear that $\sigma\{F(x_k), 1 \leq k \leq n\} \subset \sigma\{x_k, 1 \leq k \leq n\}$ and $\sigma\{F(x_k), k \geq n\} \subset \sigma\{x_k, k \geq n\}$. From here by Definition 1.4.1 we know that the mixing coefficient of $\{F(x_k)\}_{k\geq 0}$ is not bigger than that of $\{x_k\}_{k\geq 0}$. Thus the assertion (ii) holds.

The assertions for $\{x_k\}_{k\geq 0}$ defined by the linear stochastic system (1.3.44) can be obtained similarly to that for the NARX system (1.3.29). The proof is completed. □

1.5 Stationary Processes

Let $\{X_n\}_{n\geq 1}$ and $\{Y_n\}_{n\geq 1}$ be the random sequences on the probability space $(\Omega, \mathscr{F}, \mathbb{P})$, and let $f(\cdot)$ be a measurable function from $(\mathbb{R}^\infty, \mathscr{B}^\infty)$ to $(\mathbb{R}, \mathscr{B})$.

Definition 1.5.1 *Define*

$$\xi = f(\{X_n\}_{n\geq 1}, \{Y_n\}_{n\geq 1}), \tag{1.5.1}$$

$$\xi_k = f(\{X_{n+k-1}\}_{n\geq 1}, \{Y_{n+k-1}\}_{n\geq 1}) \quad \forall k \geq 1. \tag{1.5.2}$$

The random variable ξ_k is called the translate *of ξ_1 by $k-1$. If $\xi = \xi_k$ for all $k \geq 1$, then ξ is said to be* invariant.

Define $\overline{\mathscr{F}} = \sigma\{\{X_n\}_{n\geq 1}, \{Y_n\}_{n\geq 1}\}$. For $B \in \overline{\mathscr{F}}$, its translate, denoted by $\{B_k\}_{k\geq 1}$, is defined through the indicator function $I_B(\cdot)$. If $I_B(\cdot)$ is invariant, then B is said to be an *invariant set* or *invariant event*. The class $\mathscr{C}$ of all invariant events is closed under a countable set of operations and thus $\mathscr{C}$ is a σ-algebra defined by $\{X_n\}_{n\geq 1}, \{Y_n\}_{n\geq 1}$.

Define $X^n = \sum_{k=1}^n X_k$ and $Y^n = \sum_{k=1}^n Y_k$. In this section, we always assume that $Y_n > 0$ and $Y^n \to \infty$ as $n \to \infty$. We present a few examples of invariant random variables and events defined by $\{X_n\}_{n\geq 1}, \{Y_n\}_{n\geq 1}$. We have the following equalities,

$$\begin{aligned}\frac{X_2 + \cdots + X_{n+1}}{Y_2 + \cdots + Y_{n+1}} &= \frac{X^{n+1} - X_1}{Y^{n+1} - Y_1} = \frac{X^{n+1}}{Y^{n+1} - Y_1} - \frac{X_1}{Y^{n+1} - Y_1} \\ &= \frac{X^{n+1}}{Y^{n+1}} \cdot \frac{Y^{n+1}}{Y^{n+1} - Y_1} - \frac{X_1}{Y^{n+1} - Y_1}.\end{aligned} \tag{1.5.3}$$

Noticing that $Y^n \to \infty$ as $n \to \infty$, from (1.5.3) we have

$$\limsup_{n\to\infty} \frac{X_2 + \cdots + X_{n+1}}{Y_2 + \cdots + Y_{n+1}} = \limsup_{n\to\infty} \frac{X^{n+1}}{Y^{n+1}}, \tag{1.5.4}$$

$$\liminf_{n\to\infty} \frac{X_2 + \cdots + X_{n+1}}{Y_2 + \cdots + Y_{n+1}} = \liminf_{n\to\infty} \frac{X^{n+1}}{Y^{n+1}}. \tag{1.5.5}$$

Thus, $\limsup\limits_{n\to\infty}\frac{X^n}{Y^n}$ and $\liminf\limits_{n\to\infty}\frac{X^n}{Y^n}$ are invariant and the ω-sets

$$C=\left\{\liminf_{n\to\infty}\frac{X^n}{Y^n}=\limsup_{n\to\infty}\frac{X^n}{Y^n}\right\}, \tag{1.5.6}$$

$$D=\left\{\liminf_{n\to\infty}\frac{X^n}{Y^n}\neq\limsup_{n\to\infty}\frac{X^n}{Y^n}\right\}, \tag{1.5.7}$$

$$\underline{C}_a=\left\{\liminf_{n\to\infty}\frac{X^n}{Y^n}<a\right\}, \tag{1.5.8}$$

$$\overline{C}_b=\left\{\limsup_{n\to\infty}\frac{X^n}{Y^n}>b\right\} \tag{1.5.9}$$

are invariant events.

We now are in a position to present the basic inequality for invariant functions. We introduce a lemma.

Let $\{a_1,a_2,\cdots,a_{n+m}\}$ be finite numbers. The term $a_k\in\{a_1,a_2,\cdots,a_{n+m}\}$ is called *m-positive* if

$$\max_{k\le l\le\min(n+m,k+m-1)}\{a_k+\cdots+a_l\}>0. \tag{1.5.10}$$

Lemma 1.5.1 *If the m-positive terms exist for* $\{a_1,a_2,\cdots,a_{n+m}\}$, *then their sum is positive.*

Proof. See Appendix A. □

Define

$$B^m=\left\{\omega:\sup_{j\le m}\frac{X^j}{Y^j}>b\right\}. \tag{1.5.11}$$

Lemma 1.5.2 *For any* $n\ge1$, *any positive sequence* $\{Z^n\}_{n\ge1}$, *and any set* C *measurable with respect to* $\mathscr{F}$, *it follows that*

$$\sum_{k=1}^{n}\int_{B_k^mC}\left(\frac{X_k}{Z^n}-b\frac{Y_k}{Z^n}\right)\mathrm{d}\mathbb{P}+\sum_{k=n+1}^{n+m}\int_{C}\left(\frac{X_k}{Z^n}-b\frac{Y_k}{Z^n}\right)_+\mathrm{d}\mathbb{P}\ge0, \tag{1.5.12}$$

where B_k^m *is the translate of* B^m *by* $k-1$.

Proof. See Appendix A. □

We proceed to consider the sequences $\{X_n\}_{n\ge1}$, $\{Y_n\}_{n\ge1}$ and their translates $\{\xi_n\}_{n\ge1}$. Assume A is measurable with respect to $\sigma\{\{X_n\}_{n\ge1},\{Y_n\}_{n\ge1}\}$, and $\{A_k\}_{k\ge1}$ is the translate of A.

Definition 1.5.2 $\{\xi_k\}_{k\ge1}$ *is called* integral stationary *if*

$$\int_{A_k}\xi_k\mathrm{d}\mathbb{P}=\int_{A_1}\xi_1\mathrm{d}\mathbb{P}\ \ \forall\, k\ge1; \tag{1.5.13}$$

and $\{A_k\}_{k\ge1}$ *is called* probability stationary *if*

$$\mathbb{P}\{A_k\}=\mathbb{P}\{A_1\}\ \ \forall\, k\ge1. \tag{1.5.14}$$

Lemma 1.5.3 *Let $\{X_n\}_{n\geq 1}$ and $\{Y_n\}_{n\geq 1}$ be integral stationary and $E|X_1| < \infty$, $E|Y_1| < \infty$. Then for any invariant event C it holds that*

$$\int_{C\underline{C}_a} (aY_1 - X_1)\mathrm{d}\mathbb{P} \geq 0, \quad \int_{C\overline{C}_b} (X_1 - bY_1)\mathrm{d}\mathbb{P} \geq 0, \tag{1.5.15}$$

where $\underline{C}_a$ and $\overline{C}_b$ are defined by (1.5.8) and (1.5.9), respectively.

Proof. See Appendix A. □

We now present the main result of the section, the strong law of large numbers for stationary processes.

Theorem 1.5.1 *Assume $\{X_n\}_{n\geq 1}$ is integral stationary and $E|X_1| < \infty$. Then*

(i) X^n/n converges almost surely to a random variable U;

(ii) U is invariant with respect to the family $\{X_n\}_{n\geq 1}$;

(iii) $U = E(X_1|\mathscr{C})$ a.s. where $\mathscr{C}$ is the invariant σ-algebra generated by $\{X_n\}_{n\geq 1}$.

Proof. See Appendix A. □

Assume $\{X_n\}_{n\geq 1}$ is a sequence of iid random variables with $E|X_1| < \infty$. By Theorem 1.5.1 we have $\sum_{k=1}^n X_k/n \to E(X_1|\mathscr{C})$ a.s. as $n \to \infty$, where $\mathscr{C}$ is the invariant σ-algebra of $\{X_n\}_{n\geq 1}$. From Definition 1.2.2 we know that the invariant events of $\{X_n\}_{n\geq 1}$ are also tail events of $\{X_n\}_{n\geq 1}$. By Theorem 1.2.1, $\mathbb{P}\{C\} = 0$ or 1 $\forall\, C \in \mathscr{C}$. Thus the σ-algebra $\mathscr{C}$ is degenerate and $E(X_1|\mathscr{C}) = EX_1$ a.s.

Theorem 1.5.2 *(Kolmogorov Strong Law of Large Number) If $\{X_n\}_{n\geq 1}$ is iid with finite EX_1, then*

$$\frac{\sum_{k=1}^n X_k}{n} \underset{n\to\infty}{\longrightarrow} EX_1 \ \text{ a.s.} \tag{1.5.16}$$

Let us consider the linear stochastic system (1.3.44). Suppose that $\{\varepsilon_k\}_{k\geq 0}$ is iid with $E\varepsilon_k = 0$, $R_\varepsilon \triangleq E\varepsilon_k\varepsilon_k^T > 0$, $E\|\varepsilon_k\|^2 < \infty$, and the matrix F is stable, i.e., all eigenvalues of F are strictly inside the unit cycle.

Under the above conditions, the covariance function of $\{x_k\}_{k\geq 0}$ generated by (1.3.44) exists. Thus, for any $j \geq 0$ we can define $R(j) \triangleq \lim_{k\to\infty} Ex_kx_{k-j}^T$. From (1.3.44), it is seen that

$$R(j) = \sum_{k=0}^{\infty} F^{k+j}GR_\varepsilon G^T F^{kT}, \quad j \geq 0. \tag{1.5.17}$$

However, since R_ε generally is unknown, we cannot obtain $R(j)$ directly from (1.5.17). We use the following average to estimate $R(j)$

$$\frac{1}{n}\sum_{k-1}^{n} x_{k+j}x_k^T. \tag{1.5.18}$$

Theorem 1.5.3 *For the linear stochastic system (1.3.44), assume that $\{\varepsilon_k\}_{k\geq 0}$ is iid with $E\varepsilon_k = 0$, $R_\varepsilon \triangleq E\varepsilon_k\varepsilon_k^T > 0$, and the matrix F is stable. Then*

$$\frac{1}{n}\sum_{k=1}^{n} x_{k+j}x_k^T \underset{n\to\infty}{\longrightarrow} R(j) \text{ a.s.} \tag{1.5.19}$$

for all $j \geq 0$.

Proof. Here we only prove the convergence of (1.5.19) for $j \geq 1$ while for the case $j = 0$ it can similarly be proved.

From (1.3.44), we have the following identities

$$x_{k+1} = F^{k+1}x_0 + \sum_{l=1}^{k+1} F^{k+1-l}G\varepsilon_l, \tag{1.5.20}$$

$$x_{k+j} = F^j x_k + \sum_{l=k+1}^{k+j} F^{k+j-l}G\varepsilon_l, \tag{1.5.21}$$

$$x_{k+j+1} = Fx_{k+j} + G\varepsilon_{k+j+1}, \tag{1.5.22}$$

and for $j \geq 1$

$$\begin{aligned}
x_{k+j+1}x_{k+1}^T =& (Fx_{k+j} + G\varepsilon_{k+j+1})(Fx_k + G\varepsilon_{k+1})^T \\
=& Fx_{k+j}x_k^T F^T + Fx_{k+j}\varepsilon_{k+1}^T G^T + G\varepsilon_{k+j+1}x_k^T F^T + G\varepsilon_{k+j+1}\varepsilon_{k+1}^T G^T \\
=& Fx_{k+j}x_k^T F^T + F\Big(F^j x_k + \sum_{l=k+1}^{k+j} F^{k+j-l}G\varepsilon_l\Big)\varepsilon_{k+1}^T G^T \\
&+ G\varepsilon_{k+j+1}x_k^T F^T + G\varepsilon_{k+j+1}\varepsilon_{k+1}^T G^T \\
=& Fx_{k+j}x_k^T F^T + F^{j+1}x_k\varepsilon_{k+1}^T G^T + \sum_{l=k+1}^{k+j} F^{k+j+1-l}G\varepsilon_l\varepsilon_{k+1}^T G^T \\
&+ G\varepsilon_{k+j+1}x_k^T F^T + G\varepsilon_{k+j+1}\varepsilon_{k+1}^T G^T.
\end{aligned} \tag{1.5.23}$$

By stability of F, there exist constants $c > 0$ and $0 < \rho < 1$ such that $\|F^k\| \leq c\rho^k$. Then from (1.5.20), we have

$$\|x_{k+1}\|^2 = O(\rho^{2k}) + O\Big(\sum_{l=1}^{k+1} \rho^{k+1-l}\|\varepsilon_l\|^2\Big)$$

and

$$\sum_{k=1}^{n+1} \|x_{k+1}\|^2 = O(1) + O\Big(\sum_{k=1}^{n+1}\sum_{l=1}^{k+1} \rho^{k+1-l}\|\varepsilon_l\|^2\Big) = O(1) + O\Big(\sum_{l=1}^{n+1}\|\varepsilon_l\|^2\Big) = O(n). \tag{1.5.24}$$

Noticing that (1.5.23) is recursive with respect to $x_{k+j}x_k^T$, we have

$$x_{k+j+1}x_{k+1}^T = I_{1,k} + I_{2,k} + I_{3,k} + I_{4,k} + I_{5,k}, \qquad (1.5.25)$$

where

$$I_{1,k} \triangleq F^{k+1}x_j x_0^T F^{(k+1)T}, \qquad (1.5.26)$$

$$I_{2,k} \triangleq \sum_{s=0}^{k} F^{k+j+1-s}x_s\varepsilon_{s+1}^T G^T F^{(k-s)T}, \qquad (1.5.27)$$

$$I_{3,k} \triangleq \sum_{s=0}^{k}\sum_{l=s}^{s+j-1} F^{k+j-l}G\varepsilon_{l+1}\varepsilon_{s+1}^T G^T F^{(k-s)T}, \qquad (1.5.28)$$

$$I_{4,k} \triangleq \sum_{s=0}^{k} F^{k-s}G\varepsilon_{s+j+1}x_s^T F^{(k+1-s)T}, \qquad (1.5.29)$$

$$I_{5,k} \triangleq \sum_{s=0}^{k} F^{k-s}G\varepsilon_{s+j+1}\varepsilon_{s+1}^T G^T F^{(k-s)T}. \qquad (1.5.30)$$

For $I_{1,k}$, by stability of F we have

$$\frac{1}{n}\sum_{k=0}^{n} I_{1,k} = o(1). \qquad (1.5.31)$$

For $I_{2,k}$ the following identities take place,

$$\begin{aligned}
\sum_{k=0}^{n} I_{2,k} &= \sum_{k=0}^{n}\sum_{s=0}^{k} F^{k+j+1-s}x_s\varepsilon_{s+1}^T G^T F^{(k-s)T} \\
&= \sum_{s=0}^{n}\sum_{k=s}^{n} F^{k+j+1-s}x_s\varepsilon_{s+1}^T G^T F^{(k-s)T} \\
&= \sum_{s=0}^{n}\sum_{k=s}^{\infty} F^{k+j+1-s}x_s\varepsilon_{s+1}^T G^T F^{(k-s)T} \\
&\quad - \sum_{s=0}^{n}\sum_{k=n+1}^{\infty} F^{k+j+1-s}x_s\varepsilon_{s+1}^T G^T F^{(k-s)T} \\
&= \sum_{s=0}^{n}\sum_{k=0}^{\infty} F^{k+j+1}x_s\varepsilon_{s+1}^T G^T F^{kT} \\
&\quad - \sum_{s=0}^{n}\sum_{k=n+1}^{\infty} F^{k+j+1-s}x_s\varepsilon_{s+1}^T G^T F^{(k-s)T}.
\end{aligned} \qquad (1.5.32)$$

In what follows c always denotes a constant, but it may change from place to place.

By stability of F, (1.5.24), and Theorem 1.2.14, we have

$$\begin{aligned}
&\Big\|\sum_{s=0}^{n}\sum_{k=0}^{\infty}F^{k+j+1}x_s\varepsilon_{s+1}^T G^T F^{kT}\Big\| \\
&=\Big\|\sum_{k=0}^{\infty}F^{k+j+1}\Big(\sum_{s=0}^{n}x_s\varepsilon_{s+1}^T\Big)G^T F^{kT}\Big\| \\
&=O\Big(\Big(\sum_{s=0}^{n}\|x_s\|^2\Big)^{\frac{1}{2}+\eta}\Big)=O\Big(n^{\frac{1}{2}+\eta}\Big) \quad \text{a.s. } \forall\,\eta>0,
\end{aligned} \tag{1.5.33}$$

and

$$\begin{aligned}
&\Big\|\sum_{s=0}^{n}\sum_{k=n+1}^{\infty}F^{k+j+1-s}x_s\varepsilon_{s+1}^T G^T F^{(k-s)T}\Big\| \le c\sum_{s=0}^{n}\sum_{k=n+1}^{\infty}\rho^{2(k-s)}\|x_s\varepsilon_{s+1}^T\| \\
&\le c\sum_{s=0}^{n}\rho^{2(n-s)}\|x_s\varepsilon_{s+1}^T\| \le c\Big(\sum_{s=0}^{n}\|x_s\|^2\|\varepsilon_{s+1}\|^2\Big)^{\frac{1}{2}} \\
&=c\Big(\sum_{s=0}^{n}\|x_s\|^2\Big(\|\varepsilon_{s+1}\|^2-E\|\varepsilon_{s+1}\|^2\Big)+\sum_{s=0}^{n}\|x_s\|^2E\|\varepsilon_{s+1}\|^2\Big)^{\frac{1}{2}} \\
&=O\Big(\big(n^{1+\eta}+n\big)^{\frac{1}{2}}\Big)=O\Big(n^{\frac{1}{2}+\frac{\eta}{2}}\Big) \quad \text{a.s. } \forall\,\eta>0.
\end{aligned} \tag{1.5.34}$$

From (1.5.32)–(1.5.34) it follows that

$$\frac{1}{n}\sum_{k=0}^{n}I_{2,k}=o(1). \tag{1.5.35}$$

Carrying out the similar discussion as that for $I_{2,k}$ and noticing $j\ge 1$, we can prove that

$$\frac{1}{n}\sum_{k=0}^{n}I_{4,k}=o(1), \tag{1.5.36}$$

$$\frac{1}{n}\sum_{k=0}^{n}I_{5,k}=o(1). \tag{1.5.37}$$

For $I_{3,k}$, in the case $l=s$, we have

$$\begin{aligned}
&\sum_{k=0}^{n}\sum_{s=0}^{k}F^{k+j-s}G\varepsilon_{s+1}\varepsilon_{s+1}^T G^T F^{(k-s)T} \\
&=\sum_{s=0}^{n}\sum_{k=s}^{n}F^{k+j-s}G\varepsilon_{s+1}\varepsilon_{s+1}^T G^T F^{(k-s)T} \\
&=\sum_{s=0}^{n}\sum_{k=s}^{\infty}F^{k+j-s}G\varepsilon_{s+1}\varepsilon_{s+1}^T G^T F^{(k-s)T}-\sum_{s=0}^{n}\sum_{k=n+1}^{\infty}F^{k+j-s}G\varepsilon_{s+1}\varepsilon_{s+1}^T G^T F^{(k-s)T}
\end{aligned}$$

$$=\sum_{s=0}^{n}\sum_{k=0}^{\infty}F^{k+j}G\left(\varepsilon_{s+1}\varepsilon_{s+1}^{T}-R_{\varepsilon}\right)G^{T}F^{kT}+(n+1)\sum_{k=0}^{\infty}F^{k+j}GR_{\varepsilon}G^{T}F^{kT}$$
$$-\sum_{s=0}^{n}\sum_{k=n+1}^{\infty}F^{k+j-s}G\varepsilon_{s+1}\varepsilon_{s+1}^{T}G^{T}F^{(k-s)T}. \tag{1.5.38}$$

By Theorem 1.5.2 and stability of F it follows that

$$\sum_{s=0}^{n}\sum_{k=0}^{\infty}F^{k+j}G\left(\varepsilon_{s+1}\varepsilon_{s+1}^{T}-R_{\varepsilon}\right)G^{T}F^{kT}$$
$$=\sum_{k=0}^{\infty}F^{k+j}G\sum_{s=0}^{n}\left(\varepsilon_{s+1}\varepsilon_{s+1}^{T}-R_{\varepsilon}\right)G^{T}F^{kT}=o(n)\ \text{ a.s.} \tag{1.5.39}$$

By (1.5.17) it is clear that

$$(n+1)\sum_{k=0}^{\infty}F^{k+j}GR_{\varepsilon}G^{T}F^{kT}=(n+1)R(j). \tag{1.5.40}$$

Since $\sum_{s=0}^{n}\|\varepsilon_{s+1}\|^2/n \underset{n\to\infty}{\longrightarrow} E\|\varepsilon_1\|^2$ a.s., we see $\|\varepsilon_{n-k}\|^2/n \underset{n\to\infty}{\longrightarrow} 0$ a.s. for any fixed $k\geq 0$. Then we have

$$\frac{1}{n}\left\|\sum_{s=0}^{n}\sum_{k=n+1}^{\infty}F^{k+j-s}G\varepsilon_{s+1}\varepsilon_{s+1}^{T}G^{T}F^{(k-s)T}\right\|\leq c\frac{1}{n}\sum_{s=0}^{n}\sum_{k=n+1}^{\infty}\rho^{2(k-s)}\|\varepsilon_{s+1}\|^2$$
$$\leq c\frac{1}{n}\sum_{s=0}^{n}\rho^{2(n-s)}\|\varepsilon_{s+1}\|^2\leq c\frac{1}{n}\sum_{t=0}^{n_1}\rho^{2t}\|\varepsilon_{n-t}\|^2+c\frac{1}{n}\sum_{t=n_1+1}^{n}\rho^{2t}\|\varepsilon_{n-t}\|^2$$
$$\leq c\frac{1}{n}\sum_{t=0}^{n_1}\|\varepsilon_{n-t}\|^2+c\rho^{2n_1}\frac{1}{n}\sum_{t=n_1+1}^{n}\|\varepsilon_{n-t}\|^2=o(1)\ \text{ a.s.} \tag{1.5.41}$$

by letting first $n\to\infty$ and then $n_1\to\infty$.

By using Theorem 1.2.14, we have

$$\sum_{k=0}^{n}\sum_{s=0}^{k}\sum_{l=s+1}^{s+j-1}F^{k+j-l}G\varepsilon_{l+1}\varepsilon_{s+1}^{T}G^{T}F^{(k-s)T}=o(n)\ \text{ a.s. } \forall\, l>s. \tag{1.5.42}$$

Combining (1.5.38)–(1.5.42), we have proved that

$$\frac{1}{n}\sum_{k=0}^{n}I_{3,k}-R(j)=o(1). \tag{1.5.43}$$

From (1.5.31), (1.5.35), (1.5.36), (1.5.37), and (1.5.43), we derive (1.5.19) for $j\geq 1$. □

Let $\{y_k\}_{k\geq 0}$, $y_k\in\mathbb{R}^m$ be a sequence of random vectors.

A1.5.1 *$Ey_k = 0,\ k \geq 0$ and the following equality takes place*

$$R_j \triangleq \int_\Omega y_{k+j} y_k^T \mathrm{d}\mathbb{P} = \int_\Omega y_{1+j} y_1^T \mathrm{d}\mathbb{P} \ \forall\, j \geq 0. \tag{1.5.44}$$

Definition 1.5.3 *The function $\Phi_y(z)$ given by*

$$\Phi_y(z) = \sum_{k=-\infty}^{\infty} R(k) z^k \tag{1.5.45}$$

is called the power spectral density function of $\{y_k\}_{k\geq 0}$, where z in (1.5.45) is a complex variable.

Theorem 1.5.4 *For the vector sequence $\{y_k\}_{k\geq 0}$ assume that A1.5.1 holds, the power spectral density function $\Phi_y(z)$ is rational, $\Phi_y(z)|_{z=e^{iw}} > 0\ \forall\, w \in (0, 2\pi]$, and both $\Phi_y(z)$ and its inverse $\Phi_y^{-1}(z)$ is analytic on $\{z : r < |z| < R\}$ for some $0 < r < 1$ and $R > 1$. Then there exist a matrix rational function $H(z)$ and a vector sequence $\{\xi_k\}_{k\geq 0}$ with $\xi_k \in \mathbb{R}^m$, $E\xi_k = 0,\ k \geq 0$, and $E\xi_k \xi_j^T = R_\xi \delta(k, j)$ such that*

$$y_k = H(z)\xi_k, \tag{1.5.46}$$

where $R_\xi > 0$, $\delta(k, j) = 1$ if $k = j$, and $\delta(k, j) = 0$ otherwise. The representation (1.5.46) is unique in the sense that $H(0) = I$ and both $H(z)$ and $H^{-1}(z)$ are stable, i.e., their poles are strictly outside the unit cycle.

The sequence $\{\xi_k\}_{k\geq 0}$ is usually called the innovation and the formula (1.5.46) is named as the innovation representation of stationary processes with rational power spectrum.

1.6 Notes and References

In this chapter some basic concepts from probability theory and stochastic processes such as independence, martingales, mdses, Markov chains, mixing, stationarity, etc. are introduced. The results given in Chapter 1 are mostly provided with mathematical derivatives. However, for readers, who are mainly interested in the idea of the proposed approach or in the solutions to the problems discussed in the subsequent chapters, the detailed proof may be ignored and the book is still readable.

For more about probability concepts introduced in Sections 1.1 and 1.2, we refer to [30], [36], [74], and [107]. For ergodicity of Markov chains valued in general state space,we refer to [81] and [90], where in [90] the theory is formulated in a more general nonnegative operator framework. The definitions of different types of the mixing random variables and the corresponding covariance inequalities can be found in [37] and [39], the inequalities connecting ergodicity of Markov chains and the mixing coefficients are given in [33], while the almost sure convergence of mixingale series is given in [78] and [79]. Reference [109] discusses the mixing of a number of nonlinear stochastic systems while references [45], [82], and [116] discuss the strong-mixing of linear stochastic systems. For stationary random processes we refer to [77] and [98], while the innovation representation can be found in [4] and [17].

Chapter 2

Recursive Parameter Estimation

CONTENTS

A practical system may be modeled as to belong to a certain class of dynamic systems with unknown parameters, for example, the class of linear stochastic systems with input u_k and output y_k. Then, the task of system identification is to determine the unknown system coefficients and systems orders. In the linear case, i.e., in the case where the system output linearly depends upon the unknown parameters, the least-squares (LS) method is commonly used and often gives satisfactory results. This will be addressed in Chapter 3. However, in many cases a system from a class is defined by parameters nonlinearly entering the system, and the LS method may not be as convenient as for the linear case. Then the task of system identification is to estimate unknown parameters and the nonlinearity on the basis of the available system input-output data set $D_n \triangleq \{u_k,\ 0 \le k \le n-1,\ y_j,\ 0 \le j \le n\}$.

Parameter estimates may be obtained by minimizing some error criterion with data running over the data set D_n for fixed n. In this case, the estimate is nonrecursive, though the optimal estimate may be iteratively obtained while searching the optimum over the feasible set of parameters. The recursive approach suggests to derive the parameter estimate at any time k from the estimate at time $k-1$ incorporating with the information contained in the input-output data $(u_{k-1},\ y_k)$ received at time k. The advantage of recursive estimation obviously consists in simplicity of updating estimates when D_n is expanding as n increases.

In this chapter we present the recursive parameter estimation method when systems nonlinearly depend on parameters. As a matter of fact, parameters to be estimated are treated as roots of some unknown regression functions; then the problem becomes how to recursively seek the roots of unknown functions which can be observed with noise. This is the topic of SA.

2.1 Parameter Estimation as Root-Seeking for Functions

A large class of problems arising from systems and control can finally be reduced to parameter estimation problems. Let us give some examples.

(i) This is obvious for identifying linear stochastic systems, since the aim of system identification in this case is to estimate coefficients and orders of the system and variance of the driven noise which all are parameters. In the case where nonlinear functions are involved in the system, the task of identification includes estimating nonlinear functions contained in the system. The unknown nonlinear function may be expressed as a linear combination of basis functions with unknown coefficients, and identification of the nonlinear function reduces to estimating these coefficients or parameters. Even if the nonlinear function $f(\cdot)$ cannot be parameterized and we have to estimate $f(v)$ for any v in order to obtain an interpolation function from estimates, then $f(v)$ with fixed v may also be treated as a parameter required to estimate.

(ii) Adaptive regulation is an other example. Let the quality indices of an industrial product depend upon system's inputs such as temperature, pressure, etc., which serve as the control of the system. Assume the dynamics of the system is unknown or not completely known. The problem is to give control in order for the quality indices to follow the given constants. This problem is called adaptive regulation. In some cases, the optimal control appears to be a constant vector u^* and the control problem turns to be estimating the parameter u^*.

(iii) The optimal iterative learning control (ILC) at a given time of a repeated cycle usually is a constant vector for a large class of nonlinear systems. So, the problem of ILC is reduced to approaching to the optimal control as the number of iteration cycles increases, in other words, to estimating the optimal control.

(iv) There are many other problems including adaptive filtering, blind identification in signal processing, principal component analysis, consensus control for multi-agent systems, and so on, which in a certain formulation can be transformed to problems of parameter estimation.

Let us by x^0 denote the parameter to be estimated. We can always treat $x^0 \in \mathbb{R}^m$ as a root of some function $f(\cdot) \in \mathbb{R}^m \to \mathbb{R}^m$, called regression function, for example, $f(x) = x - x^0$, $f(x) = x\|x - x^0\|^2$, $f(x) = b\sin\|x - x^0\|$ with $b \in \mathbb{R}^m$, or $f(x) = x^0(x - x^0)^T Ax$, etc. Then, estimating parameter x^0 becomes seeking the root of the unknown regression function $f(\cdot)$.

Let $x_k \in \mathbb{R}^m$ be the estimate for x^0 at time k, and let the data $O_{k+1} \in \mathbb{R}^m$ be available at time $k+1$. No matter O_{k+1} actually contains information concerning x^0 or not, it can always be written as an observation of $f(x_k)$:

$$O_{k+1} = f(x_k) + \varepsilon_{k+1}, \tag{2.1.1}$$

where $\varepsilon_{k+1} \triangleq O_{k+1} - f(x_k)$ is the observation error.

Let us call $\big(f(\cdot),\ \{O_k\},\ \{\varepsilon_k\}\big)$ as the "triplet" for a root-seeking problem. Thus, we have outlined a solution route to problems from diverse areas: First, consider if the problem under consideration can be transformed to a parameter estimation problem; Second, select an appropriate regression function and form observations from available data; Third, choose a root-seeking algorithm to recursively produce estimates for roots; Finally, the most important step is to establish convergence of the estimates to the true parameters.

It is worth noting that four steps in the solution route are closely related, in particular, an appropriately chosen triplet greatly eases the convergence analysis. The triplet is usually selected according to the consideration from physics, mechanics, mathematics, engineering practice, and others. Since three elements in the triplet are related by the equation (2.1.1), only two of them can arbitrarily be selected.

Let us discuss the importance of selecting the triplet.

(i) Selection of observations. Assume a new data $y_{k+1} \in \mathbb{R}^m$ has arrived at time $k+1$. Then, at this time y_{k+1} together with the past estimates $\{x_i,\ 0 \le i \le k\}$ and the past observations $\{O_i,\ 0 \le i \le k\}$ are available, and any function with compatible dimension of these variables may serve as O_{k+1}, for example, $O_{k+1} = y_{k+1}$, $O_{k+1} = x_k^T x_k y_{k+1}$, $O_{k+1} = (x_k + O_k)^T y_{k+1} y_{k+1}$, etc. Consider a one-dimensional example. Assume the linear function $x - x^0$ is observed, and the data $y_{k+1} = x_k - x^0 + \delta_{k+1}$ at time $k+1$ are available, where $\{\delta_k\}$ is a sequence of zero-mean iid random variables. Then, it is natural to take the regression function as $f(x) = x - x^0$, observation $O_{k+1} = y_{k+1}$, and the observation error $\varepsilon_{k+1} = \delta_{k+1}$. The nice statistical properties of $\{\delta_k\}$ normally ease the convergence analysis. However, it is allowed to take, say, $O_{k+1} \triangleq (x_k + y_{k+1})y_{k+1} = (2x_k - x^0)(x_k - x^0) + \varepsilon_{k+1}$, where the regression function is $f(x) = (2x - x^0)(x - x^0)$ and $\varepsilon_{k+1} \triangleq \delta_{k+1}^2 + \delta_{k+1}(3x_k - 2x^0)$, which is not so simple as iid for convergence analysis. The observation error may even be worse if the observations are unsuccessfully chosen.

(ii) Selection of regression function. It is understandable that properties of observation errors are of significance in convergence analysis for estimates. Assume we have the observation O_{k+1} for regression function $f_1(\cdot)$: $O_{k+1} = f_1(x_k) + \varepsilon_{k+1}^{(1)}$, but it may be written as the observation for some different regression function $f_2(\cdot)$: $O_{k+1} = f_2(x_k) + \varepsilon_{k+1}^{(2)}$, where $\varepsilon_{k+1}^{(2)} \triangleq f_1(x_k) - f_2(x_k) + \varepsilon_{k+1}^{(1)}$. Sometimes, $f_1(\cdot)$ is inconvenient for analysis, but a "better" $f_2(\cdot)$ may be chosen and $\varepsilon_{k+1}^{(2)}$ is still analyzable. For many problems to be discussed in the later chapters, $f_2(\cdot)$ is often taken as a linear function, and the resulting error $\varepsilon_{k+1}^{(2)}$ may be complicated both structurally and statistically. This makes the classical root-seeking algorithms and the corresponding analysis methods not applicable and explains why we modify the classical algorithm and introduce a new analysis method.

2.2 Classical Stochastic Approximation Method: RM Algorithm

Let $\{y_k\}$ be a stationary process with $E\|y_k\|^2 < \infty$. On the basis of samples $y_1, y_2, y_3, \cdots$ we want to recursively estimate the unknown mean value $Ey_k \triangleq x^0$. Set $x_k \triangleq \frac{1}{k}\sum_{i=1}^{k} y_i$. By ergodicity of the stationary process $x_k \xrightarrow[k\to\infty]{} x^0$. We can write $\{x_k\}$ in a recursive way.

$$\begin{aligned} x_{k+1} &= \frac{1}{k+1}\sum_{i=1}^{k+1} y_i = \frac{k}{k(k+1)}\Big(y_{k+1} + \sum_{i=1}^{k} y_i\Big) \\ &= x_k + \frac{1}{k+1}\big(y_{k+1} - x_k\big). \end{aligned} \tag{2.2.1}$$

On the other hand, estimating x^0 can be viewed as a root-seeking problem. Let us write down the corresponding triplet $\big(f(\cdot),\ \{O_k\},\ \{\varepsilon_k\}\big)$:

$$f(x) \triangleq x^0 - x,\ O_{k+1} \triangleq y_{k+1} - x_k,\ \varepsilon_{k+1} \triangleq y_{k+1} - x^0,$$

which implies $O_{k+1} = f(x_k) + \varepsilon_{k+1}$ as required.

The sample average algorithm (2.2.1) actually suggests the following algorithm for the corresponding root-seeking problem

$$x_{k+1} = x_k + a_k O_{k+1},\quad a_k = \frac{1}{k+1}. \tag{2.2.2}$$

This, in fact, is the special case of SA algorithm for the linear regression function.

Consider the root-seeking problem with observation equation (2.1.1). For searching the root $x^0 : f(x^0) = 0$ Robbins and Monro proposed the following SA algorithm, which is now known as RM algorithm:

$$x_{k+1} = x_k + a_k O_{k+1} \text{ with arbitrary } x_0, \tag{2.2.3}$$

where the step-size $\{a_k\}$ is required to satisfy the following conditions

$$a_k > 0,\ a_k \xrightarrow[k\to\infty]{} 0,\ \sum_{i=1}^{\infty} a_i = \infty. \tag{2.2.4}$$

Let us explain the meaning of these requirements.

(i) For the case $m = 1$, $+a_k$ is used for the case when $f(\cdot)$ decreasingly crosses the abscissa (m=1) and it should change to $-a_k$ whenever $f(\cdot)$ increasingly crosses the abscissa. In the multidimensional case the sign for a_k is determined by the first approximation of $f(\cdot)$ in the neighborhood of x^0.

(ii) Combining (2.2.3) with (2.1.1) gives

$$x_{k+1} = x_k + a_k f(x_k) + a_k \varepsilon_{k+1}.$$

From here it is seen that for convergence of $\{x_k\}$ the noise effect must be depressed, and for this we must have $a_k \xrightarrow[k\to\infty]{} 0$ unless $\{\varepsilon_k\}$ is vanishing.

(iii) The condition $\sum_{i=1}^{\infty} a_i = \infty$ means that $\{a_k\}$ should not too fast decrease to zero. Assume the converse: $\sum_{i=1}^{\infty} a_i < \infty$. Then, in the relatively simple noise-free ($\varepsilon_k \equiv 0$) case we have $\sum\limits_{i=0}^{\infty} \|x_{i+1} - x_i\| \leq \sum\limits_{i=0}^{\infty} a_i \|f(x_i)\| < \infty$ if $f(\cdot)$ is a bounded function, and hence

$$\begin{aligned}\|x_k - x^0\| &= \|x_k - x_0 + x_0 - x^0\| = \|\sum_{i=0}^{k-1}(x_{i+1} - x_i) + x_0 - x^0\| \\ &\geq \|x_0 - x^0\| - \sum_{i=0}^{\infty} \|x_{i+1} - x_i\| > 0,\end{aligned}$$

if the initial value x_0 is far enough from x^0. This means that the estimate x_k cannot converge to x^0. Therefore, in order the estimate to avoid possible stopping at some point away from the true root x^0 by a positive distance it is necessary to require $\sum_{i=1}^{\infty} a_i = \infty$.

It is clear that the condition $a_k \xrightarrow[k\to\infty]{} 0$ is not sufficient to suppress the noise effect and to guarantee $x_k \xrightarrow[k\to\infty]{} x^0$. Let us clarify what is implied by $x_k \xrightarrow[k\to\infty]{} x^0$ and formulate it as a theorem.

Theorem 2.2.1 *For x_k defined by (2.2.3) with $O_{k+1} = f(x_k) + \varepsilon_{k+1}$ and $a_k > 0\ \forall k \geq 0$, assume $x_k \xrightarrow[k\to\infty]{} x^0$ for some sample path ω and $f(\cdot)$ is continuous at x^0, where $f(x^0) = 0$. Then, ε_{k+1} for this ω can be separated into two parts $\varepsilon_{k+1} = \varepsilon_{k+1}^{(1)} + \varepsilon_{k+1}^{(2)}$ such that*

$$\sum_{i=1}^{\infty} a_i \varepsilon_{i+1}^{(1)} < \infty \quad \textit{and} \quad \varepsilon_{k+1}^{(2)} \xrightarrow[k\to\infty]{} 0.$$

Proof. From (2.2.3) it follows that

$$\varepsilon_{k+1} = \frac{x_{k+1} - x_k}{a_k} - f(x_k) = \varepsilon_{k+1}^{(1)} + \varepsilon_{k+1}^{(2)},$$

where $\varepsilon_{k+1}^{(1)} \triangleq \frac{x_{k+1} - x_k}{a_k}$, $\varepsilon_{k+1}^{(2)} \triangleq -f(x_k)$. Since $x_k \to x^0$ and $f(\cdot)$ is continuous at x^0, we have $\varepsilon_{k+1}^{(2)} \xrightarrow[k\to\infty]{} 0$, and

$$\sum_{i=1}^{\infty} a_i \varepsilon_{i+1}^{(1)} = \sum_{i=1}^{\infty} (x_{i+1} - x_i) = x^0 - x_1 < \infty.$$

□

Probabilistic method

When the probabilistic method is applied for convergence analysis of the RM algorithm, the conditions imposed on the observation errors $\{\varepsilon_k\}$ normally guarantee that $\sum_{i=1}^{\infty} a_i \varepsilon_{i+1} < \infty$ *a.s.* For example, it is often to require that $\sum_{i=1}^{\infty} a_i^2 < \infty$ and $(\varepsilon_k, \mathscr{F}_k)$ is an mds with

$$E(\varepsilon_{k+1}|\mathscr{F}_k) = 0 \text{ and } \limsup_{k\to\infty} E(\|\varepsilon_{k+1}\|^2|\mathscr{F}_k) < \infty \;\; a.s.$$

Then by Theorem 1.2.13 we have $\sum_{i=1}^{\infty} a_i \varepsilon_{i+1} < \infty$.

Let us formulate a typical convergence theorem with such a kind of noise assumption.

Theorem 2.2.2 *Assume that for the RM algorithm (2.2.3) with observations given by (2.1.1) the following conditions hold.*

A2.2.1

$$a_i > 0, \quad \sum_{i=1}^{\infty} a_i = \infty, \text{ and } \sum_{i=1}^{\infty} a_i^2 < \infty.$$

A2.2.2 *There exists a continuously twice differentiable Lyapunov function $v(\cdot)$ $\mathbb{R}^m \to \mathbb{R}$ satisfying the following conditions:*

(i) Its second derivative is bounded;

(ii) $v(x) > 0 \, \forall \, x \neq x^0$, $v(x^0) = 0$, and $v(x) \to \infty$ as $\|x\| \to \infty$;

(iii) For any $\varepsilon > 0$ there is a $\beta_\varepsilon > 0$ such that

$$\sup_{\|x - x^0\| > \varepsilon} v_x^T(x) f(x) = -\beta_\varepsilon < 0$$

where $v_x(x)$ denotes the gradient of $v(\cdot)$.

A2.2.3 *The observation noise $(\varepsilon_k, \mathscr{F}_k)$ is an mds with*

$$E(\varepsilon_k|\mathscr{F}_{k-1}) = 0, \quad E\|\varepsilon_k\|^2 < \infty, \tag{2.2.5}$$

where $\{\mathscr{F}_k\}$ is a family of nondecreasing σ-algebras.

A2.2.4

$$\|f(x)\|^2 + E(\|\varepsilon_k\|^2|\mathscr{F}_{k-1}) < c(1 + v(x)) \quad \forall \, k \geq 0, \tag{2.2.6}$$

where c is a positive constant.

Then $x_k \xrightarrow[k\to\infty]{} x^0$ a.s. for any initial value x_0.

Here we give some remarks on Theorem 2.2.2.

(i) The elegant proof of the theorem is based on the martingale theory at the price of imposing very restrictive conditions on the observation noise $\{\varepsilon_k\}$.

(ii) In Section 2.1 we have explained that $\{\varepsilon_k\}$ may be very complicated because of arbitrariness in selection of $\{O_k\}$ and $f(\cdot)$. Consequently, Condition A2.2.3 is hardly to be satisfied in problems from practical systems.

(iii) Condition A2.2.4 implies $\|f(x)\|^2 < c(1+v(x))$ and

$$\limsup_{k\to\infty} E(\|\varepsilon_k\|^2|\mathscr{F}_{k-1}) < c(1+v(x)).$$

In the case where $v(x)$ is a quadratic function, then $\|f(x)\|^2 < c(1+v(x))$ means that as $\|x\| \to \infty$ the growth rate of $\{f(x)\}$ should not be faster than linear.

(iv) It is clear that under the conditions of the theorem we have $\sum_{i=1}^{\infty} a_i\varepsilon_{i+1} < \infty$ *a.s.*, which is stronger than the necessary conditions indicated in Theorem 2.2.1 for convergence of x_k. This gap is well accounted in the so-called ordinary differential equation (ODE) method for convergence analysis.

ODE method

The idea of the ODE method is as follows. The estimate $\{x_k\}$ generated by the RM algorithm is interpolated to a piecewise linear continuous function with interpolating length equal to $\{a_k\}$, the step-size used in the algorithm. The tail part x_t of the interpolating function is shown to satisfy the ordinary differential equation $\dot{x} = f(x)$. The sought-for root x^0 is the equilibrium of the ODE. By stability of this equation, or by assuming existence of a Lyapunov function, it is proved that $x_t \xrightarrow[t\to\infty]{} x^0$. From this, it can be deduced that $x_k \xrightarrow[k\to\infty]{} x^0$.

The noise class treated by the ODE method is much wider than that considered by the probabilistic method. Let the step-size $\{a_k\}$ satisfy (2.2.4).

Introduce the integer-valued function $m(k,T)$ for any $T > 0$ and any integer k as follows:

$$m(k,T) \triangleq \max\Big\{m : \sum_{i=k}^{m} a_i \le T\Big\}. \tag{2.2.7}$$

Noticing $a_k \xrightarrow[k\to\infty]{} 0$, we find that $m(k,T)$ diverges to infinity as $k \to \infty$ for any fixed $T > 0$. In fact, $m(k,T)$ counts the number of iterations starting from time k as long as the sum of step-sizes does not exceed T. As a matter of fact, the sum of step-sizes may be considered as a function of the upper limit of the sum, and then $m(k,T)$ is the inverse of this function.

The integer-valued function $m(k,T)$ will be used throughout the book.

From Theorem 2.2.1 it is seen that when the RM algorithm converges to the root

x^0 of $f(\cdot)$ and $f(\cdot)$ is continuous at x^0, then the noise can be separated into two parts. After removing a vanishing component from the noise ε_k the weighted sum of the rest with weights $\{a_i\}$ is convergent. This means that the key factor of the noise that effects convergence of the RM algorithm is the behavior of the weighted sum $\sum_{i=1}^{\infty} a_i\varepsilon_{i+1}$ if a vanishing ingredient of the noise is ignored. This property essential for convergence of the algorithm is well captured by the condition used in the ODE method:

$$\lim_{T\to 0}\limsup_{k\to\infty}\frac{1}{T}\left\|\sum_{i=k}^{m(k,T)} a_i\varepsilon_{i+1}\right\| = 0 \text{ and } a_k\varepsilon_{k+1}\xrightarrow[k\to\infty]{} 0, \tag{2.2.8}$$

or the slightly different one:

$$\lim_{T\to 0}\limsup_{k\to\infty}\frac{1}{T}\left\|\sum_{i=k}^{m(k,T_k)} a_i\varepsilon_{i+1}\right\| = 0 \quad \forall\, T_k \in [0,T]. \tag{2.2.9}$$

It is clear that by appropriately choosing T_k we have $\sum_{i=k}^{m(k,T_k)} a_i\varepsilon_{i+1} = a_k\varepsilon_{k+1}$, which incorporating with (2.2.9) leads to $a_k\varepsilon_{k+1}\xrightarrow[k\to\infty]{} 0$. Therefore, (2.2.8) is implied by (2.2.9).

It is also clear that if $\varepsilon_k\xrightarrow[k\to\infty]{} 0$ or $\sum_{i=1}^{\infty} a_i\varepsilon_{i+1} < \infty$, then both (2.2.8) and (2.2.9) are fulfilled.

Let us explain these conditions. Since $a_i\xrightarrow[i\to\infty]{} 0$, we have $\sum_{i=k}^{m(k,T)} a_i\xrightarrow[k\to\infty]{} T$. Hence, $\frac{1}{T}\sum_{i=k}^{m(k,T)} a_i\varepsilon_{i+1}$ asymptotically is the weighted average of ε_{i+1}. Further, "$k\to\infty$" means that we are interested in this average only for the tail part of the weighted sum. Finally, the weighted sum $\sum_{i=k}^{m(k,T)} a_i\varepsilon_{i+1}$ should asymptotically be of $o(T)$ as $T\to 0$.

Let us formulate a theorem with Condition (2.2.8) applied.

Theorem 2.2.3 *Let $\{x_k\}$ be defined by the RM algorithm (2.2.3) and (2.2.4). Assume that there exists a twice continuously differentiable Lyapunov function $v(\cdot)$ such that $v(x) > 0$, $\forall x \neq x^0$, $v(x^0) = 0$, $v(x)\to\infty$ as $\|x\|\to\infty$, and*

$$v_x^T(x)f(x) < 0 \quad \forall\, x \neq x^0,$$

and that $\{x_k\}$ is bounded and (2.2.8) is satisfied for some sample ω. Then, $x_k\xrightarrow[k\to\infty]{} x^0$ for this ω.

We limit ourselves to pointing out some observations.

(i) The strong point of Theorem 2.2.3 consists in imposing rather mild restrictions on the observation noise without involving any statistical requirements.

(ii) There is no growth rate restriction on the regression function $f(\cdot)$.

(iii) The weakness of the theorem is the boundedness assumption on $\{x_k\}$, because this is difficult to be verified before establishing its convergence.

(iv) Since ε_{i+1} may depend upon the past estimates $x_j,\ j \le i$, both (2.2.8) and (2.2.9) are trajectory-dependent conditions. Therefore, in general, they are unverifiable beforehand. However, if somehow we can show $\sum_{i=1}^{\infty} a_i \varepsilon_{i+1} < \infty$ a.s., then both (2.2.8) and (2.2.9) are satisfied.

(v) In (2.2.8) and (2.2.9) the index k runs over the whole set of integers, and hence the whole sequence $\{x_k\}$ is concerned. In what follows a weaker than (2.2.9) condition will be used. Namely, the index k need not run over the whole set of integers, but only run over indices n_k of all convergent subsequences $\{x_{n_k}\}$

$$\lim_{T\to 0} \limsup_{k\to\infty} \frac{1}{T} \left\| \sum_{i=n_k}^{m(n_k,T_k)} a_i \varepsilon_{i+1} \right\| = 0 \quad \forall\, T_k \in [0,T]. \tag{2.2.10}$$

We now explain why relaxing (2.2.9) to (2.2.10) is of significance. Consider the case where $\varepsilon_{i+1}(x_i)$ is a function of x_i. Before establishing convergence, the behavior of $\{x_i\}$ is supposed to be arbitrary, e.g., heavily fluctuating, diverging to infinity, etc. Then, $\varepsilon_{i+1}(x_i)$ may similarly behave even if ε_{i+1} linearly depends on x_i. However, if $\varepsilon_{i+1}(x_i)$ is restricted on a convergent subsequence x_{n_k}, then it is conceivable that $\{\varepsilon_{n_k+1}(x_{n_k})\}$ has a better behavior. Besides, it is natural to expect in many cases that x_i is not too far from x_{n_k} whenever $i \in [n_k,\ m(n_k,T)]$ if T is small enough.

Condition (2.2.10) plays an important role in convergence analysis for SA algorithms to be considered in subsequence.

2.3 Stochastic Approximation Algorithm with Expanding Truncations

In the last section we have introduced the RM algorithm and the probabilistic and ODE methods for its convergence analysis. In fact, there is also the weak convergence method, but no analysis method can help if the algorithm itself is divergent. Let us consider the following example.

Example 2.3.1

$$f(x) = -(x-10)^3 \ \textit{with}\ x^0 = 10, \quad \varepsilon_k \equiv 0, \quad a_k = \frac{1}{k+1}.$$

The RM algorithm recursively gives estimates $\{x_k\}$ *for* $x^0 = 10$:

$$x_{k+1} = x_k - \frac{1}{k+1}(x_k - 10)^3. \tag{2.3.1}$$

Starting from $x_0 = 0$, *the direct computation shows*

$$x_0 = 0,\ x_1 = 1000,\ x_2 = -485148500,\ x_3 \approx 3.8 \times 10^{25},$$

and $\limsup_{k\to\infty} x_k = \infty$ *and* $\liminf_{k\to\infty} x_k = -\infty$.

However, if we take $a_k = \frac{1}{k+10^3}$ instead of $\frac{1}{k+1}$, then we have $x_k \xrightarrow[k\to\infty]{} 10$. Alternatively, since the initial value for the algorithm (2.3.1) is allowed to be arbitrary, we may take x_0 such that $|x_0 - 10| < 1$. Then, we also have $x_k \xrightarrow[k\to\infty]{} 10$.

This reminds us to modify the RM algorithm (2.2.3) and (2.2.4) so that the step-size can be adaptively adjusted.

Let $(f(\cdot), \{O_k\}, \{\varepsilon_k\})$ be the triplet of a root-seeking problem, where $f(\cdot) \in \mathbb{R}^m$ and may have multi-roots. Let $J^0 \triangleq \{x : f(x) = 0\}$ be the root set of $f(\cdot)$.

Let $\{M_k\}$ be a sequence of positive numbers increasingly diverging to infinity, and let x^* be a fixed point in $\mathbb{R}^m$. Fix an arbitrary initial value x_0, and denote by x_k the estimate at time k, serving as the kth approximation to J^0. Define x_k by the following recursion:

$$x_{k+1} = (x_k + a_k O_{k+1}) I_{[\|x_k + a_k O_{k+1}\| \le M_{\sigma_k}]} + x^* I_{[\|x_k + a_k O_{k+1}\| > M_{\sigma_k}]}, \tag{2.3.2}$$

$$\sigma_k = \sum_{i=1}^{k-1} I_{[\|x_i + a_i O_{i+1}\| > M_{\sigma_i}]}, \qquad \sigma_0 = 0, \tag{2.3.3}$$

$$O_{k+1} = f(x_k) + \varepsilon_{k+1}, \tag{2.3.4}$$

where I_A is the indicator function of set $A : I_A = 1$ if A holds, and $I_A = 0$ otherwise.

We explain the algorithm. The integer σ_k is the number of truncations up to time k, while M_{σ_k} serves as the truncation bound when the $(k+1)th$ estimate is generated. From (2.3.2) it is seen that if the estimate at time $k+1$ generated by the algorithm remains in the truncation region, i.e., if $\|x_k + a_k O_{k+1}\| \le M_{\sigma_k}$, then the algorithm evolves as the RM algorithm. If $(x_k + a_k O_{k+1})$ exits from the sphere with radius M_{σ_k}, i.e., if $\|x_k + a_k O_{k+1}\| > M_{\sigma_k}$, then the estimate at time $k+1$ is pulled back to the pre-specified point x^*, and at the same time the truncation bound is enlarged from M_{σ_k} to $M_{\sigma_{k+1}} = M_{\sigma_k + 1}$.

Consequently, if it can be shown that the number of truncations is finite, or equivalently, $\{x_k\}$ generated by (2.3.2)–(2.3.4) is bounded, then in a finite number of steps the algorithm ceases truncations, i.e., becomes the RM algorithm. In fact, the key step for the convergence analysis of (2.3.2)–(2.3.4) is to show the finiteness of the number of truncations.

The convergence analysis of (2.3.2)–(2.3.4) is carried out in a deterministic way for a fixed sample (ω).

Let us first list assumptions to be imposed.

A2.3.1 $a_k > 0$, $a_k \xrightarrow[k\to\infty]{} 0$, *and* $\sum_{k=1}^{\infty} a_k = \infty$.

A2.3.2 *$f(\cdot)$ is measurable and locally bounded.*

A2.3.3 *There are a continuously differentiable function (not necessarily being nonnegative) $v(\cdot): \mathbb{R}^m \to \mathbb{R}$ and a set $J \subset \mathbb{R}^m$ such that*

$$\sup_{\delta \le d(x,J) \le \Delta} f^T(x) v_x(x) < 0 \tag{2.3.5}$$

for any $\Delta > \delta > 0$, and $v(J) \triangleq \{v(x) : x \in J\}$ is nowhere dense, where $d(x,J) = \inf_y\{\|x-y\| : y \in J\}$ and $v_x(\cdot)$ denotes the gradient of $v(\cdot)$. Further, x^* used in (2.3.2) is such that $\|x^*\| < c_0$ and $v(x^*) < \inf_{\|x\|=c_0} v(x)$ for some $c_0 > 0$.

Remark 2.3.1 *It is clear that J contains both sets $J^0 \triangleq \{x : f(x) = 0\}$, the root set of $f(\cdot)$, and $J^1 \triangleq \{x : v_x(x) = 0\}$. In real applications it is often to take $v(\cdot)$ such that $J^1 = J^0$. In this case, J in A2.3.3 may be taken to equal J^0.*

A2.3.4 *Along indices $\{n_k\}$ of any convergent subsequence $\{x_{n_k}\}$ of $\{x_k\}$*

$$\lim_{T\to 0} \limsup_{k\to\infty} \frac{1}{T} \left\| \sum_{i=n_k}^{m(n_k,T_k)} a_i \varepsilon_{i+1} \right\| = 0 \quad \forall\, T_k \in [0,T], \tag{2.3.6}$$

where $m(n_k, T_k)$ is defined by (2.2.7).

In fact, (2.3.6) coincides with (2.2.10), and it is written here just for readability. It is worth noting that (2.3.6) implies that

$$a_{n_k} \varepsilon_{n_k+1} \xrightarrow[k\to\infty]{} 0 \tag{2.3.7}$$

for any convergent subsequence $\{x_{n_k}\}$. To see this it suffices to take $T_k = a_{n_k}$ in (2.3.6).

Prior to presenting the general convergence theorem (GCT) for SAAWET, one of the main tools used in convergence analysis for various algorithms to be given in the sequel, let us first prove a lemma, which demonstrates the basic feature of SAAWET: Along any convergent subsequence $\{x_{n_k}\}$ of $\{x_k\}$ generated by (2.3.2)–(2.3.4), $\{x_k\}$ are close to each other in the neighborhood of x_{n_k} and where there is no truncation.

Lemma 2.3.1 *Let $\{x_k\}$ be generated by (2.3.2)–(2.3.4) with an initial value x_0. Assume that A2.3.1 and A2.3.2 hold, and for the sample path ω under consideration A2.3.4 holds. Let $\{x_{n_k}\}$ be a convergent subsequence of $\{x_k\}$: $x_{n_k} \xrightarrow[k\to\infty]{} \bar{x}$ at the considered ω. Then, for this ω there is $T > 0$ such that for all sufficiently large k the algorithm generating x_l with subscript l valued in $[n_k, \cdots, m(n_k,t)]$ has no truncation:*

$$x_{l+1} = x_l + a_l O_{l+1} \quad \forall l : n_k \le l \le m(n_k,t) \text{ and } \forall t \in [0,T], \tag{2.3.8}$$

and

$$\|x_{l+1} - x_{n_k}\| \le ct \quad \forall l : n_k \le l \le m(n_k,t) \text{ and } \forall t \in [0,T], \tag{2.3.9}$$

where $c > 0$ is a constant but it may depend on ω.

Proof. In what follows all random variables concerned are fixed at the considered ω.

In the case $\lim_{k\to\infty}\sigma_k < \infty$ the truncation of the algorithm ceases in a finite number of steps and $\{x_k\}$ is bounded, so the assertions of the lemma become obvious. Thus, we need only to consider the case $\sigma_k \xrightarrow[k\to\infty]{} \infty$.

(1) We first show that there are constants $T > 0$ and $M > 0$ such that for any $t \in [0,T]$ there exists $k_t > 0$ such that for any $k > k_t$

$$\Big\|\sum_{i=n_k}^{l+1} a_i O_{i+1}\Big\| \le M \quad \forall l : n_k - 1 \le l \le m(n_k,t), \tag{2.3.10}$$

where M is independent of t and k.

It is clear that A2.3.2 implies $a_{n_k} f(x_{n_k}) \xrightarrow[k\to\infty]{} 0$, which combining with (2.3.7) yields $a_{n_k} O_{n_{k+1}} \xrightarrow[k\to\infty]{} 0$. So, (2.3.10) should be proved only for $l : n_k \le l \le m(n_k,t)$.

Take $c_1 > \|\bar{x}\|$. By the convergence $x_{n_k} \xrightarrow[k\to\infty]{} \bar{x}$, there is k_c such that

$$\|x_{n_k}\| \le (c_1 + \|\bar{x}\|)/2 \quad \forall k \ge k_c. \tag{2.3.11}$$

Assume the converse: (2.3.10) is not true.

Take a sequence $\{t_j\}$ of positive real numbers such that $t_j > 0$ and $t_j \to 0$ as $j \to \infty$. Since (2.3.10) is not true by the converse assumption, for $j = 1$ there are $k_1 > k_c$ and $l_1 : n_{k_1} \le l_1 \le m(n_{k_1}, t_1)$ such that

$$\Big\|\sum_{i=n_{k_1}}^{l_1+1} a_i O_{i+1}\Big\| > (c_1 - \|\bar{x}\|)/2,$$

and for any $j > 1$ there are $k_j > k_{j-1}$ and $l_j : n_{k_j} \le l_j \le m(n_{k_j}, t_j)$ such that

$$\Big\|\sum_{i=n_{k_j}}^{l_j+1} a_i O_{i+1}\Big\| > (c_1 - \|\bar{x}\|)/2. \tag{2.3.12}$$

Without loss of generality we may assume

$$l_j = \inf\Big\{l : \Big\|\sum_{i=n_{k_j}}^{l+1} a_i O_{i+1}\Big\| > (c_1 - \|\bar{x}\|)/2\Big\}. \tag{2.3.13}$$

Then for any $l : n_{k_j} \le l \le l_j$, from (2.3.11) and (2.3.13) it follows that

$$\Big\|x_{n_{k_j}} + \sum_{i=n_{k_j}}^{l} a_i O_{i+1}\Big\| < c_1. \tag{2.3.14}$$

Since there is j_0 such that $M_{\sigma_{n_j}} > c_1 \ \forall j \ge j_0$, from (2.3.14) it follows that

$$x_{l+1} = x_l + a_l O_{l+1} \quad \forall l : n_{k_j} \le l \le l_j. \tag{2.3.15}$$

From (2.3.14) and (2.3.15) and by A2.3.2 we have

$$\|x_l\| \le c_1 \quad \text{and} \quad \|f(x_l)\| \le c_2 \quad \forall l : n_{k_j} \le l \le l_j + 1, \tag{2.3.16}$$

where c_2 is a constant.

For any fixed $t \in (0, T)$, if j is large enough, then $t_j < t$ and $l_j + 1 < m(n_{k_j}, t)$, and by A2.3.4

$$\limsup_{j\to\infty} \Big\| \sum_{i=n_{k_j}}^{l_j} a_i \varepsilon_{i+1} \Big\| = 0 \quad \text{and} \quad \limsup_{j\to\infty} \Big\| \sum_{i=n_{k_j}}^{l_j+1} a_i \varepsilon_{i+1} \Big\| = 0. \tag{2.3.17}$$

From here it follows that

$$\lim_{j\to\infty} a_{l_j+1} \varepsilon_{l_j+2} = 0. \tag{2.3.18}$$

From (2.3.15) we have

$$\|x_{l_j+1} - x_{n_{k_j}}\| \le \sum_{i=n_{k_j}}^{l_j} a_i \|f(x_i)\| + \Big\| \sum_{i=n_{k_j}}^{l_j} a_i \varepsilon_{i+1} \Big\| \xrightarrow[j\to\infty]{} 0, \tag{2.3.19}$$

where the first term at the right-hand side of the inequality tends to zero because

$$\sum_{i=n_{k_j}}^{l_j} a_i \|f(x_i)\| \le c_2 \sum_{i=n_{k_j}}^{l_j} a_i \le c_2 t_j \xrightarrow[j\to\infty]{} 0,$$

while the second term also tends to zero as shown in (2.3.17). Noticing that $a_{l_j+1} O_{l_j+2} = a_{l_j+1}\big(f(x_{l_j+1}) + \varepsilon_{l_j+2}\big) \xrightarrow[j\to\infty]{} 0$ by (2.3.16) and (2.3.18), we then by (2.3.19) have

$$\|x_{l_j+1} - x_{n_{k_j}} + a_{l_j+1} O_{l_j+2}\| \le \|x_{l_j+1} - x_{n_{k_j}}\| + \|a_{l_j+1} O_{l_j+2}\| \xrightarrow[j\to\infty]{} 0.$$

On the other hand, by (2.3.13) we have

$$\|x_{l_j+1} - x_{n_{k_j}} + a_{l_j+1} O_{l_j+2}\| = \Big\| \sum_{i=n_{k_j}}^{l_j+1} a_i O_{i+1} \Big\| > (c_1 - \|\bar{x}\|)/2 \quad \forall j \ge 1.$$

The obtained contradiction proves (2.3.10).

(2) From (2.3.10) it follows that for any $t \in [0, T]$

$$\Big\|x_{n_k} + \sum_{i=n_k}^{l} a_i O_{i+1}\Big\| \le M + \|\bar{x}\| + 1 \le M_{\sigma_k} \le M_{\sigma_{n_k}} \tag{2.3.20}$$

$$\forall l : n_k \le l \le m(n_k, t)$$

if k is large enough.

This implies (2.3.8) and

$$\|x_{l+1}\| \leq M+1+\|\bar{x}\|, \quad \|f(x_l)\| \leq c_3 \ \forall l : n_k \leq l \leq m(n_k,t), \tag{2.3.21}$$

where c_3 is a constant. The second inequality of (2.3.21) yields

$$\|\sum_{i=n_k}^{l} a_i f(x_i)\| \leq c_3 t. \tag{2.3.22}$$

From (2.3.6) we have

$$\|\sum_{i=n_k}^{m(n_k,t)} a_i \varepsilon_{i+1}\| \leq c_3 t \tag{2.3.23}$$

for large enough k and small enough T.

Combining (2.3.22) and (2.3.23) by (2.3.8) leads to

$$\|x_{l+1} - x_{n_k}\| \leq ct \quad \forall l : n_k \leq l \leq m(n_k,t)$$

for all large enough k, where $c \triangleq 2c_3$. This proves (2.3.9). □

It is worth noting that A2.3.3 is not required in Lemma 2.3.1.

We now present GCT for SAAWET.

Theorem 2.3.1 *Let $\{x_k\}$ be given by (2.3.2)–(2.3.4) with an initial value x_0. Assume A2.3.1–A2.3.3 hold. Then, $d(x_k, J) \xrightarrow[k\to\infty]{} 0$ for those sample paths ω where A2.3.4 holds.*

Proof. (1) We show that the truncation in (2.3.2)–(2.3.4) ceases in a finite number of steps.

We say that the sequence $\{v(x_{l_k}),\ldots,v(x_{m_k})\}$ crosses the interval $[\delta_1,\delta_2]$, if $v(x_{l_k}) \leq \delta_1, v(x_{m_k}) \geq \delta_2$, and $\delta_1 < v(x_i) < \delta_2 \ \forall i : l_k < i < m_k$.

We first show that any interval $[\delta_1 < \delta_2]$ with $d([\delta_1,\delta_2], v(J)) > 0$ cannot be crossed by infinitely many sequences $\{v(x_{l_k}),\ldots,v(x_{m_k})\}$, $k = 1,2,\cdots$ with $\{\|x_{l_k}\|\}$ bounded, where $v(\cdot)$ is the function figured in A2.3.3.

Assume the converse: there are infinitely many crossings $v(x_{l_k}),\ldots, v(x_{m_k})$, $k = 1,2,\ldots$, and $\{\|x_{l_k}\|\}$ is bounded.

By boundedness of $\{\|x_{l_k}\|\}$, without loss of generality, we may assume $x_{l_k} \xrightarrow[k\to\infty]{} \bar{x}$.

By setting $t = a_{l_k}$ in (2.3.9), we have

$$\|x_{l_k+1} - x_{l_k}\| \leq 2ca_{l_k} \xrightarrow[k\to\infty]{} 0. \tag{2.3.24}$$

By definition of crossing, $v(x_{l_k+1}) > \delta_1 \geq v(x_{l_k})$, so we have

$$v(x_{l_k}) \xrightarrow[k\to\infty]{} \delta_1 = v(\bar{x}) \quad \text{and} \quad d(\bar{x}, J) \triangleq \delta > 0. \tag{2.3.25}$$

From (2.3.9) we see that if we take t sufficiently small, then

$$d(x_l, J) \geq \frac{\delta}{2} \quad \forall\, l : l_k \leq l \leq m(l_k, t) \tag{2.3.26}$$

for sufficiently large k.

By (2.3.21) and (2.3.9), for large k we then have

$$\begin{aligned} v(x_{m(l_k,t)+1}) - v(x_{l_k}) &= \sum_{i=l_k}^{m(l_k,t)} a_i O_{i+1}^T v_x(\bar{x}) + o(t) \\ &= \sum_{i=l_k}^{m(l_k,t)} a_i f^T(x_i) v_x(x_i) + \sum_{i=l_k}^{m(l_k,t)} a_i f^T(x_i)(v_x(\bar{x}) - v_x(x_i)) \\ &+ \sum_{i=l_k}^{m(l_k,t)} a_i v_x^T(\bar{x}) \varepsilon_{i+1} + o(t), \end{aligned} \tag{2.3.27}$$

where $v_x(\cdot)$ denotes the gradient of $v(\cdot)$ and $o(t) \to 0$ as $t \to 0$.

Condition A2.3.6 implies that

$$\limsup_{k \to \infty} \left\| v_x^T(\bar{x}) \sum_{i=l_k}^{m(l_k,t)} a_i \varepsilon_{i+1} \right\| = o(t). \tag{2.3.28}$$

By (2.3.9) and (2.3.21) it follows that as $t \to 0$

$$\| \sum_{i=l_k}^{m(l_k,t)} a_i f^T(x_i) \big(v_x(\bar{x}) - v_x(x_i)\big) \| = o(t). \tag{2.3.29}$$

Putting (2.3.28) and (2.3.29) into (2.3.27), we have

$$v(x_{m(l_k,t)+1}) - v(x_{l_k}) = \sum_{i=l_k}^{m(l_k,t)} a_i f^T(x_i) v_x(x_i) + o(t), \tag{2.3.30}$$

which by (2.3.26) and (2.3.5) yields that there are $\alpha > 0$ and $t > 0$ such that

$$v(x_{m(l_k,t)+1}) - v(x_{l_k}) \leq -\alpha t \tag{2.3.31}$$

for all sufficiently large k.

Noticing (2.3.25), from (2.3.31) we derive

$$\limsup_{k \to \infty} v(x_{m(l_k,t)+1}) \leq \delta_1 - \alpha t. \tag{2.3.32}$$

However, by (2.3.9) we have

$$\lim_{t \to 0} \max_{l_k \leq l \leq m(l_k,t)} |v(x_{l+1}) - v(x_{l_k}))| = 0,$$

which implies that $m(l_k,t)+1<m_k$ for small enough t.

This means that $v(x_{m(l_k,t)+1})\in[\delta_1,\delta_2)$, which contradicts (2.3.32). The obtained contradiction shows impossibility of infinitely many crossings.

We are now in a position to show that the algorithm (2.3.2)–(2.3.4) ceases the truncation in a finite number of steps.

By A2.3.3, $v(J)$ is nowhere dense, and hence a nonempty interval $[\delta_1,\delta_2]$ exists such that $[\delta_1,\delta_2]\subset\big(v(x^*),\inf_{\|x\|=c_0}v(x)\big)$ and $d\big([\delta_1,\delta_2],v(J)\big)>0$. If $\sigma_k\xrightarrow[k\to\infty]{}\infty$, then x_k, starting from x^*, crosses the sphere $\{x:\|x\|=c_0\}$ infinitely many times. Consequently, $v(x_k)$ crosses $[\delta_1,\delta_2]$ infinitely often with $\{x_{l_k}\}$ bounded. But, this is impossible. Therefore, starting from some k_0, the algorithm (2.3.2)–(2.3.4) ceases the truncation in a finite number of steps and $\{x_k\}$ is bounded.

(2) We show that $v(x_k)$ converges and $d(x_k,J)\xrightarrow[k\to\infty]{}0$.

Let

$$v_1\triangleq\liminf_{k\to\infty}v(x_k)\le\limsup_{k\to\infty}v(x_k)\triangleq v_2.$$

We want to show $v_1=v_2$.

If $v_1<v_2$ and at least one of v_1 and v_2 does not belong to $v(J)$, then an interval $[\delta_1,\delta_2]\subset[v_1,v_2]$ exists such that $d([\delta_1,\delta_2],v(J))>0$ and $\delta_2>\delta_1$. But, we just have shown that this is impossible. So, if $v_1<v_2$, then both v_1 and v_2 should belong to $v(J)$ and

$$\lim_{k\to\infty}d(v(x_k),v(J))=0. \tag{2.3.33}$$

We now show that $\{v(x_k)\}$ is dense in $[v_1,v_2]$. For this it suffices to show that $x_{k+1}-x_k\xrightarrow[k\to\infty]{}0$. Assume the converse: there is a subsequence

$$\lim_{k\to\infty}\|x_{l_k+1}-x_{l_k}\|\triangleq\beta>0. \tag{2.3.34}$$

Without loss of generality, we may assume x_{l_k} converges. Otherwise, a convergent subsequence can be extracted, which is possible because $\{x_k\}$ is bounded. However, if we take $t=a_{l_k}$ in (2.3.9), we have

$$\|x_{l_k+1}-x_{l_k}\|\le 2c_2a_{l_k}\xrightarrow[k\to\infty]{}0,$$

which contradicts (2.3.34). Therefore, $\{v(x_k)\}$ is dense in $[v_1,v_2]$, and from (2.3.33) it follows that $v(J)$ is dense in $[v_1,v_2]$. However, by A2.3.3 $v(J)$ is nowhere dense. The obtained contradiction implies that it is impossible to have $v_1<v_2$, i.e., v_1 and v_2 must be the same and hence $v(x_k)$ converges.

For proving $d(x_k,J)\xrightarrow[k\to\infty]{}0$, it suffices to show that all limit points of $\{x_k\}$ belong to J.

Assume the converse: $x_{l_k}\to\bar{x}\notin J$, $d(\bar{x},J)\triangleq\delta>0$. By (2.3.9) we have

$$d(x_l,J)>\frac{\delta}{2}\quad\forall l:l_k\le l\le m(l_k,t)$$

for all large k if t is small enough. By (2.3.5) it follows that

$$v_x^T(x_l)f(x_l) < -b < 0 \quad \forall l : l_k \le l \le m(l_k,t),$$

and from (2.3.30)

$$v(x_{m(l_k,t)+1}) - v(x_{l_k}) \le -\frac{bt}{2} \tag{2.3.35}$$

for small enough t. This leads to a contradiction because $v(x_k)$ converges and the left-hand side of (2.3.35) tends to zero as $k \to \infty$. Thus, we conclude $d(x_k, J) \xrightarrow[k\to\infty]{} 0$, and the proof is completed. □

It is worth noting that any $J' \supset J$ also satisfies (2.3.5) with J replaced by J'.

Let us denote by J^* the set of limiting points of $\{x_k\}$ produced by the algorithm (2.3.2)–(2.3.4).

Theorem 2.3.2 *Under the conditions of Theorem 2.3.1, J^* is a connected subset of $\bar{J}$, the closure of J, and $d(x_k, J^*) \xrightarrow[k\to\infty]{} 0$.*

Proof. Assume the converse: i.e., J^* is disconnected. Then, there are closed sets J_1^* and J_2^* such that$J^* = J_1^* \cup J_2^*$ and $d(J_1^*, J_2^*) > 0$.

Set

$$\rho \triangleq \frac{1}{3} d(J_1^*, J_2^*).$$

Since $d(x_k, J^*) \xrightarrow[k\to\infty]{} 0$, there exists k_0 such that

$$x_k \in B(J_1^*, \rho) \cup B(J_2^*, \rho) \quad \forall k \ge k_0,$$

where $B(A, \rho)$ denotes the ρ-neighborhood of set A.

Define

$$n_0 \triangleq \inf\{k > k_0, d(x_k, J_1^*) < \rho\},$$
$$m_l \triangleq \inf\{k > n_l, d(x_k, J_2^*) < \rho\},$$
$$n_{l+1} \triangleq \inf\{k > m_l, d(x_k, J_1^*) < \rho\}.$$

It is clear that $m_l < \infty, n_l < \infty \ \forall l$, and

$$x_{n_l} \in B(J_1^*, \rho), \quad x_{n_l - 1} \in B(J_2^*, \rho).$$

Since $d(J_1^*, J_2^*) = 3\rho$, it follows that

$$\|x_{n_l} - x_{n_l - 1}\| \ge \rho. \tag{2.3.36}$$

By boundedness of $\{x_{n_l - 1}\}$, we may assume that $x_{n_l - 1}$ converges. Then, taking $t = a_{n_l - 1}$ in (2.3.9), we derive

$$\|x_{n_l} - x_{n_l - 1}\| \le c a_{n_l - 1} \xrightarrow[l\to\infty]{} 0,$$

which contradicts (2.3.36). The obtained contradiction proves the theorem. □

At beginning of the section we have explained that introducing the expanding truncations to the RM algorithm aims at avoiding its divergence caused by inadequately selecting the initial value or the step-size of the algorithm. So, it is natural to ask what do Theorems 2.3.1 and 2.3.2 imply if it is a priori known that $\{x_k\}$ generated by the RM algorithm (2.2.3) is bounded? In fact, we have the following theorem.

Theorem 2.3.3 *Assume A2.3.1, A2.3.2, and A2.3.3' hold, where A2.3.3' is a simplified version of A2.3.3:*

A2.3.3' *There are a continuously differentiable function (not necessarily being nonnegative) $v(\cdot):\mathbb{R}^m \to \mathbb{R}$ and a set $J \subset \mathbb{R}^m$ such that*

$$\sup_{\delta \le d(x,J) \le \Delta} f^T(x) v_x(x) < 0$$

for any $\Delta > \delta > 0$, and $v(J) \triangleq \{v(x) : x \in J\}$ is nowhere dense.

If $\{x_k\}$ generated by (2.2.3) is bounded, then $d(x_k, J^*) \xrightarrow[k\to\infty]{} 0$ for sample paths where A2.3.4 holds, where J^* is a connected subset of $\overline{J}$.

Proof. We need only to show $d(x_k, J) \xrightarrow[k\to\infty]{} 0$, which implies $d(x_k, J^*) \xrightarrow[k\to\infty]{} 0$ by Theorem 2.3.2. Let us check the proof for Lemma 2.3.1 and Theorem 2.3.1. Since $\{x_k\}$ is bounded, (2.3.8) and (2.3.9) become obvious. We now deal with the algorithm (2.2.3) having no truncation, so (1) in the proof of Theorem 2.3.1 is no longer needed while (2) of the proof can be carried out without any change. □

Remark 2.3.2 *Comparing Theorem 2.3.3 with Theorem 2.2.3, we find that the limiting set of the algorithm in Theorem 2.3.3 is allowed to be more than a singleton x^0, and, more importantly, the noise condition (2.3.6) (or (2.2.10)) used in Theorem 2.3.3 is weaker than (2.2.8) used in Theorem 2.2.3.*

If it is a priori known that the algorithm (2.3.2)–(2.3.4) evolves in a closed subset S of $\mathbb{R}^m$, then for convergence of the algorithm it suffices to require the corresponding conditions hold on S. In particular, A2.3.3 is modified to

A2.3.3⁰ *There are a continuously differentiable function (not necessarily being nonnegative) $v(\cdot):\mathbb{R}^m \to \mathbb{R}$ and a set $J \subset \mathbb{R}^m$ such that*

$$\sup_{\delta \le d(x,J\cap S) \le \Delta,\ x\in S} f^T(x) v_x(x) < 0$$

for any $\Delta > \delta > 0$, and $v(J \cap S)$ is nowhere dense. Further, x^* used in (2.3.2) is such that $x^* \in S$, $\|x^*\| < c_0$ for some $c_0 > 0$, and $v(x^*) < \inf_{\|x\|=c_0} v(x)\ \ \forall x \in S$.

Theorem 2.3.4 *Assume that $\{x_k\}$ produced by the algorithm (2.2.3) evolves in a closed subset S of $\mathbb{R}^m$ starting from some $x_0 \in S$ and that A2.3.1, A2.3.2, A2.3.3^0, and A2.3.4 hold. Then, $d(x_k, S \cap J^*) \xrightarrow[k\to\infty]{} 0$, where J^* is a connected subset of $\overline{J}$.*

Proof. The convergence of the algorithm can be proved along the lines as that for Theorem 2.3.1, and it is clear that all limiting points should be in $S \cap J^*$. □

In the case where J is a singleton x^0, combining Theorem 2.3.1 and Theorem 2.2.1 leads to a necessary and sufficient condition on the observation noise for convergence of $\{x_k\}$.

Theorem 2.3.5 *For $\{x_k\}$ generated by (2.3.2)–(2.3.4) assume the following conditions:*

(i) $a_k > 0$, $a_k \xrightarrow[k\to\infty]{} 0$, *and* $\sum_{k=1}^{\infty} a_k = \infty$;

(ii) $f(\cdot)$ *is measurable, continuous at* x^0, *and locally bounded;*

(iii) There is a continuously differentiable function (not necessarily being nonnegative) $v(\cdot) : \mathbb{R}^m \to \mathbb{R}$ *such that*

$$\sup_{\delta \le \|x-x^0\| \le \Delta} f^T(x) v_x(x) < 0$$

for any $\Delta > \delta > 0$ *where* $v_x(\cdot)$ *denotes the gradient of* $v(\cdot)$. *Further,* x^* *used in (2.3.2) is such that* $\|x^*\| < c_0$ *and* $v(x^*) < \inf_{\|x\|=c_0} v(x)$ *for some* $c_0 > 0$.

Then $x_k \xrightarrow[k\to\infty]{} x^0$ *if and only if the observation noise* $\{\varepsilon_{k+1}\}$ *in (2.3.4) can be decomposed into two parts* $\varepsilon_{k+1} = e_{k+1} + \nu_{k+1}$ *such that*

$$\sum_{k=1}^{\infty} a_k e_{k+1} < \infty \quad \text{and} \quad \nu_{k+1} \xrightarrow[k\to\infty]{} 0. \tag{2.3.37}$$

Proof. Sufficiency. Since (2.3.37) guarantees A2.3.4, the sufficiency of the theorem follows from Theorem 2.3.1.

Necessity. Assume $x_k \xrightarrow[k\to\infty]{} x^0$. Then, there is a k_0 possibly depending upon ω (sample) such that the truncation in (2.3.2) ceases at k_0. In other words, the algorithm becomes $x_{k+1} = x_k + a_k f(x_k) + a_k \varepsilon_{k+1}$ $\forall k \ge k_0$. Thus, we may decompose $\varepsilon_{k+1} = e_{k+1} + \nu_{k+1}$ as follows

$$e_{k+1} \triangleq \frac{x_{k+1} - x_k}{a_k} \quad \forall k \ge k_0 \ \text{ and } \ e_{k+1} \triangleq \varepsilon_{k+1} \ \forall k < k_0$$
$$\nu_{k+1} \triangleq f(x_k) \ \forall k \ge k_0 \ \text{ and } \ \nu_{k+1} \triangleq 0 \ \forall k < k_0,$$

which clearly satisfy (2.3.37). □

The regression function discussed until now is time-invariant, but the time-varying regression function $\{f_k(\cdot)\}$ can also be dealt with.

The algorithm (2.3.2)–(2.3.4) correspondingly changes to the following:

$$x_{k+1} = (x_k + a_k O_{k+1}) I_{[\|x_k + a_k O_{k+1}\| \le M_{\sigma_k}]} + x^* I_{[\|x_k + a_k O_{k+1}\| > M_{\sigma_k}]}, \tag{2.3.38}$$

$$\sigma_k = \sum_{i=1}^{k-1} I_{[\|x_i + a_i O_{i+1}\| > M_{\sigma_i}]}, \qquad \sigma_0 = 0, \tag{2.3.39}$$

$$O_{k+1} = f_k(x_k) + \varepsilon_{k+1}. \tag{2.3.40}$$

Replacing A2.3.2 and A2.3.3, we introduce the following assumptions.

A2.3.2' *The functions $f_k(\cdot)$ are measurable and uniformly locally bounded, i.e., for any constant $c \geq 0$*

$$\sup_k \sup_{\|x\|<c} \|f_k(x)\| < \infty.$$

A2.3.3" *There are a continuously differentiable function (not necessarily being nonnegative) $v(\cdot) : \mathbb{R}^m \to \mathbb{R}$ and a set $J \subset \mathbb{R}^m$ such that*

$$\limsup_{k\to\infty} \sup_{\delta \leq d(x,J) \leq \Delta} f_k^T(x) v_x(x) < 0$$

for any $\Delta > \delta > 0$, and $v(J) \triangleq \{v(x) : x \in J\}$ is nowhere dense. Further, x^* used in (2.3.38) is such that $\|x^*\| < c_0$ and $v(x^*) < \inf_{\|x\|=c_0} v(x)$ for some $c_0 > 0$.

Remark 2.3.3 *Similar to that noticed in Remark 2.3.1, J includes both*

$$J^0 \triangleq \bigcup_{j=1}^{\infty} \bigcap_{k=1}^{\infty} \{x : f_{k+j}(x) = 0\}$$

and $J^1 = \{x : \ v_x(x) = 0\}$, where J^0 is the set of common roots for the functions $\{f_k(\cdot)\}$ with a possible exception of a finite number of them.

Theorem 2.3.6 *Let $\{x_k\}$ be given by (2.3.38)–(2.3.40) with an initial value x_0. Assume A2.3.1, A2.3.2', and A2.3.3" hold. Then, $d(x_k, J^*) \xrightarrow[k\to\infty]{} 0$ for sample paths where A2.3.4 holds, where J^* is a connected subset of $\overline{J}$.*

Proof. First, replacing $f(x_k)$ with $f_k(x_k)$ in the proof of Theorem 2.3.1 we derive $d(x_k, J) \xrightarrow[k\to\infty]{} 0$. Then, applying the argument similar to that carried out in the proof for Theorem 2.3.2 we obtain the assertion of the theorem. □

Remark 2.3.4 *Theorem 2.3.5 also holds for the algorithm (2.3.38)–(2.3.40) with time-varying regression functions if conditions (ii) and (iii) used in Theorem 2.3.5 are replaced correspondingly by (ii)' and (iii)' listed below.*

(ii)' *The functions $f_k(\cdot)$ are measurable, uniformly locally bounded, i.e., for any constant $c \geq 0$*

$$\sup_k \sup_{\|x\|<c} \|f_k(x)\| < \infty,$$

and equi-continuous at x^0, i.e., $\sup_k \|f_k(x^0 + \delta) - f(x^0)\| \xrightarrow[\delta\to 0]{} 0$;

(iii)' *There is a continuously differentiable function (not necessarily being nonnegative)* $v(\cdot): \mathbb{R}^m \to \mathbb{R}$ *such that*

$$\limsup_{k\to\infty} \sup_{\delta \le d(x,J) \le \Delta} f_k^T(x) v_x(x) < 0$$

for any $\Delta > \delta > 0$. *Further,* x^* *used in (2.3.38) is such that* $\|x^*\| < c_0$ *and* $v(x^*) < \inf_{\|x\|=c_0} v(x)$ *for some* $c_0 > 0$.

2.4 SAAWET with Nonadditive Noise

For SAAWET given by (2.3.2)–(2.3.4) the noise ε_{k+1} additively enters the observation equation $O_{k+1} = f(x_k) + \varepsilon_{k+1}$. We now consider the case where the regression function is the averaged observation and the noise non-additively appears in the observation. To be precise, the observation and the regression function are as follows:

$$O_{k+1} = f(x_k, \xi_{k+1}), \quad f(\cdot) \triangleq \int_{-\infty}^{\infty} f(\cdot, y) dF(y), \tag{2.4.1}$$

where $f(\cdot,\cdot): \mathbb{R}^m \times \mathbb{R}^m \to \mathbb{R}^m$ and $F(\cdot)$ is the limiting distribution function of ξ_k.

It is clear that writing O_{k+1} in the standard form (2.3.4) we have

$$O_{k+1} = f(x_k) + \varepsilon_{k+1}, \quad \varepsilon_{k+1} \triangleq f(x_k, \xi_{k+1}) - f(x_k). \tag{2.4.2}$$

In the sequel, let us denote by $\mathscr{F}_j^k \triangleq \sigma\{\xi_i, j \le i \le k\}$ the σ-algebra generated by ξ_i, $j \le i \le k$, by $F_k(z)$ the distribution function of ξ_k, and by $F_{k+1}(z; \mathscr{F}_1^j)$ the conditional distribution of ξ_{k+1} given $\mathscr{F}_1^j$, $k \ge j$.

The following assumptions are to be imposed.

A2.4.1 $a_k > 0$, $\sum_{k=1}^{\infty} a_k = \infty$, $\sum_{k=1}^{\infty} a_k^2 < \infty$;

A2.4.2 $f(\cdot,\cdot)$ *is a measurable function and is locally Lipschitz-continuous with respect to its first argument, i.e., for any fixed* $L > 0$;

$$\|(f(x,z) - f(y,z)) I_{[\|x\| \le L, \|y\| \le L]}\| \le c_L \|x - y\| g(z), \tag{2.4.3}$$

where c_L *is a constant depending on* L *and* $g(z) > 0$;

A2.4.3 *(=A2.3.3)*

A2.4.4 *(Noise Condition)*

(i) $\{\xi_k\}$ *is an* ϕ*-mixing process with mixing coefficient* $\phi(k) \xrightarrow[k\to\infty]{} 0$;

(ii)

$$\sup_k E[(g^2(\xi_{k+1}) + \|f(0,\xi_{k+1})\|^2)|\mathscr{F}_1^k] \triangleq \mu^2 < \infty, \quad E\mu^2 < \infty \tag{2.4.4}$$

and

$$\int_{-\infty}^{\infty} (g^2(z) + \|f(0,z)\|^2) dF(z) \triangleq \lambda^2 < \infty, \tag{2.4.5}$$

where $g(\cdot)$ is defined in (2.4.3);

(iii) $\psi_k \triangleq \sup_{A \in \mathscr{B}^m} \left| P(\xi_k \in A) - \int_A dF(x) \right| \xrightarrow[k\to\infty]{} 0$, *where $\mathscr{B}^m$ denotes the Borel σ-algebra in $\mathbb{R}^m$.*

It is clear that $\psi_k \xrightarrow[k\to\infty]{} 0$ implies that $F_k(z)$ converges to $F(z)$ at all points of its continuity.

Prior to describing the theorem, let us first prove a lemma.

Lemma 2.4.1 *Let $\{x_k\}$ be generated by (2.3.2) and (2.3.3) with $O_{k+1} = f(x_k, \xi_{k+1})$. Assume A2.4.1, A2.4.2, A2.4.3, and A2.4.4 hold. Then there is an Ω_0 with $\mathbb{P}\Omega_0 = 1$ such that for any $\omega \in \Omega_0$ and any bounded subsequence $\{x_{n_k}\}$ of $\{x_k\}$ with $n_k \xrightarrow[k\to\infty]{} \infty$, the following estimate takes place*

$$\|x_i - x_{n_k}\| \le ct \quad \forall i : n_k \le i \le m(n_k, t) \quad \forall t \in [0, T] \tag{2.4.6}$$

for $\forall k \ge k_0$ starting from some integer k_0, if T is small enough, where $m(k,t)$ is given by (2.2.7) and $c = (1 + \sqrt{2}c_L L\mu)$, where, without loss of generality, it is assumed that $\|x_{n_k}\| \le L/2$ with $c_L L \ge 1$.

Proof. For any $L > 0$, set

$$g_L(z) \triangleq \sup_{\|x\| \le L} \|f(x,z)\|. \tag{2.4.7}$$

By setting $y = 0$ in (2.4.3), it is clear that

$$g_L(z) \le c_L L g(z) + \|f(0,z)\|. \tag{2.4.8}$$

From (2.4.8) and (2.4.4), it follows that

$$\begin{aligned}\sup_k E g_L^2(\xi_{k+1}) &\le \sup_k \left\{ 2c_L^2 L^2 E g^2(\xi_{k+1}) + 2E\|f(0,\xi_{k+1})\|^2 \right\} \\ &\le 2(1 \vee c_L^2 L^2) E\mu^2 \le 2c_L^2 L^2 E\mu^2,\end{aligned} \tag{2.4.9}$$

and

$$\sup_k E(g_L^2(\xi_{k+1})|\mathscr{F}_1^k) \le 2c_L^2 L^2 \mu^2, \tag{2.4.10}$$

where (and hereafter) L is taken large enough so that $c_L L \geq 1$.

Since $E(\mu^2|\mathscr{F}_1^k)$ is a convergent martingale, there is a $\nu^2 < \infty$ a.s. such that

$$E(\mu^2|\mathscr{F}_1^k) \leq \nu^2 < \infty \quad \forall\, k \geq 1. \tag{2.4.11}$$

From (2.4.10) and $\sum_{k=1}^{\infty} a_k^2 < \infty$, by Theorem 1.2.8 we have

$$\sum_{k=1}^{\infty} a_k\{g_L(\xi_{k+1}) - E(g_L(\xi_{k+1})|\mathscr{F}_1^k)\} < \infty \ a.s. \tag{2.4.12}$$

for any integer L. Denote by Ω_L the ω-set where the above series converges, and set

$$\Omega_0 \triangleq \bigcap_{L=1}^{\infty} \Omega_L \cap [\mu < \infty, \nu < \infty].$$

It is clear that $P\Omega_0 = 1$.

Let $\omega \in \Omega_L \cap [\mu < \infty, \nu < \infty]$ be fixed. Then for any integer $p \geq n_k$ by (2.4.7), (2.4.10), and Lyapunov inequality we have

$$\begin{aligned}
&\|\sum_{i=n_k}^{p} a_i f(x_i, \xi_{i+1}) I_{[\|x_i\| \leq L]}\| \leq \|\sum_{i=n_k}^{p} a_i g_L(\xi_{i+1})\| \\
&\leq \|\sum_{i=n_k}^{p} a_i [g_L(\xi_{i+1}) - E(g_L(\xi_{i+1})|\mathscr{F}_1^i)]\| + \|\sum_{i=n_k}^{p} a_i E(g_L(\xi_{i+1})|\mathscr{F}_1^i)\| \\
&\leq \|\sum_{i=n_k}^{p} a_i [g_L(\xi_{i+1}) - E(g_L(\xi_{i+1})|\mathscr{F}_1^i)]\| + \sqrt{2} c_L L \mu \sum_{i=n_k}^{p} a_i,
\end{aligned} \tag{2.4.13}$$

where the first term at the right-hand side tends to zero as $k \to \infty$ by (2.4.12).

Assume i_0 is sufficiently large such that $M_{\sigma_{k_i}} > L \ \ \forall i \geq i_0$ if $\sigma_k \xrightarrow[k \to \infty]{} \infty$, or $\lim_{k \to \infty} \sigma_k = \sigma_{k_{i_0}}$ whenever $\lim_{k \to \infty} \sigma_k < \infty$ (in this case there will be no truncation in (2.3.2) for $k > k_{i_0}$).

Assume $k > k_{i_0}$ and fix a small enough T such that $cT < L/2$, where $c = 1 + \sqrt{2} c_L L \mu$. Let $t \in [0, T]$ be arbitrarily fixed.

We prove (2.4.6) by induction. It is clear (2.4.6) is true for $i = n_k$.

Assume (2.4.6) is true and there is no truncation for $i = n_k, n_k + 1, \ldots, s < m(n_k, t)$. Noticing $\|x_i\| \leq \|x_i - x_{n_k}\| + \|x_{n_k}\| < L$, $i = n_k, \ldots, s$, by (2.4.13) we have

$$\begin{aligned}
&\|\sum_{i=n_k}^{s} a_i f(x_i, \xi_{i+1})\| = \|\sum_{i=n_k}^{s} a_i f(x_i, \xi_{i+1}) I_{[\|x_i\| \leq L]}\| \\
&\leq \|\sum_{i=n_k}^{s} a_i (g_L(\xi_{i+1}) - E(g_L(\xi_{i+1})|\mathscr{F}_1^i))\| + \sqrt{2} c_L L \mu t \\
&< (1 + \sqrt{2} c_L L \mu) t = ct < L/2,
\end{aligned}$$

if k is large enough.

This means that at time $s+1$ there is no truncation in (2.3.2), and

$$\begin{aligned}&\|x_{s+1}-x_{n_k}\| = \|\sum_{i=n_k}^{s} a_i f(x_i,\xi_{i+1})\| < ct,\\ &\|x_{s+1}\| < L/2+ct < L. \end{aligned} \tag{2.4.14}$$

□

Theorem 2.4.1 *Assume A2.4.1–A2.4.4. Then for $\{x_k\}$ generated by (2.3.2) and (2.3.3) with $O_{k+1} = f(x_k,\xi_{k+1})$ the following convergence takes place*

$$d(x_k,J^*) \xrightarrow[k\to\infty]{} 0 \text{ a.s.},$$

where J^* is a connected subset of $\overline{J}$.

Proof. It is clear that A2.4.2 implies A2.3.2. So, if we can verify A2.3.4 for $\{\varepsilon_k\}$ given by (2.4.2), then applying Theorem 2.3.1 we derive $d(x_k,J) \xrightarrow[k\to\infty]{} 0$ a.s., and then by Theorem 2.3.2 the conclusion of the theorem follows.

Let $\{x_{n_k}\}$ be a convergent subsequence of $\{x_k\}$.

We now express $\varepsilon_{i+1} \triangleq f(x_i,\xi_{i+1}) - f(x_i)$ given by (2.4.2) as follows:

$$\varepsilon_{i+1} = \sum_{s=1}^{5} \varepsilon_{i+1}^{(s)}, \tag{2.4.15}$$

where

$$\varepsilon_{i+1}^{(1)} \triangleq f(x_i,\xi_{i+1}) - E(f(x_i,\xi_{i+1})|\mathscr{F}_1^{i-j}), \tag{2.4.16}$$

$$\varepsilon_{i+1}^{(2)} \triangleq E(f(x_i,\xi_{i+1})|\mathscr{F}_1^{i-j}) - E(f(x_{n_k},\xi_{i+1})|\mathscr{F}_1^{i-j}), \tag{2.4.17}$$

$$\varepsilon_{i+1}^{(3)} \triangleq E(f(x_{n_k},\xi_{i+1})|\mathscr{F}_1^{i-j}) - Ef(x,\xi_{i+1})|_{x=x_{n_k}}, \tag{2.4.18}$$

$$\varepsilon_{i+1}^{(4)} \triangleq E[f(x,\xi_{i+1}) - f(y,\xi_{i+1})]|_{x=x_{n_k},y=x_i}, \tag{2.4.19}$$

$$\varepsilon_{i+1}^{(5)} \triangleq Ef(x,\xi_{i+1})|_{x=x_i} - f(x_i). \tag{2.4.20}$$

We now proceed to show that each $\{\varepsilon_{i+1}^{(s)},\}$ $s=1,\cdots,5$ satisfies A2.3.4. If this is done, then the proof of the theorem is completed.

Verifying A2.3.4 for $\{\varepsilon_{i+1}^{(1)}\}$

By (2.4.10), for $j>0$ we have

$$\begin{aligned}\sup_i E(g_L^2(\xi_{i+1})|\mathscr{F}_1^{i-j}) =& \sup_i E(E(g_L^2(\xi_{i+1})|\mathscr{F}_1^{i})|\mathscr{F}_1^{i-j})\\ \le& 2c_L^2 L^2 \sup_i E(\mu^2|\mathscr{F}_1^{i-j}),\end{aligned}$$

which converges to a finite limit as $i \to \infty$ by Theorem 1.2.7.

Therefore, for any integers L and j by Theorem 1.2.8

$$\sum_{i=1}^{\infty} a_i[f(x_i,\xi_{i+1})I_{[\|x_i\|\le L]} - E(f(x_i,\xi_{i+1})I_{[\|x_i\|\le L]}|\mathscr{F}_1^{i-j})] < \infty \text{ a.s.} \tag{2.4.21}$$

Consequently, there is $\Omega' \subset \Omega_0$ with $\mathbb{P}\Omega' = 1$ such that (2.4.21) holds for any integers L and j.

Let $\omega \in \Omega'$ be fixed. By Lemma 2.4.1, $\|x_i\| \le L\ \forall i : n_k \le i \le m(n_k,t)$ for small t. Then

$$\sum_{i=n_k}^{m(n_k,t)} a_i \varepsilon_{i+1}^{(1)} = \sum_{i=n_k}^{m(n_k,t)} a_i \big[f(x_i,\xi_{i+1})I_{[\|x_i\|\le L]} - E(f(x_i,\xi_{i+1})I_{[\|x_i\|\le L]}\big|\mathscr{F}_1^{i-j}\big] \xrightarrow[k\to\infty]{} 0 \tag{2.4.22}$$

for any j by (2.4.21). Hence, A2.3.4 holds for $\{\varepsilon_{i+1}^{(1)}\}$.

Verifying A2.3.4 for $\{\varepsilon_{i+1}^{(2)}\}$

By (2.4.4), (2.4.11), and Lemma 2.4.1 we have

$$\begin{aligned}
&\|\frac{1}{T}\sum_{i=n_k}^{m(n_k,t)} a_i \varepsilon_{i+1}^{(2)}\| \\
&= \frac{1}{T}\|\sum_{i=n_k}^{m(n_k,t)} a_i[E\big(f(x_i,\xi_{i+1})I_{[\|x_i\|\le L}|\mathscr{F}_1^{i-j}\big) - E(f(x_{n_k},\xi_{i+1})|\mathscr{F}_1^{i-j})]\| \\
&\le \frac{1}{T}\sum_{i=n_k}^{m(n_k,t)} a_i c_L E[\|x_i - x_{n_k}\| g(\xi_{i+1})|\mathscr{F}_1^{i-j}] \\
&\le c_L \sum_{i=n_k}^{m(n_k,t)} a_i [E((1+\sqrt{2}c_L L\mu)^2|\mathscr{F}_1^{i-j})]^{\frac{1}{2}} \cdot [E(g^2(\xi_{i+1})|\mathscr{F}_1^{i-j})]^{\frac{1}{2}} \\
&\le c_L \sum_{i=n_k}^{m(n_k,t)} a_i \big[E\big((2+4c_L^2L^2\mu^2)|\mathscr{F}_1^{i-j}\big)\big]^{\frac{1}{2}} [E(g^2(\xi_{i+1})|\mathscr{F}_1^{i-j})]^{\frac{1}{2}}.
\end{aligned}$$

Noticing (2.4.4) and (2.4.11), we then have for any $t \in [0,T]$

$$\|\frac{1}{T}\sum_{i=n_k}^{m(n_k,t)} a_i \varepsilon_{i+1}^{(2)}\| \le c_L(2+4c_L^2L^2\nu^2)^{\frac{1}{2}} \nu t \xrightarrow[T\to 0]{} 0. \tag{2.4.23}$$

Verifying A2.3.4 for $\{\varepsilon_{k+1}^{(3)}\}$

Applying the Jordan–Hahn decomposition to the signed measure

$$\mathrm{d}G_{i+1,j}(z) \triangleq \mathrm{d}F_{i+1}(z;\mathscr{F}_1^{i-j}) - \mathrm{d}F_{i+1}(z),\ i > j,$$

we know that there is a Borel set D in $\mathbb{R}^m$ such that for any Borel set A in $\mathbb{R}^m$

$$\int_A d(G_{i+1,j}(z))_+ = \int_{A\cap D^c} dG_{i+1,j}(z) \le \phi(j+1), \tag{2.4.24}$$

$$\int_A d(G_{i+1,j}(z))_- = \int_{A\cap D} dG_{i+1,j}(z) \le \phi(j+1), \tag{2.4.25}$$

where the inequalities in (2.4.24) and (2.4.25) take place because $\{\xi_k\}$ is a ϕ-mixing with mixing coefficient $\phi(i)$. Moreover, we have

$$dG_{i+1,j}(z) = d(G_{i+1,j}(z))_+ - d(G_{i+1,j}(z))_-, \tag{2.4.26}$$

$$d(G_{i+1,j}(z))_+ + d(G_{i+1,j}(z))_- = dF_{i+1}(z;\mathscr{F}_1^{i-j}) + dF_{i+1}(z). \tag{2.4.27}$$

Then, we have

$$\begin{aligned}\Big\|\sum_{i=n_k+j}^{m(n_k,t)} a_i\varepsilon_{i+1}^{(3)}\Big\| &= \Big\|\sum_{i=n_k+j}^{m(n_k,t)} a_i\big[E(f(x_{n_k},\xi_{i+1})|\mathscr{F}_1^{i-j}) - Ef(x,\xi_{i+1})|_{x=x_{n_k}}\big]\Big\|\\ &= \Big\|\sum_{i=n_k+j}^{m(n_k,t)} a_i\int_{-\infty}^{\infty} f(x_{n_k},z)\big[dF_{i+1}(z;\mathscr{F}_1^{i-j}) - dF_{i+1}(z)\big]\Big\|\\ &\le \sum_{i=n_k+j}^{m(n_k,t)} a_i\int_{-\infty}^{\infty} g_L(z)\Big(d(G_{i+1,j}(z))_+ + d(G_{i+1,j}(z))_-\Big),\end{aligned}$$

where for the last inequality (2.4.7) and (2.4.27) are invoked by noticing $\|x_{n_k}\| \le L/2$.

From here by the Hölder inequality, (2.4.24), and (2.4.25) it follows that

$$\begin{aligned}\Big\|\sum_{i=n_k+j}^{m(n_k,t)} a_i\varepsilon_{i+1}^{(3)}\Big\| \le \sum_{i=n_k}^{m(n_k,t)} a_i\phi^{\frac{1}{2}}(j+1)\Big[\Big(\int_{-\infty}^{\infty} g_L^2(z)d(G_{i+1,j}(z))_+\Big)^{\frac{1}{2}}\\ +\Big(\int_{-\infty}^{\infty} g_L^2(z)d(G_{i+1,j}(z))_-\Big)^{\frac{1}{2}}\Big],\end{aligned}$$

which by (2.4.27) implies that

$$\begin{aligned}\Big\|\sum_{i=n_k+j}^{m(n_k,t)} a_i\varepsilon_{i+1}^{(3)}\Big\| &\le 2\sum_{i=n_k+j}^{m(n_k,t)} a_i\phi^{\frac{1}{2}}(j+1)\Big[\int_{-\infty}^{\infty} g_L^2(z)\Big(dF_{i+1}(z;\mathscr{F}_1^{i-j}) + dF_{i+1}(z)\Big)\Big]^{\frac{1}{2}}\\ &= 2\sum_{i=n_k+j}^{m(n_k,t)} a_i\phi^{\frac{1}{2}}(j+1)\Big(E(g_L^2(\xi_{i+1})|\mathscr{F}_1^{i-j}) + Eg_L^2(\xi_{i+1})\Big)^{\frac{1}{2}}\\ &\le 2\sqrt{2}c_L L(\mu^2 + E\mu^2)^{\frac{1}{2}}\sum_{i=n_k+j}^{m(n_k,t)} a_i\phi^{\frac{1}{2}}(j+1),\end{aligned} \tag{2.4.28}$$

where the last inequality follows from (2.4.9) and (2.4.10).

On the other hand, for fixed $j \geq 1$ by (2.4.10) and (2.4.11), we have

$$\Big\| \sum_{i=n_k}^{n_k+j-1} a_i \big[E(f(x_{n_k}, \xi_{i+1})|\mathscr{F}_1^{i-j}) - Ef(x, \xi_{i+1})|_{x=x_{n_k}}\big] \Big\|$$
$$\leq \sum_{i=n_k}^{n_k+j-1} a_i \big[E(g_L(\xi_{i+1})|\mathscr{F}_1^{i-j}) + Eg_L(\xi_{i+1})\big]$$
$$\leq \sqrt{2} c_L L(\mu + (E\mu^2)^{\frac{1}{2}}) \sum_{i=n_k}^{n_k+j-1} a_i, \tag{2.4.29}$$

which for any fixed j tends to zero as $k \to \infty$.

Therefore, from (2.4.28) and (2.4.29) we have

$$\lim_{T\to 0} \limsup_{k\to\infty} \frac{1}{T} \Big\| \sum_{i=n_k}^{m(n_k,t)} a_i \varepsilon_{i+1}^{(3)} \Big\| \leq 2\sqrt{2} c_L L(\mu^2 + E\mu^2)^{\frac{1}{2}} \phi^{\frac{1}{2}}(j+1) \xrightarrow[j\to\infty]{} 0.$$

Verifying A2.3.4 for $\{\varepsilon_{i+1}^{(4)}\}$

By (2.4.3), Lemma 2.4.1, and (2.4.4) we have

$$\frac{1}{T} \Big\| \sum_{i=n_k}^{m(n_k,t)} a_i \varepsilon_{i+1}^{(4)} \Big\| = \frac{1}{T} \Big\| \sum_{i=n_k}^{m(n_k,t)} a_i E[f(x, \xi_{i+1}) - f(y, \xi_{i+1})]|_{x=x_{n_k}, y=x_i} \Big\|$$
$$\leq \frac{1}{T} \sum_{i=n_k}^{m(n_k,t)} a_i c_L \|x_{n_k} - x_i\| Eg(\xi_{i+1}) \leq (1 + \sqrt{2} c_L L\mu) c_L (E\mu^2)^{\frac{1}{2}} t \xrightarrow[T\to 0]{} 0.$$

Verifying A2.3.4 for $\{\varepsilon_{i+1}^{(5)}\}$

Similar to (2.4.24)–(2.4.27) applying the Jordan–Hahn decomposition to the signed measure,

$$\mathrm{d}G_{i+1}(z) \triangleq \mathrm{d}F_{i+1}(z) - \mathrm{d}F(z),$$

we know that there is a Borel set D in $\mathbb{R}^m$ such that for any Borel set A in $\mathbb{R}^m$

$$\int_A \mathrm{d}(G_{i+1}(z))_+ = \int_{A\cap D^c} \mathrm{d}G_{i+1}(z) \leq \psi_{i+1}, \tag{2.4.30}$$
$$\int_A \mathrm{d}(G_{i+1}(z))_- = \int_{A\cap D} \mathrm{d}G_{i+1}(z) \leq \psi_{i+1}, \tag{2.4.31}$$

where the inequalities follow from A2.4.4, (iii).

Moreover, we have

$$\mathrm{d}G_{i+1}(z) = \mathrm{d}(G_{i+1}(z))_+ - \mathrm{d}(G_{i+1}(z))_-,$$
$$\mathrm{d}(G_{i+1}(z))_+ + \mathrm{d}(G_{i+1}(z))_- = \mathrm{d}F_{i+1}(z) + \mathrm{d}F(z).$$

Finally, by the Hölder inequality, (2.4.30), and (2.4.31) we conclude that

$$\Big\|\sum_{i=n_k}^{m(n_k,t)} a_i\big(Ef(x,\xi_{i+1})|_{x=x_i}-f(x_i)\big)\Big\|$$
$$=\Big\|\sum_{i=n_k}^{m(n_k,t)} a_i\int_{-\infty}^{\infty} f(x_i,z)\big(\mathrm{d}F_{i+1}(z)-\mathrm{d}F(z)\big)\Big\|$$
$$\le\Big\|\sum_{i=n_k}^{m(n_k,t)} a_i\int_{-\infty}^{\infty} f(x_i,z)\big(\mathrm{d}(G_{i+1}(z))_+ +\mathrm{d}(G_{i+1}(z))_-\big)\Big\|$$
$$\le\sum_{i=n_k}^{m(n_k,t)} a_i\psi_{i+1}^{\frac{1}{2}}\Big[\Big(\int_{-\infty}^{\infty} f^2(x_i,z)\mathrm{d}(G_{i+1}(z))_+\Big)^{\frac{1}{2}}+\Big(\int_{-\infty}^{\infty} f^2(x_i,z)\mathrm{d}(G_{i+1}(z))_-\Big)^{\frac{1}{2}}\Big]$$
$$\le 2\sum_{i=n_k}^{m(n_k,t)} a_i\psi_{i+1}^{\frac{1}{2}}\Big(\int_{-\infty}^{\infty} f^2(x_i,z)(\mathrm{d}F_{i+1}(z)+\mathrm{d}F(z))\Big)^{\frac{1}{2}}$$
$$\le 2\sum_{i=n_k}^{m(n_k,t)} a_i\psi_{i+1}^{\frac{1}{2}}\Big(Eg_L^2(\xi_{i+1})+\int_{-\infty}^{\infty} g_L^2(z)\mathrm{d}F(z)\Big)^{\frac{1}{2}},$$

where for the last inequality (2.4.14) is invoked. Then by (2.4.5), (2.4.8), and (2.4.9) we have

$$\Big\|\frac{1}{T}\sum_{i=n_k}^{m(n_k,t)} a_i(Ef(x,\xi_{i+1})|_{x=x_i}-f(x_i))\Big\|$$
$$\le 2\frac{1}{T}\sum_{i=n_k}^{m(n_k,t)} a_i\psi_{i+1}^{\frac{1}{2}}(2c_L^2L^2E\mu^2+\lambda^2)^{\frac{1}{2}}$$
$$\le 2\sqrt{2}c_LL(E\mu^2+\lambda^2)^{\frac{1}{2}}\frac{1}{T}\sum_{i=n_k}^{m(n_k,t)} a_i\psi_{i+1}^{\frac{1}{2}}\xrightarrow[k\to\infty]{}0,$$

since $\psi_{i+1}\xrightarrow[i\to\infty]{}0$.

The proof of the theorem is completed. □

Remark 2.4.1 *In the proof of the theorem when we verify that the noise satisfies A2.3.4, the noise is separated into several additive terms as in (2.4.15) and the verification is carried out for each term. Satisfaction of A2.3.4 for each term is guaranteed by various reasons. This kind of treatment is typical and will be applied many times in what follows.*

2.5 Linear Regression Functions

Let $x_k \in \mathbb{R}^m$, $H \in \mathbb{R}^{m\times m}$, and $O_{k+1} \in \mathbb{R}^m$. As noted before, from the identity

$$O_{k+1} = Hx_k + \varepsilon_{k+1} \text{ with } \varepsilon_{k+1} \triangleq O_{k+1} - Hx_k,$$

it is concluded that any observation of compatible dimension may be regarded as an observation of linear regression function with noise ε_{k+1}. Further, for the linear regression function Hx if H is positive definite, then the quadratic function $x^T Hx$ may serve as the corresponding Lyapunov function $v(\cdot)$ required in A2.3.3. Also, if H is stable, then as to be shown in (2.5.5) there is a positive definite matrix P such that $H^T P + PH = -I$, and hence $x^T Px$ may serve as the corresponding Lyapunov function, where by stability of a matrix we mean that all its eigenvalues are with negative real parts. Therefore, in both cases where H is positive definite or H is stable, Condition A2.3.3 for Theorem A2.3.1 (GCT) is automatically satisfied, and for convergence of SAAWET it suffices to verify the noise condition A2.3.4. This means that considering the linear regression function case is of great importance. This will also be evidenced in coming chapters.

It is worth noting that the expanding truncations may not be necessary to be used if after choosing the linear regression function the resulting observation noise already satisfies A2.3.4. In fact, we have the following result for the RM algorithm with time-varying linear regression functions.

Let the observation $O_{k+1} \in \mathbb{R}^m$ on the time-varying linear regression function $H_k x$ be with additive noise ε_{k+1} :

$$O_{k+1} = H_k x_k + \varepsilon_{k+1}.$$

Consider the RM algorithm

$$x_{k+1} = x_k + a_k H_k x_k + a_k \varepsilon_{k+1} \tag{2.5.1}$$

with step-size satisfying the following conditions

$$a_k > 0,\ a_k \xrightarrow[k\to\infty]{} 0,\ \sum_{k=1}^{\infty} a_k = \infty.$$

Theorem 2.5.1 *Assume that the $m \times m$-matrices $\{H_k\}$ converge to a stable matrix $H : H_k \xrightarrow[k\to\infty]{} H$ and $\{x_k\}$ is given by (2.5.1). The estimate x_k converges to zero if and only if the observation noise can be decomposed into two parts $\varepsilon_{k+1} = e_{k+1} + \nu_{k+1}$ such that*

$$s \triangleq \sum_{k=0}^{\infty} a_k e_{k+1} < \infty \quad \text{and} \quad \nu_k \xrightarrow[k\to\infty]{} 0. \tag{2.5.2}$$

Proof. We first prove sufficiency. Assume (2.5.2) holds, and set

$$\Phi_{k,j} \triangleq (I + a_k H_k) \cdots (I + a_j H_j), \quad \Phi_{j,j+1} \triangleq I. \tag{2.5.3}$$

We now show that there exist constants $c_0 > 0$ and $c > 0$ such that

$$\|\Phi_{k,j}\| \le c_0 \exp[-c\sum_{i=j}^{k} a_i] \quad \forall k \ge j \text{ and } \forall j \ge 0. \tag{2.5.4}$$

Let S be an $m \times m$ negative definite matrix. Consider

$$P = -\int_0^\infty e^{H^T t} S e^{Ht} dt.$$

Since H is stable, the positive definite matrix P is well-defined. Integrating by parts, we have

$$\begin{aligned} P &= -\int_0^\infty e^{H^T t} S de^{Ht} H^{-1} \\ &= -e^{H^T t} S e^{Ht} H^{-1}\Big|_0^\infty + H^T \int_0^\infty e^{H^T t} S e^{Ht} dt H^{-1} \\ &= SH^{-1} - H^T P H^{-1}, \end{aligned}$$

which implies

$$H^T P + PH = S. \tag{2.5.5}$$

This means that if H is stable, then for any negative definite matrix S we can find a positive definite matrix P to satisfy equation (2.5.5). This fact is called the Lyapunov theorem and (2.5.5) called the Lyapunov equation. Consequently, we can find $P > 0$ such that

$$PH + H^T P = -2I,$$

where I denotes the identity matrix of compatible dimension.

Since $H_k \to H$ as $k \to \infty$, there exists k_0 such that for $\forall k \ge k_0$

$$PH_k + H_k^T P \le -I. \tag{2.5.6}$$

Consequently,

$$\begin{aligned} \Phi_{k,j}^T P \Phi_{k,j} &= \Phi_{k-1,j}^T (I + a_k H_k)^T P (I + a_k H_k) \Phi_{k-1,j} \\ &= \Phi_{k-1,j}^T (P + a_k^2 H_k^T P H_k + a_k H_k^T P + a_k P H_k) \Phi_{k-1,j} \\ &\le \Phi_{k-1,j}^T (P + a_k^2 H_k^T P H_k - a_k I) \Phi_{k-1,j} \\ &= \Phi_{k-1,j}^T P^{\frac{1}{2}} (I - a_k P^{-1} + a_k^2 P^{-\frac{1}{2}} H_k^T P H_k P^{-\frac{1}{2}}) P^{\frac{1}{2}} \Phi_{k-1,j}. \end{aligned} \tag{2.5.7}$$

Since $a_k \xrightarrow[k\to\infty]{} 0$ and $P > 0$, without loss of generality, we may assume that k_0 is sufficiently large such that for all $k \ge k_0$,

$$\|I - a_k P^{-1} + a_k^2 P^{-\frac{1}{2}} H^T P H P^{-\frac{1}{2}}\| \le 1 - 2ca_k < e^{-2ca_k} \tag{2.5.8}$$

for some constant $c > 0$. Combining (2.5.7) and (2.5.8) leads to

$$\Phi_{k,j}^T P \Phi_{k,j} \leq \big(\exp(-2c\sum_{i=j}^{k} a_i)\big) I,$$

and hence

$$\|\Phi_{k,j}\| \leq \lambda_{\min}^{-\frac{1}{2}}(P) \exp\big(-c\sum_{i=j}^{k} a_i\big), \tag{2.5.9}$$

where $\lambda_{\min}(P)$ denotes the minimum eigenvalue of P.

Paying attention to that

$$\|\Phi_{k_0-1,j}\| \leq \prod_{i=j}^{k_0-1}(1+a_i\|H_i\|) \leq \prod_{i=1}^{k_0-1}(1+a_i\|H_i\|),$$

from (2.5.9) we derive

$$\begin{aligned}\|\Phi_{k,j}\| &\leq \|\Phi_{k,k_0}\| \cdot \|\Phi_{k_0-1,j}\| \leq \lambda_{\min}^{-\frac{1}{2}}(P)\exp(-c\sum_{i=k_0}^{k} a_i) \prod_{i=1}^{k_0-1}(1+a_i\|H_i\|) \\ &\leq \lambda_{\min}^{-\frac{1}{2}}(P)\exp(c\sum_{i=j}^{k_0-1} a_i)\prod_{i=0}^{k_0-1}(1+a_i\|H_i\|)\exp(-c\sum_{i=j}^{k} a_i),\end{aligned}$$

which verifies (2.5.4).

From (2.5.1) and (2.5.3) it follows that

$$x_{k+1} = \Phi_{k,0}x_0 + \sum_{j=0}^{k} \Phi_{k,j+1} a_j (e_{j+1} + v_{j+1}). \tag{2.5.10}$$

We have to show that the right-hand side of (2.5.10) tends to zero as $k \to \infty$.

For any fixed j, $\|\Phi_{k,j}\| \to 0$ as $k \to \infty$ because of (2.5.4) and $\sum_{i=1}^{\infty} a_i = \infty$. This implies that $\Phi_{k,0}x_0 \to 0$ as $k \to \infty$ for any initial value x_0.

Since $v_k \to 0$ as $k \to \infty$, for any $\varepsilon > 0$ there exists a k_1 such that $\|v_k\| < \varepsilon$ $\forall k \geq k_1$. Then by (2.5.4) we have

$$\begin{aligned}\|\sum_{j=0}^{k}\Phi_{k,j+1}a_j v_{j+1}\| \leq c_0 \sum_{j=0}^{k_1-1}\Big[\exp(-c\sum_{i=j+1}^{k} a_i)\Big] a_j \|v_{j+1}\| \\ + \varepsilon c_0 \sum_{j=k_1}^{k}\Big(\exp(-c\sum_{i=j+1}^{k} a_i)\Big) a_j.\end{aligned} \tag{2.5.11}$$

The first term at the right-hand side of (2.5.11) tends to zero by divergence of $\sum_{i=1}^{\infty} a_i$, while the second term can be estimated as follows:

$$\begin{aligned}
&\varepsilon c_0 \sum_{j=k_1}^{k} a_j \exp\Big(-c \sum_{i=j+1}^{k} a_i\Big) \\
&\leq 2\varepsilon c_0 \sum_{j=k_1}^{k} \Big(a_j - \frac{c a_j^2}{2}\Big) \exp\Big(-c \sum_{i=j+1}^{k} a_i\Big) \\
&\leq \frac{2\varepsilon c_0}{c} \sum_{j=k_1}^{k} (1 - e^{-c a_j}) \cdot \exp(-c \sum_{i=j+1}^{k} a_i) \\
&= \frac{2\varepsilon c_0}{c} \sum_{j=k_1}^{k} \Big[\exp(-c \sum_{i=j+1}^{k} a_i) - \exp(-c \sum_{i=j}^{k} a_i)\Big] \leq \frac{2\varepsilon c_0}{c},
\end{aligned} \tag{2.5.12}$$

where the first inequality is valid for sufficiently large k_1 since $a_j \to 0$ as $j \to \infty$, and the second inequality is valid when $0 < c a_j < 1$.

Therefore, the right-hand side of (2.5.11) tends to zero as $k \to \infty$ and then $\varepsilon \to 0$.

Let us now estimate $\sum_{j=0}^{k} \Phi_{k,j+1} a_j e_{j+1}$.

Set

$$s_k = \sum_{j=0}^{k} a_j e_{j+1}, \quad s_{-1} = 0.$$

For any $\varepsilon > 0$, there exists $k_2 > k_1$ such that $\|s_j - s\| \leq \varepsilon \ \forall j \geq k_2$, since $s_n \to s < \infty$ by assumption. By partial summation, we have

$$\begin{aligned}
&\sum_{j=0}^{k} \Phi_{k,j+1} a_j e_{j+1} = \sum_{j=0}^{k} \Phi_{k,j+1}(s_j - s_{j-1}) \\
&\quad = s_k - \sum_{j=0}^{k} (\Phi_{k,j+1} - \Phi_{k,j}) s_{j-1} \\
&\quad = s_k - \sum_{j=0}^{k} (\Phi_{k,j+1} - \Phi_{k,j}) s - \sum_{j=0}^{k} (\Phi_{k,j+1} - \Phi_{k,j})(s_{j-1} - s) \\
&\quad = s_k - s + \Phi_{k,0} s - \sum_{j=0}^{k_2} (\Phi_{k,j+1} - \Phi_{k,j})(s_{j-1} - s) \\
&\quad + \sum_{j=k_2+1}^{k} \Phi_{k,j+1} a_j H_j (s_{j-1} - s),
\end{aligned} \tag{2.5.13}$$

where except for the last term, the sum of remaining terms tends to zero as $k \to \infty$ by (2.5.4) and $s_k \to s$.

Since $\|s_j - s\| \le \varepsilon$ for $j \ge k_2$ and $H_j \to H$ as $j \to \infty$, by (2.5.4) we have

$$\|\sum_{j=k_2+1}^{k} \Phi_{k,j+1} a_j H_j (s_{j-1} - s)\|$$

$$\le \varepsilon \sup_{1\le j<\infty} \|H_j\| \sum_{j=k_2+1}^{k} c_0[\exp(-c \sum_{i=j+1}^{k} a_i)]a_j,$$

which tends to zero as $k \to \infty$ and $\varepsilon \to 0$ by (2.5.12) and the fact that $\sup_{1\le j<\infty} \|H_j\| < \infty$. Thus, the right-hand side of (2.5.13) tends to zero as $k \to \infty$.

Thus, the sufficiency part of the theorem has been proved.

We now assume $x_k \xrightarrow[k\to\infty]{} 0$. The necessity part of the theorem is proved completely similar to that for Theorem 2.2.1. To see this we need only to set

$$e_{k+1} \triangleq \frac{x_{k+1} - x_k}{a_k}, \quad \nu_{k+1} \triangleq -H_k x_k.$$

The proof of the theorem is completed. □

Remark 2.5.1 *Comparing Theorem 2.5.1 with Theorem 2.3.5 and Remark 2.3.4, we find their conclusions are similar, but Theorem 2.5.1 is for the RM algorithm, i.e., the algorithm without truncations, and its proof is self-contained without referring to Theorem 2.3.1.*

Remark 2.5.2 *If the regression functions are with root equal to $x^0 \neq 0$, then instead of (2.5.1) the following algorithm takes place*

$$x_{k+1} = x_k + a_k H_k (x_k - x^0) + a_k \varepsilon_{k+1}. \tag{2.5.14}$$

Correspondingly, Theorem 2.5.1 changes to the following: Assume that the $m \times m$-matrices $\{H_k\}$ converge to a stable matrix H : $H_k \xrightarrow[k\to\infty]{} H$ and $\{x_k\}$ is given by (2.5.14). The estimate x_k converges to x^0 if and only if the observation noise can be decomposed into two parts $\varepsilon_{k+1} = e_{k+1} + \nu_{k+1}$ such that

$$\sum_{k=1}^{\infty} a_k e_{k+1} < \infty \quad \text{and} \quad \nu_k \xrightarrow[k\to\infty]{} 0.$$

Theorem 2.5.1 considers the case where H_k converges to a stable H. We now extend the theorem from stability of individual H_k to stability of $\sum_{s=k}^{k+K} H_s \ \forall k \ge k_0$ where k_0 is an integer.

Concerning step-size, in addition to the requirement

$$a_k > 0, \ a_k \xrightarrow[k\to\infty]{} 0, \ \sum_{k=1}^{\infty} a_k = \infty,$$

we assume

$$\frac{1}{a_k} - \frac{1}{a_{k-1}} \xrightarrow[k\to\infty]{} \alpha \ge 0. \tag{2.5.15}$$

Theorem 2.5.2 *Let $\{H_s\}$ be $m\times m$-matrices with $\sup_s \|H_s\| < \infty$. Assume that there are an $m\times m$-matrix $Q > 0$ and integers $k_0 > 0$ and $K > 0$ such that*

$$\sum_{s=k}^{k+K}(H_s^T Q + QH_s) \le -\beta I, \;\; \beta > 0 \;\; \forall k \ge k_0. \tag{2.5.16}$$

Then, the estimate x_k given by (2.5.1) converges to zero if and only if the observation noise can be decomposed into two parts $\varepsilon_{k+1} = e_{k+1} + \nu_{k+1}$ such that (2.5.2) holds.

Proof. From definition of $\{a_k\}$ as $k \to \infty$ we have

$$\frac{a_{k-l}}{a_k} - 1 = a_{k-l}\Big(\frac{1}{a_k} - \frac{1}{a_{k-l}}\Big) = O(a_k) \;\; \forall l \ge 1. \tag{2.5.17}$$

For any $k \ge (k_0 + K)$ and $s \ge k_0$ with $k \ge s$ we have

$$\begin{aligned}
\Phi_{k,s}^T Q\Phi_{k,s} &= \Phi_{k-1,s}^T(I + a_k H_k^T)Q(I + a_k H_k)\Phi_{k-1,s}\\
&= \Phi_{k-K,s}^T(I + a_{k-K+1}H_{k-K+1}^T)\cdots(I + a_k H_k^T)Q(I + a_k H_k)\\
&\quad \cdots(I + a_{k-K+1}H_{k-K+1})\Phi_{k-K,s}\\
&= \Phi_{k-K,s}^T\Big[Q + \Big(\sum_{l=k-K+1}^{k} a_l H_l^T Q + Q\sum_{l=k-K+1}^{k} a_l H_l\Big) + o(a_k)\Big]\Phi_{k-K,s}\\
&= \Phi_{k-K,s}^T\Big\{Q + a_k\Big[\Big(\sum_{l=k-K+1}^{k} H_l^T\Big)Q + Q\Big(\sum_{l=k-K+1}^{k} H_l\Big)\Big] + o(a_k)\Big\}\Phi_{k-K,s},
\end{aligned}$$

where $\Phi_{k,s}$ is defined by (2.5.3) and for the last equality (2.5.17) is invoked.

For any large k there is an integer r such that $k - rK \ge s$ and $K > k - rK - s$. Then, noticing $0 < a_k < 1$ by (2.5.16) we have

$$\begin{aligned}
\Phi_{k,s}^T Q\Phi_{k,s} &\le \Phi_{k-K,s}^T(Q - a_k\beta I + o(a_k))\Phi_{k-K,s}\\
&\le \Phi_{k-K,s}^T Q^{\frac{1}{2}}(I - a_k\beta Q^{-1} + o(a_k))Q^{\frac{1}{2}}\Phi_{k-K,s}\\
&\le \Phi_{k-K,s}^T Q^{\frac{1}{2}}\Big(I - \frac{\beta}{K}Q^{-1}\sum_{l=k-K+1}^{k} a_l + o(a_k)\Big)Q^{\frac{1}{2}}\Phi_{k-K,s}\\
&\le \Big(1 - \frac{\beta}{K}\lambda_{\min}(Q^{-1})\sum_{l=k-K+1}^{k} a_l + o(a_k)\Big)\Phi_{k-K,s}^T Q\Phi_{k-K,s}\\
&\le \exp\Big(-c\sum_{l=k-K+1}^{k} a_l\Big)\Phi_{k-K,s}^T Q\Phi_{k-K,s} \le \cdots\\
&\le \exp\Big(-c\sum_{l=k-rK+1}^{k} a_l\Big)\Phi_{k-rK,s}^T Q\Phi_{k-rK,s},
\end{aligned}$$

where c is a positive constant and $\lambda_{\min}(M)$ denotes the minimum eigenvalue of a matrix M.

Since $k - rK \geq s$ and $K > k - rK - s$, $||\Phi_{k-rK,s}||$ is bounded, and hence there is a positive constant $c_1 > 0$ such that

$$\Phi_{k,s}^T Q \Phi_{k,s} \leq c_1 \exp(-c \sum_{l=s}^{k} a_l) I,$$

which, by noticing $Q > 0$, implies

$$||\Phi_{k,s}|| \leq c_2 \exp(-\frac{c}{2} \sum_{l=s}^{k} a_l) \text{ with } c_2 > 0. \tag{2.5.18}$$

For any $s < k_0$, $||\Phi_{k_0-1,s}||$ is bounded and by (2.5.18) we have

$$||\Phi_{k,s}|| \leq ||\Phi_{k,k_0}|| \, ||\Phi_{k_0-1,s}|| \leq ||\Phi_{k_0-1,s}|| c_2 \exp(-\frac{c}{2} \sum_{l=k_0}^{k} a_l). \tag{2.5.19}$$

Combining (2.5.18) and (2.5.19) leads to that for all large k and any $s \geq 0$

$$||\Phi_{k,s}|| \leq c_0 \exp(-\frac{c}{2} \sum_{l=s}^{k} a_l) \tag{2.5.20}$$

for some $c_0 > 0$.

From (2.5.1), we have

$$x_{k+1} = \Phi_{k,0} x_0 + \sum_{s=0}^{t} \Phi_{k,s+1} a_s \varepsilon_{s+1}. \tag{2.5.21}$$

Comparing with Theorem 2.5.1, we find that (2.5.20) and (2.5.21) correspond to (2.5.9) and (2.5.10), respectively. Then, proving the theorem can be carried out along the lines of the proof for Theorem 2.5.1. □

2.6 Convergence Rate of SAAWET

Let the regression function $f(\cdot) \in \mathbb{R}^m \to \mathbb{R}^m$ be differentiable at its root x^0. Its Taylor expansion is as follows

$$f(x) = F(x - x^0) + \delta(x), \tag{2.6.1}$$

where $\delta(x^0) = 0$ and $\delta(x) = o(||x - x^0||)$ as $x \to x^0$.

We consider the rate of convergence of $||x_k - x^0||$ to zero in the case F in (2.6.1) is nondegenerate, where x_k is given by (2.3.2)–(2.3.4).

The following conditions are to be used.

A2.6.1 *$a_k > 0$, $a_k \xrightarrow[k\to\infty]{} 0$, $\sum_{k=1}^{\infty} a_k = \infty$, and*

$$\frac{a_k - a_{k+1}}{a_k a_{k+1}} \xrightarrow[k\to\infty]{} \alpha \geq 0. \tag{2.6.2}$$

A2.6.2 *A continuously differentiable function $v(\cdot) : \mathbb{R}^l \to \mathbb{R}$ exists such that*

$$\sup_{\Delta_1 \leq \|x - x^0\| \leq \Delta_2} f^T(x) v_x(x) < 0 \tag{2.6.3}$$

for any $\Delta_2 > \Delta_1 > 0$, and $v(x^*) < \inf_{\|x\|=c_0} v(x)$ for some $c_0 > 0$ with $c_0 > \|x^*\|$, where x^* is used in (2.3.2).

A2.6.3 *For the sample path ω under consideration the observation noise $\{\varepsilon_k\}$ in (2.3.4) can be decomposed into two parts $\varepsilon_k = \varepsilon_k' + \varepsilon_k''$ such that*

$$\sum_{k=1}^{\infty} a_k^{1-\delta} \varepsilon_{k+1}' < \infty, \quad \varepsilon_{k+1}'' = O(a_k^{\delta}) \tag{2.6.4}$$

for some $\delta \in (0, 1]$.

A2.6.4 *$f(\cdot)$ is measurable and locally bounded, and is differentiable at x^0 such that as $x \to x^0$*

$$f(x) = F(x - x^0) + \delta(x), \quad \delta(x^0) = 0, \quad \delta(x) = o(\|x - x^0\|). \tag{2.6.5}$$

The matrix F is stable (this implies nondegeneracy of F). In addition, $F + \alpha\delta I$ is also stable, where α and δ are given by (2.6.2) and (2.6.4), respectively.

Remark 2.6.1 *If $a_k = \frac{a}{k}$ with $a > 0$, then (2.6.2) holds with $\alpha = \frac{1}{a}$. Also, (2.6.2) is satisfied if $a_k = \frac{a}{k^{\frac{1}{2}+\beta}}$ with $a > 0$, $\beta \in (0, \frac{1}{2})$. In this case $\alpha = 0$.*

Remark 2.6.2 *Take $\delta > 0$ sufficiently small such that $\beta > \frac{\delta}{2(1-\delta)}$. Then $(\frac{1}{2} + \beta)2(1 - \delta) > 1$ and $\sum_{k=1}^{\infty} a_k^{2(1-\delta)} = \sum_{k=1}^{\infty} \frac{a}{k^{2(1-\delta)(\frac{1}{2}+\beta)}} < \infty$. Assume $\{\varepsilon_k, \mathscr{F}_k\}$ is an mds with $\sup_k E(\|\varepsilon_{k+1}\|^2 | \mathscr{F}_k) < \infty$. Then by Theorem 1.2.8 $\sum_{k=1}^{\infty} a_k^{(1-\delta)} \varepsilon_{k+1} < \infty$* a.s. *Therefore (2.6.4) is satisfied* a.s. *with $\varepsilon_k'' \equiv 0$.*

Theorem 2.6.1 *Assume A2.6.1–A2.6.4 hold. Then for those sample paths for which (2.6.4) holds, x_k given by (2.3.2)–(2.3.4) converges to x^0 with the following convergence rate:*

$$\|x_k - x^0\| = o(a_k^{\delta}), \tag{2.6.6}$$

where δ is the one given in (2.6.4).

Proof. We first note that by Theorem 2.3.5 $x_k \to x^0$ and there is no truncation after a finite number of steps. Without loss of generality, we may assume $x^0 = 0$.

By (2.6.2), $\frac{a_k - a_{k+1}}{a_{k+1}} \underset{k\to\infty}{\longrightarrow} 0$. Hence, by the Taylor expansion we have

$$(\frac{a_k}{a_{k+1}})^\delta = (1 + \frac{a_k - a_{k+1}}{a_{k+1}})^\delta = 1 + \delta\frac{a_k - a_{k+1}}{a_{k+1}} + O((\frac{a_k - a_{k+1}}{a_{k+1}})^2). \tag{2.6.7}$$

Write $\delta(x)$ given by (2.6.5) as follows

$$\delta(x) = \big(\delta(x)\frac{x^T}{\|x\|^2}\big)x \quad \text{or} \quad \delta(x_k) = D_k x_k \tag{2.6.8}$$

where

$$D_k \triangleq \big(\delta(x_k)\frac{x_k^T}{\|x_k\|^2}\big) \to 0 \quad \text{as} \quad k \to \infty. \tag{2.6.9}$$

By (2.6.7) and (2.6.5), for sufficiently large k we have

$$\begin{aligned}
\frac{x_{k+1}}{a_{k+1}^\delta} &= (a_k/a_{k+1})^\delta\Big[\frac{1}{a_k^\delta}(x_k + a_k(Fx_k + D_k x_k) + a_k\varepsilon_{k+1})\Big] \\
&= \Big(1 + \delta\frac{a_k - a_{k+1}}{a_{k+1}} + O((\frac{a_k - a_{k+1}}{a_{k+1}})^2)\Big)\Big(\frac{a_k x_k}{a_k^\delta}(\frac{1}{a_k}I + F \\
&\quad + D_k) + a_k^{1-\delta}\varepsilon_{k+1}\Big) \\
&= \frac{x_k}{a_k^\delta} + a_k(F + \delta\frac{a_k - a_{k+1}}{a_{k+1}}(\frac{1}{a_k}I + F) + C_k)\frac{x_k}{a_k^\delta} \\
&\quad + a_k(\frac{\varepsilon'_{k+1}}{a_k^\delta} + \frac{\varepsilon''_{k+1}}{a_k^\delta} + O(\frac{a_k - a_{k+1}}{a_{k+1}})\frac{\varepsilon_{k+1}}{a_k^\delta}),
\end{aligned} \tag{2.6.10}$$

where

$$\begin{aligned}
C_k &\triangleq \Big(\delta\frac{a_k - a_{k+1}}{a_{k+1}} + O((\frac{a_k - a_{k+1}}{a_{k+1}})^2) + 1\Big)D_k \\
&\quad + O((\frac{a_k - a_{k+1}}{a_{k+1}})^2)\Big(\frac{1}{a_k}I + F\Big) \underset{k\to\infty}{\longrightarrow} 0.
\end{aligned}$$

Since $\sum_{k=1}^{\infty} a_k^{1-\delta}\varepsilon'_{k+1} < \infty$, we have $\frac{a_k\varepsilon'_{k+1}}{a_k^\delta} \underset{k\to\infty}{\longrightarrow} 0$ and hence by (2.6.2)

$$O(\frac{a_k - a_{k+1}}{a_{k+1}})\frac{\varepsilon_{k+1}}{a_k^\delta} \underset{k\to\infty}{\longrightarrow} 0. \tag{2.6.11}$$

Set

$$\begin{aligned}
F_k &\triangleq F + \delta\frac{a_k - a_{k+1}}{a_{k+1}}(\frac{1}{a_k}I + F) + C_k, \quad z_k \triangleq \frac{x_k}{a_k^\delta}, \\
e_{k+1} &\triangleq \frac{\varepsilon'_{k+1}}{a_k^\delta} \text{ and } \nu_{k+1} \triangleq \frac{\varepsilon''_{k+1}}{a_k^\delta} + O(\frac{a_k - a_{k+1}}{a_{k+1}})\frac{\varepsilon_{k+1}}{a_k^\delta}.
\end{aligned} \tag{2.6.12}$$

Then (2.6.10) can be rewritten as

$$z_{k+1} = z_k + a_k F_k z_k + a_k(e_{k+1} + v_{k+1}).$$

Noticing that $F_k \xrightarrow[k\to\infty]{} F + \alpha\delta I$, which is stable by A2.6.4, we see that all conditions of Theorem 2.5.1 are satisfied. Therefore, $z_k \xrightarrow[k\to\infty]{} 0$, which proves the theorem. □

Remark 2.6.3 *Take* $a_k = \frac{1}{k^\alpha}$, $\alpha \in (\frac{1}{2}, 1]$ *and let* $\varepsilon''_k \equiv 0$ *in (2.6.4). Assume* $\{\varepsilon_k, \mathscr{F}_k\}$ *is an mds with* $\sup_k E(\|\varepsilon_{k+1}\|^2/\mathscr{F}_k) < \infty$. *Consider how does the convergence rate depend on* α*? In order to have* $\sum_{k=1}^{\infty} a_k^{1-\delta}\varepsilon_{k+1} < \infty$ a.s., *it suffices to require*

$$\sum_{k=1}^{\infty} E\Big(\frac{1}{k^{2\alpha(1-\delta)}}\|\varepsilon_{k+1}\|^2/\mathscr{F}_k\Big) < \infty, \tag{2.6.13}$$

which is guaranteed if $2\alpha(1-\delta) > 1$, *or* $\delta < 1 - \frac{1}{2\alpha}$. *Since* $\alpha \in (\frac{1}{2}, 1]$, *the best convergence rate* $o(\frac{1}{k^\delta})\ \forall\, \delta < \frac{1}{2}$ *is achieved at* $\alpha = 1$. *For* $\alpha \in (\frac{1}{2}, 1]$, *the convergence rate is* $o(\frac{1}{k^{\alpha\delta}})$. *Since* $\alpha\delta < \alpha(1 - \frac{1}{2\alpha}) = \alpha - \frac{1}{2}$, *the convergence rate is slowing down as* α *approaches to* $\frac{1}{2}$.

2.7 Notes and References

The stochastic approximation (SA) algorithm was first introduced by Robbins and Monro in [96], where the mean square convergence of estimates was established under the independence assumption on the observation noise. Much effort was devoted to relaxing conditions for convergence of the algorithm, see, e.g., [11], [43], and [103] among others.

Concerning methods for convergence analysis, besides the probabilistic approach [88] and the ODE approach [9], [69], and [76], there has been developed also the weak convergence method [70]. For the proof of Theorem 2.2.2 we refer to [19], [43], and [88] while for the proof of Theorem 2.2.3 we refer to [9], [69], and [76].

The initial version of SAAWET and its convergence analysis were introduced in [28] with noise condition (2.2.8) used. The improved noise condition A2.3.4 was first applied in [18]. For a detailed discussion of SAAWET, we refer to [19], where properties of SAAWET, such as pathwise convergence, convergence rate, asymptotic normality, asymptotic efficiency, robustness, etc. as well as various kinds of applications, are well summarized. However, GCT (Theorem 2.3.1) of SAAWET serving as the basic algorithm for recursive identification and parameter estimation considered in the book has been rewritten with a better presentation in comparison with the corresponding Theorem 2.2.1 given in [19]. Other material on SAAWET given here has also been revised and improved, while Theorem 2.5.2 has newly appeared in [40].

Chapter 3

Recursive Identification for ARMAX Systems

CONTENTS

In this chapter we consider identification of linear stochastic systems, by which we mean that the system output linearly depends on system inputs and the system is driven by an additive noise. When the system noise $\{w_k\}$ is uncorrelated, then the least-squares (LS) method normally gives satisfactory results. However, when the system noise is correlated, e.g., the noise is a moving average $w_k + C_1 w_{k-1} + \cdots + C_r w_{k_r+1}$, then $C_1, \cdots C_r$ together with other system coefficients have to be estimated. In this case the LS method no longer well behaves, since $w_{k-1}, \cdots, w_{k-r+1}$ cannot be observed. A natural approach is to replace them in the LS estimate with their estimates at time k. This leads to the so-called extended least-squares (ELS) estimate. However, in order for the ELS estimate to give a satisfactory result a restrictive strictly positive realness (SPR) condition usually has to be imposed. This is the weakness of ELS, but its strong point consists in that it works for systems with feedback control, and hence it may be applied to adaptive control systems.

In the case where the identification task is to build the system mathematical model, the control terms can be selected at users' disposal, for example, one may choose a sequence of iid random variables (vectors) and independent of $\{w_k\}$ to serve as system inputs. While in the time series analysis, the mathematical models are usually built simply without control terms. For both cases the SAAWET-based recursive estimates for coefficients of the system are proposed in the chapter without requiring the SPR condition. In addition, the system orders incorporated with coefficient estimation are also recursively estimated.

3.1 LS and ELS for Linear Systems

Here by the linear system we mean it in its input-output form, namely, the ARMAX process.

ARMAX.

The ARMAX system considered in the book is described as follows.

$$\begin{aligned} &y_k + A_1 y_{k-1} + \cdots + A_p y_{k-p} = B_0 u_{k-1} + \cdots + B_q u_{k-q-1} \\ &\quad + w_k + C_1 w_{k-1} + \cdots C_r w_{k-r}, \ \ k \geq 0, \\ &y_k = 0, \ \ u_k = 0, \ \ w_k = 0 \ \ \forall k < 0, \ \ p \geq 0, \ \ q \geq 0, \ \ r \geq 0, \end{aligned} \tag{3.1.1}$$

where y_k denotes the m-dimensional system output, u_k the l-dimensional system input, and w_k the m-dimensional driven noise.

If $B_0 \neq 0$, then the system (3.1.1) is with delay 1 and hence is casual. If $B_0 = \cdots = B_{d-2} = 0$, then the first term at the right-hand side of (3.1.1) is $B_{d-1} u_{k-d}$. In this case we say that the system is with time delay d.

When the control u_k at time k is defined by the outputs $y_k, y_{k-1}, \cdots, y_0$, then we say that (3.1.1) is a feedback control system.

System (3.1.1) is often written in the compact form

$$A(z) y_k = B(z) u_{k-1} + C(z) w_k, \tag{3.1.2}$$

where $A(z)$, $B(z)$, and $C(z)$ are the matrix polynomials in backward-shift operator z with $z y_k = y_{k-1}$:

$$A(z) = I + A_1 z + A_2 z^2 + \cdots + A_p z^p, \tag{3.1.3}$$

$$B(z) = B_0 + B_1 z + \cdots + B_q z^q \tag{3.1.4}$$

$$C(z) = I + C_1 z + \cdots + C_r z^r. \tag{3.1.5}$$

System (3.1.2) is called ARMAX (autoregressive moving average with exogenous inputs), where $A(z) y_k$, $C(z) w_k$, and $B(z) u_{k-1}$ are called its *AR*-part, *MA*-part, and *X*-part, respectively. Sometimes, we also call the system output $\{y_k\}$ as an ARMAX.

If $B(z) \equiv 0$, then (3.1.2) turns to be $A(z) y_k = C(z) w_k$ and $\{y_k\}$ (or the system itself) is called ARMA. If $C(z) = I$ (or $r = 0$), then the system (3.1.2) becomes $A(z) y_k = B(z) u_{k-1} + w_k$ and $\{y_k\}$ is called ARX. If $A(z) = I$ and $B(z) = 0$, then the system becomes $y_k = C(z) w_k$ and $\{y_k\}$ is called MA.

The task of system identification for ARMAX is to estimate the system orders $(p,\ q,\ r)$ and the matrix coefficients of polynomials

$$\theta^T \triangleq [-\theta_A^T\ \theta_B^T\ \theta_C^T], \tag{3.1.6}$$

$$\theta_A^T \triangleq [A_1 \cdots A_p],\ \ \theta_B^T \triangleq [B_0 \cdots B_q],\ \ \theta_C^T \triangleq [C_1 \cdots C_r] \tag{3.1.7}$$

on the basis of the available system input-output data, where M^T denotes the transpose of a matrix M. At any time n the available data set is denoted by $D_n \triangleq \{u_k,\ 0 \leq k \leq n-1,\ y_j,\ 0 \leq j \leq n\}$.

Parameter estimates may be obtained by minimizing some error criterion with data running over the data set D_n for fixed n. In this case, the estimate is derived on the basis of a data set of fixed size and hence is nonrecursive, though the optimal estimate may be iteratively obtained while searching the optimum over D_n. The recursive approach suggests to derive the parameter estimate at any time k from the estimate at time $k-1$ incorporating with the information contained in the input-output data $(u_{k-1},\ y_k)$ received at time k. The advantage of recursive estimation obviously consists in simplicity of updating estimates when D_n is expanding as n increases. However, it is clear that any recursive estimate at time n, in general, cannot be better than the estimate obtained by an optimization method based on D_n. It is worth noting that the LS estimate for ARX systems probably is an exception in the sense that it is obtained by minimizing an error criterion but it can recursively be derived. Let us consider this.

LS.

Assume the orders (p,q) are known. Define the stochastic regressor

$$\phi_k^T \triangleq [y_k^T, \cdots, y_{k-p+1}^T, u_k^T, \cdots, u_{k-q}^T] \tag{3.1.8}$$

with $y_i = 0,\ u_i = 0\ \forall i < 0$.

Then, the ARX system is written as follows

$$y_{k+1} = \theta^T \phi_k + w_{k+1}. \tag{3.1.9}$$

Combining equations (3.1.9) from $k=0$ to $k=n$ leads to a big matrix equation

$$Y_{n+1} = \Phi_n \theta + W_{n+1}, \tag{3.1.10}$$

where

$$Y_{n+1} \triangleq \begin{bmatrix} y_1^T \\ \vdots \\ y_{n+1}^T \end{bmatrix},\quad \Phi_n \triangleq \begin{bmatrix} \phi_0^T \\ \vdots \\ \phi_n^T \end{bmatrix},\quad W_{n+1} \triangleq \begin{bmatrix} w_1^T \\ \vdots \\ w_{n+1}^T \end{bmatrix}. \tag{3.1.11}$$

It is noticed that W_{n+1} cannot be observed and the best estimate θ_{n+1} for θ at time $n+1$ is naturally selected to minimize the error criterion $(Y_{n+1} - \Phi_n\theta)^T(Y_{n+1} - \Phi_n\theta)$, i.e.,

$$(Y_{n+1} - \Phi_n\theta_{n+1})^T(Y_{n+1} - \Phi_n\theta_{n+1}) = \min_{\theta}(Y_{n+1} - \Phi_n\theta)^T(Y_{n+1} - \Phi_n\theta), \tag{3.1.12}$$

where by $A \geq B$ for two square symmetric matrices of the same dimension it is meant that $A-B$ is nonnegative definite.

The estimate obtained in such a way is called the LS estimate. We now derive the explicit expression of the LS estimate. It is direct to verify that

$$\begin{aligned}&(Y_{n+1}-\Phi_n\theta)^T(Y_{n+1}-\Phi_n\theta)\\&\quad=[\theta-(\Phi_n^T\Phi_n)^{-1}\Phi_n^TY_{n+1}]^T\Phi_n^T\Phi_n[\theta-(\Phi_n^T\Phi_n)^{-1}\Phi_n^TY_{n+1}]\\&\quad\quad+Y_{n+1}^T[I-\Phi_n(\Phi_n^T\Phi_n)^{-1}\Phi_n^T]Y_{n+1},\end{aligned}\tag{3.1.13}$$

whenever $\Phi_n^T\Phi_n$ is nondegenerate. If $\Phi_n^T\Phi_n$ is singular, then (3.1.13) remains true with $(\Phi_n^T\Phi_n)^{-1}$ replaced by the pseudo-inverse $(\Phi_n^T\Phi_n)^+$. To see this, it suffices to note that

$$\Phi_n^T\Phi_n(\Phi_n^T\Phi_n)^+\Phi_n^T=\Phi_n^T\Phi_n\Phi_n^+=\Phi_n^T.$$

The last term in (3.1.13) is free of θ and is positive definite. Therefore, from (3.1.13) it is seen that the minimum of $(Y_{n+1}-\Phi_n\theta)^T(Y_{n+1}-\Phi_n\theta)$ is achieved at

$$\theta_{n+1}=(\Phi_n^T\Phi_n)^{-1}\Phi_n^TY_{n+1}=\Big(\sum_{i=0}^n\phi_i\phi_i^T\Big)^{-1}\sum_{i=0}^n\phi_iy_{i+1}^T.\tag{3.1.14}$$

This is the expression of the LS estimate for θ.

We now show that $\{\theta_k\}$ can recursively be derived.

Set $P_{k+1}\triangleq\big(\sum_{i=0}^k\phi_i\phi_i^T\big)^{-1}$. Then, by the matrix inverse identity

$$(C^{-1}+C^{-1}B^TA^{-1}BC^{-1})^{-1}=C-B^T(A+BC^{-1}B^T)^{-1}B,\tag{3.1.15}$$

we have

$$\begin{aligned}P_{k+1}&=(P_k^{-1}+\phi_k\phi_k^T)^{-1}\\&=P_k-P_k\phi_k(1+\phi_k^TP_kP_k^{-1}P_k\phi_k)^{-1}\phi_k^TP_k\\&=P_k-a_kP_k\phi_k\phi_k^TP_k,\end{aligned}\tag{3.1.16}$$

where

$$a_k\triangleq(1+\phi_k^TP_k\phi)_k^{-1}.\tag{3.1.17}$$

From (3.1.14) and (3.1.16) it follows that

$$\begin{aligned}\theta_{k+1}&=(P_k-a_kP_k\phi_k\phi_k^TP_k)\Big(\sum_{i=0}^{k-1}\phi_iy_{i+1}^T+\phi_ky_{k+1}^T\Big)\\&=\theta_k-a_kP_k\phi_k\phi_k^T\theta_k+P_k\phi_ky_{k+1}^T-a_kP_k\phi_k\phi_k^TP_k\phi_ky_{k+1}^T\\&=\theta_k-a_kP_k\phi_k\phi_k^T\theta_k+P_k\phi_k(1-a_k\phi_k^TP_k\phi_k)y_{k+1}^T\\&=\theta_k-a_kP_k\phi_k\phi_k^T\theta_k+a_kP_k\phi_ky_{k+1}^T\\&=\theta_k+a_kP_k\phi_k(y_{k+1}^T-\phi_k^T\theta_k).\end{aligned}$$

Thus, the LS estimate, being optimal in the squared errors sense, has the following recursive expressions:

$$\theta_{k+1} = \theta_k + a_k P_k \phi_k (y_{k+1}^T - \phi_k^T \theta_k), \quad a_k \triangleq (1 + \phi_k^T P_k \phi)_k^{-1}, \tag{3.1.18}$$

$$P_{k+1} = P_k - a_k P_k \phi_k \phi_k^T P_k \ \ \forall k \geq k_0, \tag{3.1.19}$$

starting from that k_0 for which $\sum_{i=0}^{k_0} \phi_i \phi_i^T$ is nonsingular.

Since in real applications k_0 may not be available, one may simply set arbitrary initial values θ_0 and $P_0 = \alpha_0 I$ with $\alpha_0 > 0$ and compute θ_k according to (3.1.18) and (3.1.19) starting from $k = 0$. By such a selection the resulting estimates are as follows

$$P_k = \Big(\sum_{i=0}^{k-1} \phi_i \phi_i^T + P_0^{-1}\Big)^{-1}, \quad \theta_k = P_k \sum_{i=0}^{k-1} \phi_i y_{i+1}^T + P_k P_0^{-1} \theta_0. \tag{3.1.20}$$

To be convinced of this it suffices to recursively express P_k and θ_k given by (3.1.20) and to check the initial conditions.

From here it is seen that the first term at the right-hand side of the expression for θ_k defined by (3.1.20) gives the standard LS estimate, while its last term represents the deviation caused by setting θ_0 and P_0 of θ_k given by (3.1.20) from the LS estimate. Later, it will be shown that for strong consistency of θ_k the minimal eigenvalue of P_k^{-1} should diverge to infinity. Therefore, the difference between θ_k and the LS estimate should be vanishing.

ELS.

We now turn to consider the coefficient estimation for ARMAX given by (3.1.2). One may replace ϕ_k given by (3.1.8) with

$$\phi_k^0 \triangleq [y_k^T, \cdots, y_{k-p+1}^T, u_k^T, \cdots, u_{k-q}^T, w_k^T, \cdots, w_{k-r+1}^T], \tag{3.1.21}$$

and similar to (3.1.9) represent ARMAX as $y_{k+1} = \theta^T \phi_k^0 + w_{k+1}$. Then, proceeding as above one might derive the LS estimate for θ given by (3.1.6). However, $\{w_k\}$ and hence $\{\phi_k^0\}$ are not available, and the LS estimate expressed in such a way cannot be used.

To overcome the difficulty arising here the natural way is to replace $\{w_k\}$ with their estimates $\{\hat{w}_k\}$. In order to express $\hat{w}_k$ in a reasonable way let us return back to (3.1.9), from which we see that for ARX it is natural to define the estimate $\hat{w}_{k+1}$ for w_{k+1} as follows

$$\hat{w}_{k+1} \triangleq y_{k+1} - \theta_{k+1}^T \phi_k, \tag{3.1.22}$$

where θ_{k+1} is the LS estimate for θ.

Replacing w_i with $\hat{w}_i$, $k - r + 1 \leq i \leq k$ in (3.1.21) we define the stochastic regressor ϕ_k for ARMAX:

$$\phi_k \triangleq [y_k^T, \cdots, y_{k-p+1}^T, u_k^T, \cdots, u_{k-q}^T, \hat{w}_k^T, \cdots, \hat{w}_{k-r+1}^T], \tag{3.1.23}$$

and apply the same algorithm (3.1.18) to estimate θ defined by (3.1.6). With arbitrary initial values $P_0 = \alpha_0 I > 0$ and θ_0 the estimate θ_k for any $k > 0$ is recursively calculated as follows:

$$\theta_{k+1} = \theta_k + a_k P_k \phi_k (y_{k+1}^T - \phi_k^T \theta_k), \; a_k \triangleq (1 + \phi_k^T P_k \phi_k)^{-1} \tag{3.1.24}$$

$$P_{k+1} = P_k - a_k P_k \phi_k \phi_k^T P_k. \tag{3.1.25}$$

Like (3.1.20) we still have

$$P_k = \Big(\sum_{i=0}^{k-1} \phi_i \phi_i^T + P_0^{-1}\Big)^{-1}.$$

Similar to this we define

$$P_k^0 = \Big(\sum_{i=0}^{k-1} \phi_i^0 \phi_i^{0T} + P_0^{0-1}\Big)^{-1} \text{ with } P_0^0 = \alpha_0^0 I > 0. \tag{3.1.26}$$

It is clear that for ARX systems $P_k \equiv P_k^0$ if $P_0 = P_0^0$.

In the next section it will be shown that the properties of θ_k heavily depend on the behavior of the matrix P_k. Let us denote by

$$\lambda_{\max}(k), \; \lambda_{\min}(k), \; \lambda_{\max}^0(k), \text{ and } \lambda_{\min}^0(k) \tag{3.1.27}$$

the maximal and minimal eigenvalues of P_k^{-1} and P_k^{0-1}, respectively.

Persistent Excitation (PE).

By the PE condition we mean that

$$\limsup_{k\to\infty} \frac{\lambda_{\max}(k)}{\lambda_{\min}(k)} < \infty \text{ and } \lambda_{\min}(k) \xrightarrow[k\to\infty]{} \infty. \tag{3.1.28}$$

The PE condition is usually used to guarantee strong consistency of LS or ELS, but it will be shown in the next section that a weaker condition is sufficient.

In the following, the norm of a matrix A is defined by $\|A\| \triangleq (\lambda_{\max}\{A^T A\})^{\frac{1}{2}}$.

3.2 Estimation Errors of LS/ELS

Strictly Positive Real (SPR) Condition.

In ARMAX the driven noise $\{w_k\}$ is usually assumed to be a sequence of uncorrelated random vectors/variables, so the MA-part $C(z)w_k$ is a sequence of correlated random vectors whenever $C(z) \neq I$. To deal with correlatedness the so-called SPR condition is often used.

Definition 3.2.1 *A matrix $H(z)$ of rational functions with real coefficients is called SPR if $H(z)$ has no poles in $|z| \leq 1$ and*

$$H(e^{i\lambda}) + H^T(e^{-i\lambda}) > 0 \;\; \forall \lambda \in [0, 2\pi]. \tag{3.2.1}$$

Lemma 3.2.1 *Assume $H(z)$ is a matrix of rational functions with real coefficients. Then, $H(z)$ is SPR if and only if $H^{-1}(z)$ is SPR, and in this case there is a constant $\varepsilon > 0$ such that*

$$\sum_{i=0}^{n} u_i^T y_i \geq \varepsilon \sum_{i=0}^{n} (\|u_i\|^2 + \|y_i\|^2) \ \forall n \geq 0, \tag{3.2.2}$$

where $\{u_i\}$ and $\{y_i\}$ are the input and output of the system with transfer function $H(z)$:

$$y_k = H(z)u_k \ \forall k \geq 0 \ \text{and} \ u_k = 0, \ y_k = 0 \ \forall k < 0.$$

Proof. Assume $H(z)$ is SPR. Noticing that $H^{-1}(e^{i\lambda})$ is the complex conjugate matrix of $H^{-T}(e^{-i\lambda})$, we have

$$\begin{aligned} & H^{-1}(e^{i\lambda}) + H^{-T}(e^{-i\lambda}) \\ & = H^{-1}(e^{i\lambda})\big(H(e^{i\lambda}) + H^T(e^{-i\lambda})\big)H^{-T}(e^{-i\lambda}) > 0 \ \forall \lambda \in [0, \ 2\pi]. \end{aligned}$$

To prove that $H^{-1}(z)$ is SPR, it remains to show that $H^{-1}(z)$ has no poles in $|z| \leq 1$. For this it suffices to show that $H(z)$ has no zeros in $|z| \leq 1$.

From (3.2.1) it follows that there is a positive number $\varepsilon > 0$ such that

$$H(e^{i\lambda}) + H^T(e^{-i\lambda}) - \varepsilon H(e^{i\lambda})H^T(e^{-i\lambda}) > 0 \ \forall \lambda \in [0, \ 2\pi],$$

which is equivalent to

$$\big(H(e^{i\lambda}) - \frac{1}{\varepsilon}I\big)\big(H(e^{i\lambda}) - \frac{1}{\varepsilon}I\big)^* < \frac{1}{\varepsilon^2}I \ \forall \lambda \in [0, \ 2\pi],$$

where M^* denotes the complex conjugate matrix of matrix M.

Therefore, for any complex vector x with $\|x\| = 1$, by Hölder inequality we have

$$\begin{aligned} \big|x^*H(e^{i\lambda})x - \frac{1}{\varepsilon}\big| &= \big|x^*\big(H(e^{i\lambda}) - \frac{1}{\varepsilon}I\big)x\big| \\ &\leq \|H(e^{i\lambda}) - \frac{1}{\varepsilon}I\| < \frac{1}{\varepsilon}. \end{aligned}$$

From here by the maximum principle for analytical functions it follows that

$$\big|x^*H(z)x - \frac{1}{\varepsilon}\big| < \frac{1}{\varepsilon} \ \forall z : |z| \leq 1.$$

This means that $H(z)$ has no zero in $\{z : \ |z| \leq 1\}$.

Noticing that $H(z) = (H^{-1}(z))^{-1}$, we conclude SPR of $H(z)$ from that of $H^{-1}(z)$.

We now show (3.2.2). Since $H(z)$ is SPR, $G(z) \triangleq H(z) - 2\varepsilon I$ is also SPR for small enough $\varepsilon > 0$. Since $G(z)$ is analytic in $|z| \leq 1$, $G(z)$ can be expanded to the series

$$G(z) = \sum_{k=0}^{\infty} G_k z^k, \ \ |z| \leq 1$$

with $\sum_{k=0}^{\infty}\|G_k\|<\infty$. Noticing that $G(e^{i\lambda})+G^T(e^{-i\lambda})>0$, we have

$$
\begin{aligned}
0 &\leq \frac{1}{2\pi}\int_{-\pi}^{\pi}\Big(\sum_{k=0}^{n}u_k e^{e^{-ik\lambda}}\Big)^T\Big(\frac{G(e^{i\lambda})+G^T(e^{-i\lambda})}{2}\Big)\Big(\sum_{l=0}^{n}u_l e^{e^{il\lambda}}\Big)\mathrm{d}\lambda\\
&= \frac{1}{2\pi}\int_{-\pi}^{\pi}\Big(\sum_{k=0}^{n}u_k e^{e^{-ik\lambda}}\Big)^T G(e^{i\lambda})\Big(\sum_{l=0}^{n}u_l e^{e^{il\lambda}}\Big)\mathrm{d}\lambda\\
&= \frac{1}{2\pi}\sum_{j=0}^{\infty}\sum_{k=0}^{n}\sum_{l=0}^{n}u_k^T G_j u_l\int_{-\pi}^{\pi}e^{i(l-k)\lambda}e^{ij\lambda}\mathrm{d}\lambda\\
&= \sum_{j=0}^{n}\sum_{k=0}^{n}\sum_{(l=0,l-k=-j)}^{n}u_k^T G_j u_l=\sum_{j=0}^{n}\sum_{k=j}^{n}u_k^T G_j u_{k-j}\\
&= \sum_{k=0}^{n}u_k^T\sum_{j=0}^{n}G_j u_{k-j}=\sum_{k=0}^{n}u_k^T(G(z)u_k)\\
&= \sum_{k=0}^{n}u_k^T(y_k-2\varepsilon u_k),
\end{aligned}
$$

which implies

$$
\sum_{i=0}^{n}u_i^T y_i \geq 2\varepsilon\sum_{i=0}^{n}\|u_i\|^2 \quad \forall n\geq 0. \tag{3.2.3}
$$

By the first part of the lemma $G^{-1}(z)$ is also SPR, an argument similar to that yielding (3.2.3) leads to

$$
\sum_{i=0}^{n}u_i^T y_i \geq 2\varepsilon\sum_{i=0}^{n}\|y_i\|^2 \quad \forall n\geq 0. \tag{3.2.4}
$$

Combining (3.2.3) and (3.2.4) yields the desired assertion. □

Noticing that a polynomial has no poles, we have the following corollary.

Corollary 3.2.1 *If $H(z)$ is a polynomial with real matrix coefficients, then it is SPR if and only if*

$$
H(e^{i\lambda})+H^T(e^{-i\lambda})>0 \;\; \forall\lambda\in[0,\,2\pi],
$$

or, equivalently,

$$
H^{-1}(e^{i\lambda})+H^{-T}(e^{-i\lambda})>0 \;\; \forall\lambda\in[0,\,2\pi] \;\; \text{and} \;\; \det H(z)\neq 0 \;\; \forall|z|\leq 1.
$$

Theorem 3.2.1 *Assume the following conditions hold:*

(i) $\{w_n,\mathscr{F}_n\}$ *is an mds with*

$$
\sup_{n\geq 0}E(\|w_{n+1}\|^{\beta}|\mathscr{F}_n)\triangleq\sigma<\infty \;\text{ a.s., } \beta\geq 2; \tag{3.2.5}
$$

(ii) The transfer function $C^{-1}(z) - \frac{1}{2}I$ is SPR, i.e.,

$$C^{-1}(e^{i\lambda}) + C^{-T}(e^{-i\lambda}) - I > 0 \ \forall \lambda \in [0, 2\pi];$$

(iii) u_n is $\mathscr{F}_n$-measurable.

Then as $n \to \infty$ the estimation error of ELS is governed as follows:

$$\|\theta_{n+1} - \theta\|^2 = O\Big(\frac{\log \lambda_{\max}(n)(\log\log \lambda_{\max}(n))^{\kappa(\beta-2)}}{\lambda_{\min}(n)}\Big), \tag{3.2.6}$$

where $\kappa(x) \triangleq 0$ if $x \neq 0$, and $\kappa(x) \triangleq c > 1$ if $x = 0$ with c being arbitrary.

Proof. Denote by $\tilde{\theta}_n$ the estimation error of ELS

$$\tilde{\theta}_n \triangleq \theta - \theta_n. \tag{3.2.7}$$

Since $P_{n+1}^{-1} \geq \lambda_{\min}(n+1)I$, it is clear that

$$\|\tilde{\theta}_{n+1}\|^2 \leq \frac{1}{\lambda_{\min}(n)} \mathrm{tr} \tilde{\theta}_{n+1}^T P_{n+1}^{-1} \tilde{\theta}_{n+1}. \tag{3.2.8}$$

Therefore, to prove (3.2.6) it suffices to show

$$\mathrm{tr}\tilde{\theta}_{n+1}^T P_{n+1}^{-1} \tilde{\theta}_{n+1} = O\big(\log \lambda_{\max}(n)(\log\log \lambda_{\max}(n))^{\kappa(\beta-2)}\big). \tag{3.2.9}$$

Set

$$\xi_{n+1} = y_{n+1} - \theta_{n+1}^T \phi_n - w_{n+1}, \tag{3.2.10}$$

where $y_{n+1} - \theta_{n+1}^T \phi_n$ actually is an estimate for w_{n+1}.

We now show that

$$C^{-1}(z)\tilde{\theta}_{n+1}^T \phi_n = \xi_{n+1}. \tag{3.2.11}$$

This is because

$$\begin{aligned} C(z)\xi_{n+1} &= y_{n+1} + (C(z) - I)(y_{n+1} - \theta_{n+1}^T \phi_n) - \theta_{n+1}^T \phi_n - C(z)w_{n+1} \\ &= -(A(z) - I)y_{n+1} + B(z)u_n + (C(z) - I)(y_{n+1} - \theta_{n+1}^T \phi_n) - \theta_{n+1}^T \phi_n \\ &= \theta^T \phi_n - \theta_{n+1}^T \phi_n = \tilde{\theta}_{n+1}^T \phi_n. \end{aligned}$$

By Condition (ii) of the theorem $C^{-1}(z) - \frac{1}{2}I$ is SPR, so by Lemma 3.2.1 there are constants $k_0 > 0$ and $k_1 \geq 0$ such that

$$s_n \triangleq \sum_{i=0}^{n} \phi_i^T \tilde{\theta}_{i+1}(\xi_{i+1} - \frac{1}{2}(1 + k_0)\tilde{\theta}_{i+1}^T \phi_i) \geq 0 \ \forall n \geq 0. \tag{3.2.12}$$

From (3.1.24) it is seen that

$$\begin{aligned} & y_{n+1}^T - \phi_n^T \theta_{n+1} \\ & \quad = y_{n+1}^T - \phi_n^T [\theta_n + a_n P_n \phi_n (y_{n+1}^T - \phi_n^T \theta_n)] \\ & \quad = (1 - a_n \phi_n^T P_n \phi_n)(y_{n+1}^T - \phi_n^T \theta_n) \\ & \quad = a_n (y_{n+1}^T - \phi_n^T \theta_n). \end{aligned} \tag{3.2.13}$$

Thus, by (3.2.10) and (3.2.13) we can rewrite (3.1.24) as

$$\tilde{\theta}_{n+1} = \tilde{\theta}_n - P_n \phi_n (\xi_{n+1}^T + w_{n+1}^T). \tag{3.2.14}$$

Noticing $P_{k+1}^{-1} = P_k^{-1} + \phi_k \phi_k^T$, from (3.2.14) we have that

$$\begin{aligned} & \operatorname{tr}\tilde{\theta}_{k+1}^T P_{k+1}^{-1} \tilde{\theta}_{k+1} = \operatorname{tr}\tilde{\theta}_{k+1}^T \phi_k \phi_k^T \tilde{\theta}_{k+1} + \operatorname{tr}\tilde{\theta}_{k+1}^T P_k^{-1} \tilde{\theta}_{k+1} \\ & \quad = \|\tilde{\theta}_{k+1}^T \phi_k\|^2 + \operatorname{tr}\left[\tilde{\theta}_k - P_k \phi_k (\xi_{k+1}^T + w_{k+1}^T)\right]^T P_k^{-1} \\ & \qquad \cdot \left[\tilde{\theta}_k - P_k \phi_k (\xi_{k+1}^T + w_{k+1}^T)\right] \\ & \quad = \|\tilde{\theta}_{k+1}^T \phi_k\|^2 - 2(\xi_{k+1}^T + w_{k+1}^T)\tilde{\theta}_k^T \phi_k + \phi_k^T P_k \phi_k \|\xi_{k+1} \\ & \qquad + w_{k+1}\|^2 + \operatorname{tr}\tilde{\theta}_k^T P_k^{-1} \tilde{\theta}_k \\ & \quad = \|\tilde{\theta}_{k+1}^T \phi_k\|^2 + \operatorname{tr}\tilde{\theta}_k^T P_k^{-1} \tilde{\theta}_k - 2(\xi_{k+1}^T + w_{k+1}^T)\left[\tilde{\theta}_{k+1} + P_k \phi_k (\xi_{k+1}^T \right. \\ & \qquad \left. + w_{k+1}^T)\right]^T \phi_k + \phi_k^T P_k \phi_k \|\xi_{k+1} + w_{k+1}\|^2 \\ & \quad \le \|\tilde{\theta}_{k+1}^T \phi_k\|^2 + \operatorname{tr}\tilde{\theta}_k^T P_k^{-1} \tilde{\theta}_k - 2\xi_{k+1}^T \tilde{\theta}_{k+1}^T \phi_k - 2 w_{k+1}^T \tilde{\theta}_{k+1}^T \phi_k \\ & \quad = \operatorname{tr}\tilde{\theta}_k^T P_k^{-1} \tilde{\theta}_k - 2\left[\phi_k^T \tilde{\theta}_{k+1} (\xi_{k+1} - \frac{1}{2}(1 + k_0)\tilde{\theta}_{k+1}^T \phi_k)\right] \\ & \qquad - k_0 \|\tilde{\theta}_{k+1}^T \phi_k\|^2 - 2 w_{k+1}^T \tilde{\theta}_{k+1}^T \phi_k. \end{aligned} \tag{3.2.15}$$

Summing up both sides of (3.2.15) from 0 to n and paying attention to (3.2.12), we derive

$$\begin{aligned} & \operatorname{tr}\tilde{\theta}_{n+1}^T P_{n+1}^{-1} \tilde{\theta}_{n+1} \\ & \quad \le \operatorname{tr}\tilde{\theta}_0^T P_0^{-1} \tilde{\theta}_0 - 2 s_n - k_0 \sum_{i=0}^n \|\tilde{\theta}_{i+1}^T \phi_i\|^2 - 2 \sum_{i=0}^n w_{i+1}^T \tilde{\theta}_{i+1}^T \phi_i \\ & \quad \le O(1) - k_0 \sum_{i=0}^n \|\tilde{\theta}_{i+1}^T \phi_i\|^2 - 2 \sum_{i=0}^n w_{i+1}^T \tilde{\theta}_{i+1}^T \phi_i. \end{aligned} \tag{3.2.16}$$

We now estimate the last term in (3.2.16).

Set

$$\eta_n \triangleq y_{n+1} - \theta_n^T \phi_n - w_{n+1}, \tag{3.2.17}$$

which is $\mathscr{F}_n$-measurable.

By (3.1.24) and (3.2.17) we have

$$\tilde{\theta}_{n+1} = \tilde{\theta}_n - a_n P_n \phi_n (\eta_n^T + w_{n+1}^T). \tag{3.2.18}$$

Then by Theorem 1.2.14, the last term in (3.2.16) is estimated as follows:

$$\begin{aligned}
&\Big|\sum_{i=0}^{n} w_{i+1}^T \tilde{\theta}_{i+1}^T \phi_i\Big| \\
&= \Big|\sum_{i=0}^{n} w_{i+1}^T (\tilde{\theta}_i^T - a_i(\eta_i + w_{i+1})\phi_i^T P_i)\phi_i\Big| \\
&\le \sum_{i=0}^{n} a_i \phi_i^T P_i \phi_i \|w_{i+1}\|^2 + \Big|\sum_{i=0}^{n} w_{i+1}^T (\tilde{\theta}_i^T - a_i \eta_i \phi_i^T P_i)\phi_i\Big| \\
&\le \sum_{i=0}^{n} a_i \phi_i^T P_i \phi_i \|w_{i+1}\|^2 + O\Big(\big(\sum_{i=0}^{n} \|(\tilde{\theta}_i^T - a_i \eta_i \phi_i^T P_i)\phi_i\|^2\big)^{\alpha}\Big) \\
&= \sum_{i=0}^{n} a_i \phi_i^T P_i \phi_i \|w_{i+1}\|^2 + O\Big(\big(\sum_{i=0}^{n} \|(\tilde{\theta}_{i+1}^T + a_i w_{i+1} \phi_i^T P_i)\phi_i\|^2\big)^{\alpha}\Big) \\
&= \sum_{i=0}^{n} a_i \phi_i^T P_i \phi_i \|w_{i+1}\|^2 + O\Big(\big(\sum_{i=0}^{n} \|\tilde{\theta}_{i+1}^T \phi_i\|^2\big)^{\alpha}\Big) \\
&\quad + O\Big(\big(\sum_{i=0}^{n} (a_i \phi_i^T P_i \phi_i)^2 \|w_{i+1}\|^2\big)^{\alpha}\Big) \text{ with } \alpha \in (\tfrac{1}{2}, 1).
\end{aligned} \tag{3.2.19}$$

Putting (3.2.19) in (3.2.16) leads to

$$\operatorname{tr} \tilde{\theta}_{k+1}^T P_{k+1}^{-1} \tilde{\theta}_{k+1} \le O(1) + \sum_{i=0}^{n} a_i \phi_i^T P_i \phi_i \|w_{i+1}\|^2 \;\; a.s. \tag{3.2.20}$$

So, the problem is reduced to estimating the last term of (3.2.20).

Noticing that the matrix $P_i \phi_i \phi_i^T$ has only one nonzero eigenvalue equal to $\phi_i^T P_i \phi_i$, we have

$$\det(I + P_i \phi_i \phi_i^T) = 1 + \phi_i^T P_i \phi_i. \tag{3.2.21}$$

From here it follows that

$$\det P_{i+1}^{-1} = \det(P_i^{-1} + \phi_i \phi_i^T) = \det P_i^{-1} \det(I + P_i \phi_i \phi_i^T),$$

or, equivalently,

$$\phi_i^T P_i \phi_i = \frac{\det P_{i+1}^{-1} - \det P_i^{-1}}{\det P_i^{-1}}. \tag{3.2.22}$$

Since $a_k \triangleq (1+\phi_k^T P_k \phi_k)^{-1}$, by (3.2.22) it follows that

$$\begin{aligned}\sum_{i=0}^{n} a_i \phi_i^T P_i \phi_i &= \sum_{i=0}^{n} \frac{\det P_{i+1}^{-1} - \det P_i^{-1}}{\det P_{i+1}^{-1}} \\ &= \sum_{i=0}^{n} \int_{\det P_i^{-1}}^{\det P_{i+1}^{-1}} \frac{dx}{\det P_{i+1}^{-1}} \leq \int_{\det P_0^{-1}}^{\det P_{n+1}^{-1}} \frac{dx}{x} \\ &= \log(\det P_{n+1}^{-1}) - \log(\det P_0^{-1}). \end{aligned} \tag{3.2.23}$$

Notice that $P_n^{-1} \geq P_0^{-1} = \frac{1}{\alpha_0} I$ and the maximal eigenvalue of P_{n+1}^{-1} is $\lambda_{\max}(n)$ while its other eigenvalues are greater than or equal to $\frac{1}{\alpha_0}$. Therefore, we have

$$\left(\frac{1}{\alpha_0}\right)^{mp+l(q+1)+mr-1} \lambda_{\max}(n) \leq \det P_{n+1}^{-1} \leq (\lambda_{\max}(n))^{mp+l(q+1)+mr},$$

and

$$\begin{aligned}&\log \lambda_{\max}(n) - (mp + l(q+1) + mr - 1)\log \alpha_0 \\ &\leq \log\det P_{n+1}^{-1} \leq (mp + l(q+1) + mr)\log \lambda_{\max}(n),\end{aligned} \tag{3.2.24}$$

where $mp + l(q+1) + mr$ is the dimension of ϕ_n.

From (3.2.23) and (3.2.24) it follows that

$$\sum_{i=0}^{n} a_i \phi_i^T P_i \phi_i = O(\log \lambda_{\max}(n)) \quad a.s. \tag{3.2.25}$$

Applying Theorem 1.2.14 with $M_i = a_i \phi_i^T P_i \phi_i$, $\xi_{i+1} = \|w_{i+1}\|^2 - E(\|w_{i+1}\|^2 | \mathscr{F}_i)$ and taking $\alpha \in [1, \min(\frac{\beta}{2}, 2)]$, we have

$$\begin{aligned}&\sum_{i=0}^{n} a_i \phi_i^T P_i \phi_i \|w_{i+1}\|^2 \\ &= \sum_{i=0}^{n} M_i \xi_{i+1} + \sum_{i=0}^{n} a_i \phi_i^T P_i \phi_i E(\|w_{i+1}\|^2 | \mathscr{F}_i) \\ &= O\Big(\big(\sum_{i=0}^{n} (M_i)^\alpha\big)^{\frac{1}{\alpha}} \log^{\frac{1}{\alpha}+\eta}\big(\sum_{i=0}^{n} M_i^\alpha + e\big)\Big) + O\big(\log \lambda_{\max}(n)\big) \\ &= O\big((\log \lambda_{\max}(n))^{\frac{1}{\alpha}} \log^{\frac{1}{\alpha}+\eta}(\log \lambda_{\max}(n) + e)\big) + O\big(\log \lambda_{\max}(n)\big) \quad \forall \eta > 0,\end{aligned} \tag{3.2.26}$$

where $\log e = 1$.

If $\beta = 2$, then $\alpha = 1$; while if $\beta > 2$, then α may be taken greater than one. Consequently, from (3.2.26) we have

$$\sum_{i=0}^{n} a_i \phi_i^T P_i \phi_i \|w_{i+1}\|^2 = O\big(\log \lambda_{\max}(n) (\log\log \lambda_{\max}(n))^{\kappa(\beta-2)}\big). \tag{3.2.27}$$

Substituting (3.2.27) into (3.2.20), we derive (3.2.9). The proof is completed. □

Remark 3.2.1 *The PE condition defined by (3.1.28) is used to guarantee strong consistency of LS/ELS estimates. From Theorem 3.2.1 it can be seen that the PE condition may be weakened for strong consistency of LS/ELS. For example, if* $\lambda_{\max}(n) = O(n^b)$, $b > 0$ *and* $(\log n)^{1+\varepsilon} = O(\lambda_{\min}(n))$, *then by Theorem 3.2.1 we have strong consistency:*

$$\|\theta_{n+1} - \theta\|^2 = O\Big(\frac{\log n(\log\log n)^{\kappa(\beta-2)}}{(\log n)^{1+\varepsilon}}\Big) \xrightarrow[n\to\infty]{} 0.$$

However, in this case the PE condition does not hold:

$$\frac{\lambda_{\max}(n)}{\lambda_{min}(n)} = O\Big(\frac{n^b}{(\log n)^{1+\varepsilon}}\Big) \xrightarrow[n\to\infty]{} \infty.$$

Theorem 3.2.1 gives a nice expression for estimation error of LS/ELS, but it does not answer the question whether or not the LS/ELS estimate converges to the true parameter as the number of data increases. For example, how does the ELS behave when $\{u_k\}$ in an ARMAX is a sequence of iid random vectors independent of $\{w_k\}$? And, how is ELS for ARMA, i.e., for the case where $u_k \equiv 0$?

As a matter of fact, under the conditions of Theorem 3.2.1, it can be shown that

$$\|\theta_{n+1} - \theta\|^2 = O\left(\frac{\log \lambda^0_{\max}(n)(\log\log \lambda^0_{\max}(n))^{\kappa(\beta-2)}}{\lambda^0_{\min}(n)}\right) \tag{3.2.28}$$

if its right-hand side tends to zero as $n \to \infty$. Further, if $\{u_k\}$ is mutually independent and is independent of $\{w_k\}$ with $\sum_{k=1}^n \|u_k\|^2/n = O(1)$, then under an identifiability condition $\lambda^0_{\min}(n) \geq cn^\alpha \ \forall\, n \geq n_0$ for some n_0 with $c > 0$, $\alpha > 0$. If, in addition, $A(z)$ is stable, then putting these estimates into (3.2.28) leads to

$$\|\theta_{n+1} - \theta\|^2 = O\left(\frac{\log n(\log\log n)^{\kappa(\beta-2)}}{n^\alpha}\right) \tag{3.2.29}$$

which tends to zero as $n \to \infty$.

3.3 Hankel Matrices Associated with ARMA

(i) Hankel Matrices for Coefficient Estimation

To identify coefficients of ARMA, the row-full-rank of the associated Hankel matrices composed of impulse responses or of correlation functions is of crucial importance.

Let us consider the following linear model

$$A(z)y_k = B(z)u_k, \tag{3.3.1}$$

where

$$A(z) = I + A_1 z + \cdots + A_p z^p \ \text{ with } A_p \neq 0 \tag{3.3.2}$$

$$B(z) = B_0 + B_1 z + \cdots + B_q z^q \ \text{ with } B_q \neq 0 \tag{3.3.3}$$

are matrix polynomials in backward-shift operator $z: zy_k = y_{k-1}$. The system output y_k and input u_k are of m- and l- dimensions, respectively.

If $\{u_k\}$ is a sequence of zero-mean uncorrelated random vectors, then (3.3.1) is an ARMA process, which, however, differs from $A(z)y_k = C(z)w_k$ discussed in the last section by the following two points: i) $B(z)$, unlike $C(z)$, may not be a square matrix polynomial; and ii) B_0, unlike I in $C(z)$, is to be estimated.

In system and control, as a rule $m \geq l$ and (3.3.1) may be a part of nonlinear systems, e.g., Hammerstein systems, Wiener systems, etc. The matrix coefficients $(A_1, \cdots, A_p, B_0, B_1 \cdots, B_q)$ are required to estimate on the basis of the input-output information or estimated inputs and outputs of the system if the linear model is a part of nonlinear systems.

Assume $A(z)$ is stable, i.e., $\det A(z) \neq 0 \ \forall |z| \leq 1$.

Stability of $A(z)$ gives the possibility to define the transfer function:

$$H(z) \triangleq A^{-1}(z)B(z) = \sum_{i=0}^{\infty} H_i z^i, \tag{3.3.4}$$

where $H_0 = B_0$, $\|H_i\| = O(e^{-ri}), r > 0, i > 1$. Then, y_k in (3.3.1) can be connected with the input $\{u_k\}$ via impulse responses:

$$y_k = \sum_{i=0}^{\infty} H_i u_{k-i}. \tag{3.3.5}$$

Let us first derive the linear equations connecting $\{A_1, \cdots, A_p, B_0, B_1, \cdots, B_q\}$ with $\{H_i\}$.

From (3.3.4), it follows that

$$\begin{aligned} &B_0 + B_1 z + \cdots + B_q z^q \\ &\quad = (I + A_1 z + \cdots + A_p z^p)(H_0 + H_1 z + \cdots + H_i z^i + \cdots). \end{aligned} \tag{3.3.6}$$

Identifying coefficients for the same degrees of z at both sides yields

$$B_i = \sum_{j=0}^{i \wedge p} A_j H_{i-j} \quad \forall\, 1 \leq i \leq q, \tag{3.3.7}$$

$$H_i = -\sum_{j=1}^{i \wedge p} A_j H_{i-j} \quad \forall\, i \geq q+1, \tag{3.3.8}$$

where $a \wedge b$ denotes $\min(a,b)$.

For H_i, $q+1 \leq i \leq q+mp$, by (3.3.8) we obtain the following linear algebraic equation

$$[A_1, A_2, \cdots, A_p]L = -[H_{q+1}, H_{q+2}, \cdots, H_{q+mp}], \tag{3.3.9}$$

where

$$L \triangleq \begin{bmatrix} H_q & H_{q+1} & \cdots & H_{q+mp-1} \\ H_{q-1} & H_q & \cdots & H_{q+mp-2} \\ \vdots & \vdots & \ddots & \vdots \\ H_{q-p+1} & H_{q-p+2} & \cdots & H_{q+(m-1)p} \end{bmatrix}, \quad (3.3.10)$$

where $H_i \triangleq 0$ for $i < 0$.

Define

$$\theta_A^T \triangleq [A_1, \cdots, A_p], \;\; G^T \triangleq -[H_{q+1}, H_{q+2}, \cdots, H_{q+mp}]. \quad (3.3.11)$$

Then, from (3.3.9) it follows that

$$\theta_A = (LL^T)^{-1}LG, \quad (3.3.12)$$

if L is of row-full-rank.

In this case, if we can obtain estimates for $\{H_i\}$, then replacing $H_i,\; i = 1,2,\cdots$ in (3.3.9) with their estimates, we derive the estimate for θ_A. Finally, with the help of (3.3.7) the estimates for $B_i,\;\; i = 0,1,\cdots,q$ can also be obtained.

From here we see that the row-full-rank of the Hankel matrix L composed of impulse responses is important for estimating the system indeed.

The well-known Yule–Walker equation connects θ_A with the Hankel matrix composed of correlation functions of the system output $\{y_k\}$.

Under the stability assumption on $A(z)$, $\{y_k\}$ is asymptotically stationary. Without loss of generality, $\{y_k\}$ may be assumed to be stationary with correlation function $R_i \triangleq E y_k y_{k-i}^T$ by appropriately choosing the initial values of $\{u_k\}$.

Multiplying $y_{k-t}^T, t \geq q+1$ at both sides of (3.3.1) from the right and taking expectation, we obtain

$$\begin{aligned} &E(y_k + A_1 y_{k-1} + \cdots + A_p y_{k-p}) y_{k-t}^T \\ &= E(B_0 u_k + B_1 u_{k-1} + \cdots + B_q u_{k-q}]) y_{k-t}^T = 0 \;\; \forall t \geq q+1, \end{aligned}$$

which yields

$$\sum_{i=0}^{p} A_i R_{q-i+s} = 0 \;\; \forall s \geq 1. \quad (3.3.13)$$

For $R_i,\; q+1 \leq i \leq q+mp$, by (3.3.13) we have the following linear algebraic equation called the Yule–Walker equation:

$$[A_1, A_2, \cdots, A_p]\Gamma = -[R_{q+1}, R_{q+2}, \cdots, R_{q+mp}], \quad (3.3.14)$$

where

$$\Gamma \triangleq \begin{bmatrix} R_q & R_{q+1} & \cdots & R_{q+mp-1} \\ R_{q-1} & R_q & \cdots & R_{q+mp-2} \\ \vdots & \vdots & \ddots & \vdots \\ R_{q-p+1} & R_{q-p+2} & \cdots & R_{q+(m-1)p} \end{bmatrix}. \quad (3.3.15)$$

Similar to (3.3.12) we can rewrite (3.3.14) as

$$\theta_A = (\Gamma\Gamma^T)^{-1}\Gamma W, \tag{3.3.16}$$

whenever Γ is of row-full-rank, where $W^T \triangleq -[R_{q+1}, R_{q+2}, \cdots, R_{q+mp}]$.

From here it is seen that the row-full-rank of the Hankel matrix Γ composed of correlation functions has the similar importance as L does.

Identifiability

For the transfer function $H(z) = \sum_{i=0}^{\infty} H_i z^i$, $A^{-1}(z)B(z)$ is called its matrix fraction description (MFD) form. It is natural to consider the uniqueness issue of the description.

Denote by $\mathcal{M}$ the totality of the matrix pairs $[X(z)\ Y(z)]$ satisfying $X^{-1}(z)Y(z) = H(z)$, where $X(z)$ is $m \times m$, stable, and monic with order less than or equal to p and $Y(z)$ is $m \times l$ with order less than or equal to q.

By (3.3.4) $[A(z)\ B(z)] \in \mathcal{M}$. We are interested in conditions guaranteeing the uniqueness of MFD. To clarify this, we first prove a lemma concerning the orders of factors in a matrix polynomial factorization. In the one-dimensional case, the order of factors of a polynomial is certainly no more than the order of the polynomial. However, in the multi-dimensional case the picture is different.

A square matrix polynomial is called unimodular if its determinant is a nonzero constant. From the definition it follows that the inverse of a unimodular matrix is also a matrix polynomial. Let the $m \times l$-matrix polynomial $B(z) = B_0 + B_1 z + \cdots + B_q z^q$ be factorized as a product of two matrix polynomials $C(z)$ and $\overline{B}(z)$: $B(z) = C(z)\overline{B}(z)$. Since $B(z) = C(z)U(z)U^{-1}(z)\overline{B}(z)$ and $U^{-1}(z)\overline{B}(z)$ remains to be a matrix polynomial for any unimodular matrix $U(z)$, the factors $C'(z) \triangleq C(z)U(z)$ and $B'(z) \triangleq U^{-1}(z)\overline{B}(z)$ in the factorization $B(z) = C'(z)B'(z)$ may be with arbitrarily high orders.

The following lemma shows that the order of $B'(z)$ in the factorization $B(z) = C'(z)B'(z)$ can be made no higher than that of $B(z)$ by appropriately choosing the unimodular matrix $U(z)$.

Lemma 3.3.1 *Assume an $m \times l$-matrix polynomial $G(z)$ of order r is factorized as $G(z) = C(z)D(z)$, where $C(z)$ and $D(z)$ are matrix polynomials of $m \times m$ and $m \times l$ dimensions, respectively. Then, an $m \times m$ unimodular matrix $U(z)$ can be chosen such that in the factorization $G(z) = C'(z)D'(z)$ with $C'(z) \triangleq C(z)U(z)$ and $D'(z) \triangleq U^{-1}(z)D(z)$ the order (degree) of $D'(z)$ denoted by* $\deg D'(z)$ *is less than or equal to r.*

Proof. It is well known that the elementary column transformations, i.e., multiplying $C(z)$ from right by the matrices corresponding to exchanging the places of its *ith* column with the *jth* column, multiplying the *ith* column of $C(z)$ by a constant, and adding its *ith* column with its *jth* column multiplied by a polynomial, may lead the the matrix polynomial $C(z)$ to a lower-triangular matrix, for which at each row the highest degree appears at its diagonal element. Denoting by $U(z)$ the unimodular matrix resulting from all the elementary transformations yielding $C(z)$ to the lower-triangular form, we have

$$C(z)U(z) = \begin{bmatrix} c'_{1,1}(z) & 0 & \cdots & 0 \\ c'_{2,1}(z) & c'_{2,2}(z) & \cdots & 0 \\ \vdots & \vdots & \ddots & \vdots \\ c'_{m,1}(z) & c'_{m,2}(z) & \cdots & c'_{m,m}(z) \end{bmatrix}, \tag{3.3.17}$$

where $\deg c'_{i,s}(z) \leq \deg c'_{i,i}(z) \ \forall s \leq i \ \forall i: \ i = 1, \cdots, m$.

We now show that this $U(z)$ is the one required by the lemma.

Let $D'(z) \triangleq U^{-1}(z)D(z) = \{d'_{i,j}(z)\}_{i=1,\cdots,m; \ j=1,\cdots,l}$ and $G(z) = \{g_{i,j}(z)\}_{i=1,\cdots,m; \ j=1,\cdots,l}$.

For the lemma it suffices to show that for any fixed $j: \ j = 1, \cdots, l$

$$\deg d'_{i,j}(z) \leq \max_{1 \leq l \leq m} \deg g_{l,j}(z) \ \ \forall i: i = 1, \cdots, m. \tag{3.3.18}$$

For $i = 1$, we have $g_{1,j}(z) = c'_{1,1}(z)d'_{1,j}(z)$, which obviously implies

$$\deg d'_{1,j}(z) \leq \max_{1 \leq l \leq m} \deg g_{l,j}(z).$$

Thus, (3.3.18) holds for $i = 1$.

Assume (3.3.18) is true for $i = 1, \cdots, s-1$. We want to show that (3.3.18) also holds for $i = s$.

Assume the converse: $\deg d'_{s,j}(z) > \max_{1 \leq l \leq m} \deg g_{l,j}(z)$.

The inductive assumption incorporated with the converse assumption implies that

$$\deg d'_{s,j}(z) > \deg d'_{i,j}(z) \ \ \forall i: \ i = 1, \cdots, s-1. \tag{3.3.19}$$

By noticing $\deg c'_{s,l}(z) \leq \deg c'_{s,s}(z) \ \ \forall l \leq s$ and $\forall s: \ s = 1, \cdots, m$, by (3.3.19) from the equality $g_{s,j}(z) = c'_{s,1}(z)d'_{1,j}(z) + \cdots + c'_{s,s-1}(z)d'_{s-1,j}(z) + c'_{s,s}(z)d'_{s,j}(z)$ we derive $\deg g_{s,j}(z) = \deg c'_{s,s}(z)d'_{s,j}(z)$. From here the converse assumption leads to a contradictory inequality:

$\deg g_{s,j}(z) \geq \deg d'_{s,j}(z) > \max_{1 \leq l \leq m} \deg g_{l,j}(z)$.

The obtained contradiction proves the lemma. □

The nonnegative integer r is called the rank of a rational polynomial matrix if (1) there exists at least one subminor of order r which does not vanish identically, and (2) all subminors of order greater than r vanish identically.

We need a fact from the linear algebra, i.e., the Smith–McMillan form for a square matrix polynomial. We formulate it as a lemma.

Lemma 3.3.2 *Let $G(z)$ be an $n \times n$ rational matrix of rank r. Then there exist two $n \times n$ unimodular matrices $U(z)$ and $V(z)$ such that*

$$\begin{aligned} G(z) &= U(z)\mathrm{diag}\left[\frac{e_1(z)}{\psi_1(z)}, \frac{e_2(z)}{\psi_2(z)}, \cdots, \frac{e_r(z)}{\psi_r(z)}, 0 \cdots, 0\right] V(z) \\ &= U(z)W(z)V(z), \end{aligned} \tag{3.3.20}$$

where

(a) *$e_k(z)$ and $\psi_k(z)$ are relatively prime polynomials with unit leading coefficients $\forall k: 1 \le k \le r$;*

(b) *Each $e_k(z)$ divides $e_{k+1}(z)$ $\forall k: 1 \le k \le r-1$, and each $\psi_j(z)$ is a factor of $\psi_{j-1}(z)$ $\forall j: 2 \le j \le r$;*

(c) *The diagonal matrix $W(z)$ appearing in (3.3.20) satisfies (a) and (b), uniquely determined by $G(z)$;*

(d) *If $G(z)$ is real, then $U(z)$, $W(z)$, and $V(z)$ may also be chosen to be real.*

Theorem 3.3.1 *The following conditions are equivalent:*

A3.3.1 *The set $\mathcal{M}$ is composed of the unique pair $[A(z)\ B(z)]$;*

A3.3.2 *$A(z)$ and $B(z)$ have no common left factor and $[A_p\ \ B_q]$ is of row-full-rank;*

A3.3.3 *There are no n-dimensional vector polynomial $d(z)$ and m-dimensional vector polynomial $c(z)$ (not both zero) with orders strictly less than p and q, respectively, such that $d^T(z)H(z) + c^T(z) = 0$.*

Proof. A3.3.2 $\Rightarrow$ A3.3.1

Assume A3.3.2 holds. Take any matrix polynomial pair $[\overline{A}(z)\ \overline{B}(z)] \in \mathcal{M}$, where

$$\overline{A}(z) = I + \overline{A}_1 z + \cdots + \overline{A}_{\overline{p}} z^{\overline{p}} \text{ with } \overline{p} \le p,$$
$$\overline{B}(z) = \overline{B}_0 + \overline{B}_1 z + \cdots + \overline{B}_{\overline{q}} z^{\overline{q}} \text{ with } \overline{q} \le q.$$

We have to show that $[\overline{A}(z)\ \overline{B}(z)] = [A(z)\ \ B(z)]$. Set $C(z) \triangleq \overline{A}(z)A^{-1}(z)$. Then, we have

$$\overline{A}(z) = C(z)A(z), \tag{3.3.21}$$
$$\overline{B}(z) = \overline{A}(z)H(z) = \overline{A}(z)A^{-1}(z)B(z) = C(z)B(z). \tag{3.3.22}$$

Since both $A(z)$ and $\overline{A}(z)$ are stable matrices, $\det C(z)$ is not identically equal to zero. So, the rank of $C(z)$ is m.

By Lemma 3.3.2 $C(z)$ can be presented in the Smith–McMillan form:

$$\begin{aligned} C(z) &= U(z)\text{diag}\Big[\frac{q_1(z)}{p_1(z)}, \frac{q_2(z)}{p_2(z)}, \cdots, \frac{q_m(z)}{p_m(z)}\Big]V(z) \\ &= U(z)P^{-1}(z)Q(z)V(z), \end{aligned} \tag{3.3.23}$$

where $U(z)$ and $V(z)$ are $m \times m$ unimodular matrices, and

$$P(z) = \text{diag}\big[p_1(z), p_2(z), \cdots, p_m(z)\big],$$
$$Q(z) = \text{diag}\big[q_1(z), q_2(z), \cdots, q_m(z)\big]$$

with $p_i(z)$ and $q_i(z)$ being coprime $\forall\, i = 1, \cdots, m$.

Putting the expression of $C(z)$ given by (3.3.23) into (3.3.21) and (3.3.22) leads to

$$Q^{-1}(z)P(z)U^{-1}(z)\overline{A}(z) = V(z)A(z), \tag{3.3.24}$$

$$Q^{-1}(z)P(z)U^{-1}(z)\overline{B}(z) = V(z)B(z). \tag{3.3.25}$$

Noticing that the right-hand sides of both (3.3.24) and (3.3.25) are matrix polynomials, we find that the *ith* rows of both $P(z)U^{-1}(z)\overline{A}(z)$ and $P(z)U^{-1}(z)\overline{B}(z)$ must be divided by $q_i(z)\ \forall\, i = 1,\cdots,m$.

Noticing that $q_i(z)$ and $p_i(z)$ are coprime $\forall i = 1,\cdots,m$, we find that $Q(z)$ must be a common left factor of $U^{-1}(z)\overline{A}(z)$ and $U^{-1}(z)\overline{B}(z)$. In other words, both $Q^{-1}(z)U^{-1}(z)\overline{A}(z)$ and $Q^{-1}(z)U^{-1}(z)\overline{B}(z)$ are matrix polynomials. Noticing that $Q^{-1}(z)$ and $P(z)$ in (3.3.24) are commutative, we find that $P(z)$ is a left-common factor of $V(z)A(z)$ and $V(z)B(z)$.

Since $A(z)$ and $B(z)$ have no common left factor, there are matrix polynomials $M(z)$ and $N(z)$ such that $A(z)M(z)+B(z)N(z) = I$, and hence $V(z)A(z)M(z)V^{-1}(z)+V(z)B(z)N(z)V^{-1}(z) = I$. This means that $V(z)A(z)$ and $V(z)B(z)$ have neither a common left factor. Consequently, $P(z)$ is unimodular. Then, from (3.3.23) it is seen that $C(z)$ is a matrix polynomial: $C(z) \triangleq C_0 + C_1 z + \cdots + C_r z^r$.

From (3.3.21) and (3.3.22) we have $[\overline{A}(z)\ \overline{B}(z)] = C(z)[A(z)\ B(z)]$. Comparing the matrix coefficients of the highest order at both its sides gives us $C_r[A_p\ B_q] = 0$. The matrix $[A_p\ B_q]$ is of row-full-rank by A3.3.2, so $C_r = 0$.

Similarly, we derive $C_i = 0,\ i \geq 1$. Therefore, $C(z)$ is a constant matrix: $C(z) = C_0$. Setting $z = 0$ in (3.3.17), we find $C(z) \equiv I$.

Then, from (3.3.21) and (3.3.22) we conclude that $\overline{A}(z) \equiv A(z)$ and $\overline{B}(z) \equiv B(z)$. Thus, A3.3.1 holds.

$A3.3.1 \Rightarrow A3.3.2$

Let $[A(z)\ \ B(z)]$ be the unique pair in $\mathcal{M}$. Assume the converse: either $[A_p\ B_q]$ is not of row-of-rank or $A(z)$ and $B(z)$ have a common-left factor.

In the case $[A_p\ B_q]$ is not of row-of-rank, there exists a nonzero vector $\alpha \in \mathbb{R}^n$ such that

$$\alpha^T[A_p\ B_q] = 0.$$

Set $G(z) \triangleq I + \beta\alpha^T z$ with an arbitrary nonzero constant vector β. Then, $G(z)A(z)$ and $G(z)B(z)$ are of orders less than or equal to p and q, respectively.

Therefore, $[G(z)A(z)\ \ G(z)B(z)] \in \mathcal{M}$. This contradicts the uniqueness of $[A(z)\ \ B(z)]$ in $\mathcal{M}$, and proves the row-full-rank of $[A_p\ B_q]$.

In the case $A(z)$ and $B(z)$ have a common-left factor $C(z)$: $A(z) = C(z)\overline{A}(z),\ B(z) = C(z)\overline{B}(z)$, then $I = C(0)\overline{A}(0)$ and we may assume that both $C(z)$ and $\overline{A}(z)$ are monic and stable.

Let $U(z)$ be the unimodular matrix defined in Lemma 3.3.1. Then, $C'(z) \triangleq C(z)U(z)$ is in the form (3.3.17) with $\deg c'_{i,s}(z) \leq \deg c'_{i,i}(z)\ \forall s \leq i\ \forall i:\ i = 1,\cdots,m$.

Define $A'(z) \triangleq U^{-1}(z)\overline{A}(z)$ and $B'(z) \triangleq U^{-1}(z)\overline{B}(z)$. Then $A(z) = C'(z)A'(z)$ and $B(z) = C'(z)B'(z)$. By Lemma 3.3.1, we have $\deg A'(z) \leq p,\ \deg B'(z) \leq q$. Therefore, $[A'(z)\ \ B'(z)] \in \mathcal{M}$. By the uniqueness of $[A(z)\ \ B(z)]$ in $\mathcal{M}$, $C(z)$ must be a unimodular matrix, and hence $A(z)$ and $B(z)$ have no common left factor. Consequently, A3.3.2 holds.

*A*3.3.1 ⇒ *A*3.3.3

Let A3.3.1 hold and let $[A(z)\ B(z)] \in \mathcal{M}$ with orders p and q, respectively. Since A3.3.1 is equivalent to A3.3.2, $[A_p\ \ B_q]$ must be of row-full-rank.

We now show A3.3.3. Assume the converse that there exist m-vector polynomial $d(z)$ and l-vector polynomial $c(z)$ (not both zero) with orders strictly less than p and q, respectively, such that $d^T(z)H(z) + c^T(z) = 0$.

Let $\xi \in \mathbb{R}^m \neq 0$ and define $\widetilde{A}(z) \triangleq A(z) + z\xi d^T(z) = I + \widetilde{A}_1 z + \cdots + \widetilde{A}_p z^p$, $\widetilde{B}(z) \triangleq B(z) - z\xi c^T(z) = \widetilde{B}_0 + \widetilde{B}_1 z + \cdots + \widetilde{B}_q z^q$.

Then, we have

$$\widetilde{A}(z)H(z) = (A(z) + z\xi d^T(z))H(z) = B(z) - z\xi c^T(z) = \widetilde{B}(z).$$

It is clear that $\widetilde{A}(z)$ remains stable if $\|\xi\| > 0$ is small enough. Therefore, $[\widetilde{A}(z)\ \ \widetilde{B}(z)] \in \mathcal{M}$. By A3.3.1 we must have $\widetilde{A}(z) = A(z)$ and $\widetilde{B}(z) = B(z)$. This means that ξ must be zero and hence contradicts $\xi \neq 0$. So, A3.3.3 holds.

*A*3.3.3 ⇒ *A*3.3.1

Let A3.3.3 hold. We now show A3.3.1. Assume the converse: there are two different matrix polynomials $[A(z)\ B(z)] \in \mathcal{M}$ and $[\overline{A}(z)\ \overline{B}(z)] \in \mathcal{M}$ with orders less than or equal to p and q, respectively.

Set $X(z) \triangleq A(z) - \overline{A}(z) = X_1 z + \cdots + X_p z^p$ and $Y(z) \triangleq B(z) - \overline{B}(z) = Y_0 + Y_1 z + \cdots + Y_q z^q$. From here it follows that

$$X(z)H(z) = Y(z). \tag{3.3.26}$$

Setting $z = 0$ in (3.3.26) we find that $Y_0 = 0$. By assumption there exists at least one nonzero row in (3.3.26). Take any nonzero row in $[X(z)\ \ Y(z)]$ and write it as $z[d^T(z) \quad -c^T(z)]$. It is clear that $[d^T(z) \quad -c^T(z)]$ is a row polynomial with orders strictly less than p and q, respectively. By (3.3.26) we have $d^T(z)H(z) + c^T(z) = 0$, which contradicts A3.3.3.

The proof of the lemma is completed. □

Row-full-rank of Hankel matrix *L*

We now give the necessary and sufficient conditions for the row-full-rank of L.

Theorem 3.3.2 *Assume $[A(z)\ B(z)] \in \mathcal{M}$ and $A^{-1}(z)B(z) = \sum_{i=0}^{\infty} H_i z^i$. Then the following condition A3.3.4 is equivalent to A3.3.1, or A3.3.2, or A3.3.3 defined in Theorem 3.3.1.*

A3.3.4 *The matrix L defined by (3.3.10) is of row-full-rank.*

Proof. By Theorem 3.3.1, A3.3.1, A3.3.2, and A3.3.3 are equivalent. So, for the theorem it suffices to show that A3.3.1 and A3.3.4 are equivalent.

*A*3.3.1 ⇒ *A*3.3.4

Assume $[A(z)\ \ B(z)]$ is the unique pair in $\mathcal{M}$ with orders p and q, respectively. Then, we have that

$$H(z) = A^{-1}(z)B(z) = \frac{\mathrm{Adj}A(z)B(z)}{a(z)} = \frac{B^*(z)}{a(z)}, \tag{3.3.27}$$

where $a(z) \triangleq \det(A(z)) = \sum_{i=0}^{mp} a_i z^i$, and $B^*(z) \triangleq \mathrm{Adj}A(z)B(z) = \sum_{j=0}^{(m-1)p+q} B_j^* z^j$.

From (3.3.27) it follows that

$$\begin{aligned}(1 + a_1 z + \cdots + a_{mp} z^{mp})(H_0 + H_1 z + \cdots + H_i z^i + \cdots)\\ = B_0^* + B_1^* z + \cdots + B_{(m-1)p+q}^* z^{(m-1)p+q}.\end{aligned} \tag{3.3.28}$$

Identifying coefficients for the same degrees of z at both sides of (3.3.28), we obtain

$$H_t = -\sum_{i=1}^{mp} a_i H_{t-i} \ \forall\, t > q + (m-1)p. \tag{3.3.29}$$

If the matrix L were not of row-full-rank, then there would exist a vector $x = (x_1^T, \cdots, x_p^T)^T \neq 0$ with $x_i \in \mathbb{R}^m$ such that $x^T L = 0$, i.e.,

$$\sum_{j=1}^{p} x_j^T H_{q-j+s} = 0 \ \forall 1 \leq s \leq mp. \tag{3.3.30}$$

In this case we show that (3.3.30) holds $\forall s \geq 1$.

Noticing (3.3.29) and (3.3.30), for $s = mp + 1$ we have

$$\begin{aligned}\sum_{j=1}^{p} x_j^T H_{q-j+mp+1} &= -\sum_{j=1}^{p} x_j^T \sum_{i=1}^{mp} a_i H_{q-j+mp+1-i}\\ &= -\sum_{i=1}^{mp} a_i \sum_{j=1}^{p} x_j^T H_{q-j+mp+1-i} = 0.\end{aligned} \tag{3.3.31}$$

Hence (3.3.30) holds for $i = mp + 1$. Carrying out the similar treatment as that done in (3.3.31), we find

$$\sum_{j=1}^{p} x_j^T H_{q-j+s} = 0 \ \forall s \geq 1. \tag{3.3.32}$$

Defining $d(z) \triangleq \sum_{i=1}^{p} x_i z^{i-1}$, we have

$$\begin{aligned}d^T(z)H(z) &= \Big(\sum_{i=1}^{p} x_i^T z^{i-1}\Big) \cdot \Big(\sum_{j=0}^{\infty} H_j z^j\Big)\\ &= \sum_{i=1}^{p} \Big(x_i^T z^{i-1} \Big(\sum_{j=0}^{q-i} H_j z^j + \sum_{j=q-i+1}^{\infty} H_j z^j\Big)\Big)\end{aligned}$$

$$=\sum_{i=1}^{p}\sum_{j=0}^{q-i}x_i^T H_j z^{i+j-1}+\sum_{k=1}^{\infty}\Big(\sum_{i=1}^{p}x_i^T H_{q-i+k}\Big)z^{q+k-1}$$

$$=\sum_{i=1}^{p}\sum_{j=0}^{q-i}x_i^T H_j z^{i+j-1}\triangleq -c^T(z). \tag{3.3.33}$$

Consequently, $d^T(z)H(z)+c^T(z)=0$ and the orders of $d(z)$ and $c(z)$ are strictly less than p and q, respectively. This contradicts A3.3.3, and hence contradicts A3.3.1 by Theorem 3.3.1.

A3.3.4 $\Rightarrow$ A3.3.1

Assume the converse: there are more than one pair in $\mathcal{M}$. Then by Theorem 3.3.1 there exist $\widetilde{d}(z)=\sum_{i=1}^{p}\widetilde{x}_i z^{i-1}$ and $\widetilde{c}(z)$ (not both zero) with orders strictly less than p and q, respectively, such that

$$\widetilde{d}^T(z)H(z)=\widetilde{c}^T(z). \tag{3.3.34}$$

Noticing $\widetilde{d}^T(z)H(z)=\sum_{j=0}^{\infty}\sum_{i=1}^{p}\widetilde{x}_i^T H_{j-i+1}z^j$, we have

$$\sum_{j=0}^{\infty}\sum_{i=1}^{p}\widetilde{x}_i^T H_{j-i+1}z^j=\widetilde{c}^T(z). \tag{3.3.35}$$

Noting the order of $\widetilde{c}^T(z)$ is less than q, from (3.3.35) we must have

$$\sum_{j=q}^{\infty}\sum_{i=1}^{p}\widetilde{x}_i^T H_{j-i+1}z^j=0,\quad \text{or}\quad \sum_{i=1}^{p}\widetilde{x}_i^T H_{q-i+s}=0\ \ \forall\, s\geq 1.$$

This means that the rows of the matrix L are linearly dependent, which contradicts A3.3.4. Consequently, A3.3.1 must be held. □

As to be shown in the coming chapters, in some cases $\{H_i\}$ can be estimated, then by Theorem 3.3.2 with the help of (3.3.12) and (3.3.7), θ_A and $\theta_B^T\triangleq[B_0,B_1,\cdots,B_q]$ can also be estimated.

Row-full-rank of Hankel matrix Γ

We now consider the row-full-rank of the Hankel matrix Γ composed of correlation functions. Assume that $\{u_k\}$ is a sequence of zero-mean uncorrelated random vectors with $Eu_k u_k^T=I$.

If Γ is of row-full-rank and $\{R_i\}$ can be estimated, then by (3.3.16), θ_A can also be estimated. As concerns the coefficients θ_B, let us set $\chi_k\triangleq B(z)u_k$.

The spectral density of χ_k is

$$\Phi^{\chi}(z)=B(z)B^T(z^{-1}),$$

while the spectral density of $\{y_k\}$ given by (3.3.1)–(3.3.3) is

$$\Phi(z)\triangleq\sum_{j=-\infty}^{\infty}R_j z^j=A^{-1}(z)B(z)B^T(z^{-1})A^{-T}(z^{-1}), \tag{3.3.36}$$

which implies

$$\Phi^{\chi}(z) = B(z)B^T(z^{-1}) = A(z)\Phi(z)A^T(z^{-1}). \tag{3.3.37}$$

Since the right-hand side of (3.3.37) is equal to

$$A(z)\Phi(z)A^T(z^{-1}) = \sum_{i=0}^{p} A_i z^i \sum_{k=-\infty}^{\infty} R_k z^k \sum_{j=0}^{p} A_j^T z^{-j}$$

$$= \sum_{i=0}^{p} \sum_{k=-\infty}^{\infty} \sum_{j=0}^{p} A_i R_k A_j^T z^{i+k-j} = \sum_{k=-\infty}^{\infty} (\sum_{i=0}^{p} \sum_{j=0}^{p} A_i R_{k+j-i} A_j^T) z^k, \tag{3.3.38}$$

we have

$$B(z)B^T(z^{-1}) = \sum_{k=-q}^{q} (\sum_{i=0}^{p} \sum_{j=0}^{p} A_i R_{k+j-i} A_j^T) z^k. \tag{3.3.39}$$

Therefore, to derive θ_B it is the matter of factorizing the right-hand side of (3.3.39).

The following Theorem 3.3.3 tells us that the row-full-rank of Γ is slightly stronger than that of L.

It is worth noting that the theorem requires no stability-like condition on $B(z)$, and $B(z)$ may not be a square matrix.

However, to prove the theorem we need a lemma on factorization of matrix polynomial.

Lemma 3.3.3 *Let $B(z) = B_0 + B_1 z + \cdots + B_q z^q$ be an $m \times l$ ($m \geq l$) matrix polynomial with rank l. Then $B(z)$ can be factorized as*

$$B(z) = B_I(z)B_P(z) \tag{3.3.40}$$

where $B_I(z)$ is an $m \times l$ matrix polynomial such that $\deg[B_I(z)] \leq \deg[B(z)]$ *and its constant term $B_I(0)$ is of column-full-rank, while $B_P(z)$ is an $l \times l$ matrix polynomial satisfying $B_P(z)B_P^T(z^{-1}) = I_l$.*

Proof. Since $B(z)$ is with rank l, any minor of order l, if it is not identically equal to zero, must be of the form: $z^x g(z)$, where $x \geq 0$ is an integer and $g(z)$ is a polynomial with a nonzero constant term. Denote the greatest common factor (GCF) of minors of order m by $z^r b(z)$. Without loss of generality, $b(z)$ may be assumed to be monic. To emphasize the degree r in the common factor $z^r b(z)$, we write $B(z)$ as $B_r(z)$.

If $r = 0$, then the GCF of $B_r(z)$ is a monic polynomial $b(z)$. Since the constant term of $b(z)$ is nonzero (=1), B_0 must be of column-full-rank. Then, we may take $B_P(z) = I$ and $B_I(z) = B(z)$, which meet the requirements of the lemma.

If $r > 0$, then the GCF of $B_r(z)$ is zero at $z = 0$. This implies that all minors of order l are zero at $z = 0$. In other words, the columns of B_0 are linearly dependent. Therefore, there exists a nonzero unit l-vector ψ such that $B_0\psi = 0$. This means that

$$B_r(z)\psi = \sum_{i=0}^{q} B_i \psi z^i = z(\sum_{i=0}^{q-1} B_{i+1} z^i)\psi.$$

Let T_r be an orthogonal matrix with ψ serving as its last column.

Define the matrix polynomial $B_{r-1}(z)$ as follows:

$$B_{r-1}(z) \triangleq B_r(z)T_r\Upsilon(z),$$

where

$$\Upsilon(z) \triangleq \begin{bmatrix} I_{m-1} & 0 \\ 0 & \frac{1}{z} \end{bmatrix}.$$

Since T_r is an $l \times l$ orthogonal matrix, the GCF of $B_r(z)T_r$ coincides with that of $B_r(z)$. Further, $B_r(z)T_r\Upsilon(z)$ differs from $B_r(z)T_r$ only at the last column by one degree of z less in comparison with the former. Therefore, the GCF of $B_{r-1}(z)$ is $z^{r-1}b(z)$, and $\deg[B_{r-1}(z)] \leq \deg[B_r(z)]$.

If $r-1>0$, as before, the columns of the constant term of $B_{r-1}(z)$ are linearly dependent. Proceeding as above for r times, we arrive at

$$B_0(z) \triangleq B_r(z)T_r\Upsilon(z)T_{r-1}\Upsilon(z)\cdots T_1\Upsilon(z).$$

It is clear that $B_0(z)$ is still of rank l with $\deg B_0(z) \leq q$, and the GCF of $B_0(z)$ is $b(z)$. So, the constant term of $B_0(z)$ is of column-full-rank.

Define

$$B_I(z) \triangleq B_0(z) \tag{3.3.41}$$

$$B_P(z) \triangleq \Upsilon^{-1}(z)T_1^T\Upsilon^{-1}(z)T_2^T\cdots\Upsilon^{-1}(z)T_r^T. \tag{3.3.42}$$

It is clear that (3.3.40) holds, and all requirements of the lemma are satisfied. □

Theorem 3.3.3 *Assume $[A(z) \;\; B(z)] \in \mathcal{M}$ and B_0 is of column-full-rank. Then, the following A3.3.5 and A3.3.6 are equivalent:*

A3.3.5 *The matrix Γ defined by (3.3.15) is of row-full-rank.*

A3.3.6 *The matrix $[A_p \;\; B_q]$ is of row-full-rank and the matrix polynomials $A(z)$ and $B(z)B^T(z^{-1})z^q$ have no common left factor.*

Proof. A3.3.6 ⇒ A3.3.5

We first note that A3.3.6 implies A3.3.2, and hence, by Theorem 3.3.1, $[A(z) \; B(z)]$ is the unique pair in $\mathcal{M}$.

By (3.3.27), we have

$$\begin{aligned} &y_k + a_1 y_{k-1} + \cdots + a_{mp} y_{k-mp} \\ &\quad = B_0^* u_k + B_1^* u_{k-1} + \cdots + B_{(m-1)p+q}^* u_{k-((m-1)p+q)}. \end{aligned} \tag{3.3.43}$$

Multiplying both sides of (3.3.43) by y_{k-t}^T from the right and taking expectation, we have

$$\begin{aligned} &E(y_k + a_1 y_{k-1} + \cdots + a_{mp} y_{k-mp}) y_{k-t}^T \\ &\quad = E(u_k + B_1^* u_{k-1} + \cdots + B_{(m-1)p+q}^* u_{k-[(m-1)p+q]}) y_{k-t}^T \\ &\quad = 0 \;\; \forall t > q + (n-1)p, \end{aligned}$$

which yields

$$R_t = -\sum_{i=1}^{mp} a_i R_{t-i} \;\; \forall t > q + (m-1)p. \tag{3.3.44}$$

If Γ were not of row-full-rank, then there would exist a vector $\eta = (\eta_1^T, \cdots, \eta_p^T)^T \neq 0$ with $\eta_i \in \mathbb{R}^m$ such that $\eta^T \Gamma = 0$, i.e.,

$$\sum_{j=1}^{p} \eta_j^T R_{q-j+s} = 0, \; 1 \leq s \leq mp. \tag{3.3.45}$$

In this case we show that (3.3.45) holds $\forall s \geq 1$.

Noticing (3.3.44) and (3.3.45), for $s = mp+1$ we have

$$\begin{aligned}\sum_{j=1}^{p} \eta_j^T R_{q-j+mp+1} &= -\sum_{j=1}^{p} \eta_j^T \sum_{i=1}^{mp} a_i R_{q-j+mp+1-i} \\ &= -\sum_{i=1}^{mp} a_i \sum_{j=1}^{p} \eta_j^T R_{q-j+mp+1-i} = 0. \end{aligned} \tag{3.3.46}$$

Hence (3.3.45) holds for $i = mp+1$. Carrying out the similar treatment as that done in (3.3.46), we find

$$\sum_{j=1}^{p} \eta_j^T R_{q-j+s} = 0 \;\; \forall s \geq 1. \tag{3.3.47}$$

Defining $\widehat{d}(z) \triangleq \sum_{i=1}^{p} \eta_i z^i$, with $\Phi(z)$ given by (3.3.36) we have

$$\widehat{d}^T(z)\Phi(z)A^T(z^{-1}) = \sum_{i=1}^{p} \sum_{j=-\infty}^{\infty} \sum_{s=0}^{p} \eta_i^T R_j A_s^T z^{i+j-s}$$

$$= \sum_{i=1}^{p} \sum_{s=0}^{p} \sum_{j=q-i+1}^{\infty} \eta_i^T R_j A_s^T z^{i+j-s} + \sum_{i=1}^{p} \sum_{s=0}^{p} \sum_{j=-\infty}^{q-i} \eta_i^T R_j A_s^T z^{i+j-s}$$

$$= \sum_{s=0}^{p} \sum_{j=1}^{\infty} \Big(\sum_{i=1}^{p} \eta_i^T R_{q-i+j}\Big) z^{q+j} A_s^T z^{-s} + \sum_{i=1}^{p} \sum_{s=0}^{p} \sum_{j=-\infty}^{q-i} \eta_i^T R_j A_s^T z^{i+j-s}$$

$$= \sum_{i=1}^{p} \sum_{s=0}^{p} \sum_{j=i-q}^{\infty} \eta_i^T R_j^T A_s^T z^{i-j-s},$$

where for the last equality (3.3.47) is invoked. We then have

$$\widehat{d}^T(z)\Phi(z)A^T(z^{-1})$$

$$= \sum_{i=1}^{p} \sum_{s=0}^{p} \sum_{j=i-q}^{q-s} \eta_i^T R_j^T A_s^T z^{i-j-s} + \sum_{i=1}^{p} \sum_{s=0}^{p} \sum_{j=q-s+1}^{\infty} \eta_i^T R_j^T A_s^T z^{i-j-s}$$

$$= \sum_{i=1}^{p}\sum_{s=0}^{p}\sum_{j=i-q}^{q-s} \eta_i^T R_j^T A_s^T z^{i-j-s} + \sum_{i=1}^{p}\eta_i^T z^i \sum_{j=1}^{\infty}(\sum_{s=0}^{p} R_{q-s+j}^T A_s^T) z^{-q-j}$$

$$= \sum_{i=1}^{p}\sum_{s=0}^{p}\sum_{j=i-q}^{q-s} \eta_i^T R_j^T A_s^T z^{i-j-s}, \tag{3.3.48}$$

where for the last equality (3.3.47) is used.

Similarly, we obtain

$$A(z)\Phi(z)\widehat{d}(z^{-1}) = \sum_{s=0}^{p}\sum_{i=1}^{p}\sum_{j=s-q}^{q-i} A_s R_j^T \eta_i z^{s-i-j} \tag{3.3.49}$$

and

$$\widehat{d}^T(z)\Phi(z)\widehat{d}(z) = \sum_{i=1}^{p}\sum_{s=1}^{p}\sum_{j=-\infty}^{\infty} \eta_i^T R_j \eta_s z^{i+j-s}$$

$$= \sum_{i=1}^{p}\sum_{s=1}^{p}\sum_{j=q-i+1}^{\infty} \eta_i^T R_j \eta_s z^{i+j-s} + \sum_{i=1}^{p}\sum_{s=1}^{p}\sum_{j=-\infty}^{q-i} \eta_i^T R_j \eta_s z^{i+j-s}$$

$$= \sum_{s=1}^{p}\sum_{j=1}^{\infty}(\sum_{i=1}^{p} \eta_i^T R_{q-i+j}) z^{q+j} \eta_s z^{-s} + \sum_{i=1}^{p}\sum_{s=1}^{p}\sum_{j=-\infty}^{q-i} \eta_i^T R_j \eta_s z^{i+j-s}$$

$$= \sum_{i=1}^{p}\sum_{s=1}^{p}\sum_{j=i-q}^{\infty} \eta_i^T R_j^T \eta_s z^{i-j-s}$$

$$= \sum_{i=1}^{p}\sum_{s=1}^{p}\sum_{j=i-q}^{q-s} \eta_i^T R_j^T \eta_s z^{i-j-s} + \sum_{i=1}^{p}\sum_{s=1}^{p}\sum_{j=q-s+1}^{\infty} \eta_i^T R_j^T \eta_s z^{i-j-s}$$

$$= \sum_{i=1}^{p}\sum_{s=1}^{p}\sum_{j=i-q}^{q-s} \eta_i^T R_j^T \eta_s z^{i-j-s} + \sum_{i=1}^{p}\eta_i^T z^i \sum_{j=1}^{\infty}(\sum_{s=1}^{p} R_{q-s+j}^T \eta_s) z^{-q-j}$$

$$= \sum_{i=1}^{p}\sum_{s=1}^{p}\sum_{j=i-q}^{q-s} \eta_i^T R_j^T \eta_s z^{i-j-s}. \tag{3.3.50}$$

Set $\widehat{A}(z) = A(z) + \beta \widehat{d}^T(z)$ with $\beta \in \mathbb{R}^m$ being an arbitrary column vector. It is clear that the order of $\widehat{A}(z)$ is less than or equal to p. From (3.3.36) and (3.3.48)–(3.3.50) it follows that

$$\begin{aligned}\widehat{A}(z)\Phi(z)\widehat{A}^T(z^{-1}) &= (A(z)+\beta\widehat{d}^T(z))\Phi(z)(A(z^{-1})+\beta\widehat{d}^T(z^{-1}))^T \\ &= A(z)\Phi(z)A^T(z^{-1}) + A(z)\Phi(z)\widehat{d}(z^{-1})\beta^T \\ &\quad + \beta\widehat{d}^T(z)\Phi(z)A^T(z^{-1}) + \beta\widehat{d}^T(z)\Phi(z)\widehat{d}(z^{-1})\beta^T\end{aligned}$$

$$
\begin{aligned}
&= B(z)B^T(z^{-1}) + \sum_{s=0}^{p}\sum_{i=1}^{p}\sum_{j=s-q}^{q-i} A_s R_j^T y_i z^{s-i-j}\beta^T \\
&+ \sum_{i=1}^{p}\sum_{s=0}^{p}\sum_{j=i-q}^{q-s} \beta y_i^T R_j^T A_s^T z^{i-j-s} \\
&+ \sum_{i=1}^{p}\sum_{s=1}^{p}\sum_{j=i-q}^{q-s} \beta y_i^T R_j^T y_s \beta^T z^{i-j-s} \triangleq F(z).
\end{aligned}
\tag{3.3.51}
$$

The degrees of z in $F(z)$ are between $-q$ and q. So, it may diverge to infinity only at $z=0$ and at z equal to infinity, and hence all its nonzero finite poles should be canceled with its zeros.

Noticing

$$
\begin{aligned}
F(z) &= \widehat{A}(z)\Phi(z)\widehat{A}^T(z^{-1}) \\
&= \widehat{A}(z)A^{-1}(z)B(z)B^T(z^{-1})A^{-T}(z^{-1})\widehat{A}^T(z^{-1}),
\end{aligned}
$$

we see that all poles of $A^{-1}(z)$ should be canceled with zeros of $F(z)$. However, A3.3.6 requires that $A(z)$ and $B(z)B^T(z^{-1})z^q$ have no common left factor. This means that any pole of $A^{-1}(z)$ cannot be canceled with zeros of $B(z)B^T(z^{-1})z^q$. By stability of $A(z)$ the poles of $A^{-1}(z)$, being outside the closed unit disk, can neither be canceled with zeros of $A^{-T}(z^{-1})$ and $\widehat{A}^T(z^{-1})$, since their zeros are either at $z=0$ or at infinity. Therefore, all poles of $A^{-1}(z)$ must be canceled with zeros of $\widehat{A}(z)$. In other words, $\widehat{A}(z)A^{-1}(z)$ must be a matrix polynomial.

Let us denote $\widehat{A}(z)A^{-1}(z) \triangleq C(z) = C_0 + C_1 z + \cdots + C_s z^s$.

Then, we have $\widehat{A}(z) = C(z)A(z)$, and $C_0 = I$ by setting $z=0$. By defining $\widehat{B}(z) \triangleq C(z)B(z) = \widehat{B}_0 + \widehat{B}_1 z + \cdots + \widehat{B}_{\widehat{q}} z^{\widehat{q}}$, we have $\widehat{B}_0 = B_0$ since $C_0 = I$. Further, we have

$$
\begin{aligned}
&\widehat{A}(z)H(z) = \widehat{A}(z)A^{-1}(z)B(z) = C(z)B(z) = \widehat{B}(z) \text{ and} \\
&\widehat{B}(z)\widehat{B}^T(z^{-1}) = F(z).
\end{aligned}
\tag{3.3.52}
$$

We now show that $\widehat{B}_s = 0 \;\forall s: q+1 \le s \le \widehat{q}$ if $\widehat{q} > q$.

If $\widehat{q} > q$ and $\widehat{B}_{\widehat{q}} \neq 0$, then comparing the matrix coefficients of the degree $\widehat{q}$ at both sides of (3.3.52) we obtain $\widehat{B}_{\widehat{q}} B_0^T = 0$, since the maximal degree of z in $F(z)$ defined by (3.3.51) is q.

Since B_0 is of column-full-rank, we find that $\widehat{B}_{\widehat{q}} = 0$. The similar treatment for $q+1 \le s \le \widehat{q}-1$ leads to $\widehat{B}_s = 0, q+1 \le s \le \widehat{q}$ in $\widehat{B}(z)$. As a consequence, $[\widehat{A}(z) \;\; \widehat{B}(z)] \in \mathcal{M}$, but $[\widehat{A}(z) \;\; \widehat{B}(z)] \neq [A(z) \;\; B(z)]$. This contradicts A3.3.6, because A3.3.1 is a consequence of A3.3.6 as pointed out at the beginning of the proof. Hence Γ is of row-full-rank.

A3.3.5 $\Rightarrow$ A3.3.6

We first show the necessity of A3.3.1 (or A3.3.2, or A3.3.3, or A3.3.4), then we show that $A(z)$ and $B(z)B(z^{-1})z^q$ have no common left factor.

If $[A(z) \quad B(z)]$ is not unique in $\mathcal{M}$, then the matrix L is not of row-full-rank by Theorem 3.3.2. This means that there exists a nonzero column vector $\widetilde{x} = [\widetilde{x}_1^T, \cdots, \widetilde{x}_p^T]^T$ such that

$$\sum_{i=1}^{p} \widetilde{x}_i^T H_{q-i+s} = 0 \ \forall\, 1 \le s \le np.$$

From here as shown in (3.3.30)–(3.3.32), we have

$$\sum_{i=1}^{p} \widetilde{x}_i^T H_{q-i+s} = 0 \ \forall\, s \ge 1.$$

Therefore, we obtain

$$\begin{aligned}\sum_{i=1}^{p} \widetilde{x}_i^T R_{q-i+s} &= \sum_{i=1}^{p} \widetilde{x}_i^T \sum_{j=0}^{\infty} H_{q-i+s+j} H_j^T \\ &= \sum_{j=0}^{\infty} \left(\sum_{i=1}^{p} \widetilde{x}_i^T H_{q-i+s+j} \right) H_j^T = 0 \ \forall\, s \ge 1,\end{aligned}$$

which means that the rows of the matrix Γ are linearly dependent. This contradicts A3.3.5.

Thus, we have shown that A3.3.1, or, equivalently, A3.3.2 holds under A3.3.5. So, it remains to show that $A(z)$ and $B(z)B(z^{-1})z^q$ have no common left factor under A3.3.5.

Assume the converse: $A(z)$ and $B(z)B(z^{-1})z^q$ have a common left factor $C(z)$ which is not unimodular:

$$[A(z)\ B(z)B^T(z^{-1})z^q] = C(z)[\widetilde{A}(z)\ \overline{D}(z)]. \tag{3.3.53}$$

By Lemma 3.3.1 we may assume that $\deg[\widetilde{A}(z)] \le p$, and the matrix $C(z) = \{c_{i,j}(z)\}_{1 \le i,j \le m}$ is lower triangular with $C_0 = I$ and the degree of $c_{i,i}(z)$ is the greatest among the entries of the ith row $\forall\, i : 1 \le i \le m$.

From (3.3.53) we have

$$B(z)B^T(z^{-1})z^q = C(z)\overline{D}(z). \tag{3.3.54}$$

Set $\widetilde{D}(z) = \overline{D}(z)C^{-T}(z^{-1})z^{-q} = \{\widetilde{d}_{i,j}(z), 1 \le i, j \le m\}$. Then, from (3.3.54) it follows that

$$\widetilde{D}(z) = C^{-1}(z)B(z)B^T(z^{-1})C^{-T}(z^{-1}), \tag{3.3.55}$$

which is equivalent to

$$D(z) \triangleq B(z)B^T(z^{-1}) = C(z)\widetilde{D}(z)C^T(z^{-1}). \tag{3.3.56}$$

For any scalar rational polynomial $g(z) = g_{-a}z^{-a} + \cdots + g_0 + \cdots + g_b z^b$ with real coefficients we introduce the operators $[\cdot]_+$ and $[\cdot]_-$ such that

$$[g(z)]_+ = g_0 + \cdots + g_b z^b \text{ and } [g(z)]_- = g_0 + g_{-1}z + \cdots + g_{-a}z^a.$$

We now show that $\deg[\widetilde{d}_{i,j}(z)]_+ \le q$, $\deg[\widetilde{d}_{i,j}(z)]_- \le q$ for $1 \le i,j \le m$ by a treatment similar to but more complicated than that used in the proof of Lemma 3.3.1.

This is proved in an inductive way starting from the first column, and in each column the proof is also carried out inductively.

Noticing that $C(z)$ is lower triangular, from (3.3.56) we have

$$D(z) = \{d_{ij}(z)\},\ d_{ij}(z) = \sum_{t=1}^{i}\sum_{s=1}^{j} c_{it}(z)\widetilde{d}_{ts}(z)c_{js}(z^{-1}) \ \forall\, i,j : 1 \le i,j \le m.$$

Starting from the first column of $\widetilde{D}(z)$, we show that $\deg[\widetilde{d}_{i1}(z)]_+ \le q$ for $1 \le i \le m$ by induction.

The $(1,1)$-element $\widetilde{d}_{11}(z)$ of $\widetilde{D}(z)$ is related to $d_{11}(z)$ as follows

$$d_{11}(z) = c_{11}(z)\widetilde{d}_{11}(z)c_{11}(z^{-1}). \tag{3.3.57}$$

Since $\deg[d_{11}(z)]_+ \le q$ and the constant term of $c_{11}(z)$ equals 1, we see that $\deg[\widetilde{d}_{11}(z)]_+ \le q$.

Assume it has been established that $\deg[\widetilde{d}_{i1}(z)]_+ \le q \ \ \forall\, i : 1 \le i \le r$. We want to show $\deg[\widetilde{d}_{(r+1)1}(z)]_+ \le q$.

Assume the converse: $\deg[\widetilde{d}_{(r+1)1}(z)]_+ > q$.

Noticing $\deg[\widetilde{d}_{i1}(z)]_+ \le q \,\forall\, i : 1 \le i \le r$ and $\deg[c_{(r+1)(r+1)}(z)] \ge \deg[c_{(r+1)t}(z)] \ \forall\, t : 1 \le t \le r$, by the converse assumption and $c_{11}(0) = 1$ we see that

$$\deg\left[c_{(r+1)(r+1)}(z)\widetilde{d}_{(r+1)1}(z)c_{11}(z^{-1})\right]_+ > \deg\left[\sum_{t=1}^{r} c_{(r+1)t}(z)\widetilde{d}_{t1}(z)c_{11}(z^{-1})\right]_+,$$

and, hence

$$\begin{aligned}&\deg\left[\sum_{t=1}^{r} c_{(r+1)t}(z)\widetilde{d}_{t1}(z)c_{11}(z^{-1}) + c_{(r+1)(r+1)}(z)\widetilde{d}_{(r+1)1}(z)c_{11}(z^{-1})\right]_+ \\ &= \deg\left[c_{(r+1)(r+1)}(z)\widetilde{d}_{(r+1)1}(z)c_{11}(z^{-1})\right]_+ > q.\end{aligned} \tag{3.3.58}$$

Since

$$d_{(r+1)1}(z) = \sum_{t=1}^{r+1} c_{(r+1)t}(z)\widetilde{d}_{t1}(z)c_{11}(z^{-1}),$$

by (3.3.58) we obtain a contradictory inequality:

$$\begin{aligned}q \ge \deg[d_{(r+1)1}(z)]_+ &= \deg\left[\sum_{t=1}^{r+1} c_{(r+1)t}(z)\widetilde{d}_{t1}(z)c_{11}(z^{-1})\right]_+ \\ &= \deg\left[c_{(r+1)(r+1)}(z)\widetilde{d}_{(r+1)1}(z)c_{11}(z^{-1})\right]_+ > q.\end{aligned}$$

Thus, we have proved $\deg[\widetilde{d}_{(r+1)1}(z)]_+ \le q$ and inductively $\deg[\widetilde{d}_{i1}(z)]_+ \le q \, \forall \, i : 1 \le i \le m$.

Similarly, we can show $\deg[\widetilde{d}_{i1}(z)]_- \le q \, \forall \, i : 1 \le i \le m$. Therefore, the assertion holds for the first column.

We now assume that the assertion is true for the first j columns, i.e.,

$$\deg[\widetilde{d}_{is}(z)]_+ \le q \text{ and } \deg[\widetilde{d}_{is}(z)]_- \le q \ \ \forall i : 1 \le i \le m \text{ and } \forall s : 1 \le s \le j.$$

We want to show it also holds for the $j+1$ column.

Observing that $\widetilde{d}_{i(j+1)}(z) = \widetilde{d}_{(j+1)i}(z^{-1}), 1 \le i \le j$, we see $\deg[\widetilde{d}_{i(j+1)}(z)]_+ = \deg[\widetilde{d}_{(j+1)i}(z)]_- \le q \, \forall \, i : 1 \le i \le j$ by the inductive assumption.

Inductively, we now assume that $\deg[\widetilde{d}_{i(j+1)}(z)]_+ \le q \, \forall \, i : 1 \le i \le r$ for some $r : r \ge j$. We want to prove $\deg[\widetilde{d}_{(r+1)(j+1)}(z)]_+ \le q$.

Assume the converse: $\deg[\widetilde{d}_{(r+1)(j+1)}(z)]_+ > q$.

Noticing $\deg[c_{(r+1)(r+1)}(z)] > \deg[c_{(r+1)t}(z)] \ \forall t : 1 \le t \le r$ and the inductive assumptions $\deg[\widetilde{d}_{i(j+1)}(z)]_+ \le q \ \forall i : 1 \le i \le r$ and $\deg[\widetilde{d}_{is}(z)]_+ \le q \ \forall i : 1 \le i \le m$ and $\forall s : 1 \le s \le j$, we find that

$$\begin{aligned}
&\deg\left[c_{(r+1)(r+1)}(z)\widetilde{d}_{(r+1)(j+1)}(z)c_{(j+1)(j+1)}(z^{-1})\right]_+ \\
&> \deg\Big[\sum_{t=1}^{r}\sum_{s=1}^{j+1} c_{(r+1)t}(z)\widetilde{d}_{ts}(z)c_{(j+1)s}(z^{-1}) \\
&\quad + \sum_{s=1}^{j} c_{(r+1)(r+1)}(z)\widetilde{d}_{(r+1)s}(z)c_{(j+1)s}(z^{-1})\Big]_+.
\end{aligned}$$

Consequently, we have

$$\begin{aligned}
&\deg\Big[\sum_{t=1}^{r}\sum_{s=1}^{j+1} c_{(r+1)t}(z)\widetilde{d}_{ts}(z)c_{(j+1)s}(z^{-1}) + \sum_{s=1}^{j} c_{(r+1)(r+1)}(z)\widetilde{d}_{(r+1)s}(z)c_{(j+1)s}(z^{-1}) \\
&+ c_{(r+1)(r+1)}(z)\widetilde{d}_{(r+1)(j+1)}(z)c_{(j+1)(j+1)}(z^{-1})\Big]_+ \\
&= \deg\left[c_{(r+1)(r+1)}(z)\widetilde{d}_{(r+1)(j+1)}(z)c_{(j+1)(j+1)}(z^{-1})\right]_+.
\end{aligned} \tag{3.3.59}$$

Since

$$d_{(r+1)(j+1)}(z) = \sum_{t=1}^{r+1}\sum_{s=1}^{j+1} c_{(r+1)t}(z)\widetilde{d}_{ts}(z)c_{(j+1)s}(z^{-1}),$$

by (3.3.36) we arrive at the following contradictory inequality:

$$\begin{aligned}
q \ge \deg[d_{(r+1)(j+1)}(z)]_+ &= \deg\left[\sum_{t=1}^{r+1}\sum_{s=1}^{j+1} c_{(r+1)t}(z)\widetilde{d}_{ts}(z)c_{(j+1)s}(z^{-1})\right]_+ \\
&= \deg\left[c_{(r+1)(r+1)}(z)\widetilde{d}_{(r+1)(j+1)}(z)c_{(j+1)(j+1)}(z^{-1})\right]_+ > q.
\end{aligned}$$

This contradiction implies that $\deg[\widetilde{d}_{(r+1)(j+1)}(z)]_+ \leq q$. As a consequence, we have proved that $\deg[\widetilde{d}_{i(j+1)}(z)]_+ \leq q$ for $1 \leq i \leq m$. Similarly, we can also show that $\deg[\widetilde{d}_{i(j+1)}(z)]_- \leq q$ for $1 \leq i \leq m$.

Therefore, the assertion holds for the $j+1$ column, i.e., $\deg[\widetilde{d}_{i(j+1)}(z)]_+ \leq q$, $\deg[\widetilde{d}_{i(j+1)}(z)]_- \leq q \; \forall \, i : 1 \leq i \leq m$.

As results, $\widetilde{D}(z)$ can be written as $\widetilde{D}(z) = \sum_{i=-q}^{q} \widetilde{D}_i z^i$ with $\widetilde{D}_{-i} = \widetilde{D}_i^T$.

From (3.3.55) it follows that $\widetilde{D}(z)$ is of rank l, and is nonnegative definite on the unit circle $|z| = 1$. Then, there exists an $m \times l$ real rational spectral factor $\overline{B}(z)$ with the poles being outside the unit circle such that $\widetilde{D}(z) = \overline{B}(z)\overline{B}^T(z^{-1})$.

Notice that the poles of $\widetilde{D}(z)$ cannot be anything but 0 and ∞. Thus ∞ is the unique pole of $\overline{B}(z)$, which implies that $\overline{B}(z)$ is a matrix polynomial.

We write $\overline{B}(z)$ as $\overline{B}(z) = \overline{B}_0 + \overline{B}_1 z + \cdots + \overline{B}_{\overline{q}} z^{\overline{q}}$. By Lemma 3.3.3, $\overline{B}(z)$ can be factored as

$$\overline{B}(z) = \widetilde{B}(z) B_P(z) \tag{3.3.60}$$

where $\widetilde{B}(z) = \widetilde{B}_0 + \widetilde{B}_1 z + \cdots + \widetilde{B}_{\overline{q}} z^{\overline{q}}$ is an $m \times l$ matrix polynomial with $\deg \widetilde{B}(z) \leq \deg \overline{B}(z) = \overline{q}$, $\widetilde{B}_0$ is of column-full-rank, and $B_P(z)$ is an $l \times l$ matrix polynomial satisfying $B_P(z) B_P^T(z^{-1}) = I$. Hence, we obtain a real polynomial factorization of $\widetilde{D}(z)$ different from $\overline{B}(z)\overline{B}^T(z^{-1})$:

$$\widetilde{D}(z) = \overline{B}(z)\overline{B}^T(z^{-1}) = \widetilde{B}(z) B_P(z) B_P^T(z^{-1}) \widetilde{B}^T(z^{-1}) = \widetilde{B}(z)\widetilde{B}^T(z^{-1}) \tag{3.3.61}$$

and $\widetilde{B}_0$ is of column-full-rank.

We now show that $\widetilde{B}_s = 0 \; \forall \, s : q+1 \leq s \leq \overline{q}$.

If $\overline{q} > q$, then comparing the matrix coefficients of $z^{\overline{q}}$ at both sides of (3.3.61) we obtain $\widetilde{B}_{\overline{q}} \widetilde{B}_0^T = 0$. Since $\widetilde{B}_0$ is of column-full-rank, we have $\widetilde{B}_{\overline{q}} = 0$. By the same argument for $s : q+1 \leq s \leq \overline{q} - 1$, we see that $\widetilde{B}_s = 0 \; \forall \, s : q+1 \leq s \leq \overline{q}$. Therefore, $\deg[\widetilde{B}(z)] \leq q$.

By (3.3.53) and (3.3.55) we have

$$\begin{aligned}
& A^{-1}(z) B(z) B^T(z^{-1}) A^{-T}(z^{-1}) \\
& = [C(z)\widetilde{A}(z)]^{-1} B(z) B^T(z^{-1}) [C(z^{-1})\widetilde{A}(z^{-1})]^{-T} \\
& = \widetilde{A}(z)^{-1} C^{-1}(z) B(z) B^T(z^{-1}) C^{-T}(z^{-1}) \widetilde{A}^{-T}(z^{-1}) \\
& = \widetilde{A}(z)^{-1} \widetilde{D}(z) \widetilde{A}^{-T}(z^{-1}) = \widetilde{A}(z)^{-1} \widetilde{B}(z) \widetilde{B}^T(z^{-1}) \widetilde{A}^{-T}(z^{-1}).
\end{aligned}$$

This means that the two linear systems $\{A(z), B(z)\}$ and $\{\widetilde{A}(z), \widetilde{B}(z)\}$ share the same spectral density, and hence they have the same correlation functions. Thus, by (3.3.14) the Yule–Walker equations corresponding to them are as follows:

$$\begin{aligned}
& [A_1, A_2, \cdots, A_p]\Gamma = -[R_{q+1}, R_{q+2}, \cdots, R_{q+mp}], \\
& [\widetilde{A}_1, \widetilde{A}_2, \cdots, \widetilde{A}_p]\Gamma = -[R_{q+1}, R_{q+2}, \cdots, R_{q+mp}],
\end{aligned}$$

which imply

$$[A_1 - \widetilde{A}_1, A_2 - \widetilde{A}_2, \cdots, A_p - \widetilde{A}_p]\Gamma = 0.$$

Since the matrix $[A_1 - \widetilde{A}_1, A_2 - \widetilde{A}_2, \cdots, A_p - \widetilde{A}_p]$ is not identically zero, the matrix Γ cannot be of row-full-rank. The obtained contradiction completes the proof of the theorem. □

(ii) Hankel Matrices for Order Estimation

In the last section we have shown that θ_A for (3.3.1) can be expressed via L and Γ by (3.3.12) and (3.3.16), respectively. However, until now the problem of order estimation has not been a concern yet.

It is noticed that $B(z)$ in (3.3.1) may not be square and B_0 in $B(z)$ is to be estimated. In what follows we consider (3.1.2) with $B(z) \equiv 0$, i.e., the ARMA process:

$$A(z)y_k = C(z)w_k, \quad y_k \in \mathbb{R}^m. \tag{3.3.62}$$

Assume that $\{w_k\}$ is a sequence of zero-mean m-dimensional iid random vectors with $Ew_k w_k^T \triangleq R_w > 0$.

If $A(z)$ is stable, i.e., $\det A(z) \neq 0 \ \forall \ |z| \leq 1$, then, as mentioned before, the sequence $\{y_k\}$ is stationary with zero-mean and covariance matrix (correlation function) $Ey_k y_{k-\tau}^T \triangleq R_\tau$.

The order estimation approach presented in the book is first to estimate $t \triangleq \max(p,r)$, then estimate p and r, respectively. For this, it is important to consider the Hankel matrix $\Gamma(\mu,\nu)$ composed of correlation functions of $\{y_k\}$:

$$\Gamma(\mu,\nu) \triangleq \begin{bmatrix} R_1 & R_2 & \cdots & R_\mu \\ R_2 & R_3 & \cdots & R_{\mu+1} \\ \vdots & & \ddots & \vdots \\ R_\nu & R_{\nu+1} & \cdots & R_{\mu+\nu-1} \end{bmatrix}. \tag{3.3.63}$$

It is noticed that $\Gamma(\mu,\nu)$ is different from Γ defined by (3.3.15). We intend to establish that rank $\Gamma(\mu,\nu) = tm \ \ \forall \mu \geq tm$ and $\forall \nu \geq t, \ \ t \triangleq \max(p,r)$, but for this we need some auxiliary results. Some of them are well-known in linear system theory, but for readability we provide a self-contained proof.

In the sequel, by $\mathbb{C}$ we denote the space of complex numbers.

Lemma 3.3.4 *Any pair of matrices of compatible dimensions* (A, B) *is controllable if and only if* $[sI - A \ \ B]$ *is of row-full-rank* $\forall s \in \mathbb{C}$. *Any pair* (A, C) *is observable if and only if* $[(sI - A)^T \ \ C^T]^T$ *is of column-full-rank* $\forall s \in \mathbb{C}$.

Proof. Since observability of (A, C) is equivalent to controllability of (A^T, C^T), it suffices to prove the first assertion.

Let $[sI - A \ \ B]$ be of row-full-rank $\forall s \in \mathbb{C}$, where $A \in \mathbb{R}^{l \times l}$. Assume the converse that (A, B) is not controllable. Then there exists a nonzero vector $\xi \in \mathbb{R}^l$ such that

$$\xi^T B = 0, \ \xi^T A^i B = 0, \ i = 1 \cdots, l-1. \tag{3.3.64}$$

We first show that without loss of generality we may assume that ξ in (3.3.64) is an eigenvector of A^T with eigenvalue z_0 with $|z_0| < 1$.

From (3.3.64) we see that the row vectors ξ^T, $\xi^T A^i$, $i = 1 \cdots, l-1$ are linearly dependent. Then, there exists a minimal $k \le l-1$ such that ξ^T, $\xi^T A^i$, $i = 1, \cdots, k-1$ are linearly independent, and there exist real numbers $a_0, a_1, \cdots, a_{k-1}$, which are not simultaneously equal to zero, such that

$$\xi^T A^k + a_{k-1}\xi^T A^{k-1} + \cdots + a_1 \xi^T A + a_0 \xi^T = 0 \text{ or } \xi^T f(A) = 0, \tag{3.3.65}$$

where $f(z) \triangleq z^k + a_{k-1}z^{k-1} + \cdots + a_1 z + a_0$.

Let z_0 be a root of $f(z)$:

$$f(z) = (z - z_0)g(z), \; g(z) \triangleq z^{k-1} + \beta_{k-2}z^{k-2} + \cdots + \beta_1 z + \beta_0. \tag{3.3.66}$$

Then, by (3.3.65) and (3.3.66) we have

$$0 = \xi^T f(A) = \zeta^T (A - z_0 I) \text{ with } \zeta^T \triangleq \xi^T g(A). \tag{3.3.67}$$

Since $f(\cdot)$ is a polynomial of the minimal order such that $\xi^T f(A) = 0$, we must have $\zeta^T = \xi^T g(A) \neq 0$. From (3.3.67) it follows that z_0 is an eigenvalue of A^T and ζ is the corresponding eigenvector.

By (3.3.64) it follows that

$$\zeta^T B = \xi^T g(A) B = 0.$$

This means that $\zeta^T A^i B = 0, \; i = 1, \cdots, l-1$, i.e., ζ satisfies the same equations as ξ does. Therefore, in (3.3.64) without loss of generality we may assume that ξ is an eigenvector of A^T with eigenvalue z_0, $|z_0| < 1$.

We then have $\xi^T [B \;\; z_0 I - A] = 0$. Since $\xi \neq 0$, this contradicts the assumption that $[sI - A \;\; B]$ is of row-full-rank $\forall s \in \mathbb{C}$. The obtained contradiction shows that (A, B) is controllable.

We now assume (A, B) is controllable, and proceed to prove that $[sI - A \;\; B]$ is of row-full-rank $\forall s \in \mathbb{C}$.

Assume the converse: there are a nonzero vector ξ and a number λ such that

$$\xi^T [\lambda I - A \;\; B] = 0.$$

Then, we have $\xi^T B = 0$, $\xi^T A = \lambda \xi^T$, and hence $\xi^T A^i B = 0$, $i = 1 \cdots, l-1$. This means that (A, B) is not controllable, which contradicts with the assumption. □

Set $t \triangleq \max(p, r)$, $A_i \triangleq 0 \;\forall i > p$, $C_j \triangleq 0 \;\forall j > r$,

$$A \triangleq \begin{bmatrix} -A_1 & I & \cdots & 0 \\ \vdots & \ddots & \ddots & \vdots \\ \vdots & & \ddots & I \\ -A_t & 0 & \cdots & 0 \end{bmatrix}, \quad K \triangleq \begin{bmatrix} -A_1 + C_1 \\ -A_2 + C_2 \\ \vdots \\ -A_t + C_t \end{bmatrix}, \quad C \triangleq \begin{bmatrix} I & 0 & \cdots & 0 \end{bmatrix},$$

where I is the $m \times m$ identity matrix, and A, K, and C are of $tm \times tm$, $tm \times m$, and $m \times tm$-matrices, respectively.

Lemma 3.3.5 *The ARMA process $A(z)y_k = C(z)w_k$ has the following state space representation*

$$\begin{cases} X_{k+1} = AX_k + Kw_k, \\ y_{k+1} = CX_{k+1} + w_{k+1}, \end{cases} \qquad X_k \in \mathbb{R}^{tm}, \tag{3.3.68}$$

where A is stable (i.e., all its eigenvalues are strictly inside the unit disk) if $A(z)$ is stable (i.e., $\det A(z) \neq 0 \ \forall |z| \leq 1$). Further, $\{X_k\}$ is stationary with $EX_k = 0$, $EX_k X_k^T \triangleq P$, and P satisfies the algebraic Lyapunov equation

$$P = APA^T + KR_w K^T.$$

Proof. To prove that (3.3.68) is a state space realization of the ARMA process it suffices to show that the transfer functions for (3.3.68) and $A(z)y_k = C(z)w_k$ are the same, i.e.,

$$C(I - Az)^{-1}K \cdot z + I = A^{-1}(z)C(z). \tag{3.3.69}$$

By definitions of A, K, and C, it is straightforward to verify that

$$\begin{aligned}
&C(I - Az)^{-1}K \cdot z + I \\
&= \begin{bmatrix} I & 0 & \cdots & 0 \end{bmatrix} \cdot \begin{bmatrix} I + A_1 z & -zI & \cdots & 0 \\ A_2 z & I & \ddots & \vdots \\ \vdots & & \ddots & -zI \\ A_t z & 0 & \cdots & I \end{bmatrix}^{-1} \cdot \begin{bmatrix} -A_1 + C_1 \\ -A_2 + C_2 \\ \vdots \\ -A_t + C_t \end{bmatrix} \cdot z + I \\
&= A^{-1}(z) \begin{bmatrix} I & zI & \cdots & z^{t-1} I \end{bmatrix} \cdot \begin{bmatrix} -A_1 + C_1 \\ -A_2 + C_2 \\ \vdots \\ -A_t + C_t \end{bmatrix} \cdot z + I \\
&= A^{-1}(z)\Big(-(A_1 z + \cdots + A_t z^t) + (C_1 z + \cdots + C_t z^t) \Big) + I \\
&= A^{-1}(z)\Big(-A(z) + C(z) \Big) + I = A^{-1}(z)C(z).
\end{aligned}$$

Thus, (3.3.69) is proved.

By the definition of A, we have

$$\begin{aligned}
&\det(sI - A) \\
&= \det \begin{bmatrix} sI + A_1 & -I & \cdots & 0 \\ A_2 & sI & \ddots & \vdots \\ \vdots & & \ddots & -I \\ A_t & 0 & \cdots & sI \end{bmatrix}
\end{aligned}$$

$$= \det \begin{bmatrix} I & 0 & \cdots & 0 \\ sI & I & \ddots & \vdots \\ \vdots & & \ddots & 0 \\ s^{t-1}I & \cdots & sI & I \end{bmatrix} \cdot \begin{bmatrix} sI+A_1 & -I & \cdots & 0 \\ A_2 & sI & \ddots & \vdots \\ \vdots & & \ddots & -I \\ A_t & 0 & \cdots & sI \end{bmatrix}$$

$$= \det \begin{bmatrix} sI+A_1 & -I & 0 & \cdots & 0 \\ A_2+A_1s+s^2I & 0 & -I & \ddots & \vdots \\ A_3+A_2s+A_1s^2+s^3I & 0 & 0 & \ddots & 0 \\ \vdots & \vdots & & \ddots & -I \\ A_t+A_{t-1}s+\cdots+A_1s^{t-1}+s^tI & 0 & \cdots & \cdots & 0 \end{bmatrix},$$

and hence

$$|\det(sI-A)| = |\det(A_t+A_{t-1}s+\cdots+A_1s^{t-1}+s^tI)|. \tag{3.3.70}$$

Since $\det A(z) \neq 0 \ \forall \ |z| \leq 1$, we see that

$$\det(A_t+A_{t-1}s+\cdots+A_1s^{t-1}+s^tI) \neq 0 \ \ \forall \ |s| \geq 1.$$

Then, by (3.3.70) it follows that all eigenvalues of A are strictly inside the unit disk, i.e., A is stable.

The remaining assertions of the lemma are clear. □

We need the following condition, which will frequently be used:

A3.3.7 *$A(z)$ and $C(z)$ have no common left factor and $[A_t \ \ C_t]$ is of row-full-rank, where* $t \triangleq \max(p,r)$, $A_i \triangleq 0 \ \forall i > p$, and $C_j \triangleq 0 \ \forall j > r$.

Lemma 3.3.6 *For the representation (3.3.68) the following assertions hold:*

(i) *(A, C) is observable;*

(ii) *(A, K) is controllable if and only if A3.3.7 holds.*

Proof. (i) By induction it is straightforward to verify that

$$CA = \begin{bmatrix} X & I & 0\cdots 0 \end{bmatrix}, \text{ and } CA^i = \begin{bmatrix} X & \cdots & X & I & 0\cdots 0 \end{bmatrix},$$

where by X we denote an $m \times m$-matrix, while by $X \cdots X$ in CA^i we denote an $m \times im$-matrix.

Then, the $tm \times tm$-matrix

$$\begin{bmatrix} C \\ CA \\ \vdots \\ CA^{t-1} \end{bmatrix} \tag{3.3.71}$$

is of full-rank. Therefore, $(A,\ C)$ is observable.

(ii) By Lemma 3.3.4 controllability of (A, K) is equivalent to

$$\operatorname{rank}[sI - A \;\; K] = tm \;\; \forall\, s \in \mathbb{C}.$$

By definitions of A and K, we have

$$\begin{aligned}
&\operatorname{rank}\,[sI - A \;\; K]\\
&=\operatorname{rank}\begin{bmatrix} sI + A_1 & -I & \cdots & \cdots & -A_1 + C_1 \\ A_2 & sI & \ddots & & -A_2 + C_2 \\ \vdots & & \ddots & -I & \vdots \\ A_t & 0 & \cdots & sI & -A_t + C_t \end{bmatrix}\\
&=\operatorname{rank}\begin{bmatrix} I & 0 & \cdots & 0 \\ sI & I & \ddots & \vdots \\ \vdots & & \ddots & 0 \\ s^{t-1}I & \cdots & sI & I \end{bmatrix} \cdot \begin{bmatrix} sI + A_1 & -I & \cdots & 0 & -A_1 + C_1 \\ A_2 & sI & \ddots & \vdots & -A_2 + C_2 \\ \vdots & & \ddots & -I & \vdots \\ A_t & 0 & \cdots & sI & -A_t + C_t \end{bmatrix}\\
&=\operatorname{rank}\begin{bmatrix} sI + A_1 & -I & 0 & \cdots & 0 & -A_1 + C_1 \\ \sum_{i=0}^{2} A_{2-i}s^i & 0 & -I & \ddots & \vdots & \vdots \\ \sum_{i=0}^{3} A_{3-i}s^i & 0 & 0 & \ddots & 0 & \vdots \\ \vdots & \vdots & & \ddots & -I & \vdots \\ \sum_{i=0}^{t} A_{t-i}s^i & 0 & \cdots & \cdots & 0 & \sum\limits_{i=1}^{t}(-A_i + C_i)s^{t-i} \end{bmatrix},
\end{aligned} \tag{3.3.72}$$

where $A_0 \triangleq I$.

Therefore, controllability of $(A,\ K)$ is equivalent to

$$\operatorname{rank}\left[\sum_{i=0}^{t} A_{t-i}s^i \quad \sum_{i=1}^{t}(-A_i + C_i)s^{t-i}\right] = m \;\; \forall\, s \in \mathbb{C}. \tag{3.3.73}$$

It is clear that for (3.3.73) to hold at $s = 0$ the necessary and sufficient condition is that $[A_t \;\; C_t - A_t]$ is of row-full-rank. Noticing that for any nonzero vector ξ of compatible dimension, $\xi^T[A_t \;\; C_t - A_t] = 0$ is equivalent to $\xi^T[A_t \;\; C_t] = 0$. Consequently, for (3.3.73) to hold at $s = 0$ the necessary and sufficient condition is that the matrix $[A_t \;\; C_t]$ is of row-full-rank.

If $s \neq 0$, replacing s with z^{-1}, $z \neq 0$, from (3.3.73) we have

$$\begin{aligned}
&\operatorname{rank}\,[A_t + A_{t-1}z^{-1} + \cdots + A_1 z^{-(t-1)} + z^{-t}I \;\; \sum_{i=1}^{t}(-A_i + C_i)z^{-(t-i)}]\\
&=\operatorname{rank} z^t[A_t + A_{t-1}z^{-1} + \cdots + A_1 z^{-(t-1)} + z^{-t}I \;\; \sum_{i=1}^{t}(-A_i + C_i)z^{-(t-i)}]\\
&=\operatorname{rank}\,[A(z) \;\; C(z) - A(z)].
\end{aligned} \tag{3.3.74}$$

So, for (3.3.73) to hold at $s \neq 0$ the necessary and sufficient condition is $\text{rank}[A(z)\ \ C(z)-A(z)] = m\ \forall z \in \mathbb{C},\ z \neq 0$.

We now show that for $\text{rank}[A(z)\ \ C(z)-A(z)] = m\ \forall z \in \mathbb{C},\ z \neq 0$ the necessary and sufficient condition is that $A(z)$ and $C(z)$ have no common left factor.

Sufficiency. Let $A(z)$ and $C(z)$ have no common left factor. We show that $\text{rank}[A(z)\ \ C(z)-A(z)] = m\ \forall z \in \mathbb{C},\ z \neq 0$.

Assume the converse: there exists $z_0 \neq 0$ and a vector $\xi \neq 0$ such that

$$\xi^T[A(z_0)\ \ C(z_0)-A(z_0)] = 0.$$

Then, we have

$$\xi^T[A(z_0)\ \ C(z_0)] = 0.$$

Since $A(z)$ and $C(z)$ have no common leftfactor, there exist $X(z)$ and $Y(z)$ such that

$$A(z)X(z)+C(z)Y(z) = I. \tag{3.3.75}$$

Replacing z with z_0 in (3.3.14) and multiplying both sides of it from the left by ξ^T yield $0 = \xi^T$, which contradicts with $\xi \neq 0$. The contradiction implies that $\text{rank}[A(z)\ \ C(z)-A(z)] = m\ \forall z \in \mathbb{C},\ z \neq 0$.

Necessity. We now assume $\text{rank}[A(z)\ \ C(z)-A(z)] = m\ \ \forall z \in \mathbb{C},\ z \neq 0$, and assume the converse that there is a common left factor $B(z)$ for $A(z)$ and $C(z)$, i.e.,

$$[A(z),\ C(z)] = B(z)[A'(z),\ C'(z)].$$

Let z_0 be a root of $\det B(z)$. It is clear that $z_0 \neq 0$, because otherwise $\det A(z)$ would have zero as its root. Then, there exists a nonzero vector ξ such that $\xi^T B(z_0) = 0$, and hence $\xi^T[A(z_0),\ C(z_0)] = 0$ and $\xi^T[A(z_0)\ \ C(z_0)-A(z_0)] = 0$. This contradicts with the assumption that $\text{rank}[A(z)\ \ C(z)-A(z)] = m\ \forall z \in \mathbb{C},\ z \neq 0$. Consequently, $A(z)$ and $C(z)$ have no common leftfactor.

Thus, we have shown a chain of equivalence: Controllability of $(A,\ K)$ is equivalent to (3.3.73), which in turn is equivalent to A3.3.7. This completes the proof of the lemma. □

Remark 3.3.1 *Lemma 3.3.6 tells us that for the ARMA process $A(z)y_k = C(z)w_k$ the state space realization (3.3.68) is observable, and it is controllable if and only if A3.3.7 holds. In other words, for (3.3.68) to be the minimal realization the necessary and sufficient condition is A3.3.7; alternatively, A3.3.7 is equivalent to the corresponding observation index that equals t and the minimal degree is mt.*

Define

$$D \triangleq APC^T + KR_w,$$

where $P = EX_kX_k^T$ satisfying the Lyapunov equation $P = APA^T + KR_wK^T$.

Lemma 3.3.7 *Let $A(z)$ be stable and let $R_w > 0$.*

(i) *If* (A, K) *is controllable and if*

$$\det C\Big(\frac{1}{z_0}\Big) \neq 0 \text{ for any root } z_0 \text{ of } \det A(z),$$

then (A, D) *is controllable.*

(ii) *If* (A, D) *is controllable, then* (A, K) *is controllable.*

Proof. It is worth noting that $|\frac{1}{z_0}| < 1$ by stability of $A(z)$, and hence the condition $\det C(\frac{1}{z_0}) \neq 0$ is implied by stability of $C(z)$.

(i) Assume the converse: (A, D) is not controllable.

Then, there is a tm-dimensional vector $\xi \neq 0$ such that

$$\xi^T D = 0, \ \ \xi^T A^j D = 0 \ \ \forall j = 1, \cdots, tm-1. \tag{3.3.76}$$

By completely the same argument as that done for (3.3.67), without loss of generality, ξ may be assumed to be an eigenvector of A^T with eigenvalue $\lambda_0 : \ |\lambda_0| < 1$.

By definition of D and P, we have

$$\begin{aligned}
\xi^T D &= \xi^T APC^T + \xi^T KR_w \\
&= \lambda_0 \xi^T \sum_{j=0}^{\infty} A^j KR_w K^T A^{jT} C^T + \xi^T KR_w \\
&= \lambda_0 \sum_{j=0}^{\infty} \lambda_0^j \xi^T KR_w K^T A^{jT} C^T + \xi^T KR_w \\
&= \lambda_0 \xi^T KR_w K^T (I - \lambda_0 A^T)^{-1} C^T + \xi^T KR_w \\
&= \xi^T KR_w \Big(K^T \big(\frac{1}{\lambda_0} I - A^T\big)^{-1} C^T + I \Big).
\end{aligned} \tag{3.3.77}$$

From (3.3.77), by (3.3.69) it follows that

$$\xi^T D = \xi^T KR_w C^T(\lambda_0) A^{-T}(\lambda_0). \tag{3.3.78}$$

Since λ_0 is an eigenvalue of A, we have $\det A(\frac{1}{\lambda_0}) = 0$. Then, by the condition of the lemma we see that $\det C(\lambda_0) \neq 0$, and hence $A^{-1}(\lambda_0) C(\lambda_0)$ is nonsingular. By $\xi^T D = 0$, from (3.3.78) it follows that $\xi^T K = 0$. Since ξ is an eigenvector of A^T, we see $\xi^T A^i K = 0, \ i = 1, \cdots, tm-1$. This contradicts with the assumption that (A, K) is controllable. The obtained contradiction shows the controllability of (A, D).

(ii) Assume (A, D) is controllable.

From (3.3.78) it is seen that if (A, K) is not controllable, then $\xi^T D = 0$ for an eigenvector ξ of A^T, and hence $\xi^T A^j D = 0 \ \forall j = 1, \cdots, tm-1$. This contradicts with the assumption that (A, D) is controllable. □

We are now in a position to answer the question stated at the beginning of the section (under which conditions $\Gamma(\mu, \nu)$ defined by (3.3.63) is of rank tm $\forall \mu \geq tm$ and $\forall \nu \geq t$). For this, in addition to A3.3.7, we introduce the following conditions.

A3.3.8 *A(z) is stable and*

$$\det C\left(\frac{1}{z_0}\right) \neq 0$$

for any root z_0 *of* $\det A(z)$.

A3.3.9 $\{w_k\}$ *is a sequence of m-dimensional iid random vectors with* $Ew_k = 0,\ Ew_k w_k^T \triangleq R_w > 0$.

Theorem 3.3.4 *Assume A3.3.8 and A3.3.9 hold. Then*

$$rank\ \Gamma(\mu,\nu) = tm\ \ \forall \mu \geq tm\ \ and\ \ \forall \nu \geq t,$$

if and only if A3.3.7 is satisfied.

Proof. From (3.3.68) it follows that

$$R_0 = Ey_k y_k^T = E(CX_k + w_k)(CX_k + w_k)^T = CPC^T + R_w, \tag{3.3.79}$$

and for any $\tau \geq 1$

$$\begin{aligned}
R_\tau =& Ey_k y_{k-\tau}^T = E(CX_k + w_k)(CX_{k-\tau} + w_{k-\tau})^T \\
=& CEX_k(CX_{k-\tau} + w_{k-\tau})^T = CE(AX_{k-1} + Kw_{k-1})(CX_{k-\tau} + w_{k-\tau})^T \\
=& CAEX_{k-1}(CX_{k-\tau} + w_{k-\tau})^T = \cdots \\
=& CA^{\tau-1}EX_{k-\tau+1}(CX_{k-\tau} + w_{k-\tau})^T \\
=& CA^{\tau-1}E(AX_{k-\tau} + Kw_{k-\tau})(CX_{k-\tau} + w_{k-\tau})^T \\
=& CA^{\tau-1}(APC^T + KR_w) = CA^{\tau-1}D.
\end{aligned} \tag{3.3.80}$$

By (3.3.80), the Hankel matrix $\Gamma(\mu,\nu)$ can be decomposed as follows

$$\begin{aligned}
\Gamma(\mu,\nu) =& \begin{bmatrix} CD & CAD & \cdots & CA^{\mu-1}D \\ CAD & CA^2D & \cdots & CA^{\mu}D \\ \vdots & & \ddots & \vdots \\ CA^{\nu-1}D & CA^{\nu}D & \cdots & CA^{\mu+\nu-2}D \end{bmatrix} \\
=& \begin{bmatrix} C \\ CA \\ \vdots \\ CA^{\nu-1} \end{bmatrix} \cdot \begin{bmatrix} D & AD & \cdots & A^{\mu-1}D \end{bmatrix}.
\end{aligned} \tag{3.3.81}$$

The first factor of the matrix product at the right-hand side of (3.3.81) being an $\nu m \times tm$-matrix, as shown in Lemma 3.3.6 for (3.3.71), is of column-full-rank for $\nu \geq t$. Consequently, the rank $\Gamma(\mu,\nu) = tm\ \ \forall \mu \geq tm$ and $\nu \geq t$ if and only if the matrix $\begin{bmatrix} D & AD & \cdots & A^{\mu-1}D \end{bmatrix}$ is of row-full-rank, i.e., $(A,\ D)$ is controllable. However, under A3.3.8 and A3.3.9 by Lemma 3.3.7 controllability of $(A,\ D)$ is equivalent to controllability of $(A,\ K)$, which in turn is equivalent to A3.3.7 by Lemma 3.3.6. □

Remark 3.3.2 *It is noticed that for (3.3.62) the assumption corresponding to A3.3.2 is as follows: $A(z)$ and $C(z)$ have no common left factor and $[A_p \; C_r]$ is of row-full-rank. It is clear that A3.3.7 is stronger than A3.3.2. Let us consider the following example:*

$$y_k + A_1 y_{k-1} + A_2 y_{k-2} = w_k + C_1 w_{k-1},$$
$$A_1 = \begin{bmatrix} 1 & 0 \\ 0 & \frac{1}{2} \end{bmatrix}, \; A_2 = \begin{bmatrix} \frac{1}{4} & 0 \\ 0 & 0 \end{bmatrix}, \; C_1 = \begin{bmatrix} 0 & 0 \\ 0 & \frac{1}{3} \end{bmatrix}.$$

It is straightforward to calculate that $\det A(z)$ is with multi-root at -2 and the root of $\det C(z)$ is -3. So, both $A(z)$ and $C(z)$ are stable. Further, $A(z)$ and $C(z)$ have no common left factor, and rank$[A_2 \, C_1] = 2$. Therefore, A3.3.2 is satisfied, but $[A_2 \, C_2] = \begin{bmatrix} \frac{1}{4} & 0 & 0 & 0 \\ 0 & 0 & 0 & 0 \end{bmatrix}$, and rank$[A_2 \, C_2] = 1 < 2$. This means that A3.3.7 does not hold, and by Theorem 3.3.4 rank$\Gamma(4, 2) < 4$, although A3.3.2 takes place.

By stability of $A(z)$ the transfer function $A^{-1}(z)C(z)$ can be expanded to the series

$$A^{-1}(z)C(z) = \sum_{i=0}^{\infty} L_i z^i, \;\; L_0 = I, \; \|L_i\| = O(e^{-\alpha i}), \; \alpha > 0 \; \forall i \geq 1. \qquad (3.3.82)$$

Let $L(\mu, \nu)$ be the Hankel matrix composed of impulse responses:

$$L(\mu, \nu) \triangleq \begin{bmatrix} L_1 & L_2 & \cdots & L_\mu \\ L_2 & L_3 & \cdots & L_{\mu+1} \\ \vdots & & \ddots & \vdots \\ L_\nu & L_{\nu+1} & \cdots & L_{\mu+\nu-1} \end{bmatrix}. \qquad (3.3.83)$$

Theorem 3.3.5 *Assume $A(z)$ is stable. Then*

$$\text{rank}\, L(\mu, \nu) = tm \;\; \forall \mu \geq tm \;\; \text{and} \;\; \forall \nu \geq t$$

if and only if A3.3.7 is satisfied.

Proof. Let us associate the transfer function $A^{-1}(z)C(z)$ with an ARMA process $A(z)y_k = C(z)w_k$, where $\{w_k\}$ is a sequence of iid m-dimensional random vectors with $Ew_k = 0$ and $Ew_k w_k^T = I$. Then, we have

$$y_k = \sum_{i=0}^{k-1} L_i w_{k-i}. \qquad (3.3.84)$$

From (3.3.84) it follows that

$$Ey_k w_{k-i}^T = L_i. \qquad (3.3.85)$$

On the other hand, from (3.3.68) we have that

$$y_k = CA^i X_{k-i} + C\sum_{j=1}^{i} A^{j-1} K w_{k-j} + w_k. \tag{3.3.86}$$

Since w_{k-i} is independent of X_{k-i}, from (3.3.86) we derive

$$E y_k w_{k-i}^T = CA^{i-1}K, \tag{3.3.87}$$

which combining with (3.3.85) implies

$$L_i = CA^{i-1}K. \tag{3.3.88}$$

Consequently, we have

$$\begin{aligned}\Gamma(\mu,\nu) &= \begin{bmatrix} CK & CAK & \cdots & CA^{\mu-1}K \\ CAK & CA^2K & \cdots & CA^{\mu}K \\ \vdots & & \ddots & \vdots \\ CA^{\nu-1}K & CA^{\nu}K & \cdots & CA^{\mu+\nu-2}K \end{bmatrix} \\ &= \begin{bmatrix} C \\ CA \\ \vdots \\ CA^{\nu-1} \end{bmatrix} \cdot \begin{bmatrix} K & AK & \cdots & A^{\mu-1}K \end{bmatrix}. \end{aligned} \tag{3.3.89}$$

Similar to the proof of Theorem 3.3.4, by noticing that the first factor of the matrix product at the right-hand side of (3.3.89) is of column-full-rank for $\nu \geq t$, we find that rank $L(\mu,\nu) = tm \ \forall\ \mu \geq tm$ and $\forall\ \nu \geq t$ if and only if $(A,\ K)$ is controllable, which, in turn, is equivalent to A3.3.7 by Lemma 3.3.6. □

3.4 Coefficient Identification of ARMAX by SAAWET

We have demonstrated the nice convergence properties of ELS estimate for ARMAX under the SPR condition $C^{-1}(e^{i\lambda}) + C^{-T}(e^{-i\lambda}) - I > 0 \ \ \forall \lambda \in [0, 2\pi]$. Multiplying this inequality by $C(e^{i\lambda})$ from left and by $C^T(e^{-i\lambda})$ from right, we obtain the following inequality:

$$\begin{aligned} I &> I - C(e^{i\lambda}) - C^T(e^{-i\lambda}) + C(e^{i\lambda})C^T(e^{-i\lambda}) \\ &= [I - C(e^{i\lambda})][I - C^T(e^{-i\lambda})]. \end{aligned}$$

Integrating both sides of this inequality from 0 to 2π yields

$$C_1C_1^T + C_2C_2^T + \cdots + C_rC_r^T < I.$$

In the extreme case $C_j = 0 \ \ \forall j = 1, \cdots, r$, the correlated noise $\{C(z)w_k\}$ becomes uncorrelated $\{w_k\}$. So, roughly speaking, the SPR condition requires $\{w_k\}$ be "not too colored".

In what follows we intend to estimate the coefficients and orders of ARMAX with the help of SAAWET without imposing SPR.

Recursive Estimation of AR-part

Let us consider the ARMAX system (3.1.2) with $B(z)u_{k-1}$ removed, i.e., consider (3.3.62).

For (3.3.62) we assume the following condition:

A3.4.1 *$A(z)$ is stable and $\{w_k\}$ is a sequence of zero-mean m-dimensional iid random vectors with $Ew_kw_k^T \triangleq R_w > 0$.*

Conditions A3.3.2 and A3.3.6 in the present case read as follows:

A3.4.2 *$A(z)$ and $C(z)$ have no common left factor and $[A_p \;\; C_r]$ is of row-full-rank.*

A3.4.3 *$A(z)$ and $C(z)C^T(z^{-1})z^r$ have no common left factor and $[A_p \;\; C_r]$ is of row-full-rank.*

Remark 3.4.1 *If both $A(z)$ and $C(z)$ are stable and if $A(z)$ and $C(z)$ have no common left factor, then $A(z)$ and $C(z)C^T(z^{-1})z^r$ have no common left factor. This is because in order for $A(z)$ and $C(z)C^T(z^{-1})z^r$ to have no common left factor, it is sufficient to require that the root sets of $\det A(z)$ and $\det C(z)\det C^T(z^{-1})z^r$ have no intersection. This takes place indeed, since by stability of $C(z)$ all roots of $\det C^T(z^{-1})z^r$ are inside the unit disk.*

The Yule–Walker equation (3.3.16) still takes place, but q should be replaced by r in Γ defined by (3.3.15) and in W. To be precise, we have

$$\theta_A = (\Gamma\Gamma^T)^{-1}\Gamma W, \tag{3.4.1}$$

where

$$\Gamma \triangleq \begin{bmatrix} R_r & R_{r+1} & \cdots & R_{r+mp-1} \\ R_{r-1} & R_r & \cdots & R_{r+mp-2} \\ \vdots & \vdots & \ddots & \vdots \\ R_{r-p+1} & R_{r-p+2} & \cdots & R_{r+(m-1)p} \end{bmatrix} \in \mathbb{R}^{mp\times m^2p}, \tag{3.4.2}$$

and $W^T \triangleq -[R_{r+1}, R_{r+2}, \cdots, R_{r+mp}]$.

Defining

$$\varphi_k^T(s) \triangleq [y_k^T, \cdots, y_{k-s+1}^T], \tag{3.4.3}$$

we have

$$Ey_k\varphi_{k-1-r}^T(mp) = -W \text{ and } E\varphi_k(p)\varphi_{k-r}^T(mp) = \Gamma. \tag{3.4.4}$$

By ergodicity of $\{y_k\}$ we have

$$\frac{1}{k}\sum_{j=1}^{k} y_j\varphi_{j-1-r}^T(mp) \xrightarrow[k\to\infty]{} -W \text{ a.s.}, \quad \frac{1}{k}\sum_{j=1}^{k} \varphi_j(p)\varphi_{j-r}^T(mp) \xrightarrow[k\to\infty]{} \Gamma \text{ a.s.} \tag{3.4.5}$$

If Γ and W were known, then θ_A could be obtained by (3.4.1). However, this is not the case. The algorithms to be proposed simultaneously estimate Γ, W, and θ_A.

Let us take a sequence of positive real numbers $\{M_k\}$ increasingly diverging to infinity.

The recursive algorithms for $\theta_{A,k}$, the estimate of θ_A, are defined as follows:

$$\Gamma_k = \Gamma_{k-1} - \frac{1}{k}(\Gamma_{k-1} - \varphi_k(p)\varphi_{k-r}^T(mp)), \tag{3.4.6}$$

$$W_k = W_{k-1} - \frac{1}{k}(W_{k-1} + y_k\varphi_{k-1-r}^T(pm)), \tag{3.4.7}$$

$$\theta_{A,k} = \big(\theta_{A,k-1} - \frac{1}{k}\Gamma_k(\Gamma_k^T\theta_{A,k-1} + W_k^T)\big) \cdot I_{[\|\theta_{A,k-1} - \frac{1}{k}\Gamma_k(\Gamma_k^T\theta_{A,k-1}+W_k^T)\| \le M_{\lambda_k}]}, \tag{3.4.8}$$

$$\lambda_k = \sum_{i=1}^{k-1} I_{[\|\theta_{A,i-1} - \frac{1}{i}\Gamma_i(\Gamma_i^T\theta_{A,i-1}+W_i^T)\| > M_{\lambda_i}]}, \quad \lambda_0 = 0 \tag{3.4.9}$$

with arbitrary initial values $\Gamma_0 \in \mathbb{R}^{mp\times m^2p}$, $W_0 \in \mathbb{R}^{m\times m^2p}$, and $\theta_{A,0} \in \mathbb{R}^{mp\times m}$.

It is worth pointing out that Γ_k, W_k given by (3.4.6), (3.4.7) are the recursive expressions of the time averages of (3.4.5). The algorithm (3.4.6)–(3.4.9) generating $\theta_{A,k}$ is the SAAWET discussed in Chapter 2.

As a matter of fact, we are facing to seek the root of the function

$$f(\theta) \triangleq \Gamma\Gamma^T\theta - \Gamma W, \quad \theta \in \mathbb{R}^{mp\times m}. \tag{3.4.10}$$

To serve as the observation of $f(\cdot)$ at time k we may take

$$\Gamma_k\Gamma_k^T\theta_{A,k-1} - \Gamma_k W_k, \tag{3.4.11}$$

which can be written in the standard form:

$$\Gamma_k\Gamma_k^T\theta_{A,k-1} - \Gamma_k W_k = f(\theta_{A,k-1}) + \varepsilon_k, \tag{3.4.12}$$

where

$$\varepsilon_k = \varepsilon_k^{(1)} + \varepsilon_k^{(2)}$$
$$\varepsilon_k^{(1)} \triangleq (\Gamma_k\Gamma_k^T - \Gamma\Gamma^T)\theta_{A,k-1}, \quad \varepsilon_k^{(2)} \triangleq \Gamma W - \Gamma_k W_k. \tag{3.4.13}$$

Theorem 3.4.1 *Assume A3.4.1 and A3.4.3 hold. Then, $\theta_{A,k}$ given by (3.4.6)–(3.4.9) converges to θ_A almost surely.*

Proof. Let $\Omega_0 \triangleq \{\omega : \Gamma_k \xrightarrow[k\to\infty]{} \Gamma, \ W_k \xrightarrow[k\to\infty]{} W\}$. By (3.4.5) it is clear that the probability of Ω_0 is one, i.e., $\mathbb{P}\Omega_0 = 1$, and for any fixed $\omega \in \Omega_0$ there is a constant c_1 such that

$$\|W_k\| + \|\Gamma_k\| < c_1 \ \forall k = 1,2,\cdots. \tag{3.4.14}$$

It is noted that the step-size a_k used in SAAWET is specified to equal $\frac{1}{k}$, and $m(k,T)$ defined by (2.2.7) is also specified to be

$$m(k,T) \triangleq \max\{m : \sum_{i=k}^{m} \frac{1}{i} \leq T\}.$$

We now show that for any $\omega \in \Omega_0$, if $\{\theta_{A,m_k}\}$ is a convergent subsequence of $\{\theta_{A,k}\}$ given by (3.4.6)-(3.4.9): $\theta_{A,m_k} \xrightarrow[k\to\infty]{} \widetilde{\theta}$, then for all large enough k and sufficiently small $T > 0$

$$\theta_{A,j} = \theta_{A,j-1} - \frac{1}{j}\Gamma_j(\Gamma_j^T\theta_{A,j-1} + W_j^T), \tag{3.4.15}$$

$$\|\theta_{A,j} - \theta_{A,m_k}\| \leq cT \quad \forall j : j = m_k + 1, \ldots, m(m_k, T) + 1, \tag{3.4.16}$$

where c is independent of k but may depend on ω (sample).

This means that in small neighborhoods of θ_{A,m_k} the algorithm (3.4.8)–(3.4.9) has no truncation and all $\theta_{A,j}$ are close to each other, although it is not clear if the entire sequence $\{\theta_{A,k}\}$ is bounded or not.

If $\{\lambda_k\}$ given by (3.4.9) is bounded, then the truncation in (3.4.8) ceases in a finite number of steps and (3.4.15) is verified.

We now assume $\lambda_k \xrightarrow[k\to\infty]{} \infty$.

For any fixed $\omega \in \Omega_0$ we have

$$\|\Gamma_k(\Gamma_k^T\theta + W_k^T)\| < c_1^2(1 + \|\theta\|).$$

Temporarily consider (3.4.8) without truncations for $j \geq m_k + 1$:

$$\theta_{A,j} = \left(\theta_{A,j-1} - \frac{1}{j}\Gamma_j(\Gamma_j^T\theta_{A,j-1} + W_j^T)\right). \tag{3.4.17}$$

Comparing $\theta_{A,j}$ with ξ_j generated by the following recursion

$$\xi_j = \xi_{j-1} + \frac{1}{j}c_1^2(1 + \xi_{j-1}), \quad \xi_{m_k} = \|\theta_{A,m_k}\| \quad j \geq m_k + 1, \tag{3.4.18}$$

we see that $\|\theta_{A,j}\| \leq \xi_j \;\; \forall j \geq m_k$.

From (3.4.18) we have

$$\int_{\xi_{j-1}}^{\xi_j} \frac{dx}{c_1^2(1+x)} \leq \frac{1}{j},$$

and

$$\int_{\xi_{m_k}}^{\xi_{m(m_k,T)+1}} \frac{dx}{c_1^2(1+x)} \leq T \quad \forall k. \tag{3.4.19}$$

Since $\xi_{m_k} \xrightarrow[k\to\infty]{} \|\widetilde{\theta}\| \triangleq \mu_1$ and

$$\int_{\mu_1}^{\infty} \frac{dx}{c_1^2(1+x)} = \infty,$$

from (3.4.19) it follows that $\xi_{m(m_k,T)}$ is bounded with respect to k. Therefore, for the fixed ω, $\|\theta_{A,j}\|$ with $\theta_{A,j}$ given by (3.4.17) is bounded by some constant α_1 for $s = m_k+1,\dots,m(m_k,T)+1$ and $\forall k \geq 1$. Since $\lambda_k \xrightarrow[k\to\infty]{} \infty$, we have $M_{\lambda_k} \to \infty$ and $M_{\lambda_k} > \alpha_1$ for sufficiently large k. This verifies (3.4.15) for the case $\lambda_k \xrightarrow[k\to\infty]{} \infty$.

By boundedness of $\|\theta_{A,j}\|, \forall j = m_k,\dots,m(m_k,T)+1$, (3.4.16) follows from (3.4.15).

To analyze convergence of the algorithm given by (3.4.8)–(3.4.9) we apply Theorem 2.3.1. For this we have to check Conditions A2.3.1–A2.3.4. Since $a_k = \frac{1}{k+1}$, A2.3.1 is satisfied. A2.3.2 also holds, because $f(\theta)$ given by (3.4.10) is a linear function.

It is noticed that by replacing $B(z)$ with $C(z)$, we see that Theorem 3.3.3 is applicable. Therefore, A3.4.3 implies the row-full-rank of Γ, and hence $\Gamma\Gamma^T > 0$. To check A2.3.3, let us take $v(\theta) \triangleq -\mathrm{tr}(\Gamma\Gamma^T\theta - \Gamma W^T)^T(\Gamma\Gamma^T\theta - \Gamma W^T)$. Therefore, the root set of $f(\cdot)$ consists of the single matrix θ_A, and A2.3.3 is fulfilled too.

It remains to verify A2.3.4. Let us write $\varepsilon_j^{(1)}$ given by (3.4.13) as follows:

$$\varepsilon_j^{(1)} \triangleq \varepsilon_j^{(1,1)} + \varepsilon_j^{(1,2)} \quad \text{for } j = m_k,\dots,m(m_k,T)+1,$$

where

$$\varepsilon_j^{(1,1)} \triangleq -(\Gamma_j\Gamma_j^T - \Gamma\Gamma^T)(\theta_{A,j-1} - \theta_{A,m_k}),$$
$$\varepsilon_j^{(1,2)} \triangleq -(\Gamma_j\Gamma_j^T - \Gamma\Gamma^T)\theta_{A,m_k}.$$

Since $\Gamma_k\Gamma_k^T - \Gamma\Gamma^T \xrightarrow[k\to\infty]{} 0$ for $\omega \in \Omega_0$, by (3.4.16) $\{\varepsilon_j^{(1,1)}\}$ satisfies A2.3.4 and by the convergence of $\{\theta_{A,m_k}\}$, $\{\varepsilon_j^{(1,2)}\}$ also fulfills A2.3.4. Further, by $\Gamma_k W_k^T - \Gamma W^T \xrightarrow[k\to\infty]{} 0$ for $\omega \in \Omega_0$, the noise condition A2.3.4 is also satisfied by $\{\varepsilon_k^{(2)}\}$.

Thus, all conditions required in Theorem 2.3.1 are fulfilled, and hence the assertion of the theorem follows. □

Recursive Estimation of MA Part

We now proceed to estimate θ_C defined by (3.1.7) and R_w. Without loss of generality, we may assume $r \geq 1$, because, otherwise, $C(z) \equiv I$ and θ_C need not be estimated.

Define

$$\zeta_k \triangleq A(z)y_k, \quad \text{or} \quad \zeta_k = C(z)w_k.$$

Under A3.4.1 the process $\{\zeta_k\}$ is stationary and ergodic with correlation function

$$S \triangleq [S(0)\vdots\cdots\vdots S(r)]^T \in \mathbb{R}^{m(r+1)\times m}, \tag{3.4.20}$$

where $S(i) \triangleq E\zeta_k\zeta_{k-i}^T \in \mathbb{R}^{m\times m}, \ i = 0,\cdots,r.$

Setting

$$\varphi_k^{wT} \triangleq [w_k^T,\dots,w_{k-r}^T], \quad \varphi_k^{\zeta T} \triangleq [\zeta_k^T,\dots,\zeta_{k-r}^T],$$

we have

$$\zeta_k = [I \ \vdots \ \theta_C^T]\varphi_k^w. \tag{3.4.21}$$

Noticing $Ew_{k-i}\zeta_k^T = R_w C_i^T$ for $1 \le i \le r$, multiplying (3.4.21) from right by $\varphi_k^{\zeta T}$ and taking expectation we obtain

$$[S(0),\dots,S(r)] = [I \ \vdots \ \theta_C^T] \cdot \begin{bmatrix} R_w & 0 & \cdots & 0 \\ R_w C_1^T & R_w & \ddots & \vdots \\ \vdots & & \ddots & 0 \\ R_w C_r^T & R_w C_{r-1}^T & \cdots & R_w \end{bmatrix}. \tag{3.4.22}$$

Write the $m(r+1) \times m$-matrix X in a sub-matrix form: $X = [X(0),\cdots,X(r)]^T \in \mathbb{R}^{m(r+1)\times m}$ with sub-matrices $X(i) \in \mathbb{R}^{m\times m}$, $i = 0,\cdots,r$. Recall that the parameter to be estimated is $X^* = [R_w, C_1,\cdots,C_r]^T$.

Then, defining matrices $A(X) \in \mathbb{R}^{m(r+1)\times m(r+1)}$, and $\Phi(X) \in \mathbb{R}^{m(r+1)\times m(r+1)}$ as follows:

$$A(X) \triangleq \begin{bmatrix} I & X(1) & \cdots & \cdots & X(r) \\ 0 & I & X(1) & \cdots & X(r-1) \\ \vdots & \ddots & \ddots & \ddots & \vdots \\ \vdots & & \ddots & \ddots & X(1) \\ 0 & \cdots & \cdots & 0 & I \end{bmatrix}, \Phi(X) \triangleq \begin{bmatrix} I & 0 & \cdots & \cdots & 0 \\ 0 & X(0) & 0 & & \vdots \\ \vdots & \ddots & \ddots & \ddots & \vdots \\ \vdots & & \ddots & \ddots & 0 \\ 0 & \cdots & \cdots & 0 & X(0) \end{bmatrix}, \tag{3.4.23}$$

we find that (3.4.22) means that X^* satisfies the nonlinear algebraic equation $A(X)\Phi(X)X = S$, or

$$\Phi(X)X = U(X)S \ \text{ with } \ U(X) \triangleq A(X)^{-1}.$$

It is straightforward to verify that $U(X)$, the inverse of $A(X)$, has the following form

$$U(X) = \begin{bmatrix} I & U_1(X) & \cdots & \cdots & U_r(X) \\ 0 & I & U_1(X) & \cdots & U_{r-1}(X) \\ \vdots & \ddots & \ddots & \ddots & \vdots \\ \vdots & & \ddots & I & U_1(X) \\ 0 & \cdots & \cdots & 0 & I \end{bmatrix}, \tag{3.4.24}$$

where the $m \times m$-sub-matrices $U_j(X),\ j = 0, \cdots, r$ are computed as follows:

$$U_0(X) \triangleq I,\ U_1(X) \triangleq -X(1),\ U_l(X) \triangleq -X(1)U_{l-1}(X) \\ -X(2)U_{l-2}(X) - \cdots - X(l),\ \ l = 2, \ldots, r. \tag{3.4.25}$$

Notice that the first $m \times m$-matrix block in $\Phi(X^*)X^*$ is R_w, which is symmetric and positive definite. Therefore, X^* should satisfy the following equation

$$\Phi(X)X = \overline{U(X)S}, \tag{3.4.26}$$

where

$$\overline{U(X)S} \triangleq \begin{bmatrix} [(\sum_{i=0}^{r} U_i(X)S^T(i))(\sum_{i=0}^{r} U_i(X)S^T(i))^T]^{\frac{1}{2}} \\ \sum_{i=1}^{r} U_{i-1}(X)S^T(i) \\ \vdots \\ S^T(r) \end{bmatrix},$$

which coincides with $U(X)S$ except the first $m \times m$-matrix block, which has been symmetrized.

Thus, to recursively estimate X^* it is equivalent to give a recursive solution to the equation (3.4.26), or to recursively search the root X^* of the matrix function $\Phi(X)X - \overline{U(X)S}$.

Denote by G the root set of this matrix function, i.e.,

$$G \triangleq \{X:\ \ \Phi(X)X - \overline{U(X)S} = 0\}.$$

It is noticed that each $X = [X(0), X(1), \cdots, X(r)]^T \in G$ corresponds to a factorization for the spectral density function $\Psi(z) \triangleq \sum_{j=-r}^{r} S(j)z^j$ of $\{C(z)w_k\}$, i.e.,

$$\Psi(z) = C(z)R_w C^T(z^{-1}) = Y(z)X(0)Y^T(z^{-1}),$$

where $Y(z) \triangleq I + X(1)z + \cdots + X(r)z^r$. Conversely, each factorization of $\Psi(z)$ corresponds to an element in G.

Remark 3.4.2 *It is clear that $X(0) \geq 0$ for any $X \in G$, but its non-singularity may not be guaranteed. As a matter of fact, $Y(z)X(0)Y^T(z) = \overline{Y}(z)X(0)\overline{Y}^T(z)$, where*

$$\overline{Y}(z) \triangleq I + \sum_{i=1}^{r} (X(i)X(0)X^+(0) + X'(i)(I - X(0)X^+(0)))z^i$$

with arbitrary $m \times m$-matrices $X'(i)$, where $X^+(0)$ denotes the pseudo-inverse of $X(0)$. In other words, the spectral function $\Psi(z)$ may have infinitely many factorizations whenever $X(0)$ is degenerate. To uniquely define a representative of X in G, instead of $X(i)$ it is natural to take $X(i)X(0)X^+(0)$, or, equivalently, to take $\Phi^+(X)\Phi(X)X$ to replace X, whenever $X(0)$ is singular.

Let us denote by X_k and S_k the estimates at time k for X and S, respectively. Then, at time k, $\Phi(X_k)X_k - \overline{U(X_k)S_k}$ is available, and it may be viewed as an observation of $\Phi(X)X - \overline{U(X)S}$ at X_k with observation noise $\Phi(X_k)X_k - \overline{U(X_k)S_k} - \Phi(X_k)X_k - \overline{U(X_k)S}$.

Noticing

$$\zeta_k = C(z)w_k = y_k + \theta_A^T \varphi_{k-1}(p), \tag{3.4.27}$$

we can estimate ζ_k by $\hat{\zeta}_k$ defined as follows

$$\hat{\zeta}_k \triangleq y_k + \theta_{A,k}^T \varphi_{k-1}(p),$$

where $\theta_{A,k}$ is the estimate for θ_A at time k given by (3.4.6)–(3.4.9), and recursively define

$$\begin{aligned} &S_{k+1}(i) = S_k(i) - \frac{1}{k+1}\left(S_k(i) - \hat{\zeta}_{k+1}\hat{\zeta}_{k+1-i}^T\right), \\ &S_0(i) = 0, \quad i = 0, 1, \cdots, r. \end{aligned} \tag{3.4.28}$$

Set $S_k \triangleq [S_k(0)\vdots\cdots\vdots S_k(r)]^T$.

Let $X_0^T \triangleq [\nu I, 0, \cdots, 0]$ with $\nu \geq 1$. Take a sequence of positive real numbers $\{M_k\}$ increasingly diverging to infinity, and fix a number $\delta \in (0,1]$.

The estimate X_k for X^* is recursively defined as follows:

$$X_{k+1} = \begin{cases} X_k - \frac{\delta}{k+1}(\Phi(X_k)X_k - \overline{U(X_k)S_k}), \\ \text{if } \|X_k - \frac{\delta}{k+1}(\Phi(X_k)X_k - \overline{U(X_k)S_k})\| \leq M_{\sigma_k}, \\ X_0 \qquad \text{otherwise}, \end{cases} \tag{3.4.29}$$

$$\sigma_k = \sum_{j=0}^{k-1} I_{[\|X_j - \frac{\delta}{j+1}(\Phi(X_j)X_j - \overline{U(X_j)S_j})\| > M_{\sigma_j}]}, \quad \sigma_0 = 0. \tag{3.4.30}$$

The recursive algorithms (3.4.6)–(3.4.9) and (3.4.28)–(3.4.30) form a complete system of estimates for an ARMA process.

Write X_k in the block matrix form: $X_k = [X_k(0)\ X_k(1)\ \cdots\ X_k(r)]^T$, where $X_k(0)$ is the kth estimate for R_w and $X_k(i)$, $i = 1, \cdots, r$ are the estimates for C_i, $i = 1, \cdots, r$.

Theorem 3.4.2 *Assume A3.4.1 and A3.4.3 hold. Then both* $\{\Phi(X_k)\}$ *and* $\{\Phi(X_k)X_k\}$ *with* $\{X_k\}$ *given by (3.4.28)–(3.4.30) converge. Further,* $\{X_k\}$ *converges to a solution to (3.4.26) almost surely, i.e.,* $X_k \xrightarrow[k\to\infty]{} \overline{X} = [\overline{X}(0)\ \overline{X}(1)\ \cdots\ \overline{X}(r)]^T \in G$, *provided* $\limsup_{k\to\infty} \det X_k(0) > 0$. *Moreover, if, in addition,* $\det C(z)$ *is stable, then* $\overline{X}(0) = R_w$, $\overline{X}(i) = C_i$, $i = 1, \cdots, r$, *whenever* $\det Y(z) \neq 0\ \forall z : |z| \leq 1$, *where* $Y(z) \triangleq I + \overline{X}(1)z + \cdots + \overline{X}(r)z^r$.

Proof. The proof is carried out by six steps.

Step 1. We show that S_k given by (3.4.28) converges to S defined by (3.4.20)

$$S_k \xrightarrow[k\to\infty]{} S \ \ a.s. \tag{3.4.31}$$

It is clear that by ergodicity of $\{\zeta_k\}$

$$\frac{1}{k}\sum_{j=1}^{k}\zeta_j\zeta_{j-i}^T \xrightarrow[k\to\infty]{} S(i),$$

while $S_k(i)$ given by (3.4.28) can be expressed as

$$S_k(i) = \frac{1}{k}\sum_{j=1}^{k}\hat{\zeta}_j\hat{\zeta}_{j-i}^T.$$

Therefore, for (3.4.31) it suffices to show

$$\frac{1}{k}\sum_{j=1}^{k}\zeta_j\zeta_{j-i}^T - \frac{1}{k}\sum_{j=1}^{k}\hat{\zeta}_j\hat{\zeta}_{j-i}^T \xrightarrow[k\to\infty]{} 0. \tag{3.4.32}$$

As a matter of fact, we have

$$\begin{aligned}
&\|\frac{1}{k}\sum_{j=1}^{k}\zeta_j\zeta_{j-i}^T - \frac{1}{k}\sum_{j=1}^{k}\hat{\zeta}_j\hat{\zeta}_{j-i}^T\| \\
&= \|\frac{1}{k}\sum_{j=1}^{k}(\zeta_j - \hat{\zeta}_j)\zeta_{j-i}^T + \frac{1}{k}\sum_{j=1}^{k}\hat{\zeta}_j(\zeta_{j-i} - \hat{\zeta}_{j-i})^T\| \\
&= \|\frac{1}{k}\sum_{j=1}^{k}(\theta_A - \theta_{A,j})^T\varphi_{j-1}(p)\zeta_{j-i}^T + \frac{1}{k}\sum_{j=1}^{k}\hat{\zeta}_j\varphi_{j-i-1}^T(p)(\theta_A - \theta_{A,j-i})\| \\
&\le (\frac{1}{k}\sum_{j=1}^{k}\|\theta_A - \theta_{A,j}\|^2\|\varphi_{j-1}(p)\|^2)^{\frac{1}{2}}(\frac{1}{k}\sum_{j=1}^{k}\|\zeta_{j-i}\|^2)^{\frac{1}{2}} \\
&+ (\frac{1}{k}\sum_{j=1}^{k}\|\hat{\zeta}_j\|^2)^{\frac{1}{2}}(\frac{1}{k}\sum_{j=1}^{k}\|\varphi_{j-i-1}(p)\|^2\|\theta_A - \theta_{A,j-i}\|^2)^{\frac{1}{2}} \xrightarrow[k\to\infty]{} 0,
\end{aligned}$$

which is because $\theta_{A,k} \xrightarrow[k\to\infty]{} \theta_A$ by Theorem 3.4.1 and for any fixed $i \ge 0$

$$\frac{1}{k}\sum_{j=1}^{k}(\|\varphi_{j-1}(p)\|^2 + \|\zeta_{j-i}\|^2 + \|\hat{\zeta}_j\|^2 + \|\varphi_{j-i-1}(p)\|^2) < \infty.$$

Set $\Omega_1 \triangleq \{\omega : W_k \xrightarrow[k\to\infty]{} W,\ \Gamma_k \xrightarrow[k\to\infty]{} \Gamma,\ S_k \xrightarrow[k\to\infty]{} S\}$. It is clear that $\mathbb{P}\Omega_1 = 1$.

Step 2. We now show that for X_k defined by (3.4.29)–(3.4.30) the properties similar to (3.4.15)–(3.4.16) also take place: For any $\omega \in \Omega_1$, if $\{X_{m_k}\}$ is a convergent subsequence of $\{X_k\}$: $X_{m_k} \xrightarrow[k\to\infty]{} \widetilde{X}$, then for all large enough k and sufficiently small $T > 0$

$$X_{j+1} = X_j - \frac{\delta}{j+1}(\Phi(X_j)X_j - \overline{U(X_j)S_j}) \tag{3.4.33}$$

and

$$\|X_{j+1} - X_{m_k}\| \le cT \ \ \forall j : j = m_k, m_k+1, \dots, m(m_k, T), \tag{3.4.34}$$

where the constant c is independent of k but may depend on ω (sample).

The idea of the proof is similar to that given for establishing (3.4.15)–(3.4.16).

If $\{\sigma_k\}$ is bounded, then the truncation in (3.4.29) ceases in a finite number of steps and (3.4.33) is verified.

We now assume $\sigma_k \xrightarrow[k\to\infty]{} \infty$.

We note that for any fixed $\omega \in \Omega_1$, $\|S_k\| < c_1$ for some $c_1 > 0 \ \forall k = 1, 2, \dots$. Then from (3.4.25) it is seen that there are constants $a_1 > 0$, $a_2 > 0$, and $a_3 > 0$ such that

$$\delta \|\overline{U(X)S_j}\| \le a_1 + a_2\|X\|^r \quad \text{and} \quad \delta\|\Phi(X)X\| \le a_3\|X\|^2.$$

Temporarily consider (3.4.29) without truncations for $j \ge m_k$. Thus, we have

$$\|X_{j+1}\| \le \|X_j\| + \frac{1}{j+1}(a_3\|X_j\|^2 + a_1 + a_2\|X_j\|^r). \tag{3.4.35}$$

Let us consider the following recursion

$$\begin{aligned} \xi_{j+1} &= \xi_j + \frac{1}{j+1}(a_3\xi_j^2 + a_1 + a_2\xi_j^r), \\ \xi_{m_k} &= \|X_{m_k}\|, \quad j \ge m_k. \end{aligned} \tag{3.4.36}$$

Then $\|X_j\| \le \xi_j \ \forall j \ge m_k$.

From (3.4.36) we have

$$\int_{\xi_j}^{\xi_{j+1}} \frac{dx}{a_3x^2 + a_1 + a_2x^r} \le \frac{1}{j+1},$$

and

$$\int_{\xi_{m_k}}^{\xi_{m(m_k,T)+1}} \frac{dx}{a_3x^2 + a_1 + a_2x^r} \le T \quad \forall k. \tag{3.4.37}$$

Since $\xi_{m_k} \xrightarrow[k\to\infty]{} \|\widetilde{X}\| \triangleq \mu_1$, we have

$$\begin{aligned} &\int_{\xi_{m_k}}^{\infty} \frac{dx}{a_3x^2 + a_1 + a_2x^r} \xrightarrow[k\to\infty]{} \int_{\mu_1}^{\infty} \frac{dx}{a_3x^2 + a_1 + a_2x^r} \\ &\triangleq \mu_2 > 0, \quad \mu_2 < \infty. \end{aligned} \tag{3.4.38}$$

Let $T > 0$ be small enough such that $T < \frac{\mu_2}{2}$. Then from (3.4.37) and (3.4.38) we conclude that $\xi_{m(m_k,T)}$ is bounded with respect to k. Therefore, for a fixed ω, $\|X_j\|$ given by (3.4.35) is bounded by some constant α_1 for $j = m_k, \dots, m(m_k, T) + 1$ and

$\forall k \geq 1$. Since $\sigma_k \xrightarrow[k\to\infty]{} \infty$, we have $M_{\sigma_k} \to \infty$ and $M_{\sigma_k} > \alpha_1$ for sufficiently large k. This verifies (3.4.33) for the case $\sigma_k \xrightarrow[k\to\infty]{} \infty$.

By boundedness of $\|X_j\|$, both $U(X_j)$ and $\Phi(X_j)$ are bounded $\forall j = m_k, \cdots, m(m_k, T)$. Then, (3.4.34) follows from (3.4.33).

Step 3. We now show that for any $\omega \in \Omega_1$, $\{X_k\}$ is bounded, i.e., the truncation in (3.4.29) ceases in a finite number of steps. This means that there is a k_0, possibly depending on ω, so that starting from k_0, $\{X_k\}$ is defined as follows

$$\begin{aligned} X_{k+1} &= X_k - \frac{\delta}{k+1}(\Phi(X_k)X_k - \overline{U(X_k)S_k}) \ \ \forall k \geq k_0, \\ X_{k_0} &= [\nu I, 0, \cdots, 0]^T. \end{aligned} \tag{3.4.39}$$

Assume the converse: $\{X_k\}$ is unbounded, or $\sigma_k \xrightarrow[k\to\infty]{} \infty$.

Define

$$V(X) \triangleq \mathrm{tr}(\Phi(X)X - \overline{U(X)S})^T(\Phi(X)X - \overline{U(X)S}). \tag{3.4.40}$$

Noticing

$$U(X_0) = I, \quad \Phi(X_0) = \begin{bmatrix} I & 0 & \cdots & & 0 \\ 0 & \nu I & 0 & & \vdots \\ \vdots & \ddots & \ddots & \ddots & \\ & & & & 0 \\ 0 & \cdots & & 0 & \nu I \end{bmatrix},$$

we see that $V(X_0) = \mathrm{tr}(X_0 - S)^T(X_0 - S) \triangleq \alpha \geq \sum_{j=1}^{r} \mathrm{tr} S^T(j)S(j) > 0$. This is because at least one of C_j, $j = 1, \cdots, r$ differs from zero as mentioned at the beginning of the subsection "recursive estimation of MA part," and hence at least one of $S(j)$, $j = 1, \cdots, r$ is nonzero.

Since $\{X_k\}$ infinitely many times returns back to X_0, there are infinitely many X_k for which $\frac{\alpha}{2} < V(X_k) \leq 2\alpha$ and with $X_k(0) > \frac{1}{2}I$. Denote them by $\{X_{m_k}\}$. It is clear that $\{X_{m_k}\}$ is bounded. Thus, without loss of generality we may assume $\{X_{m_k}\}$ with $X_{m_k}(0) > \frac{1}{2}I$ is convergent $X_{m_k} \xrightarrow[k\to\infty]{} \widetilde{X}$. Since $U(X)$ and $\Phi(X)$ are continuous,

$$\begin{aligned} V(X_{m_k}) &\to \mathrm{tr}(\Phi(\widetilde{X})\widetilde{X} - \overline{U(\widetilde{X})S})^T(\Phi(\widetilde{X})\widetilde{X} - \overline{U(\widetilde{X})S}) \\ &\triangleq \delta_1 \geq \frac{\alpha}{2} > 0 \ \text{and} \ \widetilde{X}(0) \geq \frac{1}{2}I. \end{aligned}$$

Define the quadratic function as follows:

$$V_1(X) = \mathrm{tr}(\Phi(\widetilde{X})X - \overline{U(\widetilde{X})S})^T(\Phi(\widetilde{X})X - \overline{U(\widetilde{X})S}), \tag{3.4.41}$$

and fix $\delta_2 > \delta_1$.

Since $V_1(X) \xrightarrow[k\to\infty]{} \infty$ as $\|X\| \xrightarrow[k\to\infty]{} \infty$, by the converse assumption $\sigma_k \xrightarrow[k\to\infty]{} \infty$ it is possible to take two infinite sequences $\{m_k\}$ and $\{n_k\}$ such that

$$\begin{aligned} &V_1(X_{m_k}) \le \delta_1, \quad \text{and} \quad V_1(X_{m_k+1}) > \delta_1 \\ &\delta_1 < V_1(X_i) < \delta_2 \quad \forall i : m_k < i < n_k, \quad V_1(X_{n_k}) \ge \delta_2. \end{aligned} \tag{3.4.42}$$

Then by (3.4.33), (3.4.34), and the Taylor expansion there is X' in between X_{m_k} and $X_{m(m_k,T)+1}$ such that

$$\begin{aligned} V_1(X_{m(m_k,T)+1}) - V_1(X_{m_k}) = -\text{tr} \sum_{i=m_k}^{m(m_k,T)} \frac{2\delta}{i+1} [\Phi(X_i)X_i \\ - \overline{U(X_i)S_i}]^T \Phi(\widetilde{X})[\Phi(\widetilde{X})X' - \overline{U(\widetilde{X})S}], \end{aligned}$$

and hence by noticing $\Phi(\widetilde{X}) > \frac{1}{2}I$ for large k we have

$$\begin{aligned} V_1(X_{m(m_k,T)+1}) - V_1(X_{m_k}) \le -\text{tr} \sum_{i=m_k}^{m(m_k,T)} \frac{\delta}{i+1} [\Phi(\widetilde{X})\widetilde{X} \\ - \overline{U(\widetilde{X})S}]^T [\Phi(\widetilde{X})\widetilde{X} - \overline{U(\widetilde{X})S}] + o(T), \end{aligned} \tag{3.4.43}$$

where $o(T)$ is such that $\frac{0(T)}{T} \xrightarrow[T\to 0]{} 0$.

Tending k to infinity in (3.4.43) leads to

$$\limsup_{k\to\infty} V_1(X_{m(m_k,T)+1}) \le \delta_1 - \delta_1 T + o(T) < \delta_1, \tag{3.4.44}$$

if T is small enough.

On the other hand, by (3.4.34)

$$\lim_{T\to 0} \max_{m_k \le m \le m(m_k,T)} |V_1(X_{m+1}) - V_1(X_{m_k})| = 0.$$

This means that $m(m_k,T)+1 < n_k$ if T is small enough, and hence $V_1(X_{m(m_k,T)+1}) \in [\delta_1, \delta_2)$, which contradicts (3.4.44). The obtained contradiction implies that $\{\sigma_k\}$ and hence $\{X_k\}$ are bounded, and the truncation in (3.4.29) ceases in a finite number of steps. So, (3.4.39) has been proved.

It is noticed that from (3.4.39) it follows that

$$X_{k+1} = \frac{k_0}{k+1} X_{k_0} + \frac{1}{k+1} \sum_{j=k_0}^{k} (X_j - \delta\Phi(X_j)X_j + \delta\overline{U(X_j)S_j}), \tag{3.4.45}$$

and, hence,

$$\begin{aligned} X_{k+1}(0) = \frac{k_0 \nu}{k+1} I + (1-\delta)\frac{1}{k+1} \sum_{j=k_0}^{k} X_j(0) \\ + \frac{\delta}{k+1} \sum_{j=k_0}^{k} [(\sum_{i=0}^{r} U_i(X_k)S_k^T(i))(\sum_{i=0}^{r} U_i(X_k)S_k^T(i))^T]^{\frac{1}{2}} > 0. \end{aligned}$$

From here it is seen that $\det X_k(0) > 0\ \forall k \geq 0$, but in what follows we need a stronger condition: $\limsup_{k\to\infty} \det X_k(0) > 0$ as assumed in the formulation of the theorem.

Step 4. We consider the set L of limiting points of $\{X_k\}$, which, clearly, is a bounded set by boundedness of $\{X_k\}$. From (3.4.39) it is seen that $\|X_{k+1} - X_k\| \xrightarrow[k\to\infty]{} 0$, and hence L is a connected set.

If L consists of a singleton $\overline{X}$, then $\{X_k\}$ converges to $\overline{X}$. Tending k to infinity in (3.4.45) leads to $\Phi(\overline{X})\overline{X} = U(\overline{X})S$, which means that $\overline{X} \in G$.

Consequently, we need only to consider the case where L is not a singleton but contains internal points.

We parameterize $X \in L$. Let $X(t)$ be an internal point (an $m(r+1)\times m$-matrix), and let $\{X_{m_k}\}$ be a convergent subsequence such that $X_{m_k} \xrightarrow[k\to\infty]{} X(t)$. Without loss of generality, we may assume that $\{X_{m(m_k,T)}\}$ also converges. We parameterize its limit as $X(t+T) : X_{m(m_k,T)} \xrightarrow[k\to\infty]{} X(t+T)$. By (3.4.39) for large enough k we have

$$\begin{aligned} X_{m(m_k,T)+1} - X_{m_k} &= -\delta \sum_{j=m_k}^{m(m_k,T)} \frac{1}{j+1}\left(\Phi(X_j)X_j - \overline{U(X_j)S_j}\right) \\ &= -\delta T\left(\Phi(X(t))X(t) - \overline{U(X(t))S}\right) + o(T) + o(1), \end{aligned}$$

where the last equality is derived by (3.4.31) and continuity of $\Phi(\cdot)$ and $U(\cdot)$, and $o(T)$ and $o(1)$ are such that $\frac{o(T)}{T} \xrightarrow[T\to 0]{} 0$ and $o(1) \xrightarrow[k\to\infty]{} 0$.

From here it follows that

$$\dot{X}(t) = -\delta\left(\Phi(X(t))X(t) - \overline{U(X(t))S}\right). \tag{3.4.46}$$

Thus, $\{X(t),\ t_0 \leq t \leq t_f\}$, being the parameterized L, satisfies the differential equation (3.4.46).

Step 5. We now show that $\{X_k(0)\}$ converges: $X_k(0) \xrightarrow[k\to\infty]{} R$, where R is a symmetric nonnegative definite matrix.

Consider the quadratic function

$$V^{\phi}(X) \triangleq V^{\phi}(X) \triangleq \operatorname{tr}(\Phi(X)X - \overline{U(X)S})^T \Phi(X)(\Phi(X)X - \overline{U(X)S}). \tag{3.4.47}$$

We now show that $V^{\phi}(X) \equiv 0\ \ \forall\, X \in L$.

Assume the converse: $\sup_{X\in L} V^{\phi}(X) = v > 0$.

Let us fix an internal point $X(t_1)$ $t_0 < t_1 < t_f$ in L such that $V^{\phi}(X(t_1)) \triangleq v_1 > 0$.

Consider the quadratic functions $V^1(X)$ and $V^{1\phi}(X)$, where $X \in \mathbb{R}^{m(r+1)\times m}$:

$$V^1(X) \triangleq \operatorname{tr}(\Phi(X(t_1))X - \overline{U(X(t_1))S})^T(\Phi(X(t_1))X - \overline{U(X(t_1))S}), \tag{3.4.48}$$

$$\begin{aligned} V^{1\phi}(X) \triangleq\ & \operatorname{tr}(\Phi(X(t_1))X - \overline{U(X(t_1))S})^T \\ & \cdot \Phi(X(t_1))(\Phi(X(t_1))X - \overline{U(X(t_1))S}). \end{aligned} \tag{3.4.49}$$

It is clear that $V^{1\phi}(X(t_1)) = V^{\phi}(X(t_1)) = v_1$.

Consider the curve $V^1(X(t))$, $t_0 \le t \le t_f$.

Noticing (3.4.46), we have

$$\frac{dV^1(X(t))}{dt} = 2\mathrm{tr}\Big[\Big(\frac{\mathrm{d}X(t)^T}{\mathrm{d}t}\Big)\Phi(X(t_1))(\Phi(X(t_1))X(t) - \overline{U(X(t_1))S}\Big]$$
$$= -2\delta\mathrm{tr}\Big[\Phi(X(t))X(t) - \overline{U(X(t))S}\Big)^T \Phi(X(t_1))(\Phi(X(t_1))X(t) - \overline{U(X(t_1))S}\Big].$$

Since $V^\phi(X(t_1)) = v_1 > 0$, we have $\frac{dV^1(X(t))}{dt} \neq 0$ at $t = t_1$. Therefore, $V^1(X(t))$ cannot reach its maximum at $t = t_1$. Consequently, there exists an internal point $X' \in L$ such that

$$V^1(X(t_1)) \triangleq \alpha < \beta = V^1(X').$$

Since $0 \le \Phi(X(t_1)) < cI$ for some constant $c > 0$, we have

$$\alpha = V^1(X(t_1)) \ge \frac{1}{c}V^{1\phi}(X(t_1)) \ge \frac{v_1}{c} > 0.$$

Then, there exist two infinite subsequences $\{m_k\}$ and $\{n_k\}$ such that

$$\begin{aligned} &V^1(X_{m_k}) \le \alpha, \text{ and } V^1(X_{m_k+1}) > \alpha \\ &\alpha < V^1(X_j) < \beta, \quad \forall j: m_k < j < n_k, \quad V^1(X_{n_k}) \ge \beta. \end{aligned} \tag{3.4.50}$$

By (3.4.39), $V^1(X_{m_k}) \xrightarrow[k\to\infty]{} \alpha$. Without loss of generality, we may assume X_{m_k} converges to $X(t_1)$: $X_{m_k} \xrightarrow[k\to\infty]{} X(t_1)$.

Then, we have

$$\begin{aligned} &V^1(X_{m(m_k,T)+1}) - V^1(X_{m_k}) \\ &= -\mathrm{tr} \sum_{j=m_k}^{m(m_k,T)} \frac{2\delta}{j+1}[\Phi(X_j)X_j - \overline{U(X_j)S_j}]^T \Phi(X(t_1)) \\ &\qquad [\Phi(X(t_1))X_{m_k} - \overline{U(X(t_1))S}] + o(T) \\ &\le -2\delta v_1 T + o(T) + o(1). \end{aligned} \tag{3.4.51}$$

From here it follows that

$$\limsup_{k\to\infty} V^1(X_{m(m_k,T)+1}) \le \alpha - 2\delta v_1 T + o(T) < \alpha, \tag{3.4.52}$$

if T is small enough.

On the other hand, by (3.4.34)

$$\lim_{T\to 0} \max_{m_k \le m \le m(m_k,T)} |V^1(X_{m+1}) - V^1(X_{m_k})| = 0. \tag{3.4.53}$$

This means that $m(m_k,T)+1 < n_k$ if T is small enough, and, hence, $V^1(X_{m(m_k,T)+1}) \in$

$[\alpha,\beta)$, which contradicts (3.4.52). The obtained contradiction proves that $V^{\phi}(X) \equiv 0 \ \forall X \in L$.

Noticing that $\Phi(X)X - \overline{U(X)S}$ is an $(r+1)m \times m$-matrix and the first $m \times m$ block matrix appearing in the left-upper corner of $\Phi(X)$ is an identity matrix, from $V^{\phi}(X) \equiv 0 \ \forall X \in L$ we find that the first $m \times m$-matrix in $\Phi(X)X - \overline{U(X)S}$ is identically zero for all $X \in L$. Then, from (3.4.46) it follows that its first $m \times m$-matrix differential equation is

$$\dot{X^0}(t) = 0,$$

where $X^0(t),\ t \in [t_0,\ t_f]$ are the parameterized limiting points of $\{X_k(0)\}$. This means that $X^0(t),\ t \in [t_0,\ t_f]$ actually is a constant matrix denoted by R, in other words, $X_k(0) \xrightarrow[k\to\infty]{} R$.

Step 6. We now show the convergence of $\{\Phi(X_k)X_k\}$ and $\{X_k\}$.

We have just shown that $X(0) = R \ \forall X \in L$. Therefore, for any $X \in L$

$$\Phi(X) = \begin{bmatrix} I & 0 & \cdots & & 0 \\ 0 & R & 0 & & \vdots \\ \vdots & \ddots & \ddots & \ddots & \\ & & & & 0 \\ 0 & \cdots & & 0 & R \end{bmatrix} \geq 0$$

is a constant matrix.

From (3.4.46) it follows that

$$\frac{\mathrm{d}\Phi(X)X(t)}{\mathrm{d}t} = -\delta\Phi(X)\left(\Phi(X(t))X(t) - \overline{U(X(t))S}\right),\ t \in [t_0, t_f],$$

where the right-hand side equals zero, since $V^{\phi}(X) = 0 \ \forall X \in L$. This means that $\Phi(X)X(t)$ is a constant matrix, and meanwhile proves that $\Phi(X_k)X_k$ converges as $k \to \infty$.

If, in addition, $\limsup_{k\to\infty} \det X_k(0) > 0$, then $\det X(0) > 0$ and, hence, $\Phi(X) > 0$. From $V^{\phi}(X) \equiv 0 \ \forall X \in L$ we then conclude that $\Phi(X)X - \overline{U(X)S} = 0 \ \forall X \in L$.

Consequently, from (3.4.46) it follows that

$$\frac{\mathrm{d}X(t)}{\mathrm{d}t} = 0,\ t \in [t_0, t_f].$$

This means that $X(t)$ for all $t \in [t_0, t_f]$ is a constant matrix denoted by $\overline{X}$, and $X_k \xrightarrow[k\to\infty]{} \overline{X} \in G$.

Since $\overline{X}$ corresponds to a factorization of $\Psi(z) = \sum_{j=-r}^{r} S(j)z^j$, and, hence, $\zeta_k = C(z)w_k$ can be expressed as

$$\zeta_k = w_k' + \overline{X}(1)w_{k-1}' + \cdots + \overline{X}(r)w_{k-r}'$$

with $\{w'_k\}$ being mutually uncorrelated, $Ew'_k = 0$, and $Ew'_k w'^T_k = X(0)$. Since $C(z)$ is stable, by the uniqueness stated in Theorem 1.5.4 we conclude $\overline{X}(i) = C_i$, $i = 1, \cdots, r$, and $X(0) = R_w$, whenever $\det Y(z) \neq 0\ \forall z : |z| \leq 1$. □

Recursive Estimation of X-Part

We now consider the ARMAX system (3.1.2)–(3.1.5).

Replacing A3.4.1 and A3.4.3 we use the following assumptions.

A3.4.4 *$\{w_k\}$ is a sequence of iid random vectors such that $Ew_k = 0$ and $E(w_k w_k^T) \triangleq R_w > 0$, and $\{u_k\}$ is independent of $\{w_k\}$ and is also iid with $Eu_k = 0$, and $Eu_k u_k^T = R_u > 0$.*

A3.4.5 $\det A(z) \neq 0\ \forall z : |z| \leq 1$, *and* $\det C(z) \neq 0\ \forall z : |z| \leq 1$.

A3.4.6 *$A(z)$ and $B(z)R_u B^T(z^{-1}) + C(z)R_w C^T(z^{-1})$ have no common left factor, and $[A_p \vdots B_s R_u B_0^T + C_s R_w]$ is of row-full-rank, where $s \triangleq \max(q, r)$, and $B_s \triangleq 0$ if $s > q$, and $C_s \triangleq 0$ if $s > r$.*

Lemma 3.4.1 *Assume A3.4.4–A3.4.6 hold. Then $\{y_k\}$ given by (3.1.2)–(3.1.5) can be presented as an ARMA process with the same AR-part as that in (3.1.2):*

$$A(z)y_k = H(z)\xi_k, \quad H(z) = I + H_1 z + \cdots + H_s z^s, \tag{3.4.54}$$

where $A(z)$ and $H(z)$ have no common left factor and $[A_p\ H_s]$ is of row-full-rank.

Proof. Set

$$\chi_k \triangleq y_k + \theta_A^T \varphi_{k-1}(p) = B(z)u_{k-1} + C(z)w_k, \tag{3.4.55}$$
$$\eta_{k-1} \triangleq [u_{k-1}^T, \cdots, u_{k-q}^T]^T.$$

The process $\{\chi_k\}$ is stationary and ergodic with spectral density $f(e^{-i\lambda})$, where

$$f(z) = \frac{1}{2\pi}[B(z) \vdots C(z)] \begin{bmatrix} R_u & 0 \\ 0 & R_w \end{bmatrix} [B(z^{-1}) \vdots C(z^{-1})]^T,$$

which is rational and analytic on $|z| = 1$ and is of full rank almost everywhere. Then, by Theorem 1.5.4 χ_k can uniquely be represented as

$$\chi_k = H(z)\xi_k, \tag{3.4.56}$$

where ξ_k is m-dimensional, $E\xi_k = 0$, $E\xi_k\xi_j^T = R_\xi \delta_{k,j}$ with $\delta_{k,j} = 1$ if $k = j$ and $\delta_{k,j} = 0$ if $k \neq j$ and $R_\xi > 0$, and $H(z)$ is an $m \times m$-matrix of rational functions with $H(0) = I$, and both $H(z)$ and $H^{-1}(z)$ are stable.

By stability of $H(z)$, χ_k can be represented as a moving average of infinite order:

$$\chi_k = \sum_{i=0}^{\infty} H_i \xi_{k-i}, \quad H_0 = I. \tag{3.4.57}$$

On the other hand, by stability of $H^{-1}(z)$ we have

$$\xi_k = \sum_{i=0}^{\infty} F_i \chi_{k-i} = \sum_{i=0}^{\infty} F_i (B(z)u_{k-1-i} + C(z)w_{k-i}), \quad F_0 = I.$$

Therefore, from the properties of $\{u_k\}$ and $\{w_k\}$ it follows that

$$\begin{aligned} E\chi_k \xi_{k-j}^T &= E(B(z)u_{k-1} + C(z)w_k) \\ &\cdot (\sum_{i=0}^{\infty} F_i (B(z)u_{k-j-1-i} + C(z)w_{k-j-i}))^T = 0 \;\; \forall j \geq s+1. \end{aligned}$$

This means that in (3.4.57) the summation ceases at s, i.e.,

$$\chi_k = \xi_k + H_1 \xi_{k-1} + \cdots + H_s \xi_{k-s}, \tag{3.4.58}$$

where the right-hand side is uniquely determined.

In other words,

$$A(z)y_k = H(z)\xi_k, \tag{3.4.59}$$

where $H(z) = I + H_1 z + \cdots + H_s z^s$.

Considering the spectral function of χ_k we have the following equalities

$$\begin{aligned} \Psi^{\chi}(z) &= B(z)R_u B^T(z^{-1}) + C(z)R_w C^T(z^{-1}) \\ &= H(z)R_{\xi} H^T(z^{-1}). \end{aligned}$$

By A3.4.6 it follows that $A(z)$ and $H(z)$ have no common left factor.

By (3.4.58) we have $E\chi_k \xi_{k-s}^T = H_s R_{\xi}$.

On the other hand, $\xi_{k-s} = \sum_{i=0}^{\infty} F_i (B(z)u_{k-s-i-1} + C(z)w_{k-s-i})$, and by noticing $\chi_k = B(z)u_{k-1} + C(z)w_k$ we derive that $E\chi_k \xi_{k-s}^T = B_s R_u B_0^T + C_s R_w$, where B_s and C_s are defined in A3.4.6. Consequently,

$$H_s = (B_s R_u B_0^T + C_s R_w) R_{\xi}^{-1}.$$

Thus, Condition A3.4.6 implies that the matrix $[A_p \; H_s]$ is of row-full-rank. □

Thus, under the conditions A3.4.4–A3.4.6, $\{y_k\}$ is expressed as an ARMA process (3.4.59) with A3.4.1 and A3.4.3 satisfied, where $C(z)$ and $\{w_k\}$ should be replaced by $H(z)$ and ξ_k, respectively.

Consequently, the algorithm (3.4.6)–(3.4.9) can still be applied to estimate θ_A and by Theorem 3.4.1 the estimate for θ_A remains strongly consistent.

Setting

$$\hat{\chi}_k \triangleq y_k + \theta_{A,k}^T \varphi_{k-1}(p) = y_k + A_{1,k} y_{k-1} + \cdots + A_{p,k} y_{k-p}, \tag{3.4.60}$$

$$\eta_{k-1} \triangleq [u_{k-1}^T \cdots u_{k-1-q}^T]^T, \tag{3.4.61}$$

where $A_{j,k}$ $j=1,\cdots,p$ are the estimates for A_j given by $\theta_{A,k}$, we recursively estimate θ_B by the following algorithm:

$$\theta_{B,k} = \theta_{B,k-1} - \frac{1}{k}(\theta_{B,k-1} - \eta_{k-1}\hat{\chi}_k^T) \tag{3.4.62}$$

with an arbitrary initial value $\theta_{B,0} \in \mathbb{R}^{l(q+1)\times m}$. Clearly, $\theta_{B,k}$ is the time average of $\eta_{j-1}\hat{\chi}_j^T$.

Theorem 3.4.3 Assume A3.4.4–A3.4.6 hold. Then $\theta_{B,k} \xrightarrow[k\to\infty]{} \theta_B$ a.s.

Proof. Since the iid sequences $\{u_k\}$ and $\{w_k\}$ are mutually independent, we have

$$E\chi_k u_{k-i-1}^T = E(B(z)u_{k-1} + C(z)w_k)u_{k-i-1}^T = B_i,$$
$$0 \le i \le q.$$

By ergodicity of $[\chi_k^T, u_k^T]^T$ from here it follows that

$$\frac{1}{n}\sum_{k=1}^{n} \chi_k u_{k-i-1}^T \xrightarrow[n\to\infty]{} B_i. \tag{3.4.63}$$

Notice that

$$\begin{aligned}
\frac{1}{n}\sum_{k=1}^{n}\|\hat{\chi}_k - \chi_k\|^2 &= \frac{1}{n}\sum_{k=1}^{n}\|\sum_{i=1}^{p}(A_i - A_{i,k})y_{k-i}\|^2 \\
&\le \frac{p}{n}\sum_{k=1}^{n}\sum_{i=1}^{p}\|A_i - A_{i,k}\|^2\|y_{k-i}\|^2 \\
&\le \frac{p}{n}\Big(\sum_{k=1}^{K}\sum_{i=1}^{p}\|A_i - A_{i,k}\|^2\|y_{k-i}\|^2 \\
&\quad + \sum_{k=K+1}^{n}\sum_{i=1}^{p}\|A_i - A_{i,k}\|^2\|y_{k-i}\|^2\Big) \xrightarrow[n\to\infty]{} 0 \text{ a.s.},
\end{aligned} \tag{3.4.64}$$

because the first term at the right-hand side of (3.4.64) tends to zero for any fixed K, while by Theorem 3.4.1 and ergodicity of $\{y_k\}$ the second term can be made arbitrarily small if K is sufficiently large.

From (3.4.64) and ergodicity of $\{u_k\}$ by the Schwarz inequality it follows that

$$\begin{aligned}
&\|\frac{1}{n}\sum_{k=1}^{n}\chi_k u_{k-i-1}^T - \frac{1}{n}\sum_{k=1}^{n}\hat{\chi}_k u_{k-i-1}^T\| \\
&\le \Big(\frac{1}{n}\sum_{k=1}^{n}\|\hat{\chi}_k - \chi_k\|^2\Big)^{\frac{1}{2}}\Big(\frac{1}{n}\sum_{k=1}^{n}\|u_{k-i-1}^T\|^2\Big)^{\frac{1}{2}} \xrightarrow[n\to\infty]{} 0 \text{ a.s.}
\end{aligned} \tag{3.4.65}$$

Since from (3.4.62) we have

$$\theta_{B,n} = \frac{1}{n}\sum_{k=1}^{n} \hat{\chi}_k \eta_{k-1}^T,$$

the assertion of the theorem follows from (3.4.63) and (3.4.65). □

Recursive Identification of ARMAX

With the help of Lemma 3.4.1 we are able to estimate the AR-part of an ARMAX process, and Theorem 3.4.3 gives us the estimate for the X-part of an ARMAX process. For estimating its MA-part we have to re-estimate $\hat{\zeta}_k$ in (3.4.27). In contrast to $\hat{\zeta}_k = y_k + \theta_{A,k}^T \varphi_{k-1}(p)$, we now redefine it as

$$\hat{\zeta}_k \triangleq y_k + \theta_{A,k}^T \varphi_{k-1}(p) - \theta_{B,k}^T \eta_{k-1}. \tag{3.4.66}$$

By Theorems 3.4.1 and 3.4.3 it follows that under conditions A3.4.4–A3.4.6

$$S_k \xrightarrow[k\to\infty]{} S \text{ a.s.} \tag{3.4.67}$$

We summarize the complete set of algorithms for identifying coefficients of an ARMAX process and formulate it as a theorem.

Theorem 3.4.4 *Assume A3.4.4–A3.4.6 hold. Then the algorithm given by (3.4.6)–(3.4.9) for estimating θ_A and the algorithm given by (3.4.60)–(3.4.62) for estimating θ_B converge to the true values, and $\hat{X}_k \triangleq \Phi^+(X_k)\Phi(X_k)X_k$ with $\{X_k\}$ defined by (3.4.28)–(3.4.30) with $\hat{\zeta}_k$ given by (3.4.66) converges to a solution to (3.4.26) almost surely, i.e., $\hat{X}_k \xrightarrow[k\to\infty]{} \overline{X}$ and $\overline{X} = [\overline{X}(0)\ \overline{X}(1)\ \cdots\ \overline{X}(r)]^T \in G$, provided* $\limsup_{k\to\infty} \det X_k(0) > 0$. *Further,* $\overline{X}(0) = R_w$, $\overline{X}(i) = C_i$, $i = 1,\cdots,r$, *whenever* $\det Y(z) \neq 0\ \forall z : |z| < 1$, *where* $Y(z) \triangleq I + \overline{X}(1)z + \cdots + \overline{X}(r)z^r$.

Comparison of SAAWET-based Estimates with ELS

We now compare the estimates given in this section with ELS given and discussed in Sections 3.1 and 3.2 for an ARMAX process.

Consider the following ARMAX system with $m = 2,\ l = 1,\ p = q = r = 2$:

$$y_k + A_1 y_{k-1} + A_2 y_{k-2} = B_1 u_{k-1} + B_2 u_{k-2} + w_k + C_1 w_{k-1} + C_2 w_{k-2},$$

where

$$A_1 = \begin{bmatrix} 0 & 0.5 \\ 1 & 0 \end{bmatrix},\ A_2 = \begin{bmatrix} 1.2 & 0 \\ 0 & 0.5 \end{bmatrix},\ B_1 = \begin{bmatrix} 0 \\ 1 \end{bmatrix},\ B_2 = \begin{bmatrix} 2 \\ 3 \end{bmatrix},$$

$$C_1 = \begin{bmatrix} 1.2 & 0 \\ 0 & 0.6 \end{bmatrix},\ C_2 = \begin{bmatrix} 0.36 & 0 \\ 0 & 0 \end{bmatrix},$$

and both $\{u_k\}$ and $\{w_k\}$ are iid sequences $u_k \in \mathcal{N}(0,1)$ and $w_k \in \mathcal{N}(0,R_w)$ with $R_w = \begin{bmatrix} 2 & 0 \\ 0 & 1 \end{bmatrix}$.

The matrix polynomials associated with the system are as follows:

$$A(z) = I + A_1 z + A_2 z^2 = \begin{bmatrix} 1 & 0 \\ 0 & 1 \end{bmatrix} + \begin{bmatrix} 0 & 0.5 \\ 1 & 0 \end{bmatrix} z + \begin{bmatrix} 1.2 & 0 \\ 0 & 0.5 \end{bmatrix} z^2$$

$$= \begin{bmatrix} 1+1.2z^2 & 0.5z \\ z & 1+0.5z^2 \end{bmatrix},$$

$$B(z) = B_1 z + B_2 z^2 = \begin{bmatrix} 0 \\ 1 \end{bmatrix} z + \begin{bmatrix} 2 \\ 3 \end{bmatrix} z^2 = \begin{bmatrix} 2z^2 \\ z+3z^2 \end{bmatrix},$$

$$C(z) = I + C_1 z + C_2 z^2 = \begin{bmatrix} 1+1.2z+0.36z^2 & 0 \\ 0 & 1+0.6z \end{bmatrix}.$$

It is directly verified that the roots of $\det A(z)$ and $\det C(z)$ are outside the closed unit disk.

By noticing $\frac{1}{1+1.2z+0.36z^2} + \frac{1}{1+1.2z^{-1}+0.36z^{-2}} < 1$ for $z = 1$, the following expression shows that $C^{-1}(z) - \frac{1}{2}I$ is not SPR:

$$C^{-1}(z) + C^{-T}(z^{-1}) - I = \begin{bmatrix} a & 0 \\ 0 & b \end{bmatrix},$$

where $a = \frac{1}{1+1.2z+0.36z^2} + \frac{1}{1+1.2z^{-1}+0.36z^{-2}} - 1$ and $b = \frac{1}{1+0.6z} + \frac{1}{1+0.6z^{-1}} - 1$.

The spectral function $\Psi(z) = C(z)R_\omega C^T(z^{-1})$ with

$$C(z) = \begin{bmatrix} 1+1.2z+0.36z^2 & 0 \\ 0 & 1+0.6z \end{bmatrix} \text{ and } R_w = \begin{bmatrix} 2 & 0 \\ 0 & 1 \end{bmatrix} \tag{3.4.68}$$

can be factorized in different ways.

As a matter of fact,

$$\Psi(z) = C_i(z)R_i C_i^T(z^{-1}), \; i = 1, \cdots, 5,$$

where

$$C_1(z) = \begin{bmatrix} (1+\frac{3}{5}z)(1+\frac{5}{3}z) & 0 \\ 0 & 1+\frac{3}{5}z \end{bmatrix}, \; R_1 = \begin{bmatrix} \frac{18}{25} & 0 \\ 0 & 1 \end{bmatrix}$$

$$C_2(z) = \begin{bmatrix} (1+\frac{5}{3}z)^2 & 0 \\ 0 & 1+\frac{3}{5}z \end{bmatrix}, \; R_2 = \begin{bmatrix} \frac{162}{625} & 0 \\ 0 & 1 \end{bmatrix}$$

$$C_3(z) = \begin{bmatrix} (1+\frac{3}{5}z)^2 & 0 \\ 0 & 1+\frac{5}{3}z \end{bmatrix}, \; R_3 = \begin{bmatrix} 2 & 0 \\ 0 & \frac{9}{25} \end{bmatrix}$$

$$C_4(z) = \begin{bmatrix} (1+\frac{3}{5}z)(1+\frac{5}{3}z) & 0 \\ 0 & 1+\frac{5}{3}z \end{bmatrix}, \; R_4 = \begin{bmatrix} \frac{18}{25} & 0 \\ 0 & \frac{9}{25} \end{bmatrix}$$

$$C_5(z) = \begin{bmatrix} (1+\frac{5}{3}z)^2 & 0 \\ 0 & 1+\frac{5}{3}z \end{bmatrix}, \; R_5 = \begin{bmatrix} \frac{162}{625} & 0 \\ 0 & \frac{9}{25} \end{bmatrix}.$$

Except (3.4.68), all other factorizations of $\Psi(z)$ are unstable.

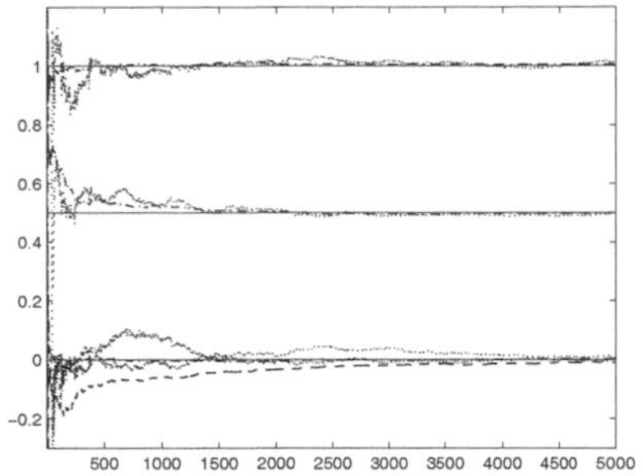

Figure 3.4.1: Estimates for A_1

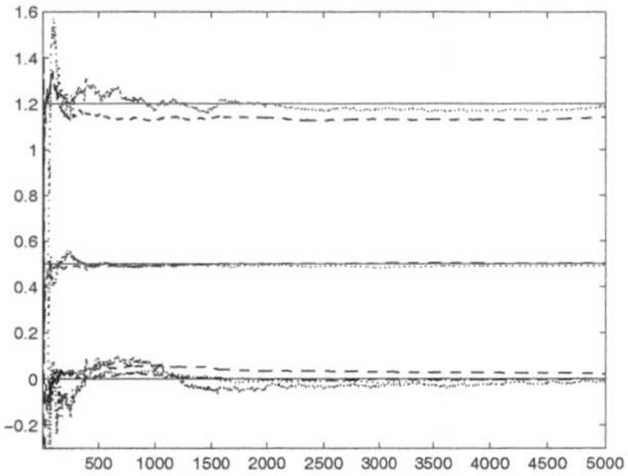

Figure 3.4.2: Estimates for A_2

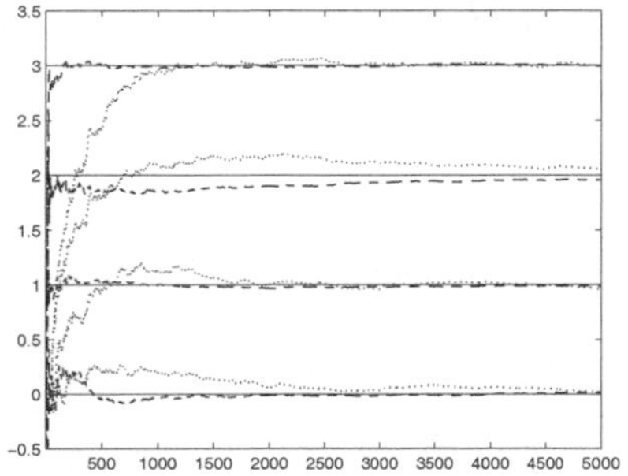

Figure 3.4.3: Estimates for B_1 and B_2

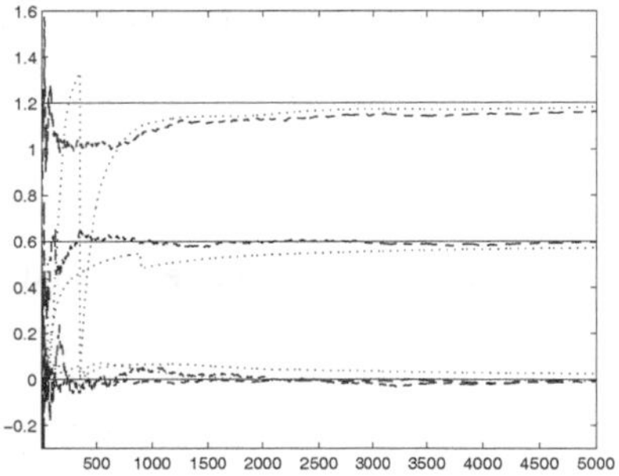

Figure 3.4.4: Estimates for C_1

The estimates given by the algorithms (3.4.6)–(3.4.9), (3.4.60), (3.4.62), and (3.4.28)–(3.4.30) are computed for 30 samples ω_i, $i = 1, \cdots, 30$. In Figures 3.4.1–3.4.6 the computer simulation results for coefficients of ARMAX and R_w are presented only for an arbitrarily chosen sample, while the simulation results for other samples are similar.

In (3.4.8)–(3.4.9) and (3.4.29)–(3.4.30) the parameters are chosen as follows: $M_k = k$, $\delta = 1$, and $\nu = 1$.

It is instructive to note that the estimates for R_w, C_1, and C_2 tend to the stable factorization of $\Psi(z)$ in all 30 simulations.

We have also estimated θ_A, θ_B, θ_C by ELS in order to compare with the estimates given by SAAWET.

In Figures 3.4.1–3.4.6 the solid lines denote the true values of the parameters, the dotted lines their estimates given by SAAWET, and the dashed lines the ELS estimates.

We see that all estimates given by SAAWET converge to the true values as time increases, while ELS gives the biased estimates for some elements. To be precise, the ELS estimates are biased for the element 0.36 in C_2 and for the element 1.2 in A_2.

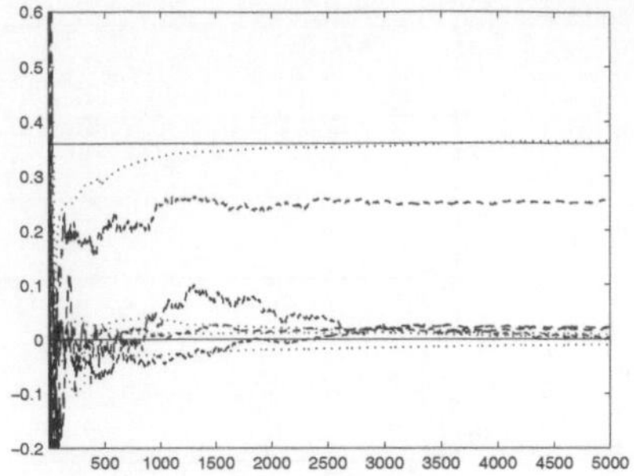

Figure 3.4.5: Estimates for C_2

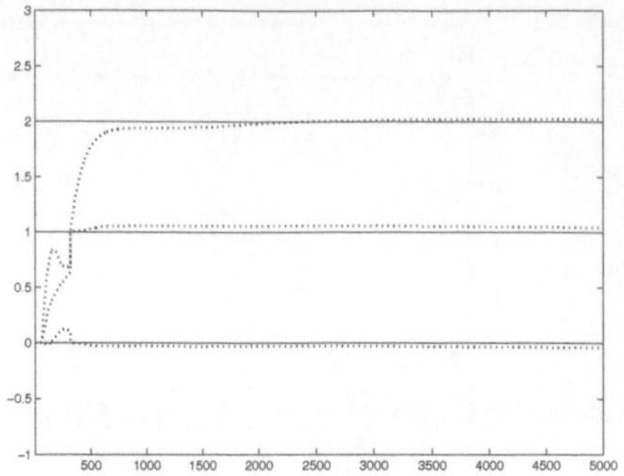

Figure 3.4.6: Estimates for R_w

3.5 Order Estimation of ARMAX

For the ARMAX system given by (3.1.2)

$$A(z)y_k = B(z)u_{k-1} + C(z)w_k,$$

we have consistently estimated the matrix coefficients in

$$\begin{aligned} A(z) &= I + A_1 z + A_2 z^2 + \cdots + A_p z^p, \\ B(z) &= B_0 + B_1 z + \cdots + B_q z^q, \\ C(z) &= I + C_1 z + \cdots + C_r z^r. \end{aligned}$$

In this section we discuss how to recursively estimate the system orders (p,q,r).

Let us start with estimating the orders (p,r) for ARMA, i.e., ARMAX with the term $B(z)u_{k-1}$ removed.

For order estimation in the existing literature the true orders (p,r) are normally assumed to belong to a known finite set M, i.e.,$(p,r) \in M$. The estimates (p_n, r_n) are then derived by minimizing a certain information criterion such as AIC, BIC, CIC, and others. For example, the estimate for orders at time n is given by

$$\begin{aligned} (p_n, r_n) &= \underset{(p',r')\in M}{\operatorname{argmin}} \left(\log \frac{\sigma_n(p',r')}{n} + \frac{a_n}{n} \right), \quad \text{or} \\ &= \underset{(p',r')\in M}{\operatorname{argmin}} \left(\sigma_n(p',r') + (p'+r')a_n \right), \end{aligned}$$

where

$$\sigma_n(p',r') = \sum_{i=0}^{n-1} \|y_{i+1} - \theta_n^T(p',r')\phi_i(p',r')\|^2$$

with $\theta_n(p',q')$ being the coefficient estimate for ARMA with orders (p',q') generated by, for example, the least squares or maximum likelihood methods, and $\{a_n\}$ is a positive, nondecreasing sequence. Under certain conditions, (p_n, r_n) converge to (p,r) almost surely as n tends to infinity. However, the estimates obtained in such a way normally are nonrecursive. Having received a new observation y_{n+1}, one has to recompute $\sigma_{n+1}(p',r')$ appearing in criteria using the entire data set $\{y_i,\ i=1,\cdots,n+1\}$ and to take the minimum with respect to p',r' running over M in order to obtain the new estimates (p_{n+1},r_{n+1}). This is quite time-consuming.

The purpose of this section is to give easily updated algorithms when new data arrive for estimating the orders of multivariate ARMA and ARMAX processes and to prove the strong consistency of the estimates.

Basic Idea of Recursive Order Estimation

Let us first list assumptions to be used in the sequel.

A3.5.1 *$A(z)$ and $C(z)$ have no common left factor, and the matrix $[A_t,C_t]$ is of row-full-rank, where $t \triangleq \max(p,r)$ and $A_t \triangleq 0$ if $t>p$ and $C_t \triangleq 0$ if $t>r$.*

A3.5.2 $\det A(z) \neq 0\ \forall\ |z| \leq 1$, *and* $\det C(z) \neq 0\ \forall\ |z|<1$.

A3.5.3 *$\{w_k\}$ is a sequence of iid random vectors such that $Ew_k=0$ with $E\|w_k\|^{2+\delta}<\infty$ for some $\delta>0$, and $Ew_kw_k^T \triangleq R_w>0$, where R_w is unknown.*

A3.5.4 *An upper bound t^* for t is available $t^*>t$.*

To avoid too heavy subscripts, in what follows instead of R_i let us write

$$R(i) \triangleq Ey_{j+i}y_j^T.$$

It is noticed that in comparison with A3.4.1 the requirement on $\{w_k\}$ is strengthened in A3.5.3. To be precise, in A3.5.3 it is additionally required that $E\|w_k\|^{2+\delta}<\infty$ for some $\delta>0$. This strengthening makes possible to derive the convergence rate of $R_k(i)$ to $R(i)$, where $R_k(i)=\frac{1}{k}\sum_{j=1}^{k}y_{j+i}y_j^T$. By the convergence rate of $R_k(i)$ we can then derive the convergence rate of the coefficient estimates.

The recursive order estimation is based on the following observation. If coefficients of an unknown polynomial can be estimated with a certain convergence rate, then the true order of the polynomial can be determined by comparing estimates for neighboring coefficients. To be precise, assume a polynomial $h(z)=z^{\mu}+h_1z^{\mu-1}+\cdots+h_{\mu}$ of order μ_0 is written as a polynomial of order $\mu \geq \mu_0$, i.e., $h_{\mu_0}\neq 0$ and $h_i=0\ \forall\ i:\mu_0<i\leq\mu$. Further, assume that the estimates $h_{j,k}$ are available for $h_j\ \forall\ j=0,1,\cdots,\mu$ $(h_{0,k}\triangleq 1)$ with convergence rate $|h_{j,k}-h_j|=O(\frac{1}{k^{\alpha}})$ with some $\alpha>0\ \forall\ j=0,1,\cdots,\mu$. Let us determine the true order μ_0 based on the coefficient estimates $h_{j,k}$.

For this we define the decision numbers

$$Q_{j,k} \triangleq \frac{|h_{j,k}|+\frac{1}{\log k}}{|h_{j+1,k}|+\frac{1}{\log k}},\ j=0,1,\cdots,\mu \text{ with } h_{\mu+1,k}\triangleq 0.$$

It is clear that $Q_{j,k}$ diverges to ∞ at those j for which $h_j\neq 0$ and $h_{j+1}=0$ and converges to finite limits for any other $j's$. Therefore, the true order can be determined as follows

$$\mu_0=\max\{j \mid Q_{j,k}\geq\varepsilon,\ j=0,\cdots,\mu\},$$

where ε is a fixed positive number greater than 1.

Noticing that by A3.5.4 the upper bound t^* for orders is available, we find that the orders of ARMA/ARMAX can similarly be determined, if the estimates for coefficients of the polynomials can be derived with the required rate of convergence.

By A3.5.1 the system under consideration is stable, and hence the influence of initial values on y_k exponentially decays as $k \to \infty$. Consequently, under A3.5.1 and A3.5.3, $\{y_k\}$ is asymptotically stationary with any given initial values. As claimed in Section 3.3, for simplicity of description, without loss of generality we may assume that the iid sequence $\{w_k\}$ is defined for $-\infty < k < \infty$, and $\{y_k\}$ is a stationary process.

We may write the ARMA system as

$$\begin{aligned} y_{k+1} + A_1 y_k + \cdots + A_{t^*} y_{k-t^*+1} &= w_{k+1} + C_1 w_k + \cdots + C_{t^*} w_{k-t^*+1} \\ A_i \triangleq 0 \ \forall\, i > p,\ C_j &\triangleq 0 \ \forall\, j > r. \end{aligned} \tag{3.5.1}$$

If $m = 1$ and the estimates for A_i, C_j, $i, j = 1, \cdots, t^*$ are available with convergence rate $O\big(\frac{1}{k^{\alpha}}\big)$, then the true orders (p, r) can be determined in a similar way. The order determination for the case $m \geq 1$ can be reduced to analyzing the one-dimensional polynomial, and the convergence rate of estimates for A_i, $i = 1, \cdots, t^*$, C_j, $j = 1, \cdots, t^*$ depends upon the convergence rate of the estimate for covariance matrix function $R(i) \triangleq E y_k y_{k-i}^T$ of $\{y_k\}$. Consequently, we may use the decision numbers as defined above to determine the orders of ARMA/ARMAX if we have estimates for $R(i)$ with the required convergence rate.

Convergence Rate of Estimate for R(i)

Let us first recursively estimate $R(i)$ by $R_k(i)$:

$$R_k(i) = \Big(1 - \frac{1}{k}\Big) R_{k-1}(i) + \frac{1}{k} y_{k+i} y_k^T \ \ \forall i \geq 0, \ \text{ or } \ R_k(i) = \frac{1}{k} \sum_{j=1}^{k} y_{j+i} y_j^T. \tag{3.5.2}$$

To establish the convergence rate of $R_k(i)$ to $R(i)$ as k tends to infinity, let us express y_k generated by $A(z) y_k = C(z) w_k$ in the following state space form different from (3.3.68):

$$\begin{cases} Z_{k+1} = F Z_k + D w_{k+1} \\ y_k = H Z_k, \end{cases} \tag{3.5.3}$$

where

$$F \triangleq \begin{bmatrix} A & B \\ 0 & 0 \end{bmatrix}$$

with

$$A \triangleq \begin{bmatrix} -A_1 & I & \cdots & 0 \\ \vdots & 0 & \ddots & \vdots \\ \vdots & \vdots & \ddots & I \\ -A_t & 0 & \cdots & 0 \end{bmatrix}, \quad B \triangleq \begin{bmatrix} 0 \\ \vdots \\ 0 \\ I \end{bmatrix}, \quad D \triangleq \begin{bmatrix} I \\ C_1 \\ C_2 \\ \vdots \\ C_t \end{bmatrix}, \ \text{ and } \ H^T \triangleq \begin{bmatrix} I \\ 0 \\ \vdots \\ 0 \\ 0 \end{bmatrix},$$

where F is a $(t+1)m \times (t+1)m$-matrix, D and H^T are $(t+1)m \times m$-dimensional matrices, and B is a $tm \times m$-matrix.

The following lemma is crucial for order estimation.

Lemma 3.5.1 *Assume conditions A3.5.2 and A3.5.3 hold. Then*

$$\|R_k(i) - R(i)\| = O(k^{-\nu}) \ \text{ a.s. } \ \forall\, \nu \in \left(0, \frac{1}{2} \wedge \frac{\delta}{2+\delta}\right), \tag{3.5.4}$$

where $\frac{1}{2} \wedge \frac{\delta}{2+\delta} \triangleq \min(\frac{1}{2}, \frac{\delta}{2+\delta})$.

Proof. It is clear that F is stable and there exist $\alpha > 0$ and $\rho \in (0,1)$ such that

$$\|F^n\| < \alpha \rho^n \ \ \forall\, n \geq 0. \tag{3.5.5}$$

Since $Z_k = F^k Z_0 + \sum_{i=1}^k F^{k-i} D w_i$, by the Hölder inequality it follows that

$$\begin{aligned}
\sum_{i=1}^n \|Z_i\|^2 &\leq \sum_{i=1}^n \Big(2\|Z_0\|^2 \alpha^2 \rho^{2i} + 2\alpha^2 \big(\sum_{j=1}^i \rho^{i-j} \|D\| \|w_j\|\big)^2\Big) \\
&\leq O(1) + \frac{2\alpha^2}{1-\rho} \sum_{j=1}^n \sum_{i=j}^n \rho^{i-j} \|D\|^2 \|w_j\|^2 \quad \text{a.s.}
\end{aligned}$$

By ergodicity, $\frac{1}{n}\sum_{j=1}^n \|w_j\|^2 = O(1)$ a.s., we then have

$$\sum_{i=1}^n \|Z_i\|^2 = O(1) + O\big(\sum_{j=1}^n \|w_j\|^2\big) = O(n) \quad \text{a.s.} \tag{3.5.6}$$

Noticing that for $j \geq 0$

$$Z_{k+j+1} = F Z_{k+j} + D w_{k+j+1} \ \text{ and } \ Z_{k+j} = F^j Z_k + \sum_{i=k}^{k+j-1} F^{k+j-1-i} D w_{i+1},$$

we have

$$\begin{aligned}
Z_{k+j+1} Z_{k+1}^T =& F Z_{k+j} Z_k^T F^T + F^{j+1} Z_k w_{k+1}^T D^T + \sum_{i=k}^{k+j-1} F^{k+j-i} D w_{i+1} w_{k+1}^T D^T \\
&+ D w_{k+j+1} Z_k^T F^T + D w_{k+j+1} w_{k+1}^T D^T,
\end{aligned}$$

and hence

$$\begin{aligned}
Z_{k+j+1} Z_{k+1}^T &= F^{k+1} Z_j Z_0^T F^{T(k+1)} \\
&+ \sum_{s=0}^k F^{k-s+j+1} Z_s w_{s+1}^T D^T F^{T(k-s)} + \sum_{s=0}^k \sum_{i=s}^{s+j-1} F^{k+j-i} D w_{i+1} w_{s+1}^T D^T F^{T(k-s)} \\
&+ \sum_{s=0}^k F^{k-s} D w_{s+j+1} Z_s^T F^{T(k+1-s)} + \sum_{s=0}^k F^{k-s} D w_{s+j+1} w_{s+1}^T D^T F^{T(k-s)}.
\end{aligned} \tag{3.5.7}$$

We now analyze each term at the right-hand side of (3.5.7).

By (3.5.5) we have

$$\frac{1}{n}\sum_{k=0}^{n} F^{k+1} Z_j Z_0^T F^{T(k+1)} = O\Big(\frac{1}{n}\Big). \tag{3.5.8}$$

Noticing that $\{w_k\}$ is iid with $Ew_k = 0$ and $Ew_k w_k^T = R_w$, by Theorem 1.2.14 we have

$$\begin{aligned}\Big\|\sum_{s=0}^{n-i} Z_s w_{s+1}^T\Big\| &= O\Big(\big(\sum_{s=0}^{n}\|Z_s\|^2\big)^{\frac{1}{2}} \log^{\frac{1}{2}+\eta}\big(\sum_{s=0}^{n}\|Z_s\|^2 + e\big)\Big)\\ &= O\Big(\big(\sum_{s=0}^{n}\|Z_s\|^2\big)^{\frac{1}{2}+\beta}\Big) \text{ a.s.,}\end{aligned}$$

where the positive numbers $\eta > 0$ and $\beta > 0$ can be chosen arbitrarily small.

By this and stability of F it follows that

$$\begin{aligned}&\Big\|\frac{1}{n}\sum_{k=0}^{n}\sum_{s=0}^{k} F^{k-s+j+1} Z_s w_{s+1}^T D^T F^{T(k-s)}\Big\|\\ &= \Big\|\frac{1}{n}\sum_{i=0}^{n} F^{i+j+1}\big(\sum_{s=0}^{n-i} Z_s w_{s+1}^T\big) D^T F^{Ti}\Big\| = O\Big(\frac{1}{n}\big(\sum_{s=0}^{n}\|Z_s\|^2\big)^{\frac{1}{2}+\beta}\Big) \quad \forall \beta > 0,\end{aligned}$$

which combining with (3.5.6) leads to

$$\Big\|\frac{1}{n}\sum_{k=0}^{n}\sum_{s=0}^{k} F^{k-s+j+1} Z_s w_{s+1}^T D^T F^{T(k-s)}\Big\| = O(n^{-(\frac{1}{2}-\beta)}) \quad \forall \beta > 0. \tag{3.5.9}$$

Since $\{w_{s+1}w_{s+1}^T - R_w\}$ is a zero-mean iid sequence with $E\|w_{s+1}w_{s+1}^T - R_w\|^{\frac{2+\delta}{2}} < \infty$ by A3.5.3 and by Theorem 1.2.4 we have

$$\frac{1}{n}\sum_{s=0}^{n}(w_{s+1}w_{s+1}^T - R_w) = o(n^{-\frac{\delta}{2+\delta}}).$$

Further, by noticing

$$\begin{aligned}&\frac{1}{n}\sum_{k=0}^{n}\sum_{s=0}^{k} F^{k+j-s} D w_{s+1} w_{s+1}^T D^T F^{T(k-s)}\\ &= \frac{1}{n}\sum_{i=0}^{n} F^{i+j} D\big(\sum_{s=0}^{n-i} w_{s+1}w_{s+1}^T\big) D^T F^{Ti}\\ &= \frac{1}{n}\sum_{i=0}^{n} F^{i+j} D \sum_{s=0}^{n-i}(w_{s+1}w_{s+1}^T - R_w) D^T F^{Ti} + \sum_{i=0}^{n} F^{i+j} D R_w D^T F^{Ti}\\ &\quad - \frac{1}{n}\sum_{i=0}^{n}(i-1) F^{i+j} D R_w D^T F^{Ti},\end{aligned}$$

we have

$$\left\|\frac{1}{n}\sum_{k=0}^{n}\sum_{s=0}^{k}F^{k+j-s}Dw_{s+1}w_{s+1}^{T}D^{T}F^{T(k-s)}-\sum_{i=0}^{\infty}F^{i+j}DR_{w}D^{T}F^{Ti}\right\|$$
$$=o(n^{-\frac{\delta}{2+\delta}})=O(n^{-\nu}). \tag{3.5.10}$$

Noticing that $\{w_{i+1}w_{s+1}\}$ is an mds for each $i,\ i=s+1,\cdots,s+j-1$, by stability of F and Theorem 1.2.14, we have

$$\left\|\frac{1}{n}\sum_{k=0}^{n}\sum_{s=0}^{k}\sum_{i=s+1}^{s+j-1}F^{k+j-i}Dw_{i+1}w_{s+1}^{T}D^{T}F^{T(k-s)}\right\|=O(n^{-(\frac{1}{2}-\beta)})\quad\forall\beta>0, \tag{3.5.11}$$

and, similarly,

$$\left\|\frac{1}{n}\sum_{k=0}^{n}\sum_{s=0}^{k}F^{k-s}Dw_{s+j+1}w_{s+1}^{T}D^{T}F^{T(k-s)}\right\|=O(n^{-(\frac{1}{2}-\beta)})\ \ \forall\beta>0\text{ and }\forall j\geq 1. \tag{3.5.12}$$

From (3.5.7)–(3.5.12) it follows that for any $\beta>0$

$$\left\|\frac{1}{n}\sum_{k=0}^{n}Z_{k+j+1}Z_{k+1}^{T}-\sum_{i=0}^{\infty}F^{i+j}DR_{w}D^{T}F^{Ti}\right\|=O(n^{-(\frac{1}{2}-\beta)})+o(n^{-\frac{\delta}{2+\delta}})=O(n^{-\nu}).$$

Since $y_k=HZ_k$, from here we have

$$R(j)=H\sum_{i=0}^{\infty}F^{i+j}DR_{w}D^{T}F^{Ti}H^{T}$$

and

$$\left\|\frac{1}{n}\sum_{k=1}^{n+1}y_{k+j}y_{k}^{T}-R(j)\right\|=O(n^{-\nu}),$$

which by (3.5.2) clearly implies (3.5.4). □

Recall the Hankel matrix $\Gamma(\mu,\nu)$ defined by (3.3.2). For simplicity of notations let us write $\Gamma(l)\triangleq\Gamma(l,l)$ and $\Gamma^{*}\triangleq\Gamma(t^{*}m)$, and denote their estimates by $\Gamma_k(l)$ and Γ_k^{*}, respectively, which are obtained by replacing $R(i)$ in $\Gamma(l)$ and Γ with its estimate $R_k(i),\ i=1,2,\cdots$.

Corollary 3.5.1 *Assume A3.5.2 and A3.5.3 hold. Then*

$$\|\Gamma_k(l)-\Gamma(l)\|=O(k^{-\nu})\ \text{ a.s. }\forall\,\nu\in\Big(0,\frac{1}{2}\wedge\frac{\delta}{2+\delta}\Big),\quad l=1,2,\cdots.$$

This is because $\Gamma(l)$ is a block matrix composed of $R(i),\ i=1,\cdots,2l-1$.

Estimation for Maximum $\mathbf{t} = \max(\mathbf{p}, \mathbf{r})$ of Orders
Define

$$\gamma_k(z) \triangleq \det(zI - \Gamma_k^T \Gamma_k) = z^{t^* m^2} + \gamma_{1,k} z^{t^* m^2 - 1} + \cdots + \gamma_{t^* m^2 - 1,k} z + \gamma_{t^* m^2,k}, \quad (3.5.13)$$

$$\gamma(z) \triangleq \det(zI - \Gamma^T \Gamma) = z^{t^* m^2} + \gamma_1 z^{t^* m^2 - 1} + \cdots + \gamma_{t^* m^2 - 1} z + \gamma_{t^* m^2}. \quad (3.5.14)$$

By Corollary 3.5.1

$$\lim_{k\to\infty} \gamma_k(z) = \gamma(z) \text{ and } |\gamma_i - \gamma_{i,k}| = O(k^{-\nu}) \ a.s. \ \forall i = 1, \cdots, t^* m^2. \quad (3.5.15)$$

Define the decision numbers

$$Q_{j,k}^{(t)} \triangleq \frac{|\gamma_{j,k}| + \frac{1}{\log k}}{|\gamma_{j+1,k}| + \frac{1}{\log k}}, \quad k \geq 1, \ j = 1, \cdots, t^* m^2,$$
$$\text{with } \gamma_{t^* m^2 + 1,k} \triangleq 0 \ \forall\, k \geq 1. \quad (3.5.16)$$

Then take a threshold $\varepsilon > 1$, say $\varepsilon = 3$, and define

$$T_k \triangleq \max\{j \mid Q_{j,k}^{(t)} \geq \varepsilon, \ j = 1, \cdots, t^* m^2\}, \quad (3.5.17)$$

if there exists some $j : 1 \leq j \leq t^* m^2$ such that $Q_{j,k}^{(t)} \geq \varepsilon$. Otherwise, define $T_k \triangleq t^* m^2$.
The estimate t_k for t is defined as

$$t_k \triangleq \left[\frac{T_k}{m}\right], \quad (3.5.18)$$

where $[a]$ denotes the integer part of a number a.
The following theorem gives the strongly consistent estimate for t.

Theorem 3.5.1 *Assume A3.5.1–A3.5.4 hold. Then*

$$\lim_{k\to\infty} t_k = t \text{ a.s.},$$

where t_k is given by (3.5.18).

Proof. By Theorem 3.3.4 the rank of $\Gamma^{*T} \Gamma^*$ is tm. This means that in (3.5.13) $\gamma_{tm} \neq 0$ and $\gamma_j = 0 \ \forall j = tm + 1, \cdots, t^* m^2$.

Noticing that $\gamma_{j,k}$ converges to γ_j faster than $\frac{1}{\log k}$ $\forall\, j = 1, \cdots, t^* m^2$ by (3.5.15), we find that

$$\lim_{k\to\infty} Q_{j,k}^{(t)} = \begin{cases} |\frac{\gamma_j}{\gamma_{j+1}}|, & \text{if } \gamma_j \neq 0, \ \gamma_{j+1} \neq 0, \\ \infty, & \text{if } \gamma_{j+1} = 0, \ \gamma_j \neq 0, \\ 0, & \text{if } \gamma_{j+1} \neq 0, \ \gamma_j = 0, \\ 1, & \text{if } \gamma_j = 0, \ \gamma_{j+1} = 0. \end{cases}$$

Therefore, $T_k \equiv tm$ for all sufficiently large k, which implies the conclusion of the theorem. □

Remark 3.5.1 *When estimating t,* $\gamma_k(z) \triangleq \det(zI - \Gamma_k^{*T}\Gamma_k^*)$ *has to be calculated for each k, but* Γ_k^* *is recursively computed. So, the quantity* $Q_{j,k}^{(t)}$ *and hence the estimate* t_k *can easily be updated as new data arrive.*

Estimation for Orders $(\mathbf{p},\mathbf{r})$

Estimating (p,r) is carried out with the help of t_k given above.

Let us define the dimension-varying matrices $G_k \in \mathbb{R}^{s_k m \times s_k m^2}$ and $W_k \in \mathbb{R}^{m \times s_k m}$:

$$G_k \triangleq \begin{bmatrix} R_k(1) & R_k(2) & \cdots & R_k(t_k m) \\ \vdots & & & \vdots \\ R_k(t_k) & R_k(t_k+1) & \cdots & R_k(t_k m + t_k - 1) \end{bmatrix},$$

$$W_k \triangleq \begin{bmatrix} R_k^T(t_k+1) \\ \vdots \\ R_k^T(t_k m + t_k) \end{bmatrix}^T.$$

Take a sequence $\{M_k\}$ of positive real numbers increasingly diverging to infinity and an arbitrary initial value $\overline{\theta}_{A,0} \in \mathbb{R}^{mt_0 \times m}$. Recursively define $\{\overline{\theta}_{A,k}\}$ by SAAWET.

If $t_{k+1} = t_k$, then define

$$\overline{\theta}_{A,k+1} = \left[\overline{\theta}_{A,k} - \frac{1}{k+1}\left(G_k G_k^T \overline{\theta}_{A,k} + G_k W_k^T\right)\right] \cdot I_{\left[\left\|\overline{\theta}_{A,k} - \frac{1}{k+1}\left(G_k G_k^T \overline{\theta}_{A,k} + G_k W_k^T\right)\right\| \le M_{\lambda_k}\right]}, \tag{3.5.19}$$

$$\lambda_k = \sum_{j=0}^{k-1} I_{\left[\left\|\overline{\theta}_{A,j} - \frac{1}{j+1}\left(G_j G_j^T \overline{\theta}_{A,j} + G_j W_j^T\right)\right\| > M_{\lambda_j}\right]}, \quad \lambda_0 = 0. \tag{3.5.20}$$

If $t_{k+1} \neq t_k$, then set $\overline{\theta}_{A,k+1} \triangleq 0 \in \mathbb{R}^{t_{k+1} m \times m}$.

As a matter of fact, $\overline{\theta}_{A,k}^T \triangleq [A_{t_k,k}, \cdots, A_{1,k}]$ is an estimate for $\overline{\theta}_A^T \triangleq [0, A_p, \cdots, A_1]$ in the case $t_k = t$, where 0 in $\overline{\theta}_A^T$ is an $m \times m(t-p)$-matrix with all elements equal to zero.

Define

$$Q_{j,k}^{(p)} \triangleq \frac{\|A_{j,k}\| + \frac{1}{\log k}}{\|A_{j+1,k}\| + \frac{1}{\log k}} \quad \text{with } A_{t_k+1,k} \triangleq 0, \;\; k \ge 1, \;\; j = 1, \cdots, t_k. \tag{3.5.21}$$

Take a threshold $\varepsilon > 1$, say $\varepsilon = 3$, and define the estimate p_k for p

$$p_k \triangleq \max\{j \mid Q_{j,k}^{(p)} \ge \varepsilon, \; j = 1, \cdots, t_k\}, \tag{3.5.22}$$

if there exists some $j : 1 \le j \le t_k$ such that $Q_{j,k}^{(p)} \ge \varepsilon$. Otherwise, define $p_k \triangleq t_k$.

Theorem 3.5.2 *Assume A3.5.1–A3.5.4 hold. Then,*

$$\lim_{k\to\infty} \overline{\theta}_{A,k} = \overline{\theta}_A \text{ and } \lim_{k\to\infty} p_k = p \text{ a.s.,}$$

where $\overline{\theta}_{A,k}$ is given by (3.5.19)–(3.5.20). Moreover,

$$\|\overline{\theta}_{A,k} - \overline{\theta}_A\| = o\Big(\frac{1}{k^\nu}\Big) \tag{3.5.23}$$

for some small enough $\nu > 0$.

Proof. Since t is an integer, by convergence of t_k to t, t_k coincides with t in a finite time. Consequently, for all sufficiently large k, the dimensions for $\overline{\theta}_{A,k}$ defined by (3.5.19)-(3.5.20) as well as for G_k and W_k will no longer change with time.

By setting

$$\varepsilon_{k+1} \triangleq (G_k G_k^T - GG^T)\overline{\theta}_{A,k} + (G_k W_k^T - GW^T) \text{ with } W \triangleq \begin{bmatrix} R^T(t+1) \\ \vdots \\ R^T(tm+t) \end{bmatrix}^T, \tag{3.5.24}$$

the algorithm defined by (3.5.19)–(3.5.20) is then rewritten as follows:

$$\begin{aligned} \overline{\theta}_{A,k+1} = & \Big[\overline{\theta}_{A,k} - \frac{1}{k+1}(GG^T\overline{\theta}_{A,k} + GW^T + \varepsilon_{k+1})\Big] \\ & \cdot I_{\big[\|\overline{\theta}_{A,k} - \frac{1}{k+1}(GG^T\overline{\theta}_{A,k} + GW^T + \varepsilon_{k+1})\| \le M_{\lambda_k}\big]}, \end{aligned} \tag{3.5.25}$$

$$\lambda_k = \sum_{j=0}^{k-1} I_{\big[\|\overline{\theta}_{A,j} - \frac{1}{j+1}(GG^T\overline{\theta}_{A,j} + GW^T + \varepsilon_{j+1})\| > M_{\lambda_j}\big]}, \quad \lambda_0 = 0. \tag{3.5.26}$$

By Lemma 3.5.1 we have $\|G_k - G\| = O\big(\frac{1}{k^\nu}\big)$, $\|W_k - W\| = O\big(\frac{1}{k^\nu}\big)$ for any $\nu \in \big(0, \frac{1}{2} \wedge \frac{\delta}{2+\delta}\big)$.

Noticing that $GG^T > 0$ and ε_k meets the conditions required by Theorem 2.5.1, we conclude $\lim_{k\to\infty} \overline{\theta}_{A,k} = \overline{\theta}_A$ a.s. by Theorem 2.5.1.

Moreover, by boundedness of $\{\overline{\theta}_{A,k}\}$, we have

$$\|\varepsilon_{k+1}\| = O\Big(\frac{1}{k^\nu}\Big) \;\; \forall\, \nu \in \Big(0, \frac{1}{2} \wedge \frac{\delta}{2+\delta}\Big).$$

So we may assume ν is small enough such that $-GG^T + \nu I < 0$. Then, taking notice of Remark 2.6.1, by Theorem 2.6.1 we conclude (3.5.23).

Notice that the polynomial

$$A_{1,k} z^{t_k - 1} + A_{2,k} z^{t_k - 2} + \cdots + A_{p,k} z^{t_k - p} + \cdots + A_{t_k,k}$$

converges to $z^{s-p}(A_1z^{p-1}+A_2z^{p-2}+\cdots+A_p+0+\cdots+0)$. Then, by (3.5.23) similar to the proof of Theorem 3.5.1 we conclude that

$$\lim_{k\to\infty} Q_{j,k}^{(p)} = \begin{cases} \frac{\|A_j\|}{\|A_{j+1}\|}, & \text{if } A_j \neq 0,\ A_{j+1} \neq 0, \\ \infty, & \text{if } A_{j+1} = 0,\ A_j \neq 0, \\ 0, & \text{if } A_{j+1} \neq 0,\ A_j = 0, \\ 1, & \text{if } A_j = 0,\ A_{j+1} = 0, \end{cases}$$

and $\lim_{k\to\infty} p_k = p$ a.s. □

Remark 3.5.2 *From the proof of Theorem 3.5.2 it is seen that ν can be any number satisfying $0<\nu<\min\left(\frac{1}{2},\frac{\delta}{2+\delta},\lambda_{\min}\{GG^T\}\right)$, where $\lambda_{\min}\{GG^T\}$ denotes the minimal eigenvalue of GG^T. It is noted that under stronger assumptions we can have the better estimates for $\|R_k(i)-R(i)\|$ and $\|\overline{\theta}_{A,k}-\overline{\theta}_A\|$. For example, if $\{w_k\}$ is iid with $E\|w_k\|^4<\infty$, then it can be shown that $\|R_k(i)-R(i)\| = O\left(\frac{(\log\log k)^{\frac{1}{2}}}{k^{\frac{1}{2}}}\right)$. Then by Theorem 2.6.1 $\|\overline{\theta}_{A,k}-\overline{\theta}_A\| = o\left(\frac{1}{k^\nu}\right)$ for any $\nu: 0<\nu<\min\left(\frac{1}{2},\lambda_{\min}\{GG^T\}\right)$.*

It is clear that Theorems 3.5.1 and 3.5.2 completely solve the order estimation problem for the ARMA process in the case where $r=t>p$. However, we have to pay attention to the possible case $p\geq r$.

We now proceed to estimate r.

Let us define

$$\zeta_k \triangleq A(z)y_k (= C(z)w_k), \tag{3.5.27}$$

$$\hat{\zeta}_k \triangleq y_k + A_{1,k}y_{k-1}+\cdots+A_{t_k,k}y_{k-t_k} \tag{3.5.28}$$

$$S_{k+1}(i) = S_k(i) - \frac{1}{k+1}\left(S_k(i)-\hat{\zeta}_{k+1}\hat{\zeta}_{k+1-i}^T\right), \quad i=0,1,\cdots,t_k. \tag{3.5.29}$$

Further, define

$$Q_{j,k}^{(r)} \triangleq \frac{\|S_k(j)\|+\frac{1}{\log k}}{\|S_k(j+1)\|+\frac{1}{\log k}} \quad \text{with } \|S_k(t_k+1)\| \triangleq 0,\ \ j=0,1,\cdots,t_k. \tag{3.5.30}$$

For a fixed $\varepsilon>1$, say, $\varepsilon=3$ the estimate for r is given by

$$r_k \triangleq \max\{j \mid Q_{j,k}^{(r)} \geq \varepsilon,\ j=0,1,\cdots,t_k\}, \tag{3.5.31}$$

if there exists some $j: 0\leq j\leq t_k$ such that $Q_{j,k}^{(r)}\geq\varepsilon$. Otherwise, set $r_k \triangleq t_k$.

Theorem 3.5.3 *Assume A3.5.1–A3.5.4 hold. Then,*

$$\lim_{k\to\infty} r_k = r \ \text{ a.s.}$$

Proof. By Theorems 3.5.1 and 3.5.2, in a finite time we have $t_k = t$ and $p_k = p$. Therefore, there exists a k_0 such that

$$S_{k+1}(i) = \frac{k_0}{k+1} S_{k_0}(i) + \frac{1}{k+1} \sum_{j=k_0}^{k} \hat{\zeta}_{j+1} \hat{\zeta}_{j+1-i}^T, \quad i = 0, 1, \cdots, t \ \forall\, k \geq k_0. \tag{3.5.32}$$

By applying Lemma 3.5.1 to the process $\zeta_k = C(z) w_k$, it follows that

$$\Big\| \frac{1}{k+1} \sum_{j=1}^{k} \zeta_{j+1} \zeta_{j+1-i}^T - S(i) \Big\| = O(k^{-\nu}) \ \forall\, \nu \in \big(0, \frac{1}{2} \wedge \frac{\delta}{2+\delta}\big) \text{ a.s.,} \tag{3.5.33}$$

$i = 0, \cdots, r$, where $S(i) \triangleq E \zeta_k \zeta_{k-i}^T$.

By (3.5.23) and (3.5.33) we obtain

$$\|S_k(i) - S(i)\| = o\big(\frac{1}{k^\nu}\big), \ i = 0, 1 \cdots, r, \quad \|S_k(i)\| = o\big(\frac{1}{k^\nu}\big), \quad r < i \leq t \tag{3.5.34}$$

for some small enough $\nu > 0$.

Therefore, we have

$$\lim_{k \to \infty} Q_{j,k}^{(r)} = \begin{cases} \frac{\|S(j)\|}{\|S(j+1)\|}, & \text{if } S(j) \neq 0,\ S(j+1) \neq 0, \\ \infty, & \text{if } S(j+1) = 0,\ S(j) \neq 0, \\ 0, & \text{if } S(j+1) \neq 0,\ S(j) = 0, \\ 1, & \text{if } S(j) = 0,\ S(j+1) = 0. \end{cases}$$

Since $C_r \neq 0$, we see that $S(r) \neq 0$, and $S(i) = 0 \ \forall i > r$ by A3.5.3. Thus, the assertion of the theorem is concluded. □

Remark 3.5.3 *Since $A_{i,k}$ and $S_k(i)$ are recursively calculated, the quantities $Q_{j,k}^{(p)}$ and $Q_{j,k}^{(r)}$ used in Theorems 3.5.2 and 3.5.3 are easily updated and hence the estimates p_k and r_k for orders are convenient to be updated too, when new data arrive.*

Extension to ARMAX Processes

We now go back to consider the ARMAX system given by (3.1.2), which, in comparison with ARMA, has an additional term $B(z)u_{k-1}$ with $u_k \in \mathbb{R}^l$, where

$$B(z) = B_0 + B_1 z + \cdots + B_q z^q \text{ with } B_q \neq 0.$$

Let $t \triangleq \max(p, q, r)$, and set $A_i = 0 \ \forall\, i > p$, $B_j = 0 \ \forall\, j > q$, and $C_l = 0 \ \forall\, l > r$. We rewrite (3.1.2) as

$$\begin{aligned} y_k + A_1 y_{k-1} + \cdots + A_t y_{k-t} &= B_0 u_{k-1} + \cdots \\ &+ B_t u_{k-t-1} + w_k + C_1 w_{k-1} + \cdots + C_t w_{k-t}. \end{aligned} \tag{3.5.35}$$

Instead of A3.5.1–A3.5.4 we now introduce the following five assumptions for the ARMAX system (3.1.2).

B3.5.0 $\{u_k\}$ *is a sequence of iid random vectors and is independent of* $\{w_k\}$ *with* $Eu_k = 0,\ E\|u_k\|^{2+\delta} < \infty$ *for some* $\delta > 0$, *and* $Eu_k u_k^T = I$.

B3.5.1 $\det A(z) \neq 0\ \forall\ |z| \leq 1$, *and* $\det C(z) \neq 0\ \forall\ |z| \leq 1$.

B3.5.2 $A(z)$ *and* $B(z)B^T(z^{-1}) + C(z)R_w C^T(z^{-1})$ *have no common left factor, and the matrix* $(A_t\ \ B_t B_0^T + C_t R_w)$ *is of row-full-rank.*

B3.5.3 $\{w_k\}$ *is a sequence of iid random vectors such that* $Ew_k = 0$ *with* $E\|w_k\|^{2+\delta} < \infty$ *for some* $\delta > 0$, *and* $Ew_k w_k^T \triangleq R_w > 0$, *where* R_w *is unknown.*

B3.5.4 *An upper bound* t^* *for* t *is available* $t^* \geq t \triangleq \max(p,q,r)$.

As before, without loss of generality, $\{y_k\}$ may be considered as a stationary process. It is noticed that B3.5.2 is reduced to A3.5.2 if $B(z) \equiv 0$.

By Lemma 3.4.1, under B3.5.0–B3.5.3 the process (3.1.2) can be represented as

$$A(z)y_k = D(z)\xi_k,\quad D(z) = I + D_1 z + \cdots + D_t z^t \tag{3.5.36}$$

with the following properties:

(i) ξ_k is m-dimensional, $E\xi_k = 0,\ E\xi_k\xi_j^T = R_\xi \delta_{k,j}$ with $\delta_{k,j} = 1$ if $k = j$ and $\delta_{k,j} = 0$ if $k \neq j$ and $R_\xi > 0$;

(ii) $D(z)$ is a stable (without root in $\{z:\ |z| \leq 1\}$) $m \times m$-matrix polynomial;

(iii) $A(z)$ and $D(z)$ have no common left factor, and $(A_t\ \ D_t)$ is of row-full-rank;

(iv) $A(z)$ and $D(z)$ are uniquely defined.

Applying Theorems 3.5.1 and 3.5.2 to the ARMA process (3.5.36) we derive the recursive and strongly consistent estimates $t_k,\ p_k$, and $A_{i,k},\ i = 1,\cdots,t$ for $t,\ p$, and $A_i,\ i = 1,\cdots,t$, respectively.

We now estimate q and θ_B.

For estimating θ_B let us define

$$\eta_{k-1} \triangleq [u_{k-1}^T,\cdots,u_{k-t-1}^T]^T,\quad \hat{\eta}_{k-1} \triangleq [u_{k-1}^T,\cdots,u_{k-t_k-1}^T]^T$$

and the estimate

$$\hat{\chi}_k \triangleq y_k + A_{1,k}y_{k-1} + \cdots + A_{t_k,k}y_{k-t_k} \tag{3.5.37}$$

for $\chi_k \triangleq y_k + A_1 y_{k-1} + \cdots + A_t y_{k-t}$.

With an arbitrary initial value $\theta_{B,0} \in \mathbb{R}^{(t_0+1)l \times m}$ we recursively define $\theta_{B,k}$ by the following algorithm:

$$\theta_{B,k+1} = \theta_{B,k} - \frac{1}{k+1}(\theta_{B,k} - \hat{\eta}_k \hat{\chi}_{k+1}^T), \tag{3.5.38}$$

whenever $t_{k+1} = t_k$. Otherwise, set $\theta_{B,k+1} \triangleq 0 \in \mathbb{R}^{(t_{k+1}+1)l \times m}$.

Clearly, if $t_j \equiv t \ \forall j \geq k_0$, then $\eta_k \equiv \hat{\eta}_k$ and

$$\theta_{B,k+1} = \frac{k_0}{k+1}\theta_{B,k_0} + \frac{1}{k+1}\sum_{j=k_0}^{k} \eta_j \hat{\chi}_{j+1}^T. \tag{3.5.39}$$

Write $\theta_{B,k}^T$ in the block form: $\theta_{B,k}^T = [B_{0,k}, \cdots, B_{t_k,k}]$, and define

$$Q_{j,k}^{(q)} \triangleq \frac{\|B_{j,k}\| + \frac{1}{\log k}}{\|B_{j+1,k}\| + \frac{1}{\log k}}, \quad k \geq 1, \quad j = 0, \cdots, t_k, \tag{3.5.40}$$

where $B_{t_k+1,k} \triangleq 0 \ \forall k \geq 1$. Take a threshold $\varepsilon > 1$, say $\varepsilon = 3$, and define the estimate q_k for q

$$q_k \triangleq \max\{j \mid Q_{j,k}^{(q)} \geq \varepsilon, \ j = 0, \cdots, t_k\}, \tag{3.5.41}$$

if there exists some $j : 0 \leq j \leq t_k$ such that $Q_{j,k}^{(q)} \geq \varepsilon$. Otherwise, define $q_k \triangleq t_k$.

For the strong consistency of $\theta_{B,k}$ and q_k, we have the following theorem.

Theorem 3.5.4 *Assume B3.5.0–B3.5.4 hold. Then,*

$$\lim_{k\to\infty} q_k = q, \quad \|B_{i,k} - B_i\| = o\left(\frac{1}{k^\nu}\right), \ i = 0 \cdots, q \tag{3.5.42}$$

for some small enough $\nu > 0$.

Proof. Since $\{u_k\}$ is iid with $Eu_k = 0$, $Eu_k u_k^T = I$ and is independent of $\{w_k\}$, we have $E\eta_k \chi_{k+1}^T = [B_0 \ B_1 \cdots B_t]^T$ where $B_i = 0, \ q < i \leq t$.

Noticing $E\|u_k\|^{2+\delta} < \infty$ for some $\delta > 0$ and applying (3.5.23) to estimating $\chi_k - \hat{\chi}_k$, we derive that $\|B_{i,k} - B_i\| = o(\frac{1}{k^\nu})$, $i = 0, \cdots, t$ for some small enough $\nu > 0$.

Finally, by the definition of $Q_{i,k}^{(q)}$ and the convergence rate of $B_{i,k}$, we conclude $q_k \to q$ a.s. □

It remains to estimate r.

Estimating r is carried out as that done in Theorem 3.5.3, but (3.5.27) and (3.5.28) should be changed to the following:

$$\zeta_k \triangleq \chi_k - B(z)u_{k-1} (= C(z)w_k),$$
$$\hat{\zeta}_k \triangleq \hat{\chi}_k - B_{0,k}u_{k-1} - \cdots - B_{q_k,k}u_{k-q_k-1}$$

with $\hat{\chi}_k$ given by (3.5.37).

We keep (3.5.29)–(3.5.31) unchanged. We then have the following theorem.

Theorem 3.5.5 *Assume B3.5.0–B3.5.4 hold. Then,*

$$\lim_{k\to\infty} r_k = r \ \text{ a.s.}$$

The proof of Theorem 3.5.5 can be carried out along the lines of that for Theorem 3.5.3 with the help of (3.5.23) and Theorem 3.5.4.

Remark 3.5.4 *With known orders of ARMAX, in Section 3.3 the recursive identification algorithms for estimating* θ_A, θ_B, θ_C, *and* R_w *have been proposed and their strong consistency has been established without imposing SPR on* $C(z)$. *So, combining Sections 3.4 and 3.5 composes a complete system of algorithms for identifying both the orders and coefficients of ARMA/ARMAX.*

Numerical Examples

Example 1. Let the ARMAX with $m = 1,\ p = q = r = 2$ be as follows

$$y_k + a_1 y_{k-1} + a_2 y_{k-2} = b_1 u_{k-1} + b_2 u_{k-2} + w_k + c_1 w_{k-1} + c_2 w_{k-2},$$

where $a_1 = -0.6,\ a_2 = 0.16,\ b_1 = 0.5,\ b_2 = 1,\ c_1 = 0.7,\ c_2 = 0.5$, and both $\{u_k\}$ and $\{w_k\}$ are iid Gaussian $w_k \in \mathcal{N}(0, 0.25),\ u_k \in \mathcal{N}(0,1)$, and they are mutually independent. It is noted that $C^{-1}(z) - \frac{1}{2}$ is not SPR, since $\frac{1}{C(e^{i\omega})} + \frac{1}{C(e^{-i\omega})}\Big|_{\omega=0} < 1$, where $C(z) = 1 + c_1 z + c_2 z^2$.

Assume an upper bound $t^* = 3$ for the unknown orders is available. We take $M_k = 2k,\ \varepsilon = 10$ in the algorithms.

More than twenty samples have been computed. As expected, for all sample paths all estimates for t, (p,q,r), a_1, a_2, b_1, and b_2 converge to the true values. The simulation results for an arbitrarily chosen sample are demonstrated by Figures 3.5.1–3.5.6. Figures 3.5.1, 3.5.3, 3.5.5, and 3.5.6 show the estimates for orders while Figures 3.5.2 and 3.5.4 show the estimates for coefficients.

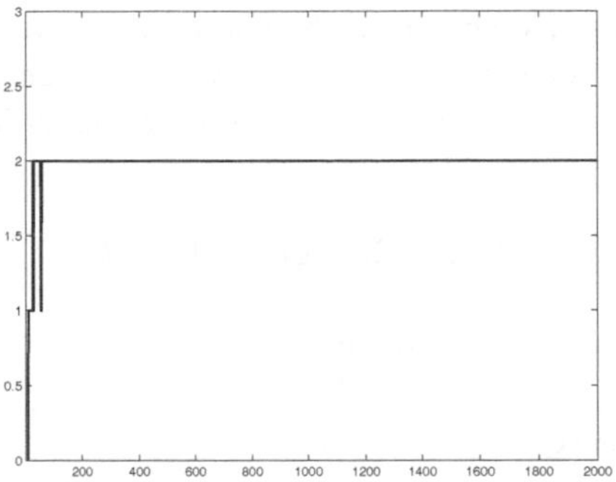

Figure 3.5.1: Estimates for t

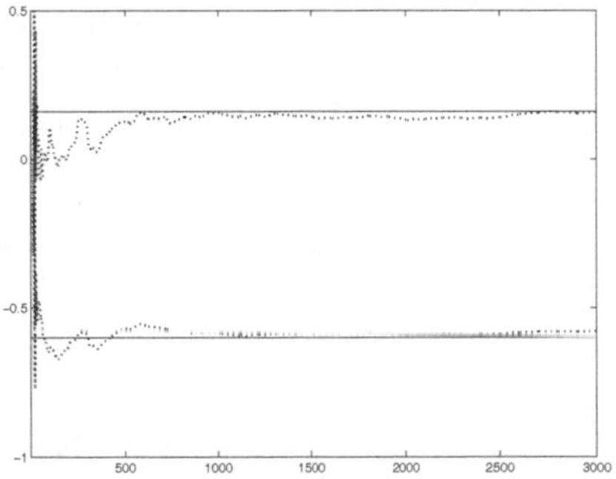

Figure 3.5.2: Estimates for a_1 and a_2

Example 2. By this example we compare the order estimates given above with those given by AIC, BIC, and CIC. Let us consider the following system

$$y_k + a_1 y_{k-1} + a_2 y_{k-2} = b_1 u_{k-1} + b_2 u_{k-2},$$

where a_1, a_2, b_1, b_2, and u_k are the same as those in Example 1.

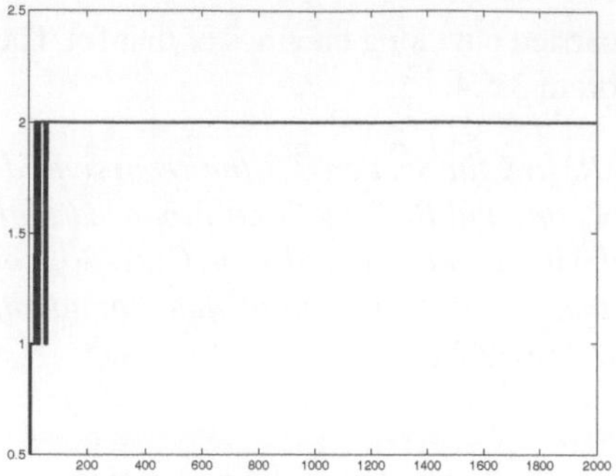

Figure 3.5.3: Estimates for p

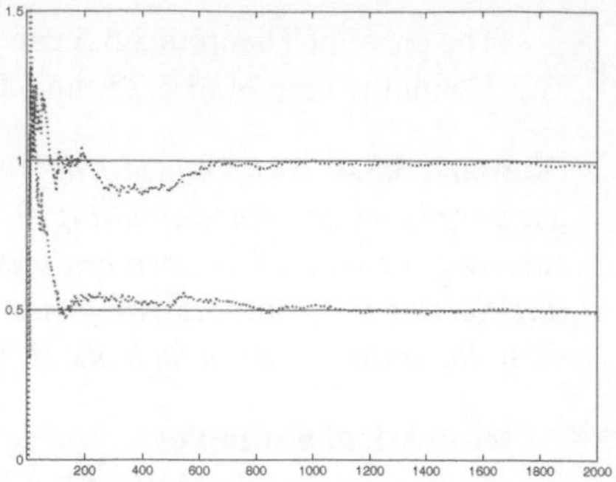

Figure 3.5.4: Estimates for b_1 and b_2

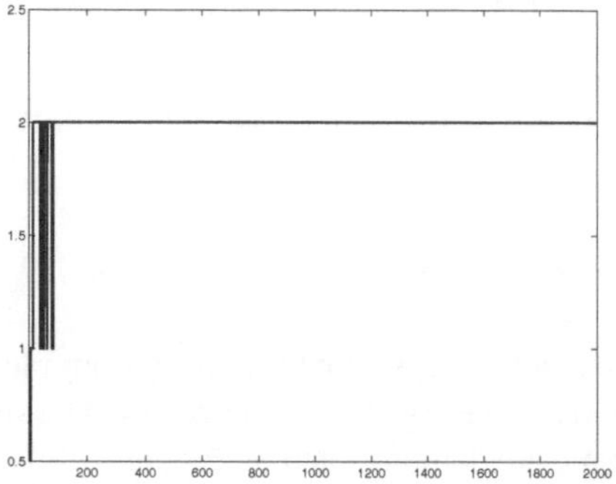

Figure 3.5.5: Estimates for q

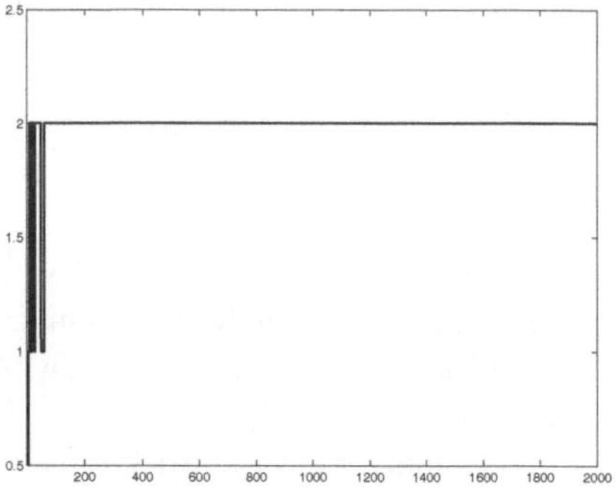

Figure 3.5.6: Estimates for r

Assume the upper bound for orders is $t^* = 3$ and take $M_k = 2k,\ \varepsilon = 10$.

Figure 3.5.7 shows the estimates for t given by the method proposed above, while Figures 3.5.8–3.5.10 present the estimates for t obtained by AIC, BIC, and CIC, respectively. To be precise, $\{t_k\}_{k=1}^N$, $N = 2000$ are plotted in Figures 3.5.8–3.5.10, where

$$t_k = \underset{1\le s\le 3}{\operatorname{argmin}}\{k\log\sigma_k(s,s) + 2d_k s\} \text{ with } d_k = 2,\ \log k, \text{ and } (\log k)^{1.5}$$

for AIC, BIC, and CIC, respectively,

$$\sigma_k(s,s) = \sum_{i=0}^{k-1} |y_{i+1} - \theta_k(s,s)^T\varphi_i(s,s)|^2,$$

$$\varphi_i(s,s) = [y_i,\cdots,y_{i+1-s},u_i,\cdots,u_{i+1-s}]^T, \text{ and}$$

$$\theta_k(s,s) = \Big(\sum_{i=0}^{k-1}\varphi_i(s,s)\varphi_i^T(s,s)\Big)^{-1}\Big(\sum_{i=0}^{k-1}\varphi_i(s,s)y_{i+1}\Big).$$

Figures 3.5.7–3.5.10 are simulated on a laptop computer with an Intel(4) 2.4GHz CPU. The computation time for plotting Figures 3.5.7–3.5.10 is listed in Table 3.5.1.

It is natural to expect that the order estimation algorithm given in this section computationally is much faster in comparison with others to run over all 2000 data, because here the recursive methods are used when new data arrive. It is also conceivable that the method given here has to reduce the initial bias step by step, and hence the AIC-like information criteria may provide a better estimate for small sampling sizes, i.e., at steps less than 300 in Figures 3.5.7–3.5.10.

Table 3.5.1: Computation time for Figures 3.5.7–3.5.10

	Figure 3.5.7	Figure 3.5.8	Figure 3.5.9	Figure 3.5.10
Computation time	1.9820 sec	85.3120 sec	85.3320 sec	87.1950 sec

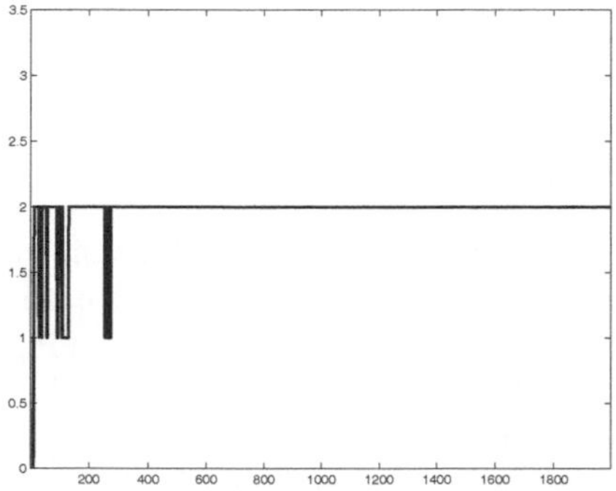

Figure 3.5.7: t_k **given in Section 3.5**

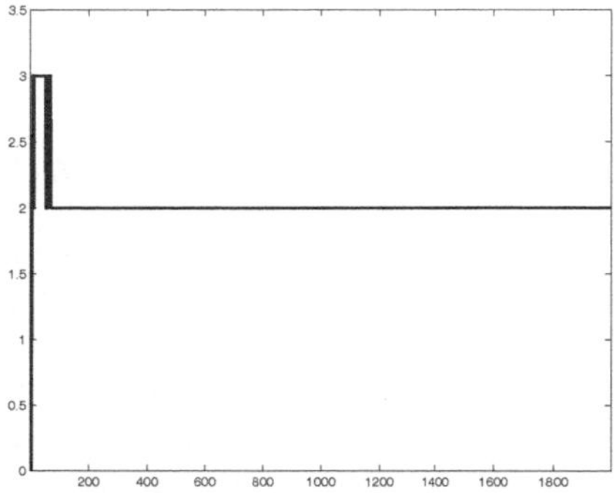

Figure 3.5.8: t_k **given by AIC**

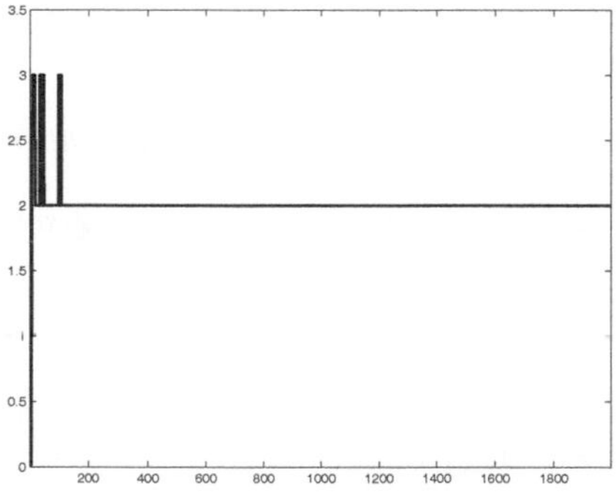

Figure 3.5.9: t_k **given by BIC**

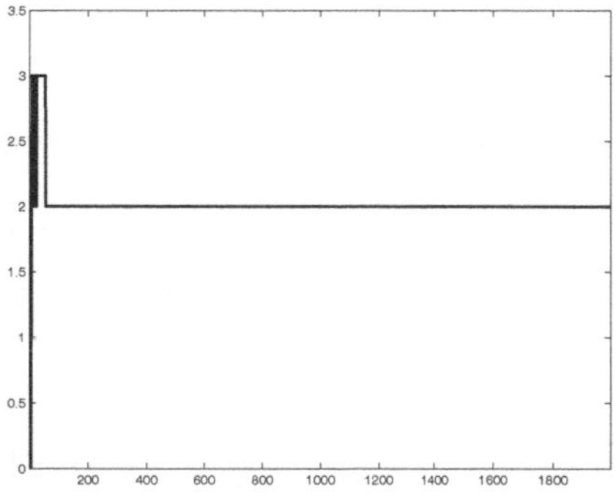

Figure 3.5.10: t_k **given by CIC**

3.6 Multivariate Linear EIV Systems

In Sections 3.4 and 3.5 the input and output of a linear system are exactly observed without errors when we identify its coefficients and orders. In this section we discuss the identification problem for linear systems when the observations on system input and output are corrupted by noises. The corresponding system is called errors-in-variables (EIV) system. We derive the recursive estimation algorithms for EIV systems and establish their strong consistency.

System Description and Estimation Algorithms

Let us consider the following MIMO EIV linear systems

$$A(z)y_k^o = B(z)u_k^o, \tag{3.6.1}$$
$$u_k = u_k^o + \eta_k, \tag{3.6.2}$$
$$y_k = y_k^o + \varepsilon_k, \tag{3.6.3}$$

with

$$A(z) = I + A_1 z + \cdots + A_p z^p, \tag{3.6.4}$$
$$B(z) = B_0 + B_1 z + B_2 z^2 + \cdots + B_q z^q, \tag{3.6.5}$$

where $A(z)$ and $B(z)$ are $m \times m$ and $m \times l$ matrix polynomials with unknown coefficients and known orders p, q, respectively, $u_k^o \in \mathbb{R}^l$ and $y_k^o \in \mathbb{R}^m$ are the true input and output of the linear systems, $u_k \in \mathbb{R}^l$ and $y_k \in \mathbb{R}^m$ are the measurement input and output corrupted by noises $\eta_k \in \mathbb{R}^l$ and $\varepsilon_k \in \mathbb{R}^m$, respectively.

The problem is recursively to estimate the matrix coefficients $\{A_1, \cdots, A_p, B_0, B_1, \cdots, B_q\}$ in (3.6.4) and (3.6.5) based on the disturbed observations $\{u_k\}$ and $\{y_k\}$.

The assumptions made on the system are as follows.

A3.6.1 *$A(z)$ is stable, i.e.,* $\det A(z) \neq 0 \;\; \forall |z| \leq 1$.

A3.6.2 *$A(z)$ and $B(z)$ have no common left factor, and $[A_p \; B_q]$ is of row-full-rank.*

A3.6.3 *The observation noises η_k and ε_k are multi-variable ARMA processes:*

$$P(z)\eta_k = Q(z)\xi_k, \quad C(z)\varepsilon_k = D(z)\zeta_k, \tag{3.6.6}$$

with

$$P(z) = I + P_1 z + P_2 z^2 + \cdots + P_{n_p} z^{n_p}, \tag{3.6.7}$$
$$Q(z) = I + Q_1 z + Q_2 z^2 + \cdots + Q_{n_q} z^{n_q}, \tag{3.6.8}$$
$$C(z) = I + C_1 z + C_2 z^2 + \cdots + C_{n_c} z^{n_c}, \tag{3.6.9}$$
$$D(z) = I + D_1 z + D_2 z^2 + \cdots + D_{n_d} z^{n_d}, \tag{3.6.10}$$

where $P(z)$ and $Q(z)$ have no common left factor, $P(z)$ is stable, so do $C(z)$ and $D(z)$, i.e., they have no common leftfactor and $C(z)$ is stable, ξ_k and ζ_k are mutually independent zero-mean iid random vectors with $E(\|\xi_k\|^\Delta) < \infty$ and $E(\|\zeta_k\|^\Delta) < \infty$ for some $\Delta > 2$. In addition, ξ_k and ζ_k have probability densities.

A3.6.4 *The input $\{u_k^o, k \geq 1\}$ is independent of $\{\xi_k\}$ and $\{\zeta_k\}$ and is a sequence of zero-mean iid random vectors with $E(u_k^o u_k^{oT}) = I$ and $E(\|u_k^o\|^{2\Delta}) < \infty$ for some $\Delta > 2$. In addition, u_k^o has probability density.*

We now derive the algebraic equations satisfied by $A_i,\ i=1,\dots,p,\ B_j,\ j=0,1,\dots,q$ and the estimation algorithms.

By A3.6.1, we have (3.3.4): $H(z) \triangleq A^{-1}(z)B(z) = \sum_{i=0}^{\infty} H_i z^i$, where $H_0 = B_0$, $\|H_i\| = O(e^{-ri}), r > 0, i > 1$.

Assuming $u_k^o = 0\ \forall k < 0$, we have

$$y_k^o = \sum_{i=0}^{k} H_i u_{k-i}^o.$$

By the independence assumption in A3.6.3 and A3.6.4 we have

$$\begin{aligned} E(y_k u_{k-i}^T) &= E(y_k^o + \varepsilon_k)(u_{k-i}^o + \eta_{k-i})^T \\ &= E y_k^o u_{k-i}^{oT} = H_i. \end{aligned} \tag{3.6.11}$$

This motivates us to estimate the impulse responses $H_i,\ i = 1,2,\dots$ by the RM algorithm for linear regression functions based on the noisy observations $\{u_k\}$ and $\{y_k\}$.

By the RM algorithm the estimates $H_{i,k}$ for $H_i,\ i = 1,2,\dots$ are generated by the following recursive algorithm:

$$H_{i,k+1} = H_{i,k} - \frac{1}{k}(H_{i,k} - y_{k+1}u_{k+1-i}^T), \tag{3.6.12}$$

or

$$H_{i,k+1} = H_{i,k} - \frac{1}{k}(H_{i,k} - H_i) + \frac{1}{k}\varepsilon_{k+1}(i), \tag{3.6.13}$$

where

$$\begin{aligned} \varepsilon_{k+1}(i) =& y_{k+1}u_{k+1-i}^T - H_i \\ =& [y_{k+1}^o u_{k+1-i}^{oT} - E y_{k+1}^o u_{k+1-i}^{oT}] + y_{k+1}^o \eta_{k+1-i}^T + \varepsilon_{k+1} u_{k+1-i}^{oT} \\ &+ \varepsilon_{k+1}\eta_{k+1-i}^T. \end{aligned} \tag{3.6.14}$$

Once the estimates $H_{i,k}$ for $H_i,\ i = 1,2,\dots$ are obtained, the matrix coefficients $\{A_1,\cdots,A_p,B_0,B_1,\cdots,B_q\}$ can be derived by the convolution relationship between them.

To be precise, replacing H_i in L defined by (3.3.10) with $H_{i,k}$ obtained from (3.6.13), we derive an estimate L_k for L:

$$L_k \triangleq \begin{bmatrix} H_{q,k} & H_{q+1,k} & \cdots & H_{q+mp-1,k} \\ H_{q-1,k} & H_{q,k} & \cdots & H_{q+mp-2,k} \\ \vdots & \vdots & \ddots & \vdots \\ H_{q-p+1,k} & H_{q-p+2,k} & \cdots & H_{q+(m-1)p,k} \end{bmatrix}. \tag{3.6.15}$$

The estimates for $\{A_1, \cdots, A_p, B_0, B_1, \cdots, B_t\}$ are naturally defined with the help of (3.3.7) and (3.3.9) as follows:

$$[A_{1,k}, A_{2,k}, \cdots, A_{p,k}]$$
$$= -[H_{q+1,k}, H_{q+2,k}, \cdots, H_{q+mp,k}] L_k^T (L_k L_k^T)^{-1}, \tag{3.6.16}$$

$$B_{i,k} = \sum_{j=0}^{i \wedge p} A_{j,k} H_{i-j,k} \;\; \forall\, 1 \le i \le q, \tag{3.6.17}$$

whenever H_k is of row-full-rank.

Strong Consistency of Estimates

To show the strong consistency of the estimates given by (3.6.16)-(3.6.17) it suffices to prove that estimates given by (3.6.12) are strongly consistent. In order to apply Theorem 2.5.1 concerning convergence of the RM algorithm for linear regression functions we have to analyze $\varepsilon_{k+1}(i)$ given by (3.6.14).

We first recall the definition and properties of α-mixing introduced in Chapter 1. Assume A3.6.1 and A3.6.4 hold, and define

$$V_{k+1} \triangleq [y_{k+1}^{oT}, y_k^{oT}, \cdots, y_{k+2-p}^{oT}, u_{k+1}^{oT}, u_k^{oT}, \cdots, u_{k+2-q}^{oT}]^T.$$

By Theorem 1.4.4 V_k is an α-mixing with mixing coefficients α_k exponentially decaying to zero:

$$\alpha_k \le c\lambda^k \;\; \forall\, k \ge 1 \tag{3.6.18}$$

for some $c > 0$ and $0 < \lambda < 1$.

Corollary 3.6.1 *Assume A3.6.3 holds. Then η_k and ε_k are all α-mixings with mixing coefficients α_k exponentially decaying to zero.*

The following lemma plays the basic role in establishing the strong consistency of the estimates.

Lemma 3.6.1 *Assume A3.6.1–A3.6.4 hold. Then the following series converge* a.s.*:*

$$\sum_{k=1}^{\infty} \frac{1}{k^{1-\nu}} [y_{k+1}^o u_{k+1-i}^{oT} - E y_{k+1}^o u_{k+1-i}^{oT}] < \infty, \tag{3.6.19}$$

$$\sum_{k=1}^{\infty} \frac{1}{k^{1-\nu}} y_{k+1}^o \eta_{k+1-i}^T < \infty, \tag{3.6.20}$$

$$\sum_{k=1}^{\infty} \frac{1}{k^{1-\nu}} \varepsilon_{k+1} u_{k+1-i}^{oT} < \infty, \tag{3.6.21}$$

$$\sum_{k=1}^{\infty} \frac{1}{k^{1-\nu}} \varepsilon_{k+1} \eta_{k+1-i}^T < \infty \tag{3.6.22}$$

for any $0 \le \nu < \frac{1}{2}$ and any integer $i \ge 1$.

Proof. For (3.6.19)–(3.6.22), it suffices to show that each entry (indexed by s,j) of the matrix series converges a.s. for $1 \le s \le m, 1 \le j \le l$.

We first prove (3.6.19); the (s,j)-element of (3.6.19) is

$$\sum_{k=1}^{\infty} \frac{1}{k^{1-\nu}} [y_{k+1}^o u_{k+1-i}^{oT} - E y_{k+1}^o u_{k+1-i}^{oT}]_{s,j}$$
$$= \sum_{k=1}^{\infty} \frac{1}{k^{1-\nu}} [y_{k+1}^o(s) u_{k+1-i}^o(j) - E y_{k+1}^o(s) u_{k+1-i}^o(j)], \tag{3.6.23}$$

where

$$y_{k+1}^o = [y_{k+1}^o(1), \cdots, y_{k+1}^o(m)]^T \text{ and } u_{k+1-i}^o = [u_{k+1-i}^o(1), \cdots, u_{k+1-i}^o(l)]^T.$$

Define $z_k^{s,j} = \frac{1}{k^{1-\nu}} [y_{k+1}^o(s) u_{k+1-i}^o(j) - E y_{k+1}^o(s) u_{k+1-i}^o(j)]$. Then $z_k^{s,j}$ is a zero-mean α-mixing with mixing coefficients exponentially decaying to zero by Theorem 1.4.4. Moreover, by the C_r-inequality and then the Hölder inequality, for any $\varepsilon > 0$ we have

$$E|z_k^{(s,j)}|^{2+\varepsilon} = E\left|\frac{1}{k^{1-\nu}} [y_{k+1}^o(s) u_{k+1-i}^o(j) - E y_{k+1}^o(s) u_{k+1-i}^o(j)]\right|^{2+\varepsilon}$$
$$\le \frac{1}{k^{(1-\nu)(2+\varepsilon)}} 2^{2+\varepsilon} E|y_{k+1}^o(s) u_{k+1-i}^o(j)|^{2+\varepsilon}$$
$$\le \frac{1}{k^{(1-\nu)(2+\varepsilon)}} 2^{2+\varepsilon} (E|y_{k+1}^o(s)|^{2(2+\varepsilon)})^{\frac{1}{2}} \cdot (E|u_{k+1-i}^o(j)|^{2(2+\varepsilon)})^{\frac{1}{2}}$$
$$= O\left(\frac{1}{k^{(1-\nu)(2+\varepsilon)}}\right),$$

which implies

$$\sum_{k=1}^{\infty} (E|z_k^{(s,j)}|^{2+\varepsilon})^{\frac{2}{2+\varepsilon}} = O\left(\sum_{k=1}^{\infty} \frac{1}{k^{2(1-\nu)}}\right) < \infty.$$

By Theorem 1.4.2 the series (3.6.23) converges a.s. for $1 \le s \le m, 1 \le j \le l$, hence (3.6.19) converges a.s.

Next, we prove (3.6.20). The (s,j)-element of (3.6.20) is

$$\sum_{k=1}^{\infty} \frac{1}{k^{1-\nu}} [y_{k+1}^o \eta_{k+1-i}^T]_{s,j} = \sum_{k=1}^{\infty} \frac{1}{k^{1-\nu}} y_{k+1}^o(s) \eta_{k+1-i}(j), \tag{3.6.24}$$

where $y_{k+1}^o = [y_{k+1}^o(1), \cdots, y_{k+1}^o(m)]^T$ and $\eta_{k+1-i} = [\eta_{k+1-i}(1), \cdots, \eta_{k+1-i}(l)]^T$.

Define $w_k^{s,j} = \frac{1}{k^{1-\nu}} y_{k+1}^o(s) \eta_{k+1-i}(j)$. Since y_k^o and η_k are mutually independent, $w_k^{s,j}$ is a zero-mean α-mixing with mixing coefficients exponentially decaying to zero by Corollary 3.6.1. Moreover, we have

$$E|w_k^{s,j}|^{2+\varepsilon} = E\left|\frac{1}{k^{1-\nu}} y_{k+1}^o(s) \eta_{k+1-i}(j)\right|^{2+\varepsilon}$$
$$= \frac{1}{k^{(1-\nu)(2+\varepsilon)}} E|y_{k+1}^o(s) \eta_{k+1-i}(j)|^{2+\varepsilon}$$

$$= \frac{1}{k^{(1-\nu)(2+\varepsilon)}} E|y_{k+1}^{o}(s)|^{2+\varepsilon} \cdot E|\eta_{k+1-i}(j)|^{2+\varepsilon}$$
$$= O(\frac{1}{k^{(1-\nu)(2+\varepsilon)}}),$$

which leads to

$$\sum_{k=1}^{\infty} (E|w_k^{(s,j)}|^{2+\varepsilon})^{\frac{2}{2+\varepsilon}} = O(\sum_{k=1}^{\infty} \frac{1}{k^{2(1-\nu)}}) < \infty.$$

Then, (3.6.24) converges a.s. for $1 \leq s \leq m, 1 \leq j \leq l$, and hence (3.6.20) converges a.s. Convergence of (3.6.21) and (3.6.22) can be proved by a treatment similar to that used for (3.6.20). □

Theorem 3.6.1 *Assume A3.6.1–A3.6.4 hold. Then, $H_{i,k}$, $i \geq 1$, defined by (3.6.12) and $A_{i,k}$, $i = 1, \ldots, p$, $B_{j,k}$, $j = 0, 1, \ldots, q$ given by (3.6.16) and (3.6.17), are strongly consistent*

$$H_{i,k} \xrightarrow[k \to \infty]{} H_i \text{ a.s., } i \geq 1,$$
$$A_{s,k} \xrightarrow[k \to \infty]{} A_s \text{ a.s., } s = 1, \cdots, p, \; B_{j,k} \xrightarrow[k \to \infty]{} B_j \text{ a.s., } j = 0, 1, \cdots, q \quad (3.6.25)$$

with convergence rates

$$\| H_{i,k} - H_i \| = o(k^{-\nu}) \text{ a.s. } \forall i \geq 1 \quad \| A_{s,k} - A_s \| = o(k^{-\nu}) \text{ a.s.,}$$
$$\| B_{j,k} - B_j \| = o(k^{-\nu}), \quad s = 1, \cdots, p, \; j = 0, 1, \cdots, q \;\; \forall \nu \in (0, \frac{1}{2}). \quad (3.6.26)$$

Proof. Applying Theorem 2.5.1 to (3.6.12) or (3.6.13) and paying attention to (3.6.14), we find that the strong consistency of $H_{i,k}$, $i \geq 1$ is guaranteed by Lemma 3.6.1 with $\nu = 0$.

For the assertion concerning convergence rates we apply Theorem 2.6.1. Comparing (3.6.13) with (2.5.14), we find that H_{ik}, $\frac{1}{k}$, and H_i in (3.6.13) correspond to x_k, a_k, and x^0 in (2.5.14), respectively, while H_k in (2.5.14) simply equals $-I$ for (3.6.13). Thus, the regression function $f(x)$ in (2.6.1) equals $-(x - H_i)$ with $\delta(x) = 0$ for the present case. It is clear that A2.6.1, A2.6.2, A2.6.4 are satisfied, while A2.6.3 also holds with $\delta = \nu$ and $\varepsilon''_{k+1} = 0$. Therefore Theorem 2.6.1 is applicable, and the convergence rate assertions of the theorem have been justified. □

Numerical Example

We present a numerical example to examine the performance of the proposed approach.

Let the MIMO EIV system be as follows:

$$y_k^o + A_1 y_{k-1}^o + A_2 y_{k-2}^o = B_1 u_{k-1}^o + B_2 u_{k-2}^o, \quad (3.6.27)$$
$$u_k = u_k^o + \eta_k,$$
$$y_k = y_k^o + \varepsilon_k.$$

where

$$A_1 = \begin{bmatrix} 1.2 & -0.4 \\ 0.2 & 0.5 \end{bmatrix}, \quad A_2 = \begin{bmatrix} 0.49 & -0.1 \\ -0.3 & 0.2 \end{bmatrix}$$

and

$$B_1 = \begin{bmatrix} -0.4 & 0.6 \\ 1 & 0 \end{bmatrix}, \quad B_2 = \begin{bmatrix} 0.5 & 0 \\ 0.3 & 0.8 \end{bmatrix}.$$

The observation noises η_k and ε_k are defined by

$$\eta_k + \begin{bmatrix} 0.3 & 0.1 \\ 0.2 & 0.4 \end{bmatrix} \eta_{k-1} = \xi_k + \begin{bmatrix} 0.2 & 0.3 \\ 0.1 & 0.1 \end{bmatrix} \xi_{k-1},$$

and

$$\begin{aligned} \varepsilon_k + \begin{bmatrix} 0.7 & -0.3 \\ 0.2 & 0.1 \end{bmatrix} \varepsilon_{k-1} + \begin{bmatrix} 0.12 & 0.2 \\ -0.1 & -0.12 \end{bmatrix} \varepsilon_{k-2} \\ = \zeta_k + \begin{bmatrix} 0.25 & 0.3 \\ -0.2 & 0.2 \end{bmatrix} \zeta_{k-1}, \end{aligned}$$

where ξ_k and ζ_k are mutually independent iid random vectors: ξ_k is uniformly distributed over $[-0.3, 0.3] \times [-0.3, 0.3]$, while $\zeta_k \in \mathcal{N}(0, 0.4^2 \times I_2)$.

The input signal u_k^o is a sequence of iid random variables uniformly distributed over $[-\sqrt{3}, \sqrt{3}]$.

It is straightforward to check that A3.6.1–A3.6.4 hold for the example defined here. Fig. 3.6.1–Fig. 3.6.4 demonstrate that the estimates converge to the true values as expected from the theoretical analysis.

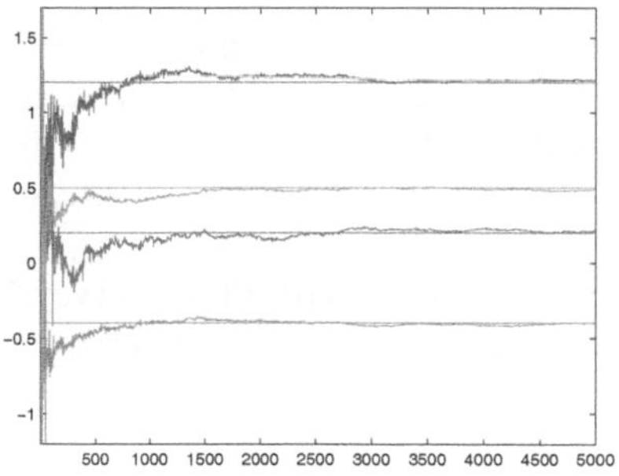

Figure 3.6.1: Estimates for A_1

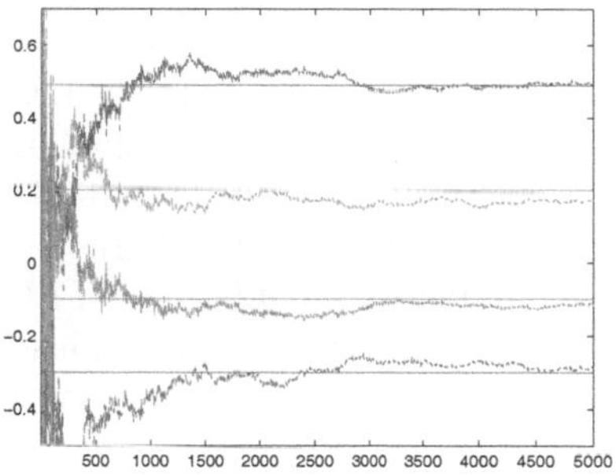

Figure 3.6.2: Estimates for A_2

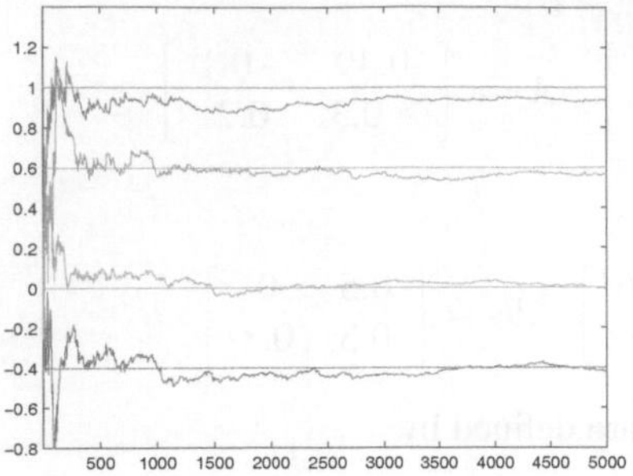

Figure 3.6.3: Estimates for B_1

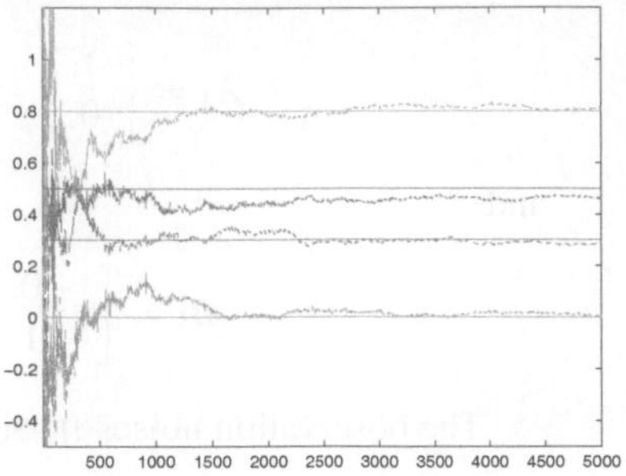

Figure 3.6.4: Estimates for B_2

3.7 Notes and References

The LS/ELS method is introduced here not only for its importance in identification of linear stochastic systems, but also for comparing with the SAAWET method presented in the book. Section 3.2 is based on [25]. For more about the ELS-based identification and adaptive control of linear stochastic systems, we refer to [26] where related concepts, properties, and techniques, such as the SPR condition, the PE condition, and the diminishingly excitation technique, are well summarized.

The rank of the Hankel matrix associated with the multivariate ARMA/ARMAX systems was discussed in [50] and [51]. We refer to [87] for the solution to the problem in a general setting. For proving (3.4.24) and (3.4.25) we use the form of inverse of block matrices given in [80]. The Marcinkiewicz–Zygmund Theorem used for proving (3.5.10) can be found in [30] among others.

The SPR condition used for the convergence of the ELS algorithm is rather restrictive [26], and there are many approaches, e.g., [15], [41], [51], [56], and [106], to overcome the difficulty. As far as the orders of ARMA/ARMAX systems are concerned, there some optimization-based criteria such as AIC [1], BIC [95], and ΦIC [26], and others [2] and [105] are proposed to estimate them. However, these approaches are nonrecursive. Based on SAAWET, the recursive identification of both coefficients and orders of multivariate ARMA/ARMAX systems is obtained in [23] and [27] and is presented in Sections 3.4 and 3.5. Figures 3.4.1–3.4.6 are reprinted with permission from [23].

For the identification of EIV systems, we refer to [102] for a survey. The sharp convergence rate of the estimates for correlation functions (law of iterative logarithm) is given in [3] and the recursive identification of EIV systems can be found in [22] and [83].

Chapter 4

Recursive Identification for Nonlinear Systems

CONTENTS

A wide class of practical systems are modeled as a linear system cascading with a static nonlinear function at its input or output: the saturation function may serve as a simplest example. If the linear subsystem is followed by a nonlinear static function, the whole system is called the Wiener system. If the static function is prior to the linear subsystem, then the system is called the Hammerstein system. These probably are the simplest but of practical importance nonlinear systems. We may also consider the cascading systems like a nonlinear static function sandwiched in between two linear subsystems (L–N–L), a linear subsystem sandwiched in between two nonlinear functions (N–L–N), and other combinations. The system L–N–L is called the Wiener–Hammerstein system, while N–L–N is called the Hammerstein–Wiener system. We call these kinds of systems "systems with static nonlinearity."

The output of an ARX system is linearly related to the system input. A direct extension of ARX is the nonlinear ARX system, where the current system output nonlinearly depends on the past system output and input.

In this chapter we discuss how to recursively estimate not only the coefficients of the linear subsystem but also the static nonlinear function contained in the system. The nonlinear function may be expressed as a linear combination of basis functions with unknown coefficients. Then, estimating the nonlinear function is equivalent to estimating the coefficients appearing in the linear combination. This is so-called the parametric approach. We may also estimate the values of the nonlinear function at any of its arguments. This corresponds to the nonparametric approach.

4.1 Recursive Identification of Hammerstein Systems

System Description

Consider the MIMO system (see Fig. 4.1.1):

$$\begin{cases} v_k = f(u_k) + \eta_k, \\ A(z)y_{k+1} = B(z)v_{k+1} + w_{k+1}, \\ z_{k+1} = y_{k+1} + \varepsilon_{k+1}, \end{cases} \tag{4.1.1}$$

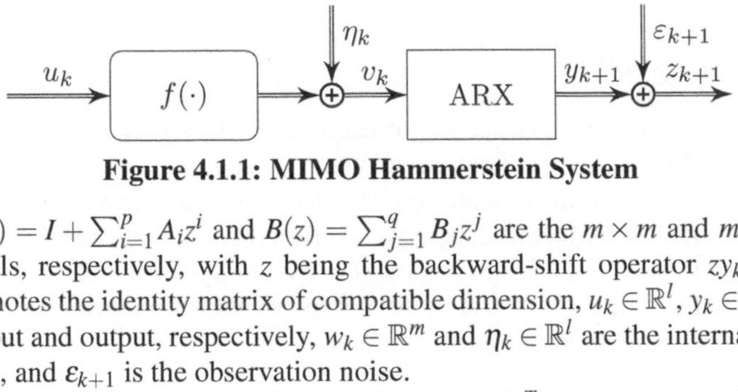

Figure 4.1.1: MIMO Hammerstein System

where $A(z) = I + \sum_{i=1}^{p} A_i z^i$ and $B(z) = \sum_{j=1}^{q} B_j z^j$ are the $m \times m$ and $m \times l$ matrix polynomials, respectively, with z being the backward-shift operator $zy_k = y_{k-1}$, I always denotes the identity matrix of compatible dimension, $u_k \in \mathbb{R}^l$, $y_k \in \mathbb{R}^m$ are the system input and output, respectively, $w_k \in \mathbb{R}^m$ and $\eta_k \in \mathbb{R}^l$ are the internal noises of the system, and ε_{k+1} is the observation noise.

The system nonlinearity is $f(\cdot) = [f_1(\cdot), \dots, f_l(\cdot)]^T$, where $f_i(\cdot) : \mathbb{R}^l \to \mathbb{R}$, $i = 1, \dots, l$.

The problem is to estimate

$$\theta_A^T \triangleq [A_1, \dots, A_p] \in \mathbb{R}^{m \times mp}, \quad \theta_B^T \triangleq [B_1, \dots, B_q] \in \mathbb{R}^{m \times lq},$$

and $f(u)$ for any given $u \in \mathbb{R}^l$ on the basis of the input-output data $\{u_k, z_k\}$.

Denote by μ_l the Lebesgue measure on $(\mathbb{R}^l, \mathscr{B}^l)$. We now list assumptions made on the system.

A4.1.1 *The input $\{u_k\}$ is a sequence of bounded iid random vectors with $Eu_k = 0$, and u_k has a probability density $p(\cdot)$ which is positive and continuous on a μ_l-positive set $U \subset \mathbb{R}^l$. Besides, $\{u_k\}$ is independent of $\Delta_k \triangleq [w_k^T, \eta_k^T, \varepsilon_k^T]^T$.*

A4.1.2 $\det A(z) \neq 0 \; \forall\, z : |z| \leq 1.$

A4.1.3 *$f(\cdot)$ is measurable and locally bounded and is continuous at u where the value $f(u)$ is estimated.*

A4.1.4 *The noises $\{w_k\}$, $\{\eta_k\}$, and $\{\varepsilon_k\}$ are sequences of iid random vectors, and they are mutually independent. Further, $E\Delta_k = 0$, $E\|\Delta_k\|^2 < \infty$, $R_w \triangleq E[w_k w_k^T] > 0$, $R_\eta \triangleq E[\eta_k \eta_k^T] \geq 0$, and $R_\varepsilon \triangleq E[\varepsilon_k \varepsilon_k^T] \geq 0$.*

A4.1.5 *$A(z)$ and $B(z)(R_v + R_\eta)B^T(z^{-1}) + R_w + A(z)R_\varepsilon A(z^{-1})$ have no common left factor, $[A_p \quad B_s(R_v + R_\eta)B_1^T + A_s R_\varepsilon]$ is of row-full-rank, where $R_v \triangleq E[f(u_k) - Ef(u_k)][f(u_k) - Ef(u_k)]^T$, $R_v + R_\eta > 0$, $s \triangleq \max\{p,q\}$; $A_i \triangleq 0$ $\forall i > p$; $B_j \triangleq 0$ $\forall j > q$ and the orders p, q of $A(z)$ and $B(z)$ are known.*

A4.1.6 *$\Upsilon \triangleq E[f(u_k)u_k^T]$ is nonsingular.*

Remark 4.1.1 *Note that $\theta_B^T P$ and $P^{-1} f(\cdot)$ with an arbitrary nonsingular P result in the same $B(z)f(\cdot)$. This means that θ_B^T and $f(\cdot)$ cannot be uniquely identified unless the indeterminate nonsingular matrix P is fixed. Let us fix $P = \Upsilon$, which is nonsingular by A4.1.6. We then have $E[(\Upsilon^{-1} f(u_k))u_k^T] = \Upsilon^{-1} E[f(u_k)u_k^T] = I$. Therefore, to uniquely define θ_B^T and $f(\cdot)$, without loss of generality, we may assume*

$$\Upsilon \triangleq E[f(u_k)u_k^T] = I. \tag{4.1.2}$$

By A4.1.2, $A(z)$ is stable. Then, by A4.1.1, A4.1.3, and A4.1.4 the processes $\{y_k\}$ and $\{z_k\}$ are asymptotically stationary. By appropriately choosing initial values they are stationary. So, without loss of generality, we may assume they are stationary. Defining $\mu \triangleq Ef(u_k) = Ev_k \in \mathbb{R}^l$, $\mu^* \triangleq Ez_k = Ey_k$, we then have

$$A(1)\mu^* = B(1)\mu, \tag{4.1.3}$$

where $A(1) \triangleq \sum_{i=0}^{p} A_i$ with $A_0 = I$, $B(1) \triangleq \sum_{j=1}^{q} B_j$.

It is clear that $\{v_k\}$ with $v_k \triangleq f(u_k) - \mu$ is an iid sequence. By setting $\zeta_k \triangleq z_k - \mu^*$, the system (4.1.1) can be written as

$$A(z)\zeta_{k+1} = B(z)(v_{k+1} + \eta_{k+1}) + w_{k+1} + A(z)\varepsilon_{k+1}. \tag{4.1.4}$$

Lemma 4.1.1 *Assume A4.1.1–A4.1.5 hold. Then the process ζ_k can be expressed as an ARMA process*

$$A(z)\zeta_k = H(z)\xi_k, \quad H(z) = I + H_1 z + \cdots + H_s z^s \tag{4.1.5}$$

with the following properties:

(1) Both $A(z)$ and $H(z)$ are stable and they have no common left factor;
(2) $[A_p \quad H_s]$ is of row-full-rank;
(3) $E\xi_k = 0$, $E\xi_k \xi_j^T = R_\xi \delta_{k,j}$.

Proof. Setting

$$\chi_{k+1} = B(z)f(u_{k+1}) + B(z)\eta_{k+1} + w_{k+1} + A(z)\varepsilon_{k+1}, \text{ or } \chi_{k+1} = A(z)z_{k+1} \tag{4.1.6}$$

we have

$$A(z)\zeta_{k+1} = \chi_{k+1} - E\chi_{k+1} \triangleq \chi^0_{k+1}.$$

By the mutual independence of v_k, η_k, w_k, and ε_k the process

$$\chi^0_{k+1} \triangleq \chi_{k+1} - E\chi_{k+1} = B(z)(v_{k+1} + \eta_{k+1}) + w_{k+1} + A(z)\varepsilon_{k+1}$$

is stationary and ergodic, and hence so is ζ_k, because $A(z)$ is stable. The spectral density of $\chi_k{}^0$ is

$$\Psi^{\chi}(z) = B(z)(R_v + R_\eta)B^T(z^{-1}) + R_w + A(z)R_\varepsilon A^T(z^{-1}).$$

Proceeding as in the proof for Lemma 3.4.1, we have the innovation representation

$$\chi^0_k = H(z)\xi_k,$$

where ξ_k is m-dimensional, $E\xi_k = 0$, $E\xi_k\xi_j^T = R_\xi\delta_{k,j}$ with $\delta_{k,j} = 1$ if $k = j$ and $\delta_{k,j} = 0$ if $k \neq j$ and $R_\xi > 0$, and $H(z)$ is an $m \times m$-matrix of rational functions with $H(0) = I$ and both $H(z)$ and $H^{-1}(z)$ are stable. This implies that

$$\chi^0_k = \sum_{i=0}^{\infty} H_i\xi_{k-i}, \;\; H_0 = I, \;\; \xi_k = \sum_{i=0}^{\infty} F_i\chi^0_{k-i} \;\; F_0 = I.$$

By A4.1.1 and A4.1.4 the completely same argument as that used in Lemma 3.4.1 leads to (4.1.5). Hence, we have

$$\chi^0_k = \xi_k + H_1\xi_{k-1} + \cdots + H_s\xi_{k-s}. \tag{4.1.7}$$

We then have

$$\begin{aligned}\Psi^{\chi}(z) &= B(z)(R_v + R_\eta)B^T(z^{-1}) + R_w + A(z)R_\varepsilon A^T(z^{-1}) \\ &= H(z)R_\xi H^T(z^{-1}).\end{aligned}$$

From A4.1.5 it follows that $A(z)$ and $B(z)(R_v + R_\eta)B^T(z^{-1}) + R_w + A(z)R_\varepsilon A^T(z^{-1})$ have no common left factor, and hence $A(z)$ and $H(z)$ have no common left factor.

Further, from (4.1.7) we have $E\chi^0_k\xi^T_{k-s} = H_sR_\xi$.

On the other hand, $\xi_{k-s} = \sum_{i=0}^{\infty} F_i\big(B(z)(\eta_{k-s-i} + v_{k-s-i}) + w_{k-s-i} + A(z)\varepsilon_{k-s-i}\big)$, and by noticing $\chi^0_k \triangleq B(z)(v_k + \eta_k) + w_k + A(z)\varepsilon_k$ we derive that $E\chi^0_k\xi^T_{k-s} = B_s(R_v + R_\eta)B_1^T + A_sR_\varepsilon$, where B_s and A_s are defined in A4.1.5. Consequently,

$$H_s = (B_s(R_v + R_\eta)B_1^T + A_sR_\varepsilon)R_\xi^{-1},$$

and, hence, by A4.1.5 $[A_p \; H_s]$ is of row-full-rank. □

Estimation Algorithms

Comparing (4.1.5) with (3.3.62), we find that they both are ARMA processes and y_k, $C(z)$, and w_k in (3.3.62) correspond to ζ_k, $H(z)$, and ξ_k, respectively, in (4.1.5).

Besides, it is clear that $\{\xi_k\}$ may be assumed as a sequence of iid random vectors in problems where only moments of order no higher than second are concerned. Therefore, to estimate coefficients of $A(z)$ in the present case, we may apply the same method as that used in Section 3.4 described by (3.4.1)–(3.4.9).

By stationarity of $\{z_k\}$ let us first recursively estimate Ez_k and $R_i \triangleq E\zeta_k\zeta_{k-i}^T$ as follows:

$$\mu_k^* = (1-\frac{1}{k})\mu_{k-1}^* + \frac{1}{k}z_k, \tag{4.1.8}$$

$$R_{i,k} = R_{i,k-1} - \frac{1}{k}\left(R_{i,k-1} - (z_k - \mu_k^*)(z_{k-i} - \mu_k^*)^T\right). \tag{4.1.9}$$

It is worth noting that (4.1.8) and (4.1.9) are the recursive expressions of

$$\mu_k^* = \frac{1}{k}\sum_{j=1}^{k} z_j \quad \text{and} \quad R_{i,k} = \frac{1}{k}\sum_{j=1}^{k}(z_j - \mu_j^*)(z_{j-i} - \mu_j^*)^T.$$

By setting

$$\varphi_k^T(t) \triangleq [\zeta_k^T, \dots, \zeta_{k-t+1}^T], \tag{4.1.10}$$

we can rewrite (4.1.5) as $\zeta_{k+1} + \theta_A^T\varphi_k(p) = H(z)\xi_{k+1}$.

Noticing $E(H(z)\xi_{k+1})\zeta_{k-t}^T = 0 \ \forall t \geq s$, we have

$$E(\zeta_{k+1} + \theta_A^T\varphi_k(p))\zeta_{k-t}^T = 0 \ \forall t \geq s.$$

Therefore, $E(\zeta_{k+1} + \theta_A^T\varphi_k(p))\varphi_{k-s}^T(mp) = 0$.

Setting

$$E\zeta_{k+1}\varphi_{k-s}^T(mp) = [R_{s+1}, \dots, R_{s+pm}] \triangleq W \in \mathbb{R}^{m\times m^2p}, \tag{4.1.11}$$

and

$$E\varphi_k(p)\varphi_{k-s}^T(mp) \triangleq \Gamma,$$

we derive the generalized Yule–Walker equation:

$$W = -\theta_A^T\Gamma, \quad \text{or} \quad \Gamma\Gamma^T\theta_A + \Gamma W^T = 0, \tag{4.1.12}$$

where

$$\Gamma \triangleq \begin{bmatrix} R(s) & R(s+1) & \cdots & R(s+mp-1) \\ R(s-1) & R(s) & \cdots & R(s+mp-2) \\ \vdots & \vdots & & \vdots \\ R(s-p+1) & R(s-p+2) & \cdots & R(s+mp-p) \end{bmatrix} \in \mathbb{R}^{mp\times m^2p}, \tag{4.1.13}$$

which coincides with (3.4.2) with r replaced by s. Therefore, Γ is of row-full-rank by Theorem 3.3.3, i.e., $\text{rank}\Gamma = mp$.

Replacing R_i in (4.1.11) and (4.1.12) with $R_{i,k}$ given by (4.1.9) leads to the estimates W_k and Γ_k for W and Γ, respectively.

Let $\{M_k\}$ be a sequence of positive real numbers increasingly diverging to infinity and let $\Gamma_0 \in \mathbb{R}^{mp\times m^2 p}$, $W_0 \in \mathbb{R}^{m\times m^2 p}$, $\theta_{A,0} \in \mathbb{R}^{mp\times m}$ be arbitrarily taken. The estimate for θ_A is defined by the SAAWET:

$$\theta_{A,k+1} = \left[\theta_{A,k} - \frac{1}{k+1}\Gamma_k(\Gamma_k^T\theta_{A,k} + W_k^T)\right]$$
$$\cdot I_{\{\|\theta_{A,k} - \frac{1}{k+1}\Gamma_k(\Gamma_k^T\theta_{A,k}+W_k^T)\|\leqslant M_{\sigma_k}\}}, \tag{4.1.14}$$

$$\sigma_k = \sum_{j=1}^{k-1} I_{\{\|\theta_{A,j} - \frac{1}{j+1}\Gamma_j(\Gamma_j^T\theta_{A,j}+W_j^T)\|> M_{\sigma_j}\}}, \tag{4.1.15}$$

with $\sigma_0 = 0$.

We now define the estimation algorithm for θ_B.

By A4.1.1 and A4.1.4 from (4.1.5) it follows that

$$E[\chi_{k+1}u_{k-i}^T] = B_i\Upsilon = B_i,\ i = 1,\dots,q. \tag{4.1.16}$$

With the help of the obtained estimates $\{\theta_{A,k} = [A_{1,k},\cdots,A_{p,k}]^T\}$ it is natural to take

$$\hat{\chi}_{k+1} = z_{k+1} + A_{1,k+1}z_k + \cdots + A_{p,k+1}z_{k+1-p}$$

as the estimate for χ_{k+1}. For any given initial values $B_{i,0}$, the estimates $B_{i,k}$ for B_i, $i = 1,\dots,q$, are recursively given by the following algorithm

$$B_{i,k+1} = B_{i,k} - \tfrac{1}{k+1}(B_{i,k} - \hat{\chi}_{k+1}u_{k-i}^T). \tag{4.1.17}$$

We now define the algorithm for estimating $f(u)$ at any fixed $u \in \mathbb{R}^l$. For this we apply the multi-variable kernel functions $\{\omega_k\}$.

To define ω_k, let $K(\cdot) : (\mathbb{R}^l, \mathscr{B}(\mathbb{R}^l)) \to (\mathbb{R}_+, \mathscr{B}(\mathbb{R}_+))$ be measurable, positive, symmetric, and satisfy the following conditions

K1 $\sup_{u\in\mathbb{R}^l}|K(u)| < \infty$,

K2 $\int_{\mathbb{R}^l}|K(u)|du < \infty$, *and* $\int_{\mathbb{R}^l}K(u)du = 1$,

K3 $\lim_{\|u\|\to\infty}\|u\|^l K(u) = 0$.

Define

$$K_h(u) \triangleq h_k^{-l}K(h_k^{-1}u^{(1)},\cdots,h_k^{-1}u^{(l)}), \tag{4.1.18}$$

where $u = [u^{(1)},\cdots,u^{(l)}]^T$, and the positive real numbers $\{h_k\}$ and $\{a_k\}$ are such that

H1 $h_k > 0,\ \lim_{k\to\infty} h_k = 0,\ \lim_{k\to\infty} kh_k^l = \infty.$

H2 $a_k > 0,\ \lim_{k\to\infty} a_k = 0,\ \sum_{k=1}^{\infty} a_k = \infty$, *and* $\sum_{k=1}^{\infty} h_k^{-l}a_k^2 < \infty$, where $\{h_k\}$ is the bandwidth of the kernel estimation and $\{a_k\}$ is the step-size used in the SA algorithms to be defined later.

As $K(u)$, for example, we may take the Gauss kernel function $K(u) = (2\pi)^{-\frac{l}{2}} e^{-\frac{1}{2}u^T u}$ with $a_k = 1/k^{\alpha}$, $h_k = 1/k^{\delta}$, $\alpha \in (0,1]$, $\delta \in (0, \frac{1}{l})$, and $2\alpha - l\delta > 1$.

Finally, define the kernel function

$$\omega_k \triangleq K_h(u - u_k) \tag{4.1.19}$$

to be used in the algorithm estimating

$$g(u) \triangleq B_1 f(u) + \sum_{i=2}^{q} B_i E f(u_{k-i}) = B_1 f(u) + B(1)\mu - B_1 \mu. \tag{4.1.20}$$

Let $\{M_k\}$ be a sequence of positive real numbers increasingly diverging to infinity, but not necessarily be the same as that used in (4.1.14).

With any initial values $g_0(u) \in \mathbb{R}^m$ and $\lambda_0(u) = 0$, the algorithm for estimating $g(u)$ is defined as follows:

$$\begin{aligned} g_{k+1}(u) &= [g_k(u) - a_k \omega_k (g_k(u) - \hat{\chi}_{k+1})] \\ &\quad \cdot I_{\{\|g_k(u) - a_k \omega_k (g_k(u) - \hat{\chi}_{k+1})\| \leqslant M_{\lambda_k(u)}\}}, \\ \lambda_k(u) &= \sum_{j=1}^{k-1} I_{\{\|g_j(u) - a_j \omega_j (g_j(u) - \hat{\chi}_{j+1})\| > M_{\lambda_j(u)}\}}. \end{aligned} \tag{4.1.21}$$

Further, define $A_k(1) \triangleq \sum_{i=0}^{p} A_{i,k}$ with $A_{0,k} = I$, $B_k(1) \triangleq \sum_{j=1}^{q} B_{j,k}$ to serve as estimates of $A(1)$ and $B(1)$, respectively.

Putting $A_k(1)$, $B_k(1)$, and μ_k^* given by (4.1.8) into (4.1.3) leads the estimate for μ:

$$\mu_k = B_k^+(1) A_k(1) \mu_k^*, \tag{4.1.22}$$

where $B_k^+(1)$ denotes the pseudo-inverse of $B_k(1)$.

Finally, from (4.1.20) we obtain the estimate for $f(u)$ as follows:

$$f_k(u) = B_{1,k}^+ \left(g_k(u) - A_k(1)\mu_k^* + B_{1,k}\mu_k \right). \tag{4.1.23}$$

Strong Consistency

We now proceed to show that the estimates μ_k^* given by (4.1.8), $\theta_{A,k}$ given by (4.1.14)–(4.1.15), $B_{i,k}$ given by (4.1.17), μ_k given by (4.1.22), $g_k(u)$ given by (4.1.21), and $f_k(u)$ given by (4.1.23) all are strongly consistent.

Theorem 4.1.1 *Assume A4.1.1–A4.1.6 hold. Then,*

$$\theta_{A,k} \xrightarrow[k \to \infty]{} \theta_A \text{ a.s.}, \quad B_{i,k} \xrightarrow[k \to \infty]{} B_i, \text{ a.s.}, \quad i = 1, \dots, q. \tag{4.1.24}$$

Proof. By ergodicity of $\{z_k\}$ we have $\mu_k^* \xrightarrow[k\to\infty]{} \mu^*$ a.s. and $R_{i,k} \xrightarrow[k\to\infty]{} R_i$ a.s., and hence $W_k \xrightarrow[k\to\infty]{} W$ a.s. and $\Gamma_k \xrightarrow[k\to\infty]{} \Gamma$ a.s. Consequently, the proof of Theorem 3.4.1 can be applied to the present case to derive $\theta_{A,k} \xrightarrow[k\to\infty]{} \theta_A$ a.s.

By ergodicity of $\chi_{k+1}u_{k-i}^T$ and the convergence of $\theta_{A,k}$ to θ_A from (4.1.16) and the definition of $\hat{\chi}_k$ we derive that

$$\frac{1}{k}\sum_{j=1}^{k}\hat{\chi}_{j+1}u_{j-i}^T \xrightarrow[k\to\infty]{} B_i.$$

Notice that (4.1.17) is merely the recursive expression of $\frac{1}{k}\sum_{j=1}^{k}\hat{\chi}_{j+1}u_{j-i}^T$, so from here the assertion concerning $B_{i,k}$ follows. □

Prior to estimating $f(\cdot)$ we prove two lemmas.

Lemma 4.1.2 *Under A4.1.1 and A4.1.3, if the kernel function K satisfies K1–K3 and $\{h_k\}$ satisfies H1, then*

$$E\omega_k \xrightarrow[k\to\infty]{} p(u),\; E[(h_k^{l/2}\omega_k)^2] \xrightarrow[k\to\infty]{} p(u)\kappa_{02}, \tag{4.1.25}$$

$$E[\omega_k f(u_k)] \xrightarrow[k\to\infty]{} p(u)f(u),\; E[\omega_k\|f(u_k)\|] \xrightarrow[k\to\infty]{} p(u)\|f(u)\|, \tag{4.1.26}$$

$$E[\|h_k^{l/2}\omega_k f(u_k)\|^2] \xrightarrow[k\to\infty]{} p(u)\|f(u)\|^2\kappa_{02}, \tag{4.1.27}$$

where $\kappa_{02} \triangleq \int_{\mathbb{R}^l} K^2(u)du$.

Proof. We have

$$\begin{aligned}
E[K_h(u-u_k)] &= \int_{\mathbb{R}^l}\frac{1}{h_k^l}K\left(\frac{u-v}{h_k}\right)p(v)dv\\
&= \int_{\mathbb{R}^l}\frac{1}{h_k^l}K\left(\frac{v}{h_k}\right)p(u-v)dv\\
&= \int_{\mathbb{R}^l}K(x)p(u-h_kx)dx \xrightarrow[k\to\infty]{} p(u)\int_{\mathbb{R}^l}K(x)dx = p(u).
\end{aligned}$$

This proves the first assertion in (4.1.25). The remaining assertions of the lemma can similarly be proved. □

Lemma 4.1.3 *Under A4.1.1, A4.1.3, and A4.1.4, if $K(\cdot)$ satisfies K1–K3, $\{h_k\}$ and $\{a_k\}$ satisfy H1 and H2, then*

$$\sum_{k=1}^{\infty}a_k(\omega_k - E\omega_k) < \infty \text{ a.s.}, \tag{4.1.28}$$

$$\sum_{k=1}^{\infty}a_k(\omega_k\|f(u_{k-j})\| - E(\omega_k\|f(u_{k-j}))\|) < \infty \quad \text{a.s. } \forall j\ge 0, \tag{4.1.29}$$

$$\sum_{k=1}^{\infty}a_k\omega_k(\|w_{k+1-j}\| - E\|w_{k+1-j}\|) < \infty \text{ a.s. } \forall j\ge 0, \tag{4.1.30}$$

$$\sum_{k=1}^{\infty} a_k \omega_k (\|\eta_{k-j}\| - E\|\eta_{k-j}\|) < \infty \text{ a.s. } \forall j \geq 0, \tag{4.1.31}$$

$$\sum_{k=1}^{\infty} a_k \omega_k (\|\varepsilon_{k+1-j}\| - E\|\varepsilon_{k+1-j}\|) < \infty \text{ a.s. } \forall j \geq 0. \tag{4.1.32}$$

The series (4.1.29)–(4.1.32) with $\|f(u_{k-j})\|, \|w_{k+1-j}\|, \|\eta_{k-j}\|$, and $\|\varepsilon_{k+1-j}\|$ replaced by $f(u_{k-j}), w_{k+1-j}, \eta_{k-j}$, and ε_{k+1-j}, respectively, are still convergent.

Proof. Let us first show (4.1.28). By A4.1.1 $\{h_k^{l/2}\omega_k\}$ is a sequence of mutually independent random vectors, which have bounded second moments by (4.1.25).

Since $\sum_{k=1}^{\infty} a_k^2/h_k^l < \infty$, by Theorem 1.2.8 it follows that

$$\sum_{k=1}^{\infty} a_k(\omega_k - E\omega_k) = \sum_{k=1}^{\infty} \frac{a_k}{h_k^{l/2}} \left(h_k^{l/2}\omega_k - h_k^{l/2} E\omega_k\right) < \infty, \quad \text{a.s.} \tag{4.1.33}$$

The assertion (4.1.29) is similarly proved.

Since ω_k, w_{k+1}, η_{k-j}, and ε_{k+1-i} are mutually independent for $j \geq 0$ and $i \geq 0$, and $E\|\Delta_k\|^2 < \infty$, we have

$$\sum_{k=1}^{\infty} a_k^2 E[\omega_k^2] E[\|\Delta_k\|^2] = O\Big(\sum_{k=1}^{\infty} \frac{a_k^2}{h_k^l} E[(h_k^{l/2}\omega_k)^2]\Big) < \infty.$$

Again, by Theorem 1.2.8, (4.1.30)–(4.1.32) hold. □

Lemma 4.1.4 *Assume A4.1.1–A4.1.6 hold. If $K(\cdot)$ satisfies K1–K3, $\{h_k\}$ and $\{a_k\}$ satisfy H1 and H2, then there exists an Ω_1 with $\mathbb{P}\{\Omega_1\} = 1$ such that for any $\omega \in \Omega_1$ and any convergent subsequence $g_{n_k}(u)$ of $\{g_k(u)\}$: $g_{n_k}(u) \xrightarrow[k\to\infty]{} \bar{g}(u)$, for sufficient large k and small enough $T > 0$ it takes place that*

$$g_{s+1}(u) = g_s(u) - a_s\omega_s(g_s(u) - \chi_{s+1}) \tag{4.1.34}$$

and

$$\|g_{s+1}(u) - g_{n_k}(u)\| \leq cT \ \ \forall s : n_k \leq s \leq m(n_k, T), \tag{4.1.35}$$

where c is a constant which is independent of k but may depend on ω.

Proof. By (4.1.25) and (4.1.28) we have

$$\begin{aligned}\sum_{i=n_k}^{s} a_i\omega_i &= \sum_{i=n_k}^{s} a_i(\omega_i - E\omega_i) + \sum_{i=n_k}^{s} a_i E\omega_i \\ &= O(T) \ \ \forall s : n_k \leq s \leq m(n_k, T)\end{aligned} \tag{4.1.36}$$

as $k \to \infty$ and $T \to 0$.

Similarly, by (4.1.29) it follows that

$$\sum_{i=n_k}^{s} a_i\omega_i\|f(u_{i-j})\| = \sum_{i=n_k}^{s} a_i\big(\omega_i\|f(u_{i-j})\| - E(\omega_i\|f(u_{i-j})\|)\big)$$
$$+\sum_{i=n_k}^{s} a_iE(\omega_i\|f(u_{i-j})\|) = O\big(\sum_{i=n_k}^{s} a_iE(\omega_i\|f(u_{i-j})\|)\big),$$

which is of $O(T)$ as $k\to\infty$ and $T\to 0$ because of (4.1.26) and (4.1.25), respectively, for $j=0$ and for $j\geq 1$ by noticing $E(\omega_i\|f(u_{i-j})\|) = E\omega_iE\|f(u_1)\|$.

A similar treatment by (4.1.30)–(4.1.32) leads to

$$\sum_{i=n_k}^{s} a_i\omega_i\big(\|\eta_{i-j}\| + \|w_{i+1-j}\| + \|\varepsilon_{i+1-j}\|\big) = O(T) \ \ \forall j\geq 0$$

as $k\to\infty$ and $T\to 0$.

Therefore, by definition (4.1.6), we derive

$$\sum_{i=n_k}^{s} a_i\omega_i\|\chi_{i+1-j}\| = O(T) \quad \forall\, s: n_k\leq s\leq m(n_k,T) \ \ \forall j\geq 0 \tag{4.1.37}$$

as $k\to\infty$ and $T\to 0$.

From here, by Theorem 4.1.1, it follows that

$$\sum_{i=n_k}^{s} a_i\omega_i\|\hat{\chi}_{i+1-j}\| = O(T) \quad \forall\, s: n_k\leq s\leq m(n_k,T) \ \ \forall j\geq 0 \tag{4.1.38}$$

as $k\to\infty$ and $T\to 0$.

Since $g_{n_k}(u) \xrightarrow[k\to\infty]{} \bar{g}(u)$, by (4.1.36) and (4.1.38) we see that the algorithm (4.1.21) is without truncations for several successive steps starting from n_k, i.e., (4.1.34) takes place for several successive $s\geq n_k$ if k is sufficiently large. Define

$$\Omega_{k,j} \triangleq (1-a_k\omega_k)\cdots(1-a_j\omega_j), \quad k\leq j, \quad \Omega_{j,j+1}=1.$$

By (4.1.37) we see

$$\log\Omega_{s,n_k} = O\big(\sum_{i=n_k}^{s} a_i\omega_i\big) \ \text{ and } \ \Omega_{s,n_k} = 1+O(T) \ \ \forall s: n_k\leq s\leq m(n_k,T)$$

as $k\to\infty$ and $T\to 0$.

This implies

$$\left\|\sum_{i=n_k}^{s}\Omega_{s,i+1}a_i\omega_i\hat{\chi}_{i+1}\right\| \leq \sum_{i=n_k}^{s} a_i\omega_i\|\hat{\chi}_{i+1}\| = O(T) \ \ \forall s: n_k\leq s\leq m(n_k,T) \tag{4.1.39}$$

as $k \to \infty$ and $T \to 0$.

Consequently, we have

$$\begin{aligned} g_{s+1}(u) =& \Omega_{s,n_k} g_{n_k}(u) + \sum_{i=n_k}^{s} \Omega_{s,i+1} a_i \omega_i \hat{\chi}_{i+1} \\ =& g_{n_k}(u) + O(T) \ \ \forall s : n_k \le s \le m(n_k, T) \end{aligned}$$

as $k \to \infty$ and $T \to 0$.

Thus, we have shown not only (4.1.35) but also the fact that there is no truncation in (4.1.21) $\forall s : n_k \le s \le m(n_k, T)$ if k is sufficiently large and T is small enough. The proof is completed. □

Theorem 4.1.2 *Assume A4.1.1–A4.1.6 hold. Then for any $u \in U$*

$$g_k(u) \xrightarrow[k\to\infty]{} g(u) \ \text{a.s.}, \ \ f_k(u) \xrightarrow[k\to\infty]{} f(u) \ \text{a.s.}$$

Proof. The algorithm (4.1.21) is rewritten as

$$\begin{aligned} g_{k+1}(u) =& [g_k(u) - a_k p(u)(g_k(u) - g(u)) - a_k e_{k+1}(u)] \\ & \cdot I_{\{\|g_k(u) - a_k p(u)(g_k(u) - g(u)) - a_k e_{k+1}(u)\| \leqslant M_{\lambda_k(u)}\}} \end{aligned} \tag{4.1.40}$$

with $e_{k+1}(u) = \sum_{i=1}^{6} e_{k+1}^{(i)}(u)$, where

$$\begin{aligned} e_{k+1}^{(1)}(u) &= (\omega_k - p(u))\, g_k(u), \\ e_{k+1}^{(2)}(u) &= \omega_k (\chi_{k+1} - \hat{\chi}_{k+1}), \\ e_{k+1}^{(3)}(u) &= -\left[\omega_k B(z) f(u_{k+1}) - p(u)(B_1 f(u) + B(1)\mu - B_1 \mu)\right], \\ e_{k+1}^{(4)}(u) &= -\omega_k w_{k+1}, \\ e_{k+1}^{(5)}(u) &= -\omega_k B(z) \eta_{k+1}, \\ e_{k+1}^{(6)}(u) &= -\omega_k A(z) \varepsilon_{k+1}. \end{aligned}$$

Then the problem becomes seeking the root of the linear function $h(x) = p(u)(x - g(u))$ with root equal to $g(u) = B_0 f(u) + B(1)\mu - B_0 \mu$.

By Theorem 2.3.1 for $g_k(u) \xrightarrow[k\to\infty]{} g(u)$ a.s. it suffices to verify the noise condition: For any fixed $\omega \in \Omega_0$ with $\mathbb{P}\Omega_0 = 1$ along any convergent subsequence $\{g_{n_k}(u)\}$ $\left(g_{n_k}(u) \xrightarrow[k\to\infty]{} \bar{g}(u)\right)$ of $\{g_k(u)\}$ the following limit takes place for $i = 1, \ldots, 6$

$$\lim_{T\to 0} \limsup_{k\to\infty} \frac{1}{T} \left\| \sum_{j=n_k}^{m(n_k, T_k)} a_j e_{j+1}^{(i)}(u) \right\| = 0, \ \forall T_k \in [0, T], \tag{4.1.41}$$

where $m(n_k, T_k) \triangleq \max\left\{m : \sum_{j=n_k}^{m} a_j \le T_k\right\}$.

For $i = 1$, we have

$$\sum_{j=n_k}^{m(n_k,T_k)} a_j(\omega_j - p(u))g_j(u)$$
$$= \sum_{j=n_k}^{m(n_k,T_k)} a_j(\omega_j - E\omega_j)(g_j(u) - \bar{g}(u)) + \bar{g}(u) \sum_{j=n_k}^{m(n_k,T_k)} a_j(\omega_j - E\omega_j)$$
$$+ \sum_{j=n_k}^{m(n_k,T_k)} a_j(E\omega_j - p(u))g_j(u). \tag{4.1.42}$$

By (4.1.28) the second term at the right-hand side of (4.1.41) tends to zero as $k \to \infty$, while the third term tends to zero by (4.1.25) and (4.1.35).

It remains to prove that the first term at the right-hand side of (4.1.42) tends to zero as $k \to \infty$. By (4.1.35) and also by the fact $g_{n_k}(u) \xrightarrow[k\to\infty]{} \bar{g}(u)$, we have

$$\lim_{T\to 0} \limsup_{k\to\infty} \frac{1}{T} \left\| \sum_{j=n_k}^{m(n_k,T_k)} a_j(\omega_j - E\omega_j)(g_j(u) - \bar{g}(u)) \right\|$$
$$= \lim_{T\to 0} \limsup_{k\to\infty} \frac{1}{T} O(T) \sum_{j=n_k}^{m(n_k,T_k)} a_j E|\omega_j - E\omega_j|$$
$$= \lim_{T\to 0} \limsup_{k\to\infty} O(1) \sum_{j=n_k}^{m(n_k,T_k)} a_j E\omega_j = 0,$$

where the last equality follows from (4.1.25). Thus, we have proved (4.1.41) for $i = 1$.

We now show (4.1.41) for $i = 2$.

By A4.1.1–A4.1.4, without loss of generality, we may assume the processes $\{y_k\}$ and $\{z_k\}$ are stationary.

Define

$$\Phi = \begin{bmatrix} -A_1 & I & \cdots & 0 \\ -A_2 & 0 & \ddots & \vdots \\ \vdots & \vdots & \ddots & \mathrm{I} \\ -A_p & 0 & \cdots & 0 \end{bmatrix}, \ H = \begin{bmatrix} I \\ 0 \\ \vdots \\ 0 \end{bmatrix},$$

where $\Phi \in \mathbb{R}^{mp\times mp}$, $H \in \mathbb{R}^{mp\times m}$, and $A_j \triangleq 0$ for $j > p$.

Then, from (4.1.6) it follows that

$$\begin{cases} x_{k+1} = \Phi x_k + H\chi_{k+1}, \\ z_{k+1} = H^T x_{k+1}, \quad k \geq 0, \end{cases} \tag{4.1.43}$$

where $x_k \in \mathbb{R}^{mp}$.

Therefore, we have

$$z_{k+1} = H^T\big(\Phi^{k+1}x_0 + \sum_{i=0}^{k}\Phi^{k-i}H^T\chi_{i+1}\big).$$

By stability of $A(z)$, there are constants $\rho_1 > 0$ and $\rho \in (0,1)$ such that $\|\Phi^k\| < \rho_1\rho^k \ \ \forall k \geq 0$. Consequently, we have

$$\|z_{k+1}\| < c_1\rho^{k+1} + c_2\sum_{i=0}^{k}\rho^{k-i}\|\chi_{i+1}\|. \tag{4.1.44}$$

We now show that for any fixed integer $j \geq 0$

$$\sum_{i=n_k}^{s} a_i\omega_i\|z_{i+1-j}\| = O(T) \ \ \forall s : n_k \leq s \leq m(n_k,T) \tag{4.1.45}$$

as $k \to \infty$ and $T \to 0$.

By (4.1.44) we have

$$\begin{aligned}\sum_{i=n_k}^{s} a_i\omega_i\|z_{i+1-j}\| &\leq \sum_{i=n_k}^{s} a_i\omega_i\big(c_1\rho^{i+1-j} + c_2\sum_{\kappa=0}^{i}\rho^{i-j-\kappa}\|\chi_{\kappa+1}\|\big)\\ &= \sum_{i=n_k}^{s} a_i\omega_i c_1\rho^{i-j+1} + c_2\sum_{i=n_k}^{s} a_i\omega_i\sum_{t=j}^{i}\rho^{t-j}\|\chi_{i-t+1}\|.\end{aligned} \tag{4.1.46}$$

Since $\rho \in (0,1)$, by (4.1.36) for any fixed integer j we have

$$\lim_{k\to\infty}\sum_{i=n_k}^{s} a_i\omega_i\rho^{i+1-j} \leq \lim_{k\to\infty}\rho^{n_k-j+1}\sum_{i=n_k}^{s} a_i\omega_i = 0. \tag{4.1.47}$$

By (4.1.37) it follows that as $T \to 0$

$$\begin{aligned}&\lim_{k\to\infty} c_2\sum_{i=n_k}^{s} a_i\omega_i\Big(\sum_{t=j}^{n_k-1} + \sum_{t=n_k}^{i}\Big)\rho^{t-j}\|\chi_{i-t+1}\|\\ &= \lim_{k\to\infty} c_2\sum_{t=j}^{n_k-1}\rho^{t-j}\sum_{i=n_k}^{s} a_i\omega_i\|\chi_{i-t+1}\|\\ &+ \lim_{k\to\infty} c_2\sum_{t=n_k}^{s}\rho^{t-j}\sum_{i=t}^{s} a_i\omega_i\|\chi_{i-t+1}\| = O(T) \ \ \forall s : n_k \leq s \leq m(n_k,T).\end{aligned} \tag{4.1.48}$$

Combining (4.1.46)–(4.1.47) leads to (4.1.45).

By Theorem 4.1.1 and (4.1.45) we have

$$\lim_{k\to\infty}\frac{1}{T}\|\sum_{j=n_k}^{m(n_k,T_k)} a_j\omega_j(\chi_{j+1}-\hat{\chi}_{j+1})\|$$
$$\leq \sum_{\kappa=0}^{p}\lim_{k\to\infty}\frac{1}{T}\sum_{j=n_k}^{m(n_k,T_k)} a_i\omega_i\|A_\kappa - A_{\kappa,j+1}\|\|z_{j+1-\kappa}\| = 0, \tag{4.1.49}$$

where $A_0 \triangleq I$, $A_{0,j}=I\ \forall j\geq 0$. Thus, we have proved (4.1.41) for $i=2$.

For $i=3$ we have

$$\sum_{j=n_k}^{m(n_k,T_k)} a_j\Big[\omega_j B(z)f(u_{j+1})-p(u)\big(B_1f(u)+B(1)\mu - B_1\mu\big)\Big]$$
$$= \sum_{j=n_k}^{m(n_k,T_k)} a_j\big(\omega_jB_1f(u_j)-B_1E[\omega_jf(u_j)]\big)$$
$$+ \sum_{j=n_k}^{m(n_k,T_k)} a_j\big(B_1E[\omega_jf(u_j)]-p(u)B_1f(u)\big)$$
$$+ \sum_{j=n_k}^{m(n_k,T_k)} a_j\big(\sum_{\kappa=2}^{q}\omega_jB_\kappa f(u_{j-\kappa+1})-E\omega_j\sum_{\kappa=2}^{q}B_\kappa Ef(u_1)\big)$$
$$+ \sum_{j=n_k}^{m(n_k,T_k)} a_j\big(E\omega_j\sum_{\kappa=2}^{q}B_\kappa Ef(u_\kappa)-p(u)B(1)\mu+p(u)B_1\mu\big). \tag{4.1.50}$$

At the right-hand side of (4.1.50) the second and the fourth terms tend to zero as $k\to\infty$ by (4.1.25) and (4.1.26), and the first and the third terms tend to zero as $k\to\infty$ by (4.1.29) and (4.1.30), where (4.1.25)–(4.1.30) are used with norm removed as indicated in Lemma 4.1.3.

Convergence to zero of $e^{(i)}_{k+1}(u)$, $i=4,5,6$ as $k\to\infty$ follows from (4.1.31) and (4.1.32) with norm removed. This completes the proof. □

Numerical Example

Consider the following system with $m=3$, $l=2$, $p=2$, and $q=2$:

$$y_{k+1}+A_1y_k+A_2y_{k-1}=B_1v_k+B_2v_{k-1}+w_{k+1},$$

where $v_k=f(u_k)+\eta_k$, $f(u_k)=\begin{bmatrix}\frac{5}{16}u_k^{(1)3}-\frac{5}{16}u_k^{(2)2}+\frac{25}{8}\\ -\frac{5}{16}u_k^{(1)2}+\frac{5}{16}u_k^{(2)3}+\frac{5}{2}\end{bmatrix}$,

$$A_1=\begin{bmatrix}0&0.5&0\\1&0&0\\0&0&-0.8\end{bmatrix},\quad A_2=\begin{bmatrix}1.5&0&0\\0&0.5&0\\0&0&0.8\end{bmatrix},$$

$$B_1=\begin{bmatrix}-0.5&0.5\\1.5&0\\0&1\end{bmatrix},\quad B_2=\begin{bmatrix}0&-0.5\\0.8&-1\\0&1.5\end{bmatrix}.$$

It is clear that $A(z)$ is stable, and A4.1.3 and A4.1.5 are satisfied.

Let the components $\{u_k^{(1)}\}$ and $\{u_k^{(2)}\}$ of $u_k = [u_k^{(1)}, u_k^{(2)}]^T$ be iid uniformly distributed over the interval $[-2,2]$, and they are mutually independent. It is clear that $\Upsilon = E[f(u_k)u_k^T] = I$. Further, let $w_k \in \mathcal{N}(0,R_w)$, $\eta_k \in \mathcal{N}(0,R_\eta)$, and $\varepsilon_k \in \mathcal{N}(0,R_\varepsilon)$ be iid and independent of $\{u_k\}$ with covariance matrices $R_w = \begin{bmatrix} 2 & 0.2 & 0.8 \\ 0 & 1 & -0.3 \\ 0.6 & 0 & 1.5 \end{bmatrix}$, $R_\eta = \begin{bmatrix} 0.2 & -0.1 \\ 0.3 & 0.1 \end{bmatrix}$, $R_\varepsilon = \begin{bmatrix} 1.2 & 0 & 0.5 \\ 0.4 & 1 & 0 \\ 0 & 0.2 & 0.8 \end{bmatrix}$.

The algorithms have run $N = 5000$ steps. The estimates for A_1, A_2, B_1, and B_2 are presented in Figs. 4.1.2–4.1.5, respectively.

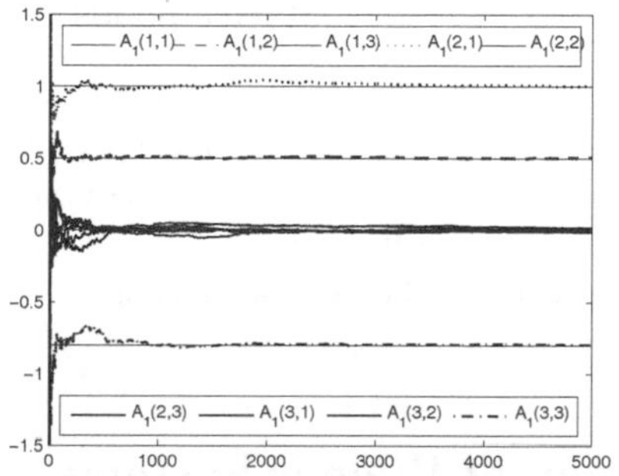

Figure 4.1.2: Estimates for A_1

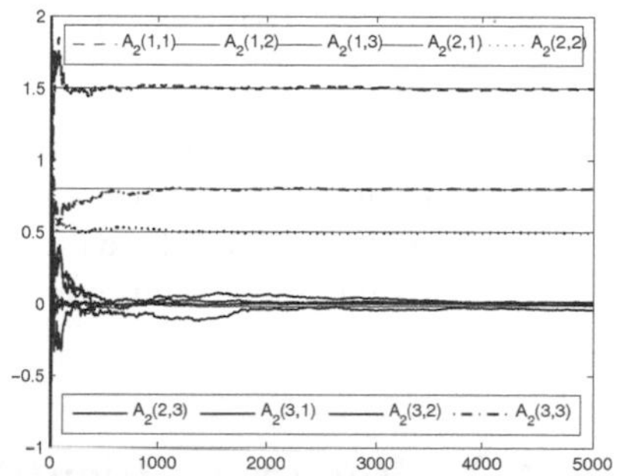

Figure 4.1.3: Estimates for A_2

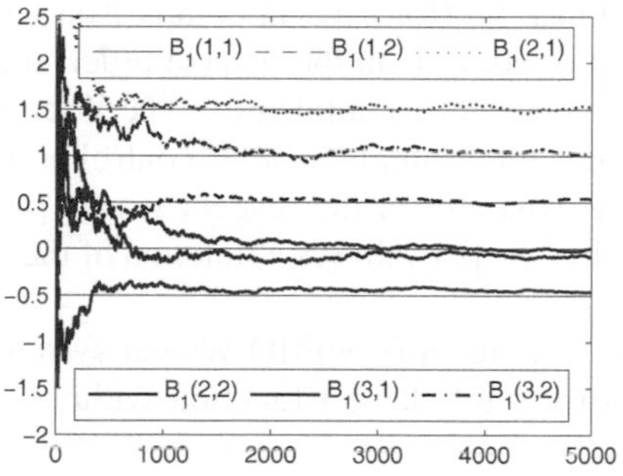

Figure 4.1.4: Estimates for B_1

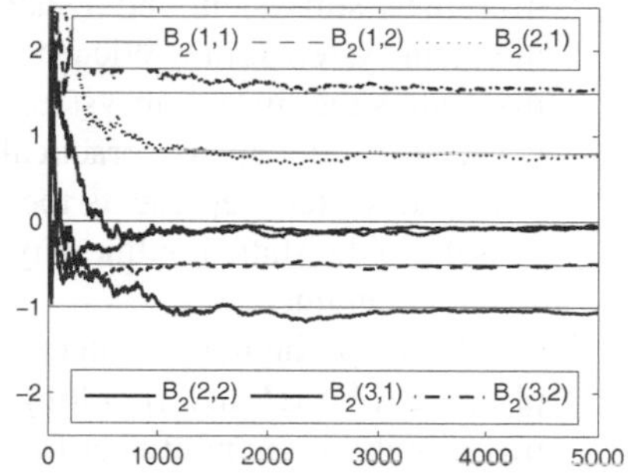

Figure 4.1.5: Estimates for B_2

For any $u \in U$, the estimates of $f_1(u)$ at $N = 5000$ are presented in Fig. 4.1.6, while the estimates of $f_2(u)$ are similar and not presented here. In the figures the solid lines denote the true surface and the dash-dotted lines the estimated surface. The estimation errors are presented by the gray surface.

When estimating $f(u)$ the domain $U = [-2,2] \times [-2,2]$ is separated into 20×20 grids and the Gauss kernel is used with $h_k = 1/k^\delta$ and $\delta = 0.18$. In the algorithm (4.1.21) we take $a_k = 1/k^\alpha$ and $\alpha = 0.7$. The estimation errors become bigger at the boundary of the domain, but they may be corrected by statistical methods.

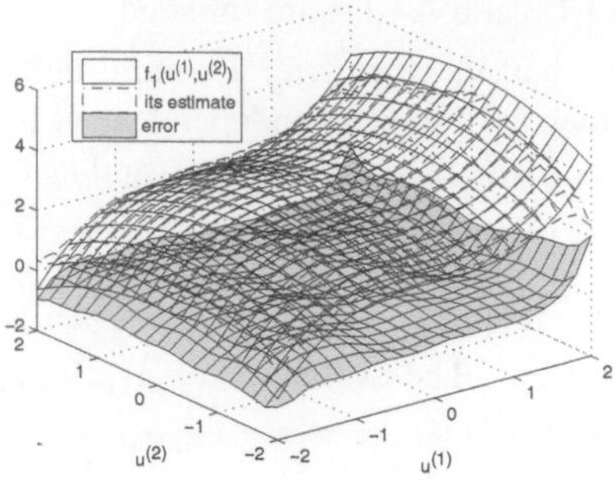

Figure 4.1.6: Estimate for $f_1(u^{(1)}, u^{(2)})$

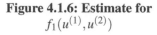

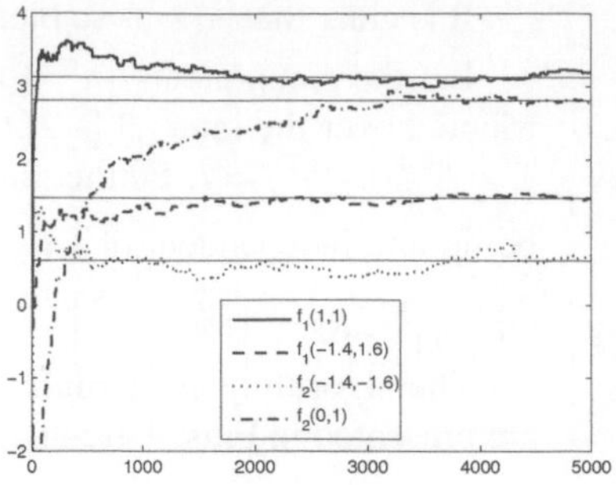

Figure 4.1.7: Estimates for f at fixed points

To demonstrate the convergence rate of estimating $f_1(\cdot,\cdot)$ and $f_2(\cdot,\cdot)$ we present estimates for $f_1(1,1)$, $f_1(-1.4,1.6)$ and $f_2(0,1)$, $f_2(-1.4,-1.6)$ vs computational steps k in Fig. 4.1.7, where the solid lines are their true values.

The simulation results are consistent with the theoretical analysis.

4.2 Recursive Identification of Wiener Systems

Like Hammerstein systems, the Wiener system is also a cascading system composed of a linear subsystem and a static nonlinearity. However, in comparison with Hammerstein systems, in a Wiener system cascading is in the reverse order: the control first enters the linear subsystem and then the static nonlinearity. The Wiener system is used to model diverse practical systems. For example, in a pH control problem, the linear subsystem represents the mixing dynamics of the reagent stream in a stirred vessel and the static nonlinearity describes the pH value as a function of the chemical species contained.

We adopt the nonparametric approach to identify MIMO Wiener systems by applying SAAWET incorporating a multi-variable kernel function without requiring invertibility of the nonlinear function.

The system is expressed as follows:

$$A(z)v_k = B(z)u_k, \quad y_k = f(v_k), \quad z_k = y_k + \varepsilon_k, \tag{4.2.1}$$

where $A(z) = I + A_1 z + \cdots + A_p z^p$ and $B(z) = B_1 z + B_2 z^2 + \cdots + B_q z^q$ are the $m \times m$ and $m \times l$ matrix-valued polynomials with unknown coefficients but with known orders p and q, respectively, and z is the backward-shift operator: $zy_{k+1} = y_k$. Moreover, $u_k \in \mathbb{R}^l$, $y_k \in \mathbb{R}^m$, and $z_k \in \mathbb{R}^m$ are the system input, output, and observation, respectively; $\varepsilon_k \in \mathbb{R}^m$ is the observation noise. The unknown nonlinear function is denoted by $f(\cdot) = [f_1, f_2, \cdots, f_m]^T$, where $f_i : \mathbb{R}^m \to \mathbb{R}$. The problem is to recursively estimate A_i, $i = 1, \cdots, p$, B_j, $j = 1, \cdots, q$, and $f(v)$ at any fixed v based on the available input-output data $\{u_k, z_k\}$.

It is important to clarify whether or not the system can uniquely be defined by the input-output $\{u_k, y_k\}$ of the system. In general, the answer is negative. To see this, let P be an $m \times m$ nonsingular matrix, and set $\tilde{A}(z) \triangleq PA(z)P^{-1}$, $\tilde{B}(z) \triangleq PB(z)$, $\tilde{f}(x) \triangleq f(P^{-1}x)$, and $\tilde{v}_k \triangleq Pv_k$. Then, the following system

$$\tilde{A}(z)\tilde{v}_k = \tilde{B}(z)u_k,\ y_k = \tilde{f}(Pv_k) = f(v_k),\ z_k = y_k + \varepsilon_k, \tag{4.2.2}$$

and the systems (4.2.1) share the same input-output $\{u_k, y_k\}$, but they have different linear and nonlinear parts. So, in order for the MIMO Wiener system to be uniquely defined, the nonsingular matrix P should be fixed in advance.

If $A(z)$ is stable, then we have (3.3.4):

$$H(z) \triangleq A^{-1}(z)B(z) = \sum_{i=1}^{\infty} H_i z^i \text{ with } H_1 = B_1,$$

where $\|H_i\| = O(e^{-ri})\ \forall i > 1$ for some $r > 0$.

If further assuming $u_k = 0\ \forall k < 0$, then similar to (3.3.5) we have

$$v_k = \sum_{i=1}^{k} H_i u_{k-i}. \tag{4.2.3}$$

If $\{u_k,\ k \geq 0\}$ is a sequence of iid Gaussian random vectors $u_k \in \mathcal{N}(0, I_{l \times l})$, then $v_k \in \mathcal{N}(0, \Sigma_k)$ and $\Sigma_k = \sum_{i=1}^{k} H_i H_i^T \xrightarrow[k \to \infty]{} \Sigma \triangleq \sum_{i=1}^{\infty} H_i H_i^T$.

Before further discussing, let us introduce assumptions to be used later on.

A4.2.1 *The input $\{u_k,\ k \geq 0\}$ is a sequence of iid Gaussian random vectors $u_k \in \mathcal{N}(0, I_{l \times l})$ and is independent of $\{\varepsilon_k\}$.*

A4.2.2 *$A(z)$ and $B(z)$ have no common left factor, $[A_p, B_q]$ is of row-full-rank, and $A(z)$ is stable:* $\det A(z) \neq 0\ \forall\ |z| \leq 1$.

A4.2.3 *$f(\cdot)$ is a measurable vector-valued function satisfying the following condition:*

$$|f_i(x_1, \cdots, x_m)| \leq c(M + |x_1|^{\nu_1} + \cdots + |x_m|^{\nu_m}), 1 \leq i \leq m, \tag{4.2.4}$$

where $c > 0$, $M > 0$, and $\nu_j \geq 0, 1 \leq j \leq m$ are constants.

A4.2.4 *$\{\varepsilon_k\}$ is a sequence of iid random vectors with $E\varepsilon_k = 0$ and $E[\|\varepsilon_k\|^2] < \infty$.*

A4.2.5 *The matrix Q is nonsingular, where $Q \triangleq E(f(v)v^T)$ with $v \in \mathcal{N}(0, \Sigma)$ and $\Sigma = \sum_{i=1}^{\infty} H_i H_i^T$.*

We now proceed to fix P.

Under Condition A4.2.2, by Theorem 3.3.2, the first m rows of L are of row-full-rank, where L is defined by (3.3.10). Therefore, Σ_k is nonsingular for all sufficiently large k.

By A4.2.3, $Q_{k+1} \triangleq E(f(v_{k+1})v_{k+1}^T)$ is meaningful, and $Q_k \xrightarrow[k\to\infty]{} Q$.

In order to fix coefficients of the system to be identified, let us choose P in (4.2.2) to equal $Q\Sigma^{-1}$. Then, $\tilde{P} \triangleq \tilde{Q}\tilde{\Sigma}^{-1} = I$ for system (4.2.1).

Therefore, under A4.2.5, without loss of generality, we may assume that $Q\Sigma^{-1} = I$ for system (4.2.1).

In the sequel, we will use a property of the stable linear system: If the δth absolute moment of the system's input is bounded, then the δth absolute moment of its output is also bounded.

Lemma 4.2.1 *Let $\{x_k\}$ be a sequence of random vectors with $\sup_k E\|x_k\|^\delta < \infty$ for some $\delta \geq 2$, and let L_i with $\|L_i\| = O(e^{-ri})$, $r > 0$ be a sequence of real matrices. Then, the process*

$$X_k = \sum_{i=1}^{k} L_i x_{k-i} \tag{4.2.5}$$

has the bounded δth absolute moment:

$$\sup_k E\|X_k\|^\delta < \infty. \tag{4.2.6}$$

Proof. Applying the Hölder inequality, we have

$$\begin{aligned}
\sup_k E\|X_k\|^\delta &\leq \sup_k E\Big[\sum_{i=1}^{k} \|L_i\|^{\frac{1}{2}} \|L_i\|^{\frac{1}{2}} \|x_{k-i}\|\Big]^\delta \\
&\leq \sup_k E\Big[\Big(\sum_{i=1}^{k} \|L_i\|^{\frac{\delta}{2(\delta-1)}}\Big)^{\frac{\delta-1}{\delta}} \cdot \Big(\sum_{i=1}^{k} \|L_i\|^{\frac{\delta}{2}} \|x_{k-i}\|^\delta\Big)^{\frac{1}{\delta}}\Big]^\delta \\
&= \sup_k \Big[\Big(\sum_{i=1}^{k} \|L_i\|^{\frac{\delta}{2(\delta-1)}}\Big)^{\delta-1} \cdot \Big(\sum_{i=1}^{k} \|L_i\|^{\frac{\delta}{2}} E\|x_{k-i}\|^\delta\Big)\Big] < \infty.
\end{aligned}$$

□

Estimation of $\{\mathbf{A_1}, \cdots, \mathbf{A_p}, \mathbf{B_0}, \cdots, \mathbf{B_q}\}$

We first estimate the impulse responses H_i and then the coefficients of $A(z)$ and $B(z)$.

Lemma 4.2.2 *Assume A4.2.1–A4.2.5 hold. Then*

$$E z_{k+1} u_{k+1-i}^T \xrightarrow[k\to\infty]{} H_i \ \forall\, i \geq 1. \tag{4.2.7}$$

Proof. By noticing $v_{k+1} \in \mathcal{N}(0, \Sigma_{k+1})$, $\Sigma_{k+1} \triangleq \sum_{i=1}^{k+1} H_i H_i^T$, the Gaussian vector $g_{k,j} \triangleq [v_{k+1}^T, u_{k+1-j}^T]^T$, $j \geq 1$ is zero-mean with covariance matrix

$$G_{k,j} \triangleq \begin{bmatrix} \Sigma_{k+1} & H_j \\ H_j^T & I \end{bmatrix}. \tag{4.2.8}$$

It is straightforward to verify that $G_{k,j}$ can be factorized $G_{k,j} = L_{k,j}L_{k,j}^T$ with

$$L_{k,j} \triangleq \begin{bmatrix} \Sigma_{k+1}^{\frac{1}{2}} & 0 \\ H_j^T\Sigma_{k+1}^{-\frac{1}{2}} & (I_{l\times l} - H_j^T\Sigma_{k+1}^{-1}H_j)^{\frac{1}{2}} \end{bmatrix}. \tag{4.2.9}$$

It then follows that $L_{k,j}^{-1}G_{k,j}L_{k,j}^{-T} = I_{(m+l)\times(m+l)}$ and $l_{k,j} = [l_{k,j}(1)^T, l_{k,j}(2)^T]^T \triangleq L_{k,j}^{-1}g_{k,j}$ is a Gaussian vector, $l_{k,j} \in \mathcal{N}(0, I_{(m+l)\times(m+l)})$, and hence the components of $g_{k,j} = L_{k,j}l_{k,j}$ can be expressed as follows:

$$v_{k+1} = \Sigma_{k+1}^{\frac{1}{2}} l_{k,j}(1), \tag{4.2.10}$$

$$u_{k+1-j} = H_j^T\Sigma_{k+1}^{-\frac{1}{2}} l_{k,j}(1) + (I_{l\times l} - H_j^T\Sigma_{k+1}^{-1}H_j)^{\frac{1}{2}} l_{k,j}(2). \tag{4.2.11}$$

Noting that the components of $l_{k,j}$ are orthogonal to each other, we obtain

$$\begin{aligned} &Ef(v_{k+1})u_{k+1-j}^T \\ &= Ef(\Sigma_{k+1}^{\frac{1}{2}} l_{k,j}(1))\left[H_j^T\Sigma_{k+1}^{-\frac{1}{2}} l_{k,j}(1) + (I_{l\times l} - H_j^T\Sigma_{k+1}^{-1}H_j)^{\frac{1}{2}} l_{k,j}(2)\right]^T \\ &= Ef(\Sigma_{k+1}^{\frac{1}{2}} l_{k,j}(1))[\Sigma_{k+1}^{\frac{1}{2}} l_{k,j}(1)]^T\Sigma_{k+1}^{-1}H_j = Q_{k+1}\Sigma_{k+1}^{-1}H_j, \end{aligned} \tag{4.2.12}$$

where $Q_{k+1} = E(f(v_{k+1})v_{k+1}^T)$.

By A4.2.1 and (4.2.11), we have

$$\begin{aligned} Ez_{k+1}u_{k+1-i}^T =& E(y_{k+1} + \varepsilon_{k+1})u_{k+1-i}^T = Ef(v_{k+1})u_{k+1-i}^T \\ =& Q_{k+1}\Sigma_{k+1}^{-1}H_i \xrightarrow[k\to\infty]{} H_i, \end{aligned}$$

since $\Sigma_k \xrightarrow[k\to\infty]{} \Sigma$, $Q_k \xrightarrow[k\to\infty]{} Q$, and $Q\Sigma^{-1} = I$.

The proof is completed. □

Based on (4.2.7), we apply SAAWET to recursively estimate H_i:

$$\begin{aligned} H_{i,k+1} =& [H_{i,k} - \frac{1}{k}(H_{i,k} - z_{k+1}u_{k+1-i}^T)] \\ & \cdot I_{[\|H_{i,k} - \frac{1}{k}(H_{i,k} - z_{k+1}u_{k+1-i}^T)\| \le M_{\delta_{i,k}}]}, \end{aligned} \tag{4.2.13}$$

$$\delta_{i,k} = \sum_{j=1}^{k-1} I_{[\|H_{i,j} - \frac{1}{j}(H_{i,j} - z_{j+1}u_{j+1-i}^T)\| > M_{\delta_{i,j}}]}, \tag{4.2.14}$$

where $\{M_k\}$ is an arbitrarily chosen sequence of positive real numbers increasingly diverging to infinity, $H_{i,0}$ is an arbitrary initial value, and I_A denotes the indicator function of a set A.

From (3.3.4), we have (3.3.5)–(3.3.12). To ease reading here we copy the basic linear equation (3.3.9) connecting θ_A with $\{H_i\}$:

$$[A_1, A_2, \cdots, A_p]L = -[H_{q+1}, H_{q+2}, \cdots, H_{q+mp}], \tag{4.2.15}$$

where L is given by (3.3.10).

In Theorem 4.2.1 it will be shown that $H_{i,k} \to H_i$ a.s. as $k \to \infty$. As a consequence, since under A4.2.2 L is of row-full-rank, by Theorem 3.3.2

$$L_k \triangleq \begin{bmatrix} H_{q,k} & H_{q+1,k} & \cdots & H_{q+mp-1,k} \\ H_{q-1,k} & H_{q,k} & \cdots & H_{q+mp-2,k} \\ \vdots & \vdots & \ddots & \vdots \\ H_{q-p+1,k} & H_{q-p+2,k} & \cdots & H_{q+(m-1)p,k} \end{bmatrix}$$

is also of row-full-rank when k is sufficiently large. Thus, L_k can serve as the kth estimate of L with $H_{i,k} = 0$ for $i \le 0$.

The estimates for $\{A_1, \cdots, A_p, B_1, \cdots, B_q\}$ are naturally defined as follows:

$$[A_{1,k}, A_{2,k}, \cdots, A_{p,k}] = -[H_{q+1,k}, H_{q+2,k}, \cdots, H_{q+mp,k}] L_k^T (L_k L_k^T)^{-1}, \tag{4.2.16}$$

$$B_{i,k} = \sum\nolimits_{j=0}^{i \wedge p} A_{j,k} H_{i-j,k} \ \ \forall\, 1 \le i \le q. \tag{4.2.17}$$

Estimation of $f(\cdot)$

We now recursively estimate $f(y)$ for any fixed $y \in \mathbb{R}^m$. By using the estimates obtained for the coefficients of the linear subsystem we can estimate the internal signals v_k on the basis of the state space representations of the linear subsystem. Then, applying SAAWET incorporated with a multi-variable kernel function we obtain the estimates for $f(y)$. Let us start with estimating v_k.

Define

$$C \triangleq \begin{bmatrix} -A_1 & I & \cdots & 0 \\ \vdots & & \ddots & \vdots \\ \vdots & & & I \\ -A_s & 0 & \cdots & 0 \end{bmatrix}, \quad D \triangleq \begin{bmatrix} B_1 \\ B_2 \\ \vdots \\ B_s \end{bmatrix}, \quad H \triangleq \begin{bmatrix} I \\ 0 \\ \vdots \\ 0 \end{bmatrix},$$

and $s \triangleq \max(p,q)$. Then, the linear part of (4.2.1) can be presented in the state space form

$$x_{k+1} = C x_k + D u_k, \quad v_{k+1} = H^T x_{k+1}, \tag{4.2.18}$$

where C is an $sm \times sm$ matrix, D is an $sm \times l$ matrix, and H is an $sm \times m$ matrix, $A_k = 0$ for $k > p$, and $B_k = 0$ for $k > q$.

Replacing A_i and B_j in C and D with $A_{i,k}$ and $B_{j,k}$ given by (4.2.16) and (4.2.17), respectively, $i = 1, \cdots, s,\ j = 1, \cdots, s$, we obtain the estimates C_k and D_k for C and D at time k. Thus, we define the estimate $\widehat{v}_{k+1}$ for v_{k+1} as follows:

$$\widehat{x}_{k+1} = C_{k+1} \widehat{x}_k + D_{k+1} u_k, \ \widehat{v}_{k+1} = H^T \widehat{x}_{k+1} \tag{4.2.19}$$

with an arbitrary initial value $\widehat{x}_0$.

To estimate $f(y)$, let us introduce the kernel function ω_k and its estimate $\widehat{\omega}_k$ as follows:

$$\begin{aligned}\omega_k(y) &= \frac{1}{(2\pi)^{\frac{m}{2}}(b_k)^m} e^{-\frac{(y-v_k)^T(y-v_k)}{2b_k^2}},\\ \widehat{\omega}_k(y) &= \frac{1}{(2\pi)^{\frac{m}{2}}(b_k)^m} e^{-\frac{(y-\widehat{v}_k)^T(y-\widehat{v}_k)}{2b_k^2}},\end{aligned} \tag{4.2.20}$$

where b_k is the window width of the kernel function ω_k, and $b_k = \frac{1}{k^a}$, $a > 0$ by setting.

We apply SAAWET incorporated with the kernel function to estimate $f(y)$.

Take $m_k = k^b$, where $b > 0$ is such that $(m+2)a + b < \frac{1}{2}$. With $\Delta_0(y) = 0$ and arbitrary $\mu_0(y)$ we recursively estimate $f(y)$ as follows.

$$\begin{aligned}\mu_{k+1}(y) = &\ [\mu_k(y) - \tfrac{1}{k}\widehat{\omega}_k(\mu_k(y) - z_k)]\\ &\cdot I_{[\|\mu_k(y)-\frac{1}{k}\widehat{\omega}_k(\mu_k(y)-z_k)\| \le m_{\Delta_k}(y)]},\\ \Delta_k(y) = &\ \sum_{j=1}^{k-1} I_{[\|\mu_j(y)-\frac{1}{j}\widehat{\omega}_j(\mu_j(y)-z_j)\| > m_{\Delta_j}(y)]}.\end{aligned} \tag{4.2.21}$$

Consistency of Estimates for Linear Subsystem

We now proceed to prove the strong consistency of the estimates given by (4.2.13) and (4.2.14).

Lemma 4.2.3 *Assume A4.2.1–A4.2.5 hold. For any* $0 \le \nu < \frac{1}{2}$ *the following series are convergent:*

$$\sum_{k=1}^{\infty} \frac{1}{k^{1-\nu}} [E(z_{k+1}u_{k+1-i}^T) - H_i] < \infty \ \text{ for } i \ge 1, \tag{4.2.22}$$

$$\sum_{k=1}^{\infty} \frac{1}{k^{1-\nu}} \left(E(z_{k+1}u_{k+1-i}^T) - z_{k+1}u_{k+1-i}^T\right) < \infty \ \text{ for } i \ge 1. \tag{4.2.23}$$

Proof. By noticing $Ef(v)v^T\Sigma^{-1} = Q\Sigma^{-1} = I$, we have

$$Ez_{k+1}u_{k+1-i}^T - H_i = (I_{1,k+1} + I_{2,k+1})H_i,$$

where

$$I_{1,k+1} = Ef(v_{k+1})v_{k+1}^T(\Sigma_{k+1}^{-1} - \Sigma^{-1})$$

and

$$I_{2,k+1} = E(f(v_{k+1})v_{k+1}^T - f(v)v^T)\Sigma^{-1}.$$

Noticing $v_{k+1} \in \mathcal{N}(0, \Sigma_{k+1})$, we have

$$E(f(v_{k+1})v_{k+1}^T) = \int_{\mathbb{R}^m} f(x)x^T \frac{1}{(2\pi)^{\frac{m}{2}}|\Sigma_{k+1}|^{\frac{1}{2}}} e^{-\frac{x^T\Sigma_{k+1}^{-1}x}{2}} \mathrm{d}x$$

and

$$E(f(v)v^T) = \int_{\mathbb{R}^m} f(x)x^T \frac{1}{(2\pi)^{\frac{m}{2}}|\Sigma|^{\frac{1}{2}}} e^{-\frac{x^T\Sigma^{-1}x}{2}} \mathrm{d}x.$$

It then follows that

$$\begin{aligned}
&\left[E(f(v_{k+1})v_{k+1}^T - f(v)v^T)\right] = E(f(v_{k+1})v_{k+1}^T) - E(f(v)v^T) \\
&= (|\Sigma_{k+1}|^{-\frac{1}{2}} - |\Sigma|^{-\frac{1}{2}}) \int_{\mathbb{R}^m} f(x)x^T \frac{1}{(2\pi)^{\frac{m}{2}}} e^{-\frac{x^T\Sigma_{k+1}^{-1}x}{2}} \mathrm{d}x \\
&+ \int_{\mathbb{R}^m} f(x)x^T \frac{1}{(2\pi)^{\frac{m}{2}}|\Sigma|^{\frac{1}{2}}} [e^{-\frac{x^T\Sigma_{k+1}^{-1}x}{2}} - e^{-\frac{x^T\Sigma^{-1}x}{2}}]\mathrm{d}x \\
&= (|\Sigma_{k+1}|^{-\frac{1}{2}} - |\Sigma|^{-\frac{1}{2}}) \int_{\mathbb{R}^m} f(x)x^T \frac{1}{(2\pi)^{\frac{m}{2}}} e^{-\frac{x^T\Sigma_{k+1}^{-1}x}{2}} \mathrm{d}x \\
&- \int_{\mathbb{R}^m} f(x)x^T \frac{1}{(2\pi)^{\frac{m}{2}}|\Sigma|^{\frac{1}{2}}} e^{-\frac{x^T\Sigma^{-1}x}{2}} (1 - e^{\frac{x^T(\Sigma^{-1}-\Sigma_{k+1}^{-1})x}{2}}).
\end{aligned}$$

Noticing that $\| \sum_{j=k+2}^{\infty} H_j H_j^T \| = O(\lambda^k)$ for some $0 < \lambda < 1$, we have

$$\begin{aligned}
&(|\Sigma_{k+1}|^{-\frac{1}{2}} - |\Sigma|^{-\frac{1}{2}}) \int_{\mathbb{R}^m} f(x)x^T \frac{1}{(2\pi)^{\frac{m}{2}}} e^{-\frac{x^T\Sigma^{-1}x}{2}} \mathrm{d}x \\
&= O\Big(\Big\| \sum_{j=k+2}^{\infty} H_j H_j^T \Big\|\Big) = O(\lambda^k).
\end{aligned}$$

Since

$$\begin{aligned}
&\int_{\mathbb{R}^n} f(x)x^T \frac{1}{(2\pi)^{\frac{m}{2}}|\Sigma|^{\frac{1}{2}}} e^{-\frac{x^T\Sigma^{-1}x}{2}} (1 - e^{\frac{x^T(\Sigma^{-1}-\Sigma_{k+1}^{-1})x}{2}}) \\
&= \int_{\|x\|\le k} f(x)x^T \frac{1}{(2\pi)^{\frac{m}{2}}|\Sigma|^{\frac{1}{2}}} e^{-\frac{x^T\Sigma^{-1}x}{2}} (1 - e^{\frac{x^T(\Sigma^{-1}-\Sigma_{k+1}^{-1})x}{2}}) \\
&+ \int_{\|x\|>k} f(x)x^T \frac{1}{(2\pi)^{\frac{m}{2}}|\Sigma|^{\frac{1}{2}}} e^{-\frac{x^T\Sigma^{-1}x}{2}} (1 - e^{\frac{x^T(\Sigma^{-1}-\Sigma_{k+1}^{-1})x}{2}}),
\end{aligned}$$

it follows that

$$\begin{aligned}
&\int_{\|x\|\le k} f(x)x^T \frac{1}{(2\pi)^{\frac{m}{2}}|\Sigma|^{\frac{1}{2}}} e^{-\frac{x^T\Sigma^{-1}x}{2}} (1 - e^{\frac{x^T(\Sigma^{-1}-\Sigma_{k+1}^{-1})x}{2}}) \\
&\le k^2 \|\Sigma^{-1} - \Sigma_{k+1}^{-1}\| \int_{\|x\|\le k} f(x)x^T \frac{1}{(2\pi)^{\frac{m}{2}}|\Sigma|^{\frac{1}{2}}} e^{-\frac{x^T\Sigma^{-1}x}{2}} \mathrm{d}x \\
&= O(\eta^k)
\end{aligned}$$

and

$$\int_{\|x\|>k} f(x)x^T \frac{1}{(2\pi)^{\frac{m}{2}}|\Sigma|^{\frac{1}{2}}} e^{-\frac{x^T\Sigma^{-1}x}{2}}(1-e^{\frac{x^T(\Sigma^{-1}-\Sigma_{k+1}^{-1})x}{2}})$$
$$\leq \frac{1}{(2\pi)^{\frac{m}{2}}|\Sigma|^{\frac{1}{2}}}\int_{\|x\|>k} f(x)x^T e^{-\frac{x^T\Sigma^{-1}x}{2}} = O(\eta^k)$$

for some $0<\eta<1$. Thus, we have $I_{2,k+1}=O(\rho^k)$ for some $0<\rho<1$.

For $I_{1,k+1}$, we have

$$I_{1,k+1} = Ef(v_{k+1})v_{k+1}^T\Sigma^{-1}(\sum_{j=k+2}^{\infty} H_jH_j^T)\Sigma_{k+1}^{-1} = O(\sum_{j=k+2}^{\infty} H_jH_j^T) = O(\lambda^k).$$

Therefore, (4.2.22) takes place.

For (4.2.23) with $i\geq 1$ we have

$$\sum_{k=1}^{\infty}\frac{1}{k^{1-\nu}}(E(z_{k+1}u_{k+1-i}^T)-z_{k+1}u_{k+1-i}^T)$$
$$=\sum_{k=1}^{\infty}\frac{1}{k^{1-\nu}}[Ef(v_{k+1})u_{k+1-i}^T - f(v_{k+1})u_{k+1-i}^T]$$
$$-\sum_{k=1}^{\infty}\frac{1}{k^{1-\nu}}\varepsilon_{k+1}u_{k+1-i}^T < \infty \ a.s. \tag{4.2.24}$$

Define $z_k^{(1)} \triangleq \frac{1}{k^{1-\nu}}[Ef(v_{k+1})u_{k+1-i}^T - f(v_{k+1})u_{k+1-i}^T]$. By Theorem 1.4.4, $z_k^{(1)}$ is a zero-mean α-mixing sequence with mixing coefficients tending to zero exponentially. By the C_r-inequality and then the Hölder inequality, for any $\varepsilon>0$ we have

$$E\|z_k^{(1)}\|^{2+\varepsilon} = O(\frac{1}{k^{(1-\nu)(2+\varepsilon)}}),$$

which implies $\sum_{k=1}^{\infty}(E\|z_k^{(1)}\|^{2+\varepsilon})^{\frac{2}{2+\varepsilon}} = O(\sum_{k=1}^{\infty}\frac{1}{k^{2(1-\nu)}})<\infty$. Then the first term at the right-hand side of (4.2.24) converges by Theorem 1.4.2.

Since $\{u_k\}$ and $\{\varepsilon_k\}$ are mutually independent and $E\|\varepsilon_k\|^2<\infty$, we have

$$\sum_{k=1}^{\infty}E\|\frac{1}{k^{1-\nu}}\varepsilon_{k+1}u_{k+1-i}^T\|^2 = \sum_{k=1}^{\infty}O(\frac{1}{k^{2(1-\nu)}})<\infty.$$

By Theorem 1.2.8, the last term in (4.2.24) converges too. □

Theorem 4.2.1 *Assume A4.2.1–A4.2.5 hold. Then, $H_{i,k}$, $i\geq 1$ defined by (4.2.13)–(4.2.14) satisfy*

$$\| H_{i,k}-H_i \| = o(k^{-\nu}) \text{ a.s. } \forall\, \nu\in(0,1/2). \tag{4.2.25}$$

As consequences, $A_{i,k}, i = 1,\cdots,p$ and $B_{j,k}, j = 1,\cdots,q$ converge and have the following convergence rate:

$$\| A_{i,k} - A_i \| = o(k^{-\nu}) \text{ a.s. } \forall \nu \in (0,1/2) \; i = 1,\cdots,p,$$
$$\| B_{j,k} - B_j \| = o(k^{-\nu}) \text{ a.s. } \forall \nu \in (0,1/2) \; j = 1,\cdots,q.$$

Proof. We rewrite (4.2.13) as

$$H_{i,k+1} = [H_{i,k} - \frac{1}{k}(H_{i,k} - H_i) - \frac{1}{k}\varepsilon_{k+1}(i)]$$
$$\cdot I_{[\|H_{i,k} - \frac{1}{k}(H_{i,k}-H_i) - \frac{1}{k}\varepsilon_{k+1}(i)\| \le M_{\delta_{i,k}}]},$$

where

$$\varepsilon_{k+1}(i) = H_i - z_{k+1}u_{k+1-i}^T$$
$$= [H_i - Ez_{k+1}u_{k+1-i}^T] + [Ez_{k+1}u_{k+1-i}^T - z_{k+1}u_{k+1-i}^T].$$

Since H_i is the single root of the linear function $-(x - H_i)$, by Theorem 2.6.1, it suffices to prove

$$\sum_{k=1}^{\infty} \frac{1}{k^{1-\nu}} \varepsilon_{k+1}(i) < \infty \text{ a.s. } i \ge 1 \; \forall \nu \in (0,1/2). \tag{4.2.26}$$

By Lemma 4.2.3, we find that (4.2.26) is true, and hence (4.2.25) holds, while the rest can be derived straightforwardly from (4.2.25). □

Consistency of the Estimates for Nonlinearity

Lemma 4.2.4 *Assume A4.2.1–A4.2.3 hold. The following limits take place*

$$\frac{\| u_k \|}{k^c} \xrightarrow[k\to\infty]{\text{a.s.}} 0, \; \frac{\| f(v_k) \|}{k^c} \xrightarrow[k\to\infty]{\text{a.s.}} 0 \; \forall c > 0\,. \tag{4.2.27}$$

Proof. By noting

$$P[\frac{\| u_k \|}{k^c} > \varepsilon] = P[\frac{\| u_k \|^{2/c}}{k^2} > \varepsilon^{2/c}] < \frac{1}{\varepsilon^{2/c}k^2} E \| u_k \|^{2/c}$$

for any given $\varepsilon > 0$, it follows that $\sum_{k=1}^{\infty} P[\frac{\|u_k\|}{k^c} > \varepsilon] < \infty$. Hence, by Theorem 1.2.11 we derive that $\frac{\|u_k\|}{k^c} \xrightarrow[k\to\infty]{\text{a.s.}} 0$. By the growth rate restriction on $f(\cdot)$, the second assertion of the lemma can be proved in a similar way since v_k is Gaussian with variance $\Sigma_k \xrightarrow[k\to\infty]{} \Sigma$. □

Lemma 4.2.5 *Assume A4.2.1–A4.2.4 hold. Then*

$$E[\omega_k] \xrightarrow[k\to\infty]{} \rho(y), \; E[\omega_k f(v_k)] \xrightarrow[k\to\infty]{} \rho(y)f(y), \tag{4.2.28}$$

$$E[\omega_k \|f(v_k)\|] \xrightarrow[k\to\infty]{} \rho(y)\|f(y)\|, \tag{4.2.29}$$

$$E|\omega_k|^\delta = O\left(1/b_k^{m(\delta-1)}\right) \; \forall \delta \ge 2, \tag{4.2.30}$$

$$E\|\omega_k f(v_k)\|^\delta = O\left(1/b_k^{m(\delta-1)}\right) \; \forall \delta \ge 2, \tag{4.2.31}$$

where $\rho(y) = \frac{1}{(2\pi)^{\frac{m}{2}}|\Sigma|^{\frac{1}{2}}} e^{-\frac{y^T\Sigma^{-1}y}{2}}$ *and* $\Sigma = \sum_{i=1}^{\infty} H_i H_i^T$.

Further, the following series are convergent almost surely:

$$\sum_{k=1}^{\infty} \frac{1}{k}(\omega_k - E\omega_k) < \infty, \tag{4.2.32}$$

$$\sum_{k=1}^{\infty} \frac{1}{k}(|\omega_k - E\omega_k| - E|\omega_k - E\omega_k|) < \infty, \tag{4.2.33}$$

$$\sum_{k=1}^{\infty} \frac{1}{k}(\omega_k \parallel y_k \parallel - E[\omega_k \parallel y_k \parallel]) < \infty, \tag{4.2.34}$$

$$\sum_{k=1}^{\infty} \frac{1}{k}(\omega_k y_k - E\omega_k y_k) < \infty, \tag{4.2.35}$$

$$\sum_{k=1}^{\infty} \frac{1}{k}\omega_k \varepsilon_k < \infty, \sum_{k=1}^{\infty} \frac{1}{k}\omega_k (\|\varepsilon_k\| - E\|\varepsilon_k\|) < \infty. \tag{4.2.36}$$

Proof. We prove the first limit in (4.2.28).

Noting $v_k \in \mathcal{N}(0, \Sigma_k)$ with $\Sigma_k = \sum_{i=1}^{k} H_i H_i^T$, we have

$$
\begin{aligned}
E\omega_k &= \int_{\mathbb{R}^m} \frac{1}{(2\pi)^m b_k^m |\Sigma_k|^{\frac{1}{2}}} e^{-\left(\frac{(y-x)^T(y-x)}{2b_k^2} + \frac{x\Sigma_k^{-1}x}{2}\right)} dx \\
&= \int_{\mathbb{R}^m} \frac{1}{(2\pi)^m |\Sigma_k|^{\frac{1}{2}}} e^{-\left(\frac{x^Tx}{2} + \frac{(y-b_kx)^T\Sigma_k^{-1}(y-b_kx)}{2}\right)} dx \\
&\xrightarrow[k\to\infty]{} \frac{1}{(2\pi)^{\frac{m}{2}}|\Sigma|^{\frac{1}{2}}} e^{-\frac{y^T\Sigma^{-1}y}{2}} \int_{\mathbb{R}^m} \frac{1}{(2\pi)^{\frac{m}{2}}} e^{-\left(\frac{x^Tx}{2}\right)} dx = \rho(y).
\end{aligned}
$$

The second limit in (4.2.28) and (4.2.29)–(4.2.31) can be proved in a similar manner.

Define $z_k^{(2)} \triangleq \frac{1}{k}(\omega_k - E\omega_k)$. Then, $z_k^{(2)}$ is a zero-mean α-mixing sequence with the mixing coefficient tending to zero exponentially by Theorem 1.4.4.

Noticing (4.2.30), we have $\sup_k E(k^{-am(1+\varepsilon)}|\omega_k|^{2+\varepsilon}) < a_1$ for some constant a_1, and by the C_r-inequality, we have

$$
\begin{aligned}
&\sum_{k=1}^{\infty} (E|z_k^{(2)}|^{2+\varepsilon})^{\frac{2}{2+\varepsilon}} = \sum_{k=1}^{\infty} \frac{1}{k^2}(E|\omega_k - E\omega_k|^{2+\varepsilon})^{\frac{2}{2+\varepsilon}} \\
&\le \sum_{k=1}^{\infty} \frac{4}{k^2}(E|\omega_k|^{2+\varepsilon})^{\frac{2}{2+\varepsilon}} = \sum_{k=1}^{\infty} \frac{4}{k^{2(1-\frac{am(1+\varepsilon)}{2+\varepsilon})}}(E(k^{-am(1+\varepsilon)}|\omega_k|^{2+\varepsilon}))^{\frac{2}{2+\varepsilon}} \\
&< \sum_{k=1}^{\infty} \frac{a_2}{k^{a_3}} < \infty,
\end{aligned}
$$

where $a_2 \triangleq 4a_1^{\frac{2}{2+\varepsilon}}$, $a_3 \triangleq 2(1-\frac{am(1+\varepsilon)}{2+\varepsilon}) > 1$, and $0 < a < \frac{1}{2(m+2)}$. Hence (4.2.32) converges by Theorem 1.4.2, and (4.2.33)–(4.2.35) can be verified by the similar treatment. For (4.2.36) Theorem 1.2.8 is applied by noticing that u_k and ε_k are mutually independent and $E\|\varepsilon_k\|^2 < \infty$, and hence $\sum_{k=1}^{\infty}\frac{1}{k^2}E(\omega_k^2\|\varepsilon_k\|^2) = O(\sum_{k=1}^{\infty}\frac{1}{k^{2-ma}}) < \infty$. □

Lemma 4.2.6 *Assume that A4.2.1–A4.2.5 hold. There exists a constant $c > 0$ with $\frac{1}{2}-(m+2)a-b-3c > 0$ such that*

$$\| v_k - \widehat{v}_k \| = o\left(\frac{1}{k^{1/2-2c}}\right), \ |\omega_k - \widehat{\omega}_k| = o\left(\frac{1}{k^{1/2-(m+2)a-2c}}\right). \tag{4.2.37}$$

Proof. From (4.2.18) and (4.2.19), we have

$$\begin{aligned}\widehat{x}_{k+1} - x_{k+1} =& C_{k+1}\widehat{x}_k - Cx_k + (D_{k+1}-D)u_k\\ =& C_{k+1}(\widehat{x}_k - x_k) + (C_{k+1}-C)x_k + (D_{k+1}-D)u_k.\end{aligned}$$

Since C is stable and $C_k \to C$, there exists a $\lambda \in (0,1)$ such that

$$\|\widehat{x}_{k+1} - x_{k+1}\| \le N_1\lambda^{k+1}\|\widehat{x}_0 - x_0\| + S(\lambda,k),$$

where

$$S(\lambda,k) = N_2\sum_{j=1}^{k+1}\lambda^{k-j+1}(\|C_j - C\|\cdot\|x_{j-1}\| + \|D_j - D\|\cdot\|u_{j-1}\|)$$

with $N_1 > 0$ and $N_2 > 0$ being constants.

Since $\|C_k - C\| = o\left(\frac{1}{k^{\frac{1}{2}-c}}\right)$ and $\|D_k - D\| = o\left(\frac{1}{k^{\frac{1}{2}-c}}\right)$, by Theorem 4.2.1 and Lemma 4.2.4, we have

$$\begin{aligned}S(\lambda,k) &= \sum_{j=1}^{k+1}\lambda^{k-j+1}o\Big(\frac{1}{j^{1/2-2c}}\Big)\\ &= o\Big(\frac{1}{k^{1/2-2c}}\Big)\sum_{j=1}^{k+1}\lambda^{k-j+1}\Big(\frac{k}{j}\Big)^{1/2-2c}\\ &= o\Big(\frac{1}{k^{1/2-2c}}\Big)\sum_{j=0}^{k}\lambda^{j}\Big(\frac{k}{k-j+1}\Big)^{1/2-2c}\\ &= o\Big(\frac{1}{k^{1/2-2c}}\Big)\Big(\sum_{j=0}^{[\frac{k}{2}]}+\sum_{j=[\frac{k}{2}]+1}^{k}\Big)\lambda^{j}\Big(\frac{k}{k-j+1}\Big)^{1/2-2c}\\ &= o\Big(\frac{1}{k^{1/2-2c}}\Big)\Big(\frac{2}{1-\lambda}+\lambda^{[\frac{k}{2}]+1}k^{\frac{3}{2}-2c}\Big) = o\Big(\frac{1}{k^{\frac{1}{2}-2c}}\Big).\end{aligned}$$

This implies that

$$\|\widehat{x}_k - x_k\| = o\Big(\frac{1}{k^{1/2-2c}}\Big),$$

because the first term at the right-hand side of (4.2.38) decays exponentially.

Since $\|v_k - \widehat{v}_k\|$ and $\|\widehat{x}_k - x_k\|$ are of the same order, we have

$$\begin{aligned}|\widehat{\omega}_k - \omega_k| =& o\Big(\frac{1}{b_k^{(m+2)}}\|\widehat{v}_k - v_k\|\Big) = o\Big(\frac{1}{b_k^{(m+2)}}\|\widehat{x}_k - x_k\|\Big)\\ =& o\Big(\frac{1}{k^{1/2-(m+2)a-2c}}\Big).\end{aligned}$$

□

Theorem 4.2.2 *Assume that A4.2.1–A4.2.5 hold. Then $\mu_k(y)$ defined by (4.2.21) is strongly consistent:*

$$\mu_k(y) \xrightarrow[k\to\infty]{} f(y) \text{ a.s., } y \in \mathbb{R}^m. \tag{4.2.38}$$

Proof. The algorithm (4.2.21) can be rewritten as

$$\begin{aligned}\mu_{k+1}(y) =& [\mu_k(y) - \frac{1}{k}\rho(y)(\mu_k(y) - f(y)) - \frac{1}{k}\bar{\varepsilon}_{k+1}(y)]\\ &\cdot I_{[\|\mu_k(y) - \frac{1}{k}\rho(y)(\mu_k(y)-f(y)) - \frac{1}{k}\bar{\varepsilon}_{k+1}(y)\| \le m_{\Delta_k}(y)]},\end{aligned}$$

where $\bar{\varepsilon}_{k+1}(y) = \widehat{\omega}_k(\mu_k(y) - z_k) - \rho(y)(\mu_k(y) - f(y))$.

Since $f(y)$ is the unique root of $-\rho(y)(x - f(y))$, by Theorem 2.3.1 for (4.2.38) it suffices to prove

$$\lim_{T\to 0}\limsup_{k\to\infty}\frac{1}{T}\left\|\sum_{j=n_k}^{m(n_k,T_k)}\frac{1}{j}\bar{\varepsilon}_{j+1}(y)\right\| = 0 \;\; \forall T_k \in [0, T] \tag{4.2.39}$$

for any convergent subsequence $\mu_{n_k}(y)$, where

$$m(k,T) \triangleq \max\{m : \sum_{j=k}^{m}\frac{1}{j} \le T\}.$$

Write $\varepsilon_{k+1}(y)$ as $\bar{\varepsilon}_{k+1}(y) = \sum_{i=1}^{4}\bar{\varepsilon}_{k+1}^{(i)}(y)$, where

$$\begin{aligned}&\bar{\varepsilon}_{k+1}^{(1)}(y) = (\widehat{\omega}_k - \omega_k)(\mu_k(y) - z_k), \;\; \bar{\varepsilon}_{k+1}^{(2)}(y) = (\omega_k - \rho(y))\mu_k(y),\\ &\bar{\varepsilon}_{k+1}^{(3)}(y) = \rho(y)f(y) - \omega_k f(v_k), \;\; \bar{\varepsilon}_{k+1}^{(4)}(y) = -\omega_k\varepsilon_k.\end{aligned}$$

We now prove (4.2.39) with $\bar{\varepsilon}_{j+1}(y)$ replaced, respectively, by $\bar{\varepsilon}_{j+1}^{(i)}(y)$, $i = 1,2,3,4$.

For $i = 1$, from (4.2.21) we have $\|\mu_k(y)\| \le k^b$, which together with (4.2.36) implies

$$\sum_{k=1}^{\infty}\frac{1}{k}\|(\widehat{\omega}_k - \omega_k)\mu_k(y)\| = o\left(\sum_{k=1}^{\infty}\frac{1}{k^{\frac{3}{2}-(m+2)a-b-2c}}\right) < \infty. \tag{4.2.40}$$

By (4.2.37) and the second limit in (4.2.27) we have

$$\sum_{k=1}^{\infty}\frac{1}{k}\|(\widehat{\omega}_k-\omega_k)f(v_k)\| \le \sum_{k=1}^{\infty}\frac{1}{k^{1-c}}|(\widehat{\omega}_k-\omega_k)|\cdot\frac{\|f(v_k)\|}{k^c}$$
$$=o\left(\sum_{k=1}^{\infty}\frac{1}{k^{\frac{3}{2}-(m+2)a-3c}}\right)<\infty. \tag{4.2.41}$$

Again by (4.2.37) it is seen that

$$\sum_{k=1}^{\infty}\frac{1}{k}\|(\widehat{\omega}_k-\omega_k)\varepsilon_k\| \le \sum_{k=1}^{\infty}\frac{1}{k}|(\widehat{\omega}_k-\omega_k)|\cdot\|\varepsilon_k\|$$
$$=o\left(\sum_{k=1}^{\infty}\frac{(\|\varepsilon_k\|-E\|\varepsilon_k\|)+E\|\varepsilon_k\|}{k^{\frac{3}{2}-(n+2)a-2c}}\right)<\infty. \tag{4.2.42}$$

Combining (4.2.40)–(4.2.42), we have

$$\sum_{k=1}^{\infty}\frac{1}{k}(\widehat{w}_k-\omega_k)(\mu_k(y)-z_k)<\infty. \tag{4.2.43}$$

Hence, (4.2.39) holds for $\bar{\varepsilon}_{j+1}^{(1)}(y)$.

For $\bar{\varepsilon}_{j+1}^{(3)}(y)$ we have

$$\bar{\varepsilon}_{k+1}^{(3)}(y)=\rho(y)f(y)-\omega_k f(v_k)=[\rho(y)f(y)-E\omega_k f(v_k)]$$
$$+[E\omega_k f(v_k)-\omega_k f(v_k)]. \tag{4.2.44}$$

By the second limit in (4.2.28) and by convergence of the series in (4.2.35) it follows that (4.2.39) holds for $i=3$.

Convergence of the first series in (4.2.36) assures that (4.2.39) holds for $i=4$.

It remains to show (4.2.39) for $i=2$. For this we first show a result similar to Lemma 4.1.4. Namely, there exists an Ω_0 with $P\Omega_0=1$ such that for any fixed sample path $\omega\in\Omega_0$ if $\mu_{n_k}(y)$ is a convergent subsequence of $\{\mu_k(y)\}:\mu_{n_k}(y)\xrightarrow[k\to\infty]{}\bar{\mu}(y)$, then for all large enough k and sufficiently small $T>0$

$$\mu_{i+1}(y)=\mu_i(y)-\frac{1}{i}\widehat{\omega}_i(\mu_i(y)-z_i), \tag{4.2.45}$$

and

$$\|\mu_{i+1}(y)-\mu_{n_k}(y)\|\le cT,\ \ i=n_k,n_k+1,\cdots,m(n_k,T), \tag{4.2.46}$$

where $c>0$ is a constant, which is independent of k but may depend on sample path ω.

Consider the recursion (4.2.45) starting from $\mu_{n_k}(y)$.

Let

$$\Phi_{i,j} \triangleq (1-\frac{1}{i}\omega_i)\cdots(1-\frac{1}{j}\omega_j),\ i\geq j,\ \Phi_{j,j+1}=1.$$

By $\lim_{k\to\infty} E\omega_k = \rho(y)$ and the convergence of the series in (4.2.32) it is clear that $\forall i \in [n_k,\cdots,m(n_k,T)]$

$$\sum_{j=n_k}^{i}\frac{1}{j}\omega_j = \sum_{j=n_k}^{i}\frac{1}{j}(\omega_j - E\omega_j)+\sum_{j=n_k}^{i}\frac{1}{j}E\omega_j = O(T), \tag{4.2.47}$$

which implies

$$\log\Phi_{i,n_k} = O(\sum_{j=n_k}^{i}\frac{1}{j}\omega_j)$$

and

$$\Phi_{i,n_k} = 1+O(T)\ \forall i\in[n_k,\cdots,m(n_k,T)] \tag{4.2.48}$$

as $k\to\infty$ and $T\to 0$.

By convergence of the second series in (4.2.36), and by (4.2.47) and (4.2.48) it is seen that

$$\left\|\sum_{j=n_k}^{i}\Phi_{i,j+1}\frac{1}{j}\omega_j\varepsilon_j\right\| \leq \sum_{j=n_k}^{i}\Phi_{i,j+1}\frac{1}{j}\omega_j(\|\varepsilon_j\|-E\|\varepsilon_j\|)$$
$$+\sum_{j=n_k}^{i}\Phi_{i,j+1}\frac{1}{j}\omega_j E\|\varepsilon_j\| = O(T)\ \forall i\in[n_k,\cdots,m(n_k,T)]. \tag{4.2.49}$$

By (4.2.29), (4.2.34), and (4.2.48) it follows that

$$\left\|\sum_{j=n_k}^{i}\Phi_{i,j+1}\frac{1}{j}\omega_j f(v_j)\right\| \leq \sum_{j=n_k}^{i}\Phi_{i,j+1}\frac{1}{j}(\omega_j\|f(v_j)\|$$
$$-E\omega_j\|f(v_j)\|+E\omega_j\|f(v_j)\|)=O(T) \tag{4.2.50}$$

$\forall i\in[n_k,\cdots,m(n_k,T)]$ as $k\to\infty$ and $T\to 0$, and by (4.2.40)–(4.2.42)

$$\left\|\sum_{j=n_k}^{i}\Phi_{i,j+1}\frac{1}{j}(\omega_j-\widehat{w}_j)(\mu_j(y)-z_j)\right\| = O(T)\quad \forall i\in[n_k,\cdots,m(n_k,T)]. \tag{4.2.51}$$

From the recursion (4.2.45) we have

$$\begin{aligned}\mu_{i+1}(y) =& \mu_i(y)-\frac{1}{i}\omega_i(\mu_i(y)-z_i)+\frac{1}{i}(\omega_i-\widehat{\omega}_i)(\mu_i(y)-z_i)\\ =&\Phi_{i,n_k}\mu_{n_k}(y)+\sum_{j=n_k}^{i}\Phi_{i,j+1}\frac{1}{j}\omega_j z_i+\sum_{j=n_k}^{i}\Phi_{i,j+1}\frac{1}{j}(\omega_j-\widehat{\omega}_j)(\mu_j(y)-z_i),\end{aligned}$$

which incorporating with (4.2.48)-(4.2.51) yields

$$\mu_{i+1}(y) = \mu_{n_k}(y) + O(T) \ \forall i \in [n_k, \cdots, m(n_k, T)]. \tag{4.2.52}$$

This means that the algorithm (4.2.21) has no truncation for $i \in [n_k, \cdots, m(n_k, T)]$, when k is large enough and $T > 0$ is sufficiently small. Since the possibly exceptional set is with probability zero, we may take Ω_0 with $P\Omega_0 = 1$ such that for all $\omega \in \Omega_0$ (4.2.45) and (4.2.46) are true for large k and small $T > 0$.

We then have

$$\sum_{j=n_k}^{m(n_k,T_k)} \frac{1}{j}\bar{\varepsilon}_{j+1}^{(2)}(y) = \sum_{j=n_k}^{m(n_k,T_k)} \frac{1}{j}(\mu_j(y) - \bar{\mu}(y))(\omega_j - E\omega_j)$$
$$+ \bar{\mu}(y) \sum_{j=n_k}^{m(n_k,T_k)} \frac{1}{j}(\omega_j - E\omega_j) + \sum_{j=n_k}^{m(n_k,T_k)} \frac{1}{j}(E\omega_j - \rho(y))\mu_j(y). \tag{4.2.53}$$

At the right-hand side of the equality (4.2.53), the second term tends to zero as $k \to \infty$ by (4.2.32), the last term tends to zero as $k \to \infty$ by the first limit in (4.2.28) and (4.2.46), while the first term is analyzed as follows. By (4.2.28), (4.2.33), (4.2.46), and $\mu_{n_k}(y) \xrightarrow[k\to\infty]{} \bar{\mu}(y)$ we have

$$\lim_{T\to 0} \limsup_{k\to\infty} \frac{1}{T} \sum_{j=n_k}^{m(n_k,T_k)} \frac{1}{j}(\mu_j(y) - \bar{\mu}(y))(\omega_j - E\omega_j)$$
$$= \lim_{T\to 0} \limsup_{k\to\infty} \frac{1}{T} O(T) \sum_{j=n_k}^{m(n_k,T_k)} \frac{1}{j} E|\omega_j - E\omega_j|$$
$$= \lim_{T\to 0} \limsup_{k\to\infty} O(1) \sum_{j=n_k}^{m(n_k,T_k)} \frac{1}{j} E\omega_j = 0. \tag{4.2.54}$$

Thus, (4.2.39) is also valid for $i = 2$. The proof is completed. □

Numerical Example

Let the linear subsystem be given as follows

$$v_k + A_1 v_{k-1} + A_2 v_{k-2} = B_1 u_{k-1} + B_2 u_{k-2}, \tag{4.2.55}$$

where $A_1 = \begin{bmatrix} 0 & 0.5 \\ -0.4 & 0 \end{bmatrix}$, $A_2 = \begin{bmatrix} 0.3 & 0 \\ 0 & -0.5 \end{bmatrix}$, $B_1 = \begin{bmatrix} 0 & \sqrt{0.85} \\ -\sqrt{0.6} & 0 \end{bmatrix}$, and $B_2 = \begin{bmatrix} -0.7 & 0 \\ 0 & 0.9 \end{bmatrix}$, and let the nonlinear function be given by $y_k = f(v_k)$ with $f(x,y) = \begin{bmatrix} f_1(x,y) \\ f_2(x,y) \end{bmatrix} = \begin{bmatrix} x + 0.3y^2 + 2 \\ 0.3x^2 + y + 1 \end{bmatrix}$.

Let the observation noise $\varepsilon_k \in \mathcal{N}(0, I_2)$ be iid, and let $u_k \in \mathcal{N}(0, I_2)$ be also iid and be independent of $\{\varepsilon_k\}$. It is noted that P for system (4.2.55) equals I, so what to be estimated coincides with those expressed above.

The estimates of A_1, A_2, B_1, B_2 are given by Fig. 4.2.1–4.2.4, respectively. The estimated curve for $f_1(\cdot)$ is given by Fig. 4.2.5, where the lower surface represents the estimation errors. The estimates for $f_1(\cdot)$ and $f_2(\cdot)$ at some particular points, namely, the estimates for $f_1(0,0.8)$, $f_1(0.8,-0.4)$, and $f_2(-0.8,0), f_2(-2,-1.6)$, are presented in Fig. 4.2.6.

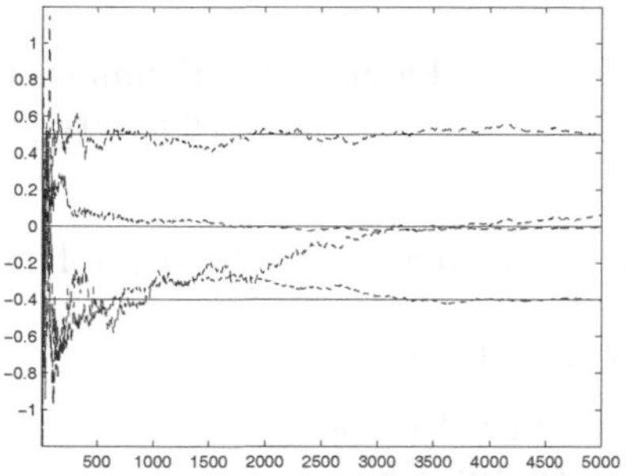

Figure 4.2.1: Estimates for A_1

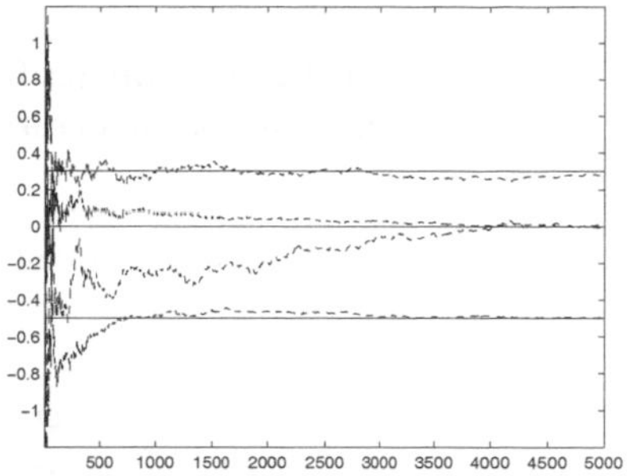

Figure 4.2.2: Estimates for A_2

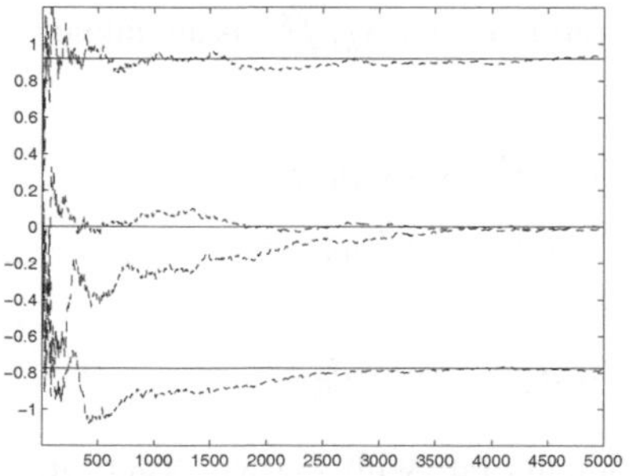

Figure 4.2.3: Estimates for B_1

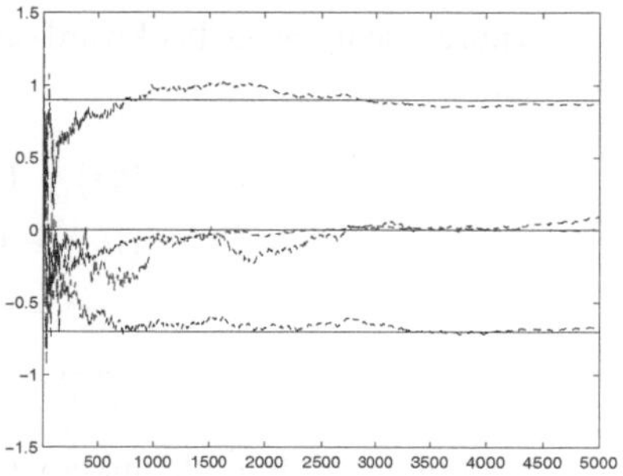

Figure 4.2.4: Estimates for B_2

4.3 Recursive Identification of Wiener–Hammerstein Systems

We now consider identification of the SISO Wiener–Hammerstein system consisting of two linear subsystems with a static nonlinearity $f(\cdot)$ in between as shown by Figure 4.3.1.

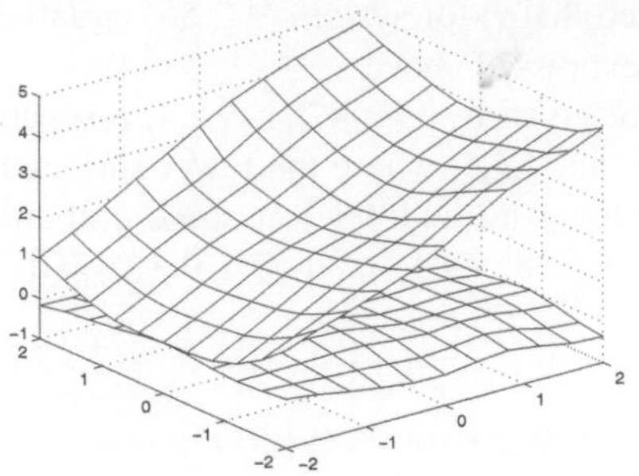

Figure 4.2.5: Estimates for f_1 and estimation errors

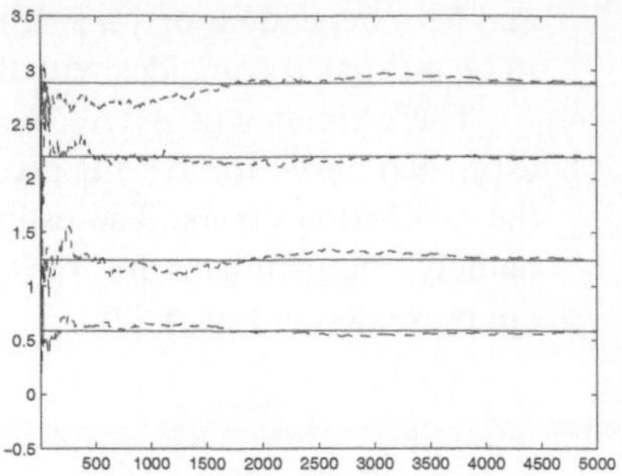

Figure 4.2.6: Estimates for f at fixed points

The system input u_k, output y_k, and observation z_k are related as follows

$$P(z)v_{k+1} = Q(z)u_k \tag{4.3.1}$$

$$\varphi_k = f(v_k), \quad \varphi_k^e = \varphi_k + \eta_k \tag{4.3.2}$$

$$C(z)y_{k+1} = D(z)\varphi_k^e + \xi_{k+1} \tag{4.3.3}$$

$$z_k = y_k + \varepsilon_k, \tag{4.3.4}$$

where z denotes the backward-shift operator $zy_{k+1} = y_k$, $f(\cdot)$ is an unknown function, and

$$P(z) = 1 + p_1 z + p_2 z^2 + \cdots + p_{n_p} z^{n_p}, \tag{4.3.5}$$

$$Q(z) = 1 + q_1 z + q_2 z^2 + \cdots + q_{n_q} z^{n_q}, \tag{4.3.6}$$

$$C(z) = 1 + c_1 z + c_2 z^2 + \cdots + c_{n_c} z^{n_c}, \tag{4.3.7}$$

$$D(z) = 1 + d_1 z + d_2 z^2 + \cdots + d_{n_d} z^{n_d} \tag{4.3.8}$$

are polynomials with unknown coefficients but with known orders n_p, n_q, n_c, n_d. These kinds of systems can model practical phenomena arising from diverse areas including sensor systems, electromechanical systems in robotics, mechatronics, biological systems, and chemical systems.

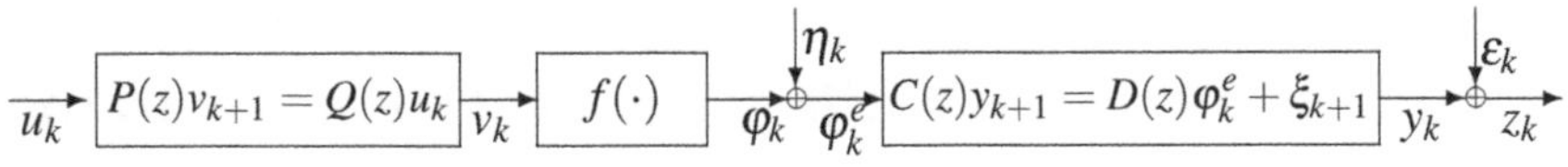

Figure 4.3.1: Wiener–Hammerstein System

If $P(z) \equiv Q(z) \equiv 1$, then the system under consideration turns to a Hammerstein system, while for the case $C(z) \equiv D(z) \equiv 1$ it becomes a Wiener system.

One may ask is it necessary to independently consider Wiener–Hammerstein systems? In other words, is it possible to reduce the Wiener–Hammerstein system to a Wiener or to a Hammerstein system with some appropriate linear subsystem and nonlinearity? The results presented in the section can be interpreted as a negative answer to this question, because they imply that the system cannot be decomposed into two sub-problems and solved by the existing Hammerstein and Wiener identification algorithms.

Remark 4.3.1 *It is worth noting that setting the first coefficients in $P(z), Q(z), C(z)$, and $D(z)$ all equal to one is not a restriction, because there is a flexibility in definition of the nonlinear function $f(\cdot)$ to be identified. Let us explain this. First, without loss of generality, we may assume the linear parts of the system are of relative degree equal to one and are given by $P(z), Q(z), C(z)$, and $D(z)$, where $P(z)$ and $C(z)$ are given by (4.3.5) and (4.3.7), respectively, while $Q(z)$ and $D(z)$ are written as*

$$Q(z) = q_0 + q_1 z + \cdots + q_{n_q} z^{n_q}, \quad D(z) = d_0 + d_1 z + \cdots + d_{n_d} z^{n_d}$$

with $q_0 \neq 0$ and $d_0 \neq 0$ but not necessarily being equal to 1. Define

$$\tilde{Q}(z) \triangleq 1 + \frac{q_1}{q_0} z + \cdots + \frac{q_{n_q}}{q_0} z^{n_q}, \quad \tilde{D}(z) \triangleq 1 + \frac{d_1}{d_0} \cdots + \frac{d_{n_d}}{d_0} z^{n_d},$$

$$\tilde{v}_k \triangleq \frac{1}{q_0} v_k, \quad \tilde{f}(x) \triangleq d_0 f(q_0 x), \quad \text{and} \quad \tilde{\varphi}_k \triangleq \tilde{f}(\tilde{v}_k).$$

Then it is clear that the system (4.3.1)–(4.3.4) and the following system

$$P(z)\tilde{v}_{k+1} = \tilde{Q}(z) u_k, \quad \tilde{\varphi}_k = \tilde{f}(\tilde{v}_k), \quad \tilde{\eta}_k \triangleq d_0 \eta_k,$$
$$C(z) y_{k+1} = \tilde{D}(z) \tilde{\varphi}_k^e + \xi_{k+1}, \quad \tilde{\varphi}_k^e \triangleq \tilde{\varphi}_k + \tilde{\eta}_k$$

are with the same input–output data $\{u_k, y_k\}$. It is noted that all polynomials $P(z), \tilde{Q}(z)$, $C(z)$, and $\tilde{D}(z)$ in the above system are monic, where the nonlinear function to be estimated is $\tilde{f}(x)$.

In the system described by (4.3.1)–(4.3.8), the first linear subsystem is an ARX system without noise, while the second linear subsystem is also an ARX but with state noise $\{\xi_k\}$. The static nonlinearity of the system is denoted by $f(\cdot)$, while η_k is the internal noise of the system and ε_k is the observation noise.

The problem is to recursively estimate the nonlinear function $f(\cdot)$ and the unknown coefficients $\{p_1, \cdots, p_{n_p}, q_1, \cdots, q_{n_q}, c_1, \cdots, c_{n_c}, \text{and } d_1, \cdots, d_{n_d}\}$ on the basis of the observation z_k and the designed input u_k.

Let us first list assumptions to be imposed on the system.

A4.3.1 *The input $\{u_k, k \geq 0\}$ is a sequence of iid Gaussian random variables $u_k \in \mathcal{N}(0, \vartheta^2)$ with $\vartheta > 0$ and is independent of $\{\eta_k\}$, $\{\xi_k\}$, and $\{\varepsilon_k\}$.*

A4.3.2 *$P(z)$ and $Q(z)$ are coprime and $P(z)$ is stable: $P(z) \neq 0 \ \forall \, |z| \leq 1$.*

A4.3.3 *$C(z)$ and $D(z)$ are coprime and both are stable: $C(z) \neq 0$ and $D(z) \neq 0$ $\forall\, |z| \leq 1$.*

By A4.3.2 and A4.3.3, we have

$$L(z) \triangleq \frac{Q(z)}{P(z)} = \sum_{i=0}^{\infty} l_i z^i, \tag{4.3.9}$$

and

$$H(z) \triangleq \frac{D(z)}{C(z)} = \sum_{i=0}^{\infty} h_i z^i, \tag{4.3.10}$$

where $|l_i| = O(e^{-r_1 i}), r_1 > 0, i \geq 1$ and $|h_i| = O(e^{-r_2 i}), r_2 > 0, i \geq 1$, and $l_0 = 1$ and $h_0 = 1$ since all polynomials (4.3.5)-(4.3.8) are monic.

A4.3.4 *$f(\cdot)$ is a measurable function and continuous at y where the value $f(y)$ is estimated. The growth rate of $f(y)$ as $|y| \to \infty$ is not faster than a polynomial. Further, at least one of the parameters ρ and κ is nonzero, where*

$$\rho \triangleq \frac{1}{\sqrt{2\pi}\sigma^5\vartheta} \int_{\mathbb{R}} (y^2 - \sigma^2\vartheta^2)\, f(y) e^{-\frac{y^2}{2\sigma^2\vartheta^2}} \mathrm{d}y, \tag{4.3.11}$$

$$\kappa \triangleq \frac{1}{\sqrt{2\pi}\sigma^7\vartheta} \int_{\mathbb{R}} (y^3 - 3\sigma^2\vartheta^2 y)\, f(y) e^{-\frac{y^2}{2\sigma^2\vartheta^2}} \mathrm{d}y, \tag{4.3.12}$$

where $\sigma^2 = \sum_{i=0}^{\infty} l_i^2$.

A4.3.5 *$\{\eta_k\}$, $\{\xi_k\}$, and $\{\varepsilon_k\}$ are mutually independent, and each of them is an iid sequence: $E\eta_k = 0, E\xi_k = 0, E\varepsilon_k = 0$, $E|\eta_k|^\Delta < \infty, E|\xi_k|^\Delta < \infty$, and $E|\varepsilon_k|^\Delta < \infty$ for some $\Delta > 3$. Besides, ξ_k and ε_k are with probability densities.*

Before proceeding further, let us introduce another parameter τ

$$\tau \triangleq \frac{1}{\sqrt{2\pi}\sigma^3\vartheta} \int_{\mathbb{R}} y f(y) e^{-\frac{y^2}{2\sigma^2\vartheta^2}} \mathrm{d}y, \tag{4.3.13}$$

which, in the case it is nonzero, may replace κ or ρ to estimate the second linear subsystem in order to have simpler calculation.

Remark 4.3.2 *The growth rate restriction required in A4.3.4 implies that there are constants $\alpha > 0$ and $\beta \geq 1$ such that*

$$|f(y)| \leq \alpha(1 + |y|^\beta) \;\; \forall y \in \mathbb{R}. \tag{4.3.14}$$

Therefore, the finiteness of integrals (4.3.11)–(4.3.12) is guaranteed.

We now show that for various nonlinear functions either ρ or κ or even both are nonzero indeed. In many cases τ is also nonzero.

(i) Let $f(\cdot)$ be a monic polynomial with arbitrary coefficients.

If $f(x)=x^2+ax+b$, then $\tau=a\vartheta^2$, $\rho=2\vartheta^4>0$, and $\kappa=0$.

If $f(x)=x^3+ax^2+bx+c$, then $\tau=(3\sigma^2\vartheta^2+b)\vartheta^2$, $\rho=2a\vartheta^4$, and $\kappa=6\sigma^6\vartheta^6>0$.

If $f(x)=x^4+ax^3+bx^2+cx+d$, then $\tau=(3\sigma^2\vartheta^2a+c)\vartheta^2$, $\rho=2\vartheta^4(6\sigma^2\vartheta^2-b)$, and $\kappa=6\vartheta^6a$. Only in the special case where $6\sigma^2\vartheta^2=b$ and $a=0$, then both ρ and κ are zero. It is worth noting that the equality $6\sigma^2\vartheta^2=b$ can easily be avoided by slightly changing the input variance ϑ^2.

The higher order polynomials can be discussed in a similar manner.

(ii) Let $f(\cdot)$ be one of the blocks that often appear in practical systems, for instance, the dead-zone, saturation, pre-load, and so on.

Let $f(\cdot)$ be a dead-zone described as follows:

$$f(x)=\begin{cases} b(x-a), & x>a \\ 0, & -a\le x\le a, \\ b(x+a), & x<-a, \end{cases} \tag{4.3.15}$$

where both a and b are greater than 0.

Then, $\tau=\frac{2\vartheta^2bI_a}{\sqrt{2\pi}}>0$, $\rho=0$, and $\kappa=\frac{2ab\vartheta^3}{\sqrt{2\pi}\sigma^3}>0$, where $I_a=\int_{\frac{a}{\sigma\vartheta}}^{\infty}e^{-\frac{x^2}{2}}\,\mathrm{d}x>0$.

Let $f(\cdot)$ be a saturation function described as follows:

$$f(x)=\begin{cases} ba, & x>a \\ bx, & -a\le x\le a, \\ -ba, & x<-a, \end{cases} \tag{4.3.16}$$

where both a and b are greater than 0.

Then, $\tau=\frac{2\vartheta^2bI_a}{\sqrt{2\pi}}>0$, $\rho=0$, and $\kappa=-\frac{2ab\vartheta^3}{\sqrt{2\pi}\sigma^3}<0$, where $I_a=\int_0^{\frac{a}{\sigma\vartheta}}e^{-\frac{x^2}{2}}\,\mathrm{d}x>0$.

Let $f(\cdot)$ be a pre-load function described as follows:

$$f(x)=\begin{cases} bx+a, & x>0 \\ 0, & x=0, \\ bx-a, & x<0, \end{cases} \tag{4.3.17}$$

where both a and b are greater than 0.

Then, $\tau=\vartheta^2b+\frac{2\vartheta a}{\sqrt{2\pi}\sigma}>0$, $\rho=0$, and $\kappa=-\frac{\vartheta^3a}{\sqrt{2\pi}\sigma^3}<0$.

From the above discussion we see that A4.3.4 is not too restrictive. It is worth noting that the constants ρ, κ, and τ are not simultaneously required to be nonzero. To be precise, we need that κ and τ are nonzero when $f(\cdot)$ is odd. Actually, in this case we need only to require κ be nonzero since τ is positive. If $f(\cdot)$ is even, then ρ is required to be nonzero. While for a general $f(\cdot)$, i.e., $f(\cdot)$ is neither odd nor even, it is only needed that either both κ and τ are nonzero or ρ is nonzero. The assumption $\rho\neq 0$ ($\tau\neq 0$) excludes $f(\cdot)$ from being an odd (even) function. These are rather restrictive conditions and are to be weakened.

From (4.3.1) and (4.3.9) it is seen that $v_k = zL(z)u_k$, and from (4.3.3) and (4.3.10) $y_k = zF(z)\varphi_k^e + \frac{1}{C(z)}\xi_k$. Therefore, both linear subsystems are physically realizable.

Assuming $u_k = 0\ \forall k < 0$, we have

$$v_k = \sum_{i=0}^{k-1} l_i u_{k-i-1}. \tag{4.3.18}$$

Define

$$y_k^0 \triangleq C^{-1}(z)D(z)\varphi_{k-1} = \sum_{i=0}^{k-1} h_i \varphi_{k-i-1}, \tag{4.3.19}$$

$\bar{\eta}_k \triangleq C^{-1}(z)D(z)\eta_{k-1} = \sum_{i=0}^{k-1} h_i \eta_{k-i-1}$, and $\bar{\xi}_k \triangleq C^{-1}(z)\xi_k$.

Then, by (4.3.4) we have

$$\begin{aligned} y_{k+1} &= C^{-1}(z)D(z)\varphi_k + C^{-1}(z)D(z)\eta_k + C^{-1}(z)\xi_{k+1} \\ &= y_{k+1}^0 + \bar{\eta}_{k+1} + \bar{\xi}_{k+1}. \end{aligned} \tag{4.3.20}$$

In fact, y_k^0 is the output of the system without noises ξ_k, η_k, and ε_k.

Estimation for $\{\mathbf{p_1}, \cdots, \mathbf{p_{n_p}}, \mathbf{q_1}, \cdots, \mathbf{q_{n_q}}\}$

We first estimate the impulse responses l_i of the first linear subsystem, and then the coefficients of polynomials $P(z)$ and $Q(z)$ by using their relationship with l_i.

The impulse responses l_i are estimated with the help of the following lemma.

Lemma 4.3.1 *Assume A4.3.1–A4.3.5 hold. Then the following limits take place:*

$$\lim_{k\to\infty} Ez_{k+1}(u_{k-1}^2 - \vartheta^2) = \rho, \tag{4.3.21}$$

$$\lim_{k\to\infty} Ez_{k+1}u_{k-1}u_{k-i-1} = \rho l_i \ \ \forall i \geq 1, \tag{4.3.22}$$

$$\lim_{k\to\infty} Ez_{k+1}(u_{k-1}^3 - 3\vartheta^2 u_{k-1}) = \kappa, \tag{4.3.23}$$

$$\lim_{k\to\infty} Ez_{k+1}(u_{k-1}^2 - \vartheta^2)u_{k-i-1} = \kappa l_i \ \ \forall i \geq 1, \tag{4.3.24}$$

$$\lim_{k\to\infty} Ez_{k+1}(u_{k-i-1}^2 - \vartheta^2) = \rho \sum\nolimits_{j=0}^{i} h_j l_{i-j}^2 \ \ \forall i \geq 0, \tag{4.3.25}$$

$$\lim_{k\to\infty} Ez_{k+1}(u_{k-i-1}^3 - 3\vartheta^2 u_{k-i-1}) = \kappa \sum\nolimits_{j=0}^{i} h_j l_{i-j}^3 \ \ \forall i \geq 0. \tag{4.3.26}$$

Proof. From (4.3.18) it is seen that $v_k \in \mathcal{N}(0, \sigma_k^2\vartheta^2)$, $\sigma_k^2 = \sum_{i=0}^{k-1} l_i^2$.

We apply the treatment similar to that used in Lemma 4.2.2. The Gaussian random vector $g_{k,j} \triangleq [v_k, u_{k-1}, u_{k-j-1}]^T$, $j \geq 1$, is zero-mean with covariance matrix

$$G_{k,j} \triangleq \begin{bmatrix} \vartheta^2\sigma_k^2 & \vartheta^2 & \vartheta^2 l_j \\ \vartheta^2 & \vartheta^2 & 0 \\ \vartheta^2 l_j & 0 & \vartheta^2 \end{bmatrix}.$$

It is straightforward to verify that $G_{k,j}$ can be factorized $G_{k,j} = \Gamma_{k,j}\Gamma_{k,j}^T$ with

$$\Gamma_{k,j} \triangleq \begin{bmatrix} \sigma_k\vartheta & 0 & 0 \\ \frac{\vartheta}{\sigma_k} & \frac{\vartheta\sqrt{\sigma_k^2-1}}{\sigma_k} & 0 \\ \frac{\vartheta l_j}{\sigma_k} & \frac{-\vartheta l_j}{\sigma_k\sqrt{\sigma_k^2-1}} & \frac{\vartheta\sqrt{\sigma_k^2-1-l_j^2}}{\sqrt{\sigma_k^2-1}} \end{bmatrix}.$$

From here it follows that $\Gamma_{k,j}^{-1}G_{k,j}\Gamma_{k,j}^{-T} = I$ and $\gamma_{k,j} = [\gamma_{k,j}(1), \gamma_{k,j}(2), \gamma_{k,j}(3)]^T \triangleq \Gamma_{k,j}^{-1}g_{k,j}$ is a Gaussian vector $\gamma_{k,j} \in \mathcal{N}(0,I)$, and hence the components of $g_{k,j} = \Gamma_{k,j}\gamma_{k,j}$ can be expressed as follows:

$$v_k = \sigma_k\vartheta\gamma_{k,j}(1), \quad u_{k-1} = \frac{\vartheta}{\sigma_k}\gamma_{k,j}(1) + \frac{\vartheta\sqrt{\sigma_k^2-1}}{\sigma_k}\gamma_{k,j}(2)$$

$$u_{k-j-1} = \frac{\vartheta l_j}{\sigma_k}\gamma_{k,j}(1) - \frac{\vartheta l_j}{\sigma_k\sqrt{\sigma_k^2-1}}\gamma_{k,j}(2) + \vartheta\sqrt{\frac{\sigma_k^2-1-l_j^2}{\sigma_k^2-1}}\gamma_{k,j}(3).$$

The proof of (4.3.21)–(4.3.26) is essentially based on these expressions and the fact that the components of $\gamma_{k,j}$ are orthogonal to each other. To demonstrate the method of the proof we prove (4.3.21)–(4.3.23) only; the rest is proved similarly. Since

$$\begin{aligned} E(f(v_k)u_{k-1}^2) =& \frac{1}{\sigma_k^2}E\Bigg(f(\sigma_k\vartheta\gamma_{k,j}(1))\Big(\vartheta^2\gamma_{k,j}^2(1) \\ &+ (\sigma_k^2-1)\vartheta^2\gamma_{k,j}^2(2) + 2\vartheta^2\sqrt{\sigma_k^2-1}\gamma_{k,j}(1)\gamma_{k,j}(2)\Big)\Bigg) \\ =& \frac{1}{\sigma_k^2}\Big(E(f(v_k)\vartheta^2\gamma_{k,j}^2(1)) + (\sigma_k^2-1)\vartheta^2 Ef(v_k)\Big), \end{aligned}$$

we have

$$E\left(f(v_k)(u_{k-1}^2 - \vartheta^2)\right) = \frac{1}{\sigma_k^4}E(f(v_k)v_k^2) - \frac{\vartheta^2}{\sigma_k^2}Ef(v_k) \triangleq \rho_k. \tag{4.3.27}$$

Similarly, we obtain

$$\begin{aligned} E(f(v_k)u_{k-1}u_{k-j-1}) =& E\Bigg(f(\sigma_k\vartheta\gamma_{k,j}(1))\Big(\frac{\vartheta}{\sigma_k}\gamma_{k,j}(1) + \frac{\vartheta\sqrt{\sigma_k^2-1}}{\sigma_k}\gamma_{k,j}(2)\Big) \\ &\cdot\Big(\frac{\vartheta l_j}{\sigma_k}\gamma_{k,j}(1) - \frac{\vartheta l_j}{\sigma_k\sqrt{\sigma_k^2-1}}\gamma_{k,j}(2) + \vartheta\sqrt{\frac{\sigma_k^2-1-l_j^2}{\sigma_k^2-1}}\gamma_{k,j}(3)\Big)\Bigg) \\ =& E\big(f(\sigma_k\vartheta\gamma_{k,j}(1))\vartheta^2\gamma_{k,j}^2(1)\frac{l_j}{\sigma_k^2}\big) - E\big(f(\sigma_k\vartheta\gamma_{k,j}(1))\vartheta^2\gamma_{k,j}^2(2)\frac{l_j}{\sigma_k^2}\big) \end{aligned}$$

$$= \Big(\frac{1}{\sigma_k^4}E(f(v_k)v_k^2) - \frac{\vartheta^2}{\sigma_k^2}Ef(v_k)\Big)l_j = \rho_k l_j, \tag{4.3.28}$$

and

$$E(f(v_k)(u_{k-1}^3 - 3\vartheta^2 u_{k-1})) = E\Bigg(f(\sigma_k\vartheta\gamma_{k,j}(1))\Big(\frac{3\vartheta^3\sqrt{\sigma_k^2-1}}{\sigma_k^3}\gamma_{k,j}^2(1)\gamma_{k,j}(2)$$

$$+\frac{3\vartheta^3(\sigma_k^2-1)}{\sigma_k^3}\gamma_{k,j}(1)\gamma_{k,j}^2(2) + \frac{\vartheta^3}{\sigma_k^3}\gamma_{k,j}^3(1) + \frac{\vartheta^3(\sigma_k^2-1)\sqrt{\sigma_k^2-1}}{\sigma_k^3}\gamma_{k,j}^3(2)$$

$$-\frac{3\vartheta^2}{\sigma_k}\gamma_{k,j}(1) - \frac{3\vartheta^2\sqrt{\sigma_k^2-1}}{\sigma_k}\gamma_{k,j}(2)\Big)\Bigg)$$

$$= \frac{1}{\sigma_k^6}E(f(v_k)v_k^3) + \frac{3\vartheta^2(\sigma_k^2-1)}{\sigma_k^4}E(f(v_k)v_k) - \frac{3\vartheta^2}{\sigma_k^2}E(f(v_k)v_k)$$

$$= \frac{1}{\sigma_k^6}E(f(v_k)v_k^3) - \frac{3\vartheta^2}{\sigma_k^4}E(f(v_k)v_k) \triangleq \kappa_k. \tag{4.3.29}$$

Noticing $\sigma_k \xrightarrow[k\to\infty]{} \sigma$, from (4.3.27)–(4.3.29) we derive (4.3.21)–(4.3.23).

The proof of the lemma is completed. □

Prior to estimating l_i we first estimate ρ and κ by SAAWET on the basis of (4.3.21) and (4.3.23), respectively.

$$\theta_{k+1}^{(0,\rho)} = \Big[\theta_k^{(0,\rho)} - \frac{1}{k}\big(\theta_k^{(0,\rho)} - z_{k+1}(u_{k-1}^2 - \vartheta^2)\big)\Big] \cdot I_{\Big[\big|\theta_k^{(0,\rho)} - \frac{1}{k}\big(\theta_k^{(0,\rho)} - z_{k+1}(u_{k-1}^2-\vartheta^2)\big)\big| \le M_{\delta_k^{(0,\rho)}}\Big]}, \tag{4.3.30}$$

$$\delta_k^{(0,\rho)} = \sum_{j=1}^{k-1} I_{\Big[\big|\theta_j^{(0,\rho)} - \frac{1}{j}\big(\theta_j^{(0,\rho)} - z_{j+1}(u_{j-1}^2-\vartheta^2)\big)\big| > M_{\delta_j^{(0,\rho)}}\Big]}, \tag{4.3.31}$$

$$\theta_{k+1}^{(0,\kappa)} = \Big[\theta_k^{(0,\kappa)} - \frac{1}{k}\big(\theta_k^{(0,\kappa)} - z_{k+1}(u_{k-1}^3 - 3\vartheta^2 u_{k-1})\big)\Big] \cdot I_{\Big[\big|\theta_k^{(0,\kappa)} - \frac{1}{k}\big(\theta_k^{(0,\kappa)} - z_{k+1}(u_{k-1}^3-3\vartheta^2 u_{k-1})\big)\big| \le M_{\delta_k^{(0,\kappa)}}\Big]}, \tag{4.3.32}$$

$$\delta_k^{(0,\kappa)} = \sum_{j=1}^{k-1} I_{\Big[\big|\theta_j^{(0,\kappa)} - \frac{1}{j}\big(\theta_j^{(0,\kappa)} - z_{j+1}(u_{j-1}^3-3\vartheta^2 u_{j-1})\big)\big| > M_{\delta_j^{(0,\kappa)}}\Big]}, \tag{4.3.33}$$

where $\{M_k\}$ is an arbitrarily chosen sequence of positive real numbers increasingly diverging to infinity, $\theta_0^{(0,\rho)}$ and $\theta_0^{(0,\kappa)}$ are the arbitrary initial values, and I_A denotes the indicator function of a set A.

In order to obtain the estimates for $\{l_i, i \ge 1\}$, we design a switching mechanism by comparing the values between $|\theta_{k+1}^{(0,\rho)}|$ and $|\theta_{k+1}^{(0,\kappa)}|$ based on (4.3.30)–(4.3.33) because we only know that at least one of the constants ρ and κ is nonzero by A4.3.4.

If $|\theta_{k+1}^{(0,\rho)}| \geq |\theta_{k+1}^{(0,\kappa)}|$, then the following algorithm based on (4.3.22) is used to derive the estimates for $\{\rho l_i,\ i \geq 1\}$:

$$\theta_{k+1}^{(i,\rho)} = \left[\theta_k^{(i,\rho)} - \frac{1}{k}\left(\theta_k^{(i,\rho)} - z_{k+1}u_{k-1}u_{k-i-1}\right)\right] \cdot I_{\left[\left|\theta_k^{(i,\rho)} - \frac{1}{k}\left(\theta_k^{(i,\rho)} - z_{k+1}u_{k-1}u_{k-i-1}\right)\right| \leq M_{\delta_k^{(i,\rho)}}\right]}, \tag{4.3.34}$$

$$\delta_k^{(i,\rho)} = \sum_{j=1}^{k-1} I_{\left[\left|\theta_j^{(i,\rho)} - \frac{1}{j}\left(\theta_j^{(i,\rho)} - z_{j+1}u_{j-1}u_{j-i-1}\right)\right| > M_{\delta_j^{(i,\rho)}}\right]}. \tag{4.3.35}$$

Here $\theta_k^{(i,\rho)}$ is obtained from the previous step of the recursion if $|\theta_k^{(0,\rho)}| \geq |\theta_k^{(0,\kappa)}|$; otherwise, i.e., if $|\theta_k^{(0,\rho)}| < |\theta_k^{(0,\kappa)}|$, then $\theta_k^{(i,\rho)}$ has not been computed in accordance with (4.3.34)-(4.3.35). In this case $\theta_k^{(i,\rho)}$ in (4.3.34) is set to equal $\theta_k^{(0,\rho)} l_{i,k}$. After having the estimates for ρ and ρl_i, the estimates for the impulse responses $\{l_i, i \geq 1\}$ at time $k+1$ are given by

$$l_{i,k+1} \triangleq \begin{cases} \dfrac{\theta_{k+1}^{(i,\rho)}}{\theta_{k+1}^{(0,\rho)}}, & \text{if } \theta_{k+1}^{(0,\rho)} \neq 0, \\ 0, & \text{if } \theta_{k+1}^{(0,\rho)} = 0. \end{cases} \tag{4.3.36}$$

If $|\theta_{k+1}^{(0,\rho)}| < |\theta_{k+1}^{(0,\kappa)}|$, then based on (4.3.24), $\{\kappa l_i,\ i \geq 1\}$ are estimated by the following algorithm:

$$\theta_{k+1}^{(i,\kappa)} = \left[\theta_k^{(i,\kappa)} - \frac{1}{k}\left(\theta_k^{(i,\kappa)} - z_{k+1}(u_{k-1}^2 - \vartheta^2)u_{k-i-1}\right)\right] \cdot I_{\left[\left|\theta_k^{(i,\kappa)} - \frac{1}{k}\left(\theta_k^{(i,\kappa)} - z_{k+1}(u_{k-1}^2 - \vartheta^2)u_{k-i-1}\right)\right| \leq M_{\delta_k^{(i,\kappa)}}\right]}, \tag{4.3.37}$$

$$\delta_k^{(i,\kappa)} = \sum_{j=1}^{k-1} I_{\left[\left|\theta_j^{(i,\kappa)} - \frac{1}{j}\left(\theta_j^{(i,\kappa)} - z_{j+1}(u_{j-1}^2 - \vartheta^2)u_{j-i-1}\right)\right| > M_{\delta_j^{(i,\kappa)}}\right]}. \tag{4.3.38}$$

Similar to the previous case, $\theta_k^{(i,\kappa)}$ is derived from the previous step of the recursion if $|\theta_k^{(0,\rho)}| < |\theta_k^{(0,\kappa)}|$; otherwise, i.e., if $|\theta_k^{(0,\rho)}| \geq |\theta_k^{(0,\kappa)}|$, then $\theta_k^{(i,\kappa)}$ has not been computed in accordance with (4.3.37)–(4.3.38). In this case $\theta_k^{(i,\kappa)}$ in (4.3.37) is set to equal $\theta_k^{(0,\kappa)} l_{i,k}$. After having the estimates for κ and κl_i, the estimates for the impulse responses $\{l_i, i \geq 1\}$ at time $k+1$ are derived by

$$l_{i,k+1} \triangleq \begin{cases} \dfrac{\theta_{k+1}^{(i,\kappa)}}{\theta_{k+1}^{(0,\kappa)}}, & \text{if } \theta_{k+1}^{(0,\kappa)} \neq 0, \\ 0, & \text{if } \theta_{k+1}^{(0,\rho)} = 0. \end{cases} \tag{4.3.39}$$

It is important to note that the strong consistency of $\theta_k^{(0,\rho)}$ and $\theta_k^{(0,\kappa)}$ as to be shown later on guarantees that switching between the algorithms (4.3.34)–(4.3.36) and (4.3.37)–(4.3.39) ceases in a finite number of steps, because by A4.3.4 at least one of ρ and κ is nonzero.

Carrying out the similar operation as that done in Section 3.3, we obtain the following linear algebraic equation:

$$L[p_1, p_2, \cdots, p_{n_p}]^T = -[l_{n_q+1}, l_{n_q+2}, \cdots, l_{n_q+n_p}]^T, \tag{4.3.40}$$

where

$$L \triangleq \begin{bmatrix} l_{n_q} & l_{n_q-1} & \cdots & l_{n_q-n_p+1} \\ l_{n_q+1} & l_{n_q} & \cdots & l_{n_q-n_p+2} \\ \vdots & \vdots & \ddots & \vdots \\ l_{n_q+n_p-1} & l_{n_q+n_p-2} & \cdots & l_{n_q} \end{bmatrix}. \tag{4.3.41}$$

Noticing that the matrix L is nonsingular under A4.3.2 by Theorem 3.3.2 and that $l_{i,k} \xrightarrow[k\to\infty]{} l_i$ a.s. as to be shown by Theorem 4.3.1, we see that L_k is nonsingular when k is sufficiently large, where

$$L_k \triangleq \begin{bmatrix} l_{n_q,k} & l_{n_q-1,k} & \cdots & l_{n_q-n_p+1,k} \\ l_{n_q+1,k} & l_{n_q,k} & \cdots & l_{n_q-n_p+2,k} \\ \vdots & \vdots & \ddots & \vdots \\ l_{n_q+n_p-1,k} & l_{n_q+n_p-2,k} & \cdots & l_{n_q,k} \end{bmatrix} \tag{4.3.42}$$

serving as the kth estimate for L with $l_{i,k} = 0$ for $i < 0$.

The estimates for $\{p_1, \cdots, p_{n_p}, q_1, \cdots, q_{n_q}\}$ are naturally to be defined as follows:

$$[p_{1,k}, p_{2,k}, \cdots, p_{n_p,k}]^T \triangleq -L_k^{-1}[l_{n_q+1,k}, l_{n_q+2,k}, \cdots, l_{n_q+n_p,k}]^T, \tag{4.3.43}$$

$$q_{i,k} \triangleq l_{i,k} + \sum_{j=1}^{i\wedge n_p} p_{j,k} l_{i-j,k}, \quad i = 1, 2, \cdots, n_q. \tag{4.3.44}$$

Estimation for $\{\mathbf{c}_1, \cdots, \mathbf{c}_{\mathbf{n_c}}, \mathbf{d}_1, \cdots, \mathbf{d}_{\mathbf{n_d}}\}$

In order to estimate the coefficients of the polynomials $C(z)$ and $D(z)$ in the second linear subsystem, we first estimate the impulse responses h_i.

From (4.3.21)–(4.3.24) it is seen that l_i $\forall i \geq 1$ can be estimated provided at least one of ρ and κ is nonzero, as demonstrated by (4.3.30)–(4.3.38). Further, from (4.3.25)–(4.3.26) it is seen that h_j $\forall j \geq 1$ can also be estimated. However, if τ defined by (4.3.13) is nonzero, then by using its estimate incorporated with the obtained estimates for l_i we can also derive estimates for h_j, $j \geq 1$. The advantage of using the estimate for τ consists in its simpler computation and faster rate of convergence.

Similar to Lemma 4.3.1 we have the following lemma.

Lemma 4.3.2 *Assume A4.3.1–A4.3.5. Then*

$$\lim_{k\to\infty} E z_{k+1} u_{k-i-1} = \tau \sum_{j=0}^{i} h_j l_{i-j}, \ i \geq 0. \tag{4.3.45}$$

Proof. Noticing that $\{u_k\}$ is Gaussian iid with zero-mean, from

$$Eu_{k-i-1}v_k = E\big(u_{k-i-1}\sum_{j=0}^{k-1} l_j u_{k-j-1}\big) = \vartheta^2 l_i,$$

we see that

$$E\Big[\big(u_{k-i-1} - \frac{l_i v_k}{\sigma_k^2}\big)v_k\Big] = 0,$$

which implies that $u_{k-i-1} - \frac{l_i v_k}{\sigma_k^2}$ is uncorrelated with and hence is independent of v_k, since for Gaussian random variables independence is equivalent to uncorrelatedness. Therefore, we have

$$E\big(u_{k-i-1} - \frac{l_i v_k}{\sigma_k^2}|v_k\big) = 0,$$

which implies

$$E(u_{k-i-1}|v_k) = \frac{l_i v_k}{\sigma_k^2},$$

and hence

$$E(u_{k-i-1}\varphi_k) = E(E(u_{k-i-1}\varphi_k|v_k)) = E\big(\varphi_k \frac{l_i v_k}{\sigma_k^2}\big) = \tau_k l_i,$$

where $\tau_k \triangleq \frac{1}{\sigma_k^2}E(f(v_k)v_k)$.

Noticing that $\varphi_{k-j}, j > i$ is independent of u_{k-i-1}, by (4.3.4), (4.3.19), and (4.3.20) we have

$$\begin{aligned}
\lim_{k\to\infty} Ez_{k+1}u_{k-i-1} &= \lim_{k\to\infty} E(y_{k+1}^0 + \bar{\eta}_{k+1} + \bar{\xi}_{k+1} + \varepsilon_{k+1})u_{k-i-1} \\
&= \lim_{k\to\infty} Ey_{k+1}^0 u_{k-i-1} = \lim_{k\to\infty} \sum_{j=0}^{k} h_j E\varphi_{k-j}u_{k-i-1} \\
&= \lim_{k\to\infty} \sum_{j=0}^{i} h_j \tau_{k-j} l_{i-j} = \tau \sum_{j=0}^{i} h_j l_{i-j}.
\end{aligned}$$

This completes the proof. □

Based on (4.3.45), we now introduce the following algorithms to estimate $\tau\sum_{j=0}^{i} h_j l_{i-j}$:

$$\begin{aligned}
\lambda_{k+1}^{(i,\tau)} = &\Big[\lambda_k^{(i,\tau)} - \frac{1}{k}\big(\lambda_k^{(i,\tau)} - z_{k+1}u_{k-i-1}\big)\Big] \\
&\cdot I_{\Big[\big|\lambda_k^{(i,\tau)} - \frac{1}{k}\big(\lambda_k^{(i,\tau)} - z_{k+1}u_{k-i-1}\big)\big| \le M_{\sigma_k^{(i,\tau)}}\Big]},
\end{aligned} \tag{4.3.46}$$

$$\sigma_k^{(i,\tau)} = \sum_{j=1}^{k-1} I_{\Big[\big|\lambda_j^{(i,\tau)} - \frac{1}{j}\big(\lambda_j^{(i,\tau)} - z_{j+1}u_{j-i-1}\big)\big| > M_{\sigma_j^{(i,\tau)}}\Big]}, \quad i \ge 0. \tag{4.3.47}$$

It is noticed that $\tau\sum_{j=0}^{i} h_j l_{i-j} = \tau$ when $i = 0$. So, the algorithm (4.3.46)–(4.3.47) with $i = 0$ gives the estimate for τ.

Similar to $\{\theta_k^{(0,\rho)}\}$ and $\{\theta_k^{(0,\kappa)}\}$ calculated above, we also need to compute $\{\lambda_k^{(0,\tau)}\}$ and compare the values $|\lambda_{k+1}^{(0,\tau)}|$, $|\theta_{k+1}^{(0,\rho)}|$, and $|\theta_{k+1}^{(0,\kappa)}|$.

(i) If $|\lambda_{k+1}^{(0,\tau)}| \geq \max(|\theta_{k+1}^{(0,\rho)}|, |\theta_{k+1}^{(0,\kappa)}|)$, then the estimates for $\{h_i, i \geq 1\}$ are given by

$$h_{i,k+1} \triangleq \frac{\lambda_{k+1}^{(i,\tau)}}{\lambda_{k+1}^{(0,\tau)}} - \sum_{j=0}^{i-1} h_{j,k+1} l_{i-j,k+1}. \tag{4.3.48}$$

In (4.3.46)–(4.3.47), $\lambda_k^{(i,\tau)}$ is obtained from the previous step of the recursion if $|\lambda_k^{(0,\tau)}| \geq \max(|\theta_k^{(0,\rho)}|, |\theta_k^{(0,\kappa)}|)$; otherwise, they are set to equal $\lambda_k^{(0,\tau)} \sum_{j=0}^{i} h_{j,k} l_{i-j,k}$.

(ii) If $|\theta_{k+1}^{(0,\rho)}| > \max(|\lambda_{k+1}^{(0,\tau)}|, |\theta_{k+1}^{(0,\kappa)}|)$, then based on (4.3.25) we introduce the following algorithms to estimate $\rho\sum_{j=0}^{i} h_j l_{i-j}^2$:

$$\begin{aligned} \lambda_{k+1}^{(i,\rho)} &= \Big[\lambda_k^{(i,\rho)} - \frac{1}{k}\big(\lambda_k^{(i,\rho)} - z_{k+1}(u_{k-i-1}^2 - \vartheta^2)\big)\Big] \\ &\quad \cdot I_{\left[\left|\lambda_k^{(i,\rho)} - \frac{1}{k}\left(\lambda_k^{(i,\rho)} - z_{k+1}(u_{k-i-1}^2 - \vartheta^2)\right)\right| \leq M_{\sigma_k^{(i,\rho)}}\right]}, \end{aligned} \tag{4.3.49}$$

$$\sigma_k^{(i,\rho)} = \sum_{j=1}^{k-1} I_{\left[\left|\lambda_j^{(i,\rho)} - \frac{1}{j}\left(\lambda_j^{(i,\rho)} - z_{j+1}(u_{j-i-1}^2 - \vartheta^2)\right)\right| > M_{\sigma_j^{(i,\rho)}}\right]}, \quad i \geq 1, \tag{4.3.50}$$

where $\lambda_k^{(i,\rho)}$ are obtained from the previous step of the recursion if $|\theta_k^{(0,\rho)}| > \max(|\lambda_k^{(0,\tau)}|, |\theta_k^{(0,\kappa)}|)$; otherwise, they are set to equal $\theta_k^{(0,\rho)} \sum_{j=0}^{i} h_{j,k} l_{i-j,k}^2$. In this case the estimates for $\{h_i, i \geq 1\}$ are defined by

$$h_{i,k+1} \triangleq \frac{\lambda_{k+1}^{(i,\rho)}}{\theta_{k+1}^{(0,\rho)}} - \sum_{j=0}^{i-1} h_{j,k+1} l_{i-j,k+1}^2. \tag{4.3.51}$$

(iii) Finally, if $|\theta_{k+1}^{(0,\kappa)}| > \max(|\lambda_{k+1}^{(0,\tau)}|, |\theta_k^{(0,\rho)}|)$, then based on (4.3.26) we introduce the following algorithms to estimate $\kappa\sum_{j=0}^{i} h_j l_{i-j}^3$:

$$\begin{aligned} \lambda_{k+1}^{(i,\kappa)} &= \Big[\lambda_k^{(i,\kappa)} - \frac{1}{k}\big(\lambda_k^{(i,\kappa)} - z_{k+1}(u_{k-i-1}^3 - 3\vartheta^2 u_{k-i-1})\big)\Big] \\ &\quad \cdot I_{\left[\left|\lambda_k^{(i,\kappa)} - \frac{1}{k}\left(\lambda_k^{(i,\kappa)} - z_{k+1}(u_{k-i-1}^3 - 3\vartheta^2 u_{k-i-1})\right)\right| \leq M_{\sigma_k^{(i,\kappa)}}\right]}, \end{aligned} \tag{4.3.52}$$

$$\sigma_k^{(i,\kappa)} = \sum_{j=1}^{k-1} I_{\left[\left|\lambda_j^{(i,\kappa)} - \frac{1}{j}\left(\lambda_j^{(i,\kappa)} - z_{j+1}(u_{j-i-1}^3 - 3\vartheta^2 u_{j-i-1})\right)\right| > M_{\sigma_j^{(i,\kappa)}}\right]}, i \geq 1, \tag{4.3.53}$$

where $\lambda_k^{(i,\kappa)}$ are obtained from the previous step of the recursion if $|\theta_k^{(0,\kappa)}| >$

$\max(|\lambda_k^{(0,\tau)}|, |\theta_k^{(0,\rho)}|)$; otherwise, they are set to equal $\theta_k^{(0,\kappa)} \sum_{j=0}^{i} h_{j,k} l_{i-j,k}^3$. In this case the estimates for $\{h_i, i \geq 1\}$ are defined by

$$h_{i,k+1} \triangleq \frac{\lambda_{k+1}^{(i,\kappa)}}{\theta_{k+1}^{(0,\kappa)}} - \sum_{j=0}^{i-1} h_{j,k+1} l_{i-j,k+1}^3. \tag{4.3.54}$$

If $|\theta_k^{(0,\rho)}| = |\lambda_k^{(0,\tau)}| = |\theta_k^{(0,\kappa)}| = 0$, then define $h_{i,k} \triangleq 0$.

As mentioned before, switching among the algorithms (4.3.36)–(4.3.38), (4.3.49)–(4.3.51), and (4.3.52)–(4.3.54) ceases in a finite number of steps if the estimates $\theta_k^{(0,\rho)}$, $\theta_k^{(0,\kappa)}$, and $\lambda_k^{(0,\tau)}$ are strongly consistent.

The estimates for $\{c_1, \cdots, c_{n_c}, d_1, \cdots, d_{n_d}\}$ are obtained in the same way as that used for estimating coefficients of the first linear subsystem.

For this define

$$H \triangleq \begin{bmatrix} h_{n_d} & h_{n_d-1} & \cdots & h_{n_d-n_c+1} \\ h_{n_d+1} & h_{n_d} & \cdots & h_{n_d-n_c+2} \\ \vdots & \vdots & \ddots & \vdots \\ h_{n_d+n_c-1} & h_{n_d+n_c-2} & \cdots & h_{n_d} \end{bmatrix} \tag{4.3.55}$$

and

$$H_k \triangleq \begin{bmatrix} h_{n_d,k} & h_{n_d-1,k} & \cdots & h_{n_d-n_c+1,k} \\ h_{n_d+1,k} & h_{n_d,k} & \cdots & h_{n_d-n_c+2,k} \\ \vdots & \vdots & \ddots & \vdots \\ h_{n_d+n_c-1,k} & h_{n_d+n_c-2,k} & \cdots & h_{n_d,k} \end{bmatrix} \tag{4.3.56}$$

with $h_{i,k} = 0$ for $i < 0$.

The matrix H is nonsingular under A4.3.3 by Theorem 3.3.2, and hence H_k is nonsingular for sufficiently large k since $h_{i,k}$ converges to h_i a.s. as to be shown by Theorem 4.3.2.

Similar to (4.3.43)–(4.3.44), the estimates for $\{c_1, \cdots, c_{n_c}, d_1, \cdots, d_{n_d}\}$ are given as follows:

$$[c_{1,k}, \cdots, c_{n_c,k}]^T \triangleq -H_k^{-1} [h_{n_d+1,k}, \cdots, h_{n_d+n_c,k}]^T, \tag{4.3.57}$$

$$d_{i,k} \triangleq h_{i,k} + \sum_{j=1}^{i \wedge n_c} c_{j,k} h_{i-j,k}, \quad i = 1, 2, \cdots, n_d. \tag{4.3.58}$$

Nonparametric Estimation for f$(\cdot)$

We now recursively estimate $f(y)$, where y is an arbitrary point at the real axis. By using the estimates obtained for coefficients of the linear subsystems we can estimate the internal signals v_k and φ_k, the input and output of the nonlinear block, on the basis of the state space representations of the linear subsystems. Then, applying SAAWET incorporated with kernel function we obtain the estimates for $f(y)$.

From (4.3.3) and (4.3.4), we have

$$C(z)z_{k+1} = D(z)(\varphi_k + \eta_k) + C(z)\varepsilon_{k+1} + \xi_{k+1}. \tag{4.3.59}$$

By defining

$$v_k \triangleq D^{-1}(z)C(z)z_{k+1},\ \phi_k \triangleq D^{-1}(z)\xi_{k+1},\ \text{and}\ \chi_k \triangleq D^{-1}(z)C(z)\varepsilon_{k+1}, \tag{4.3.60}$$

(4.3.59) can be rewritten as

$$\begin{aligned}\varphi_k &= D^{-1}(z)C(z)z_{k+1} - D^{-1}(z)C(z)\varepsilon_{k+1} - D^{-1}(z)\xi_{k+1} - \eta_k \\ &= v_k - \chi_k - \phi_k - \eta_k.\end{aligned} \tag{4.3.61}$$

Let us start with estimating v_k.

Define

$$P \triangleq \begin{bmatrix} -p_1 & 1 & \cdots & 0 \\ \vdots & & \ddots & \vdots \\ \vdots & & & 1 \\ -p_s & 0 & \cdots & 0 \end{bmatrix},\quad Q \triangleq \begin{bmatrix} 1 \\ q_1 \\ \vdots \\ q_{s-1} \end{bmatrix},\quad G \triangleq \begin{bmatrix} 1 \\ 0 \\ \vdots \\ 0 \end{bmatrix},$$

where $s \triangleq \max(n_p, n_q + 1)$. Then, (4.3.1) can be presented in the state space form

$$\begin{cases} x_{k+1} = & Px_k + Qu_k \\ v_{k+1} = & G^T x_{k+1}, \end{cases} \tag{4.3.62}$$

where P is an $s \times s$ matrix, Q and G are $s \times 1$ vectors, $p_k = 0$ for $k > n_p$, and $q_k = 0$ for $k > n_q$.

Replacing p_i and q_j in P and Q with $p_{i,k}$ and $q_{j,k}$ given by (4.3.43) and (4.3.44), respectively, $i = 1, \cdots, s,\ j = 1, \cdots, s-1$, we obtain the estimates Q_k and P_k for Q and P at time k, and, hence, the estimate $\widehat{v}_{k+1}$ for v_{k+1}:

$$\widehat{x}_{k+1} = P_{k+1}\widehat{x}_k + Q_{k+1}u_k,\ \widehat{v}_{k+1} = G^T \widehat{x}_{k+1} \tag{4.3.63}$$

with an arbitrary initial value $\widehat{x}_0$.

Similarly, we can obtain the estimate for φ_k with the help of the state space representation of the second linear subsystem.

Set

$$\bar{s} \triangleq n_d \vee (n_c + 1) \triangleq \begin{cases} n_d, & \text{if } n_d \geq n_c + 1, \\ n_c + 1, & \text{if } n_d < n_c + 1. \end{cases}$$

The first system in (4.3.60) can be expressed in the state space form:

$$\begin{cases} t_{k+1} = & Dt_k + Cz_{k+1} \\ v_k = & \bar{G}^T t_{k+1}, \end{cases} \tag{4.3.64}$$

where

$$D \triangleq \begin{bmatrix} -d_1 & 1 & \cdots & 0 \\ \vdots & & \ddots & \vdots \\ \vdots & & & 1 \\ -d_{\bar{s}} & 0 & \cdots & 0 \end{bmatrix}, C \triangleq \begin{bmatrix} 1 \\ c_1 \\ \vdots \\ c_{\bar{s}-1} \end{bmatrix},$$

and $\bar{G} \triangleq \begin{bmatrix} 1 & 0 & \cdots & 0 \end{bmatrix}^T$. Notice that D is an $\bar{s} \times \bar{s}$ matrix, C and $\bar{G}$ are $\bar{s} \times 1$ vectors, and $d_k = 0$ for $k > n_d$, $c_k = 0$ for $k > n_c$ by definition.

Let C_k and D_k be obtained from C and D with entries replaced by their estimates given by (4.3.57) and (4.3.58). The estimate $\widehat{v}_k$ for v_k is recursively given by the following algorithm with an arbitrary initial value $\widehat{t}_0$:

$$\widehat{t}_{k+1} = D_{k+1}\widehat{t}_k + C_{k+1}z_{k+1}, \ \widehat{v}_k = \bar{G}^T\widehat{t}_{k+1}. \tag{4.3.65}$$

To estimate $f(y)$, let us introduce the kernel function ω_k and its estimate $\widehat{\omega}_k$ as follows:

$$\omega_k = \frac{1}{\sqrt{2\pi}b_k}e^{-\frac{(v_k-y)^2}{2b_k^2}}, \widehat{\omega}_k = \frac{1}{\sqrt{2\pi}b_k}e^{-\frac{(\widehat{v}_k-y)^2}{2b_k^2}}, \tag{4.3.66}$$

where $b_k = \frac{1}{k^a}$ with $a > 0$ being a fixed constant. Let $m_k = k^b, b > 0$, and $3a + b < \frac{1}{2}$. Then, $f(y)$ is recursively estimated by the following algorithm with arbitrary initial value $\mu_0(y)$ and $\Delta_0(y) = 0$,

$$\begin{aligned} \mu_{k+1}(y) = \ & [\mu_k(y) - \tfrac{1}{k}\widehat{\omega}_k(\mu_k(y) - \widehat{v}_k)] \\ & \cdot I_{[|\mu_k(y) - \frac{1}{k}\widehat{\omega}_k(\mu_k(y)-\widehat{v}_k)| \le m_{\Delta_k}(y)]}, \\ \Delta_k(y) = \ & \sum_{j=1}^{k-1} I_{[|\mu_j(y) - \frac{1}{j}\widehat{\omega}_j(\mu_j(y)-\widehat{v}_j)| > m_{\Delta_j}(y)]}. \end{aligned} \tag{4.3.67}$$

Strong Consistency of Estimates

We are now proceeding to prove that all estimates given above in this section are strongly consistent. For this we are planning to verify conditions guaranteeing convergence of all the corresponding SAAWETs.

Lemma 4.3.3 *Assume A4.3.1–A4.3.5 hold. Then for any* $0 \le v < 1/2$, *the following series are convergent:*

$$\sum_{k=1}^{\infty} \frac{1}{k^{1-v}}(E(z_{k+1}(u_{k-1}^2 - \vartheta^2)) - \rho) < \infty, \tag{4.3.68}$$

$$\sum_{k=1}^{\infty} \frac{1}{k^{1-v}}(E(z_{k+1}u_{k-1}u_{k-i-1}) - \rho l_i) < \infty, \ \text{for } i \ge 1, \tag{4.3.69}$$

$$\sum_{k=1}^{\infty} \frac{1}{k^{1-v}}(Ez_{k+1}(u_{k-1}^3 - 3\vartheta^2 u_{k-1}) - \kappa) < \infty, \tag{4.3.70}$$

$$\sum_{k=1}^{\infty} \frac{1}{k^{1-v}}(Ez_{k+1}(u_{k-1}^2 - \vartheta^2)u_{k-i-1} - \kappa l_i) < \infty, \ \text{for } i \ge 1, \tag{4.3.71}$$

$$\sum_{k=1}^{\infty}\frac{1}{k^{1-\nu}}\Big(Ez_{k+1}(u_{k-i-1}^2-\vartheta^2)-\rho\sum\nolimits_{j=0}^{i}h_jl_{i-j}^2\Big)<\infty,\ \text{for } i\geq 0, \quad (4.3.72)$$

$$\sum_{k=1}^{\infty}\frac{1}{k^{1-\nu}}\Big(Ez_{k+1}(u_{k-i-1}^3-3\vartheta^2u_{k-i-1})-\kappa\sum\nolimits_{j=0}^{i}h_jl_{i-j}^3\Big)<\infty,\ \text{for } i\geq 0, \quad (4.3.73)$$

$$\sum_{k=1}^{\infty}\frac{1}{k^{1-\nu}}\Big(E(z_{k+1}u_{k-i-1})-\tau\sum\nolimits_{j=0}^{i}h_jl_{i-j}\Big)<\infty,\ \text{for } i\geq 0. \quad (4.3.74)$$

Proof. By A4.3.1, A4.3.3, and A4.3.5 we have

$$Ez_{k+1}(u_{k-1}^2-\vartheta^2)=EC^{-1}(z)D(z)\varphi_k(u_{k-1}^2-\vartheta^2)=E\varphi_k(u_{k-1}^2-\vartheta^2).$$

Hence, by (4.3.27) and $v_k\in\mathcal{N}(0,\sigma_k^2\vartheta^2)$ it follows that

$$Ez_{k+1}(u_{k-1}^2-\vartheta^2)-\rho=\rho_k-\rho=I_{k,1}-I_{k,2}, \quad (4.3.75)$$

where

$$\begin{aligned}I_{k,1}=&\frac{1}{\sqrt{2\pi}}\Big(\Big(\frac{1}{\sigma_k^5\vartheta}-\frac{1}{\sigma^5\vartheta}\Big)\int_{\mathbb{R}}y^2f(y)e^{-\frac{y^2}{2\sigma_k^2\vartheta^2}}\mathrm{d}y\\&+\frac{1}{\sigma^5\vartheta}\int_{\mathbb{R}}y^2f(y)\big(e^{-\frac{y^2}{2\sigma_k^2\vartheta^2}}-e^{-\frac{y^2}{2\sigma^2\vartheta^2}}\big)\mathrm{d}y\Big),\end{aligned}$$

and

$$\begin{aligned}I_{k,2}=&\frac{1}{\sqrt{2\pi}}\Big(\Big(\frac{\vartheta}{\sigma_k^3}-\frac{\vartheta}{\sigma^3}\Big)\int_{\mathbb{R}}f(y)e^{-\frac{y^2}{2\sigma_k^2\vartheta^2}}\mathrm{d}y\\&+\frac{\vartheta}{\sigma^3}\int_{\mathbb{R}}f(y)\big(e^{-\frac{y^2}{2\sigma_k^2\vartheta^2}}-e^{-\frac{y^2}{2\sigma^2\vartheta^2}}\big)\mathrm{d}y\Big),\end{aligned}$$

where the integrals are finite by (4.3.14).

Since $\sigma^2-\sigma_k^2=\sum_{i=k}^{\infty}l_i^2$, $|l_i|=O(e^{-r_1i})$ with $r_1>0$, we have $0<\sigma^2-\sigma_k^2=O(\rho^k)$ for some $\rho\in(0,1)$, which implies

$$\Big(\frac{1}{\sigma_k^5\vartheta}-\frac{1}{\sigma^5\vartheta}\Big)\int_{\mathbb{R}}y^2f(y)e^{-\frac{y^2}{2\sigma_k^2\vartheta^2}}\mathrm{d}y=O(\rho^k).$$

By the mean theorem there is an $x\in[\sigma_k^2,\sigma^2]$ such that

$$\int_{\mathbb{R}}y^2f(y)\big(e^{-\frac{y^2}{2\sigma_k^2\vartheta^2}}-e^{-\frac{y^2}{2\sigma^2\vartheta^2}}\big)\mathrm{d}y=\int_{\mathbb{R}}\frac{1}{2x^2\vartheta^2}y^4f(y)e^{-\frac{y^2}{2x\vartheta^2}}\mathrm{d}y(\sigma_k^2-\sigma^2),$$

where x may depend on y, but $e^{-\frac{y^2}{2x\vartheta^2}}\leq e^{-\frac{y^2}{2\sigma^2\vartheta^2}}$ uniformly with respect to y and $\frac{1}{x^2}\leq 1$.

Therefore, we obtain

$$\int_{\mathbb{R}} y^2 f(y)\big(e^{-\frac{y^2}{2\sigma_k^2\vartheta^2}} - e^{-\frac{y^2}{2\sigma^2\vartheta^2}}\big)\mathrm{d}y = O(\rho^k),$$

and hence

$$I_{k,1} = O(\rho^k). \tag{4.3.76}$$

Similarly, we can show

$$I_{k,2} = O(\rho^k). \tag{4.3.77}$$

Thus, (4.3.68) is verified by noticing (4.3.75), (4.3.76), and (4.3.77), while (4.3.69)–(4.3.74) can be shown in a similar way. □

Lemma 4.3.4 *Assume A4.3.1–A4.3.5 hold. For any* $0 \le \nu < 1/2$ *the following series are convergent:*

$$\sum_{k=1}^{\infty} \frac{1}{k^{1-\nu}}\big(E(y_{k+1}^0(u_{k-1}^2 - \vartheta^2)) - y_{k+1}^0(u_{k-1}^2 - \vartheta^2)\big) < \infty \text{ a.s.,} \tag{4.3.78}$$

$$\sum_{k=1}^{\infty} \frac{1}{k^{1-\nu}}\big(E(y_{k+1}^0 u_{k-1} u_{k-i-1}) - y_{k+1}^0 u_{k-1} u_{k-i-1}\big) < \infty \text{ a.s. for } i \ge 1, \tag{4.3.79}$$

$$\sum_{k=1}^{\infty} \frac{1}{k^{1-\nu}}\big(E(y_{k+1}^0(u_{k-1}^3 - 3\vartheta^2 u_{k-1})) - y_{k+1}^0(u_{k-1}^3 - 3\vartheta^2 u_{k-1})\big) < \infty \text{ a.s.,} \tag{4.3.80}$$

$$\sum_{k=1}^{\infty} \frac{1}{k^{1-\nu}}\big(E(y_{k+1}^0(u_{k-1}^2 - \vartheta^2)u_{k-i-1}) - y_{k+1}^0(u_{k-1}^2 - \vartheta^2)u_{k-i-1}\big) < \infty \text{ a.s. for } i \ge 1, \tag{4.3.81}$$

$$\sum_{k=1}^{\infty} \frac{1}{k^{1-\nu}}\big(E(y_{k+1}^0(u_{k-i-1}^2 - \vartheta^2)) - y_{k+1}^0(u_{k-i-1}^2 - \vartheta^2)\big) < \infty \text{ a.s. for } i \ge 0, \tag{4.3.82}$$

$$\sum_{k=1}^{\infty} \frac{1}{k^{1-\nu}}\big(E(y_{k+1}^0(u_{k-i-1}^3 - 3\vartheta^2 u_{k-i-1})) - y_{k+1}^0(u_{k-i-1}^3 - 3\vartheta^2 u_{k-i-1})\big) < \infty \text{ a.s. for } i \ge 0, \tag{4.3.83}$$

$$\sum_{k=1}^{\infty} \frac{1}{k^{1-\nu}}\big(E(y_{k+1}^0 u_{k-i-1}) - y_{k+1}^0 u_{k-i-1}\big) < \infty \text{ a.s. for } i \ge 0. \tag{4.3.84}$$

Proof. Since $y_k^0 = \sum_{i=0}^{k-1} h_i \varphi_{k-i-1} = \sum_{i=0}^{k-1} h_i f(v_{k-i-1})$, we have

$$
\begin{aligned}
&E(y_{k+1}^0 u_{k-i-1}) - y_{k+1}^0 u_{k-i-1} \\
&= E\Big(\big(\sum_{j=0}^{k} h_j \varphi_{k-j}\big) u_{k-i-1}\Big) - \big(\sum_{j=0}^{k} h_j \varphi_{k-j}\big) u_{k-i-1} \\
&= E\Big[\big(\sum_{j=0}^{i} h_j \varphi_{k-j} + \sum_{j=i+1}^{k} h_j \varphi_{k-j}\big) u_{k-i-1}\Big] - \big(\sum_{j=0}^{i} h_j \varphi_{k-j} + \sum_{j=i+1}^{k} h_j \varphi_{k-j}\big) u_{k-i-1} \\
&= E\big(\sum_{j=0}^{i} h_j \varphi_{k-j} u_{k-i-1}\big) - \sum_{j=0}^{i} h_j \varphi_{k-j} u_{k-i-1} \\
&\quad + E\big(\sum_{j=i+1}^{k} h_j \varphi_{k-j} u_{k-i-1}\big) - \sum_{j=i+1}^{k} h_j \varphi_{k-j} u_{k-i-1} \\
&= E\big(\sum_{j=0}^{i} h_j \varphi_{k-j} u_{k-i-1}\big) - \sum_{j=0}^{i} h_j \varphi_{k-j} u_{k-i-1} - \sum_{j=i+1}^{k} h_j \varphi_{k-j} u_{k-i-1}.
\end{aligned}
$$

Consequently, we have

$$
\begin{aligned}
&\sum_{k=1}^{\infty} \frac{1}{k^{1-\nu}} \big(E y_{k+1}^0 u_{k-i-1} - y_{k+1}^0 u_{k-i-1}\big) \\
&= \sum_{k=1}^{\infty} \frac{1}{k^{1-\nu}} \Big[E \sum_{j=0}^{i} h_j \varphi_{k-j} u_{k-i-1} - \big(\sum_{j=0}^{i} h_j \varphi_{k-j}\big) u_{k-i-1} - \sum_{j=i+1}^{k} h_j \varphi_{k-j} u_{k-i-1}\Big] \\
&= \sum_{k=1}^{\infty} \frac{1}{k^{1-\nu}} \big(E \varphi_{k-j} u_{k-i-1} - \varphi_{k-j} u_{k-i-1}\big) \\
&\quad - \sum_{k=1}^{\infty} \frac{1}{k^{1-\nu}} \sum_{j=i+1}^{k} h_j \varphi_{k-j} u_{k-i-1}.
\end{aligned}
\tag{4.3.85}
$$

We now prove convergence of the first term at the right-hand side of (4.3.85).

Set $z_k^{(1)} \triangleq \frac{1}{k^{1-\nu}} \big(Ef(v_{k-j}) u_{k-i-1} - f(v_{k-j}) u_{k-i-1}\big)$ for fixed $0 \le j \le i$. By Theorem 1.4.4, $z_k^{(1)}$ is a zero-mean α-mixing with mixing coefficient decaying exponentially to zero. Besides, by the C_r-inequality and then the Hölder inequality, for any $\delta > 2$ we have

$$
\begin{aligned}
E|z_k^{(1)}|^{\delta} &= \frac{1}{k^{\delta(1-\nu)}} E|E u_{k-i-1} f(v_{k-j}) - u_{k-i-1} f(v_{k-j})|^{\delta} \\
&\le \frac{2^{\delta}}{k^{\delta(1-\nu)}} E|u_{k-i-1} f(v_{k-j})|^{\delta} \le \frac{2^{\delta}}{k^{\delta(1-\nu)}} (E|u_{k-i-1}|^{2\delta})^{\frac{1}{2}} (E|f(v_{k-j})|^{2\delta})^{\frac{1}{2}} \\
&\le \frac{2^{\delta}}{k^{\delta(1-\nu)}} (E|u_{k-i-1}|^{2\delta})^{\frac{1}{2}} (E|\alpha(1+|v_{k-j}|^{\beta})|^{2\delta})^{\frac{1}{2}} < \infty,
\end{aligned}
\tag{4.3.86}
$$

where (4.3.14) is invoked. From here it is seen that

$$\sum_{k=1}^{\infty}(E|z_k^{(1)}|^{\delta})^{\frac{2}{\delta}} \le O(\sum_{k=1}^{\infty}\frac{1}{k^{2(1-\nu)}}) < \infty.$$

By Theorem 1.4.2 the first term at the right-hand side of (4.3.85) is convergent. We now show that the last term of (4.3.85) converges too.

Define $\mathcal{F}_k^u \triangleq \sigma(u_i, 0 \le i \le k)$. Then, $v_k \in \mathcal{F}_{k-1}^u$ and $\frac{1}{k^{1-\nu}}(\sum_{j=i+1}^{k} h_j\varphi_{k-j}) \in \mathcal{F}_{k-i-2}^u$. Noticing $\{u_k\}$ is iid, we have

$$E\Big(\frac{1}{k^{1-\nu}}(\sum_{j=i+1}^{k} h_j\varphi_{k-j})u_{k-i-1}|\mathcal{F}_{k-i-2}^u\Big)$$
$$= \frac{1}{k^{1-\nu}}(\sum_{j=i+1}^{k} h_j\varphi_{k-j})E(u_{k-i-1}|\mathcal{F}_{k-i-2}^u) = 0.$$

Therefore, $\Big\{\frac{1}{k^{1-\nu}}(\sum_{j=i+1}^{k} h_j\varphi_{k-j})u_{k-i-1}, \mathcal{F}_{k-i-1}^u\Big\}$ is an mds.

By Lemma 4.2.1, we have $\sup_k E(\sum_{j=i+1}^{k} h_j\varphi_{k-j})^2 < \infty$ for any fixed $i: 0 \le i \le k-1$, and, hence,

$$\sum_{k=1}^{\infty} E\Big((\frac{1}{k^{1-\nu}}\sum_{j=i+1}^{k} h_j\varphi_{k-j})u_{k-i-1}\Big)^2$$
$$= \sum_{k=1}^{\infty}\frac{1}{k^{2(1-\nu)}}E(\sum_{j=i+1}^{k} h_j\varphi_{k-j})^2 Eu_{k-i-1}^2 = \sum_{k=1}^{\infty} O(\frac{1}{k^{2(1-\nu)}}) < \infty,$$

which implies

$$\sum_{k=1}^{\infty} E\Big[\frac{1}{k^{1-\nu}}(\sum_{j=i+1}^{k} h_j\varphi_{k-j})u_{k-i-1})^2|\mathcal{F}_{k-i-2}\Big] < \infty \text{ a.s.}$$

Then, the last term of (4.3.85) converges by Theorem 1.2.8.

Thus, we have proved (4.3.84), and (4.3.78)–(4.3.83) can be shown in a similar manner. □

Lemma 4.3.5 *Assume A4.3.1–A4.3.5 hold. For any* $0 \le \nu < \frac{1}{2}$ *the following series* *are convergent:*

$$\sum_{k=1}^{\infty}\frac{1}{k^{1-v}}\big(E(z_{k+1}(u_{k-1}^2-\vartheta^2))-z_{k+1}(u_{k-1}^2-\vartheta^2)\big)<\infty,\ \text{a.s.,} \tag{4.3.87}$$

$$\sum_{k=1}^{\infty}\frac{1}{k^{1-v}}\big(E(z_{k+1}u_{k-1}u_{k-i-1})\\ -z_{k+1}u_{k-1}u_{k-i-1}\big)<\infty\ \text{a.s. for } i\geq 1, \tag{4.3.88}$$

$$\sum_{k=1}^{\infty}\frac{1}{k^{1-v}}\big(E(z_{k+1}(u_{k-1}^3-3\vartheta^2u_{k-1}))\\ -z_{k+1}(u_{k-1}^3-3\vartheta^2u_{k-1})\big)<\infty\ \text{a.s.,} \tag{4.3.89}$$

$$\sum_{k=1}^{\infty}\frac{1}{k^{1-v}}\big(E(z_{k+1}(u_{k-1}^2-\vartheta^2)u_{k-i-1})\\ -z_{k+1}(u_{k-1}^2-\vartheta^2)u_{k-i-1}\big)<\infty\ \text{a.s. for } i\geq 1, \tag{4.3.90}$$

$$\sum_{k=1}^{\infty}\frac{1}{k^{1-v}}\big(E(z_{k+1}(u_{k-i-1}^2-\vartheta^2))\\ -z_{k+1}(u_{k-i-1}^2-\vartheta^2)\big)<\infty\ \text{a.s. for } i\geq 0, \tag{4.3.91}$$

$$\sum_{k=1}^{\infty}\frac{1}{k^{1-v}}\big(E(z_{k+1}(u_{k-i-1}^3-3\vartheta^2u_{k-i-1}))\\ -z_{k+1}(u_{k-i-1}^3-3\vartheta^2u_{k-i-1})\big)<\infty\ \text{a.s. for } i\geq 0, \tag{4.3.92}$$

$$\sum_{k=1}^{\infty}\frac{1}{k^{1-v}}\big(E(z_{k+1}u_{k-i-1})-z_{k+1}u_{k-i-1}\big)<\infty\ \text{a.s. for } i\geq 0. \tag{4.3.93}$$

Proof. We now prove (4.3.93), while (4.3.87)–(4.3.92) are similarly proved.

Note that

$$\sum_{k=1}^{\infty}\frac{1}{k^{1-v}}(Ez_{k+1}u_{k-i-1}-z_{k+1}u_{k-i-1})\\ =\sum_{k=1}^{\infty}\frac{1}{k^{1-v}}(Ey_{k+1}^0u_{k-i-1}-y_{k+1}^0u_{k-i-1})-\sum_{k=1}^{\infty}\frac{1}{k^{1-v}}(\bar{\eta}_{k+1}u_{k-i-1})\\ -\sum_{k=1}^{\infty}\frac{1}{k^{1-v}}(\bar{\xi}_{k+1}u_{k-i-1})-\sum_{k=1}^{\infty}\frac{1}{k^{1-v}}\varepsilon_{k+1}u_{k-i-1}, \tag{4.3.94}$$

where $\bar{\xi}_{k+1}$ and $\bar{\eta}_{k+1}$ are defined by (4.3.20).

The first term at the right-hand side of (4.3.94) is convergent by (4.3.84).

Define the σ-algebra $\mathcal{F}_k\triangleq\sigma\{\eta_j,\xi_j,\varepsilon_{j+1},u_{j-i-2},0\leq j\leq k\}$. Then we have $\frac{1}{k^{1-v}}\bar{\eta}_{k+1}\in\mathcal{F}_k$, and

$$E(\frac{1}{k^{1-v}}\bar{\eta}_{k+1}u_{k-i-1}|\mathcal{F}_k)=\frac{1}{k^{1-v}}\bar{\eta}_{k+1}E(u_{k-i-1}|\mathcal{F}_k)=0,$$

i.e., $\{\frac{1}{k^{1-\nu}}\bar{\eta}_{k+1}u_{k-i-1}, \mathcal{F}_{k+1}\}$ is an mds.

By Lemma 4.2.1, we have $\sup_k E(\bar{\eta}_{k+1})^2 < \infty$, and hence

$$\sum_{k=1}^{\infty} E\Big(\frac{1}{k^{1-\nu}}\bar{\eta}_{k+1}u_{k-i-1}\Big)^2$$
$$= \sum_{k=1}^{\infty} \frac{1}{k^{2(1-\nu)}} E\big(\bar{\eta}_{k+1}\big)^2 E u_{k-i-1}^2 = \sum_{k=1}^{\infty} O\Big(\frac{1}{k^{2(1-\nu)}}\Big) < \infty,$$

which implies

$$\sum_{k=1}^{\infty} E\Big[\Big(\frac{1}{k^{1-\nu}}\bar{\eta}_{k+1}u_{k-i-1}\Big)^2 \big| \mathcal{F}_k\Big] < \infty \text{ a.s.}$$

Then, the second term at the right-hand side of (4.3.94) is convergent by Theorem 1.2.8. Convergence of the two remaining terms at the right-hand side of (4.3.94) can also be established by Theorem 1.2.8.

The proof is completed. □

Theorem 4.3.1 *Assume that A4.3.1–A4.3.5 hold. Then the estimates* $l_{t,k}$, $t \geq 1$; $p_{i,k}$, $i = 1, \cdots, n_p$; $q_{j,k}$, $j = 1, \cdots, n_q$ *are strongly consistent with convergence rates*

$$|l_{t,k} - l_t| = o(k^{-\nu}) \text{ a.s. } \forall \nu \in (0, 1/2) \;\; t \geq 1, \tag{4.3.95}$$
$$|p_{i,k} - p_i| = o(k^{-\nu}) \text{ a.s. } \forall \nu \in (0, 1/2) \;\; i = 1, \cdots, n_p, \tag{4.3.96}$$
$$|q_{j,k} - q_j| = o(k^{-\nu}) \text{ a.s. } \forall \nu \in (0, 1/2) \;\; j = 1, \cdots, n_q. \tag{4.3.97}$$

Proof. We first prove that $\theta_k^{(0,\rho)}$ and $\theta_k^{(0,\kappa)}$ defined, respectively, by (4.3.30)–(4.3.31) and (4.3.32)–(4.3.33) converge a.s. with the following convergence rate:

$$\theta_k^{(0,\rho)} \xrightarrow[k\to\infty]{} \rho \text{ a.s., } \Big|\theta_k^{(0,\rho)} - \rho\Big| = o(k^{-\nu}) \text{ a.s. } \forall \nu \in (0, 1/2), \tag{4.3.98}$$
$$\theta_k^{(0,\kappa)} \xrightarrow[k\to\infty]{} \kappa \text{ a.s., } \Big|\theta_k^{(0,\kappa)} - \kappa\Big| = o(k^{-\nu}) \text{ a.s. } \forall \nu \in (0, 1/2). \tag{4.3.99}$$

We rewrite (4.3.30) as

$$\theta_{k+1}^{(0,\rho)} = \Big[\theta_k^{(0,\rho)} - \frac{1}{k}\Big(\theta_k^{(0,\rho)} - \rho\Big) - \frac{1}{k}\varepsilon_{k+1}^{(0,\rho)}\Big] \cdot I_{\Big[\Big|\theta_k^{(0,\rho)} - \frac{1}{k}\Big(\theta_k^{(0,\rho)} - \rho l_i\Big) - \frac{1}{k}\varepsilon_{k+1}^{(0,\rho)}\Big| \leq M_{\delta_k}^{(0,\rho)}\Big]},$$

where

$$\varepsilon_{k+1}^{(0,\rho)} = \rho - z_{k+1}(u_{k-1}^2 - \vartheta^2)$$
$$= (\rho - E z_{k+1}(u_{k-1}^2 - \vartheta^2)) + (E z_{k+1}(u_{k-1}^2 - \vartheta^2) - z_{k+1}(u_{k-1}^2 - \vartheta^2)).$$

Since ρ is the single root of the linear function $-(y - \rho)$, by Theorem 2.6.1, it suffices to prove

$$\sum_{k=1}^{\infty} \frac{1}{k^{1-\nu}} \varepsilon_{k+1}^{(0,\rho)} < \infty \text{ a.s. } \forall \nu \in (0, 1/2). \tag{4.3.100}$$

By (4.3.68) and (4.3.87) we find that (4.3.100) is true, and hence (4.3.98) holds. Similarly, (4.3.99) can be proved by using (4.3.70) and (4.3.89).

After establishing the convergence of $\theta_k^{(0,\rho)}$ and $\theta_k^{(0,\kappa)}$, by A4.3.4 at least one of ρ and κ is nonzero, so switching between (4.3.34)–(4.3.36) and (4.3.37)–(4.3.39) may happen only a finite number of times. Therefore, for (4.3.95) it suffices to show $\forall \nu \in (0,1/2)$,

$$\theta_k^{(i,\rho)} \xrightarrow[k\to\infty]{} \rho l_i \text{ a.s.}, \quad \left|\theta_k^{(i,\rho)} - \rho l_i\right| = o(k^{-\nu}) \text{ a.s.}, \quad i \geq 1, \tag{4.3.101}$$

$$\theta_k^{(i,\kappa)} \xrightarrow[k\to\infty]{} \kappa l_i \text{ a.s.}, \quad \left|\theta_k^{(i,\kappa)} - \kappa l_i\right| = o(k^{-\nu}) \text{ a.s.}, \quad i \geq 1. \tag{4.3.102}$$

Noticing the corresponding convergent series in Lemmas 4.3.3 and 4.3.5, we can prove (4.3.101) and (4.3.102) in the similar way as that used for proving (4.3.98).

From (4.3.95) it directly follows that (4.3.96) and (4.3.97) are derived from the definitions (4.3.43) and (4.3.44) for $p_{i,k}$ and $q_{j,k}$. □

Theorem 4.3.2 *Assume A4.3.1–A4.3.5 hold. Then, $c_{i,k}$ and $d_{j,k}$ defined by (4.3.57) and (4.3.58) are strongly consistent with convergence rates:*

$$|c_{i,k} - c_i| = o(k^{-\nu}) \text{ a.s. } \forall \nu \in (0,1/2), \quad i = 1,\cdots,n_c, \tag{4.3.103}$$

$$|d_{j,k} - d_j| = o(k^{-\nu}) \text{ a.s. } \forall \nu \in (0,1/2), \quad j = 1,\cdots,n_d. \tag{4.3.104}$$

Proof. In order to prove (4.3.103) and (4.3.104), by the definitions of $c_{i,k}$ and $d_{j,k}$, it suffices to show that

$$|h_{t,k} - h_t| = o(k^{-\nu}) \text{ a.s. } \forall \nu \in (0,1/2), \quad t \geq 1. \tag{4.3.105}$$

Let us first show that $\lambda_k^{(i,\tau)}$ defined by (4.3.46)–(4.3.47) converges with the following convergence rate: $\forall\ 0 < \nu < 1/2, \ i \geq 0$

$$\lambda_k^{(i,\tau)} \xrightarrow[k\to\infty]{} \tau\sum_{j=0}^{i} h_j l_{i-j} \text{ a.s.}, \quad \left|\lambda_k^{(i,\tau)} - \tau\sum_{j=0}^{i} h_j l_{i-j}\right| = o(k^{-\nu}) \text{ a.s.} \tag{4.3.106}$$

We rewrite (4.3.46) as

$$\begin{aligned}\lambda_{k+1}^{(i,\tau)} &= \Big[\lambda_k^{(i,\tau)} - \frac{1}{k}\big(\lambda_k^{(i,\tau)} - \tau\sum_{j=0}^{i} h_j l_{i-j}\big) - \frac{1}{k} e_{k+1}^{(i,\tau)}\Big] \\ &\quad \cdot I_{\Big[\big|\lambda_k^{(i,\tau)} - \frac{1}{k}\big(\lambda_k^{(i,\tau)} - \tau\sum_{j=0}^{i} h_j l_{i-j}\big) - \frac{1}{k} e_{k+1}^{(i,\tau)}\big| \leq M_{\sigma_k^{(i,\tau)}}\Big]},\end{aligned} \tag{4.3.107}$$

where $e_{k+1}^{(i,\tau)} = \tau\sum\limits_{j=0}^{i} h_j l_{i-j} - z_{k+1} u_{k-i-1}$.

Since $\tau\sum_{j=0}^{i} h_j l_{i-j}$ is the single root of the linear function $-(x - \tau\sum_{j=0}^{i} h_j l_{i-j})$, by Theorem 2.6.1 for proving (4.3.106) it suffices to show

$$\sum_{k=1}^{\infty} \frac{1}{k^{1-\nu}} e_{k+1}^{(i,\tau)} < \infty \text{ a.s. } \forall \nu \in (0,1/2), \quad i \geq 0. \tag{4.3.108}$$

Write $e_{k+1}^{(i,\tau)}$ as

$$e_{k+1}^{(i,\tau)} = (\tau \sum_{j=0}^{i} h_j l_{i-j} - E z_{k+1} u_{k-i-1}) + (E z_{k+1} u_{k-i-1} - z_{k+1} u_{k-i-1}).$$

By (4.3.74) and (4.3.93), we find that (4.3.108) is true, and hence (4.3.106) holds.

Combining (4.3.106) with the convergence of $\theta_k^{(0,\rho)}$ and $\theta_k^{(0,\kappa)}$ established in Theorem 4.3.1, by A4.3.4 we find that switching among the algorithms (4.3.46)–(4.3.48), (4.3.49)–(4.3.51), and (4.3.52)–(4.3.54) ceases in a finite number of steps. Consequently, for (4.3.105) it suffices to show $\forall \nu \in (0, 1/2)$,

$$\lambda_k^{(i,\rho)} \xrightarrow[k\to\infty]{} \rho \sum_{j=0}^{i} h_j l_{i-j}^2 \text{ a.s.}, \left| \lambda_k^{(i,\rho)} - \rho \sum_{j=0}^{i} h_j l_{i-j}^2 \right| = o(k^{-\nu}) \text{ a.s.}, \quad i \geq 1, \tag{4.3.109}$$

$$\lambda_k^{(i,\kappa)} \xrightarrow[k\to\infty]{} \kappa \sum_{j=0}^{i} h_j l_{i-j}^3 \text{ a.s.}, \left| \lambda_k^{(i,\kappa)} - \kappa \sum_{j=0}^{i} h_j l_{i-j}^3 \right| = o(k^{-\nu}) \text{ a.s.}, \quad i \geq 1. \tag{4.3.110}$$

Noticing the corresponding convergent series in Lemmas 4.3.3 and 4.3.5, we can prove (4.3.109) and (4.3.110) in the similar way as that used for proving (4.3.106). □

Lemma 4.3.6 *Assume A4.3.1–A4.3.5 hold. The following limits take place*

$$\frac{u_k}{k^c} \xrightarrow[k\to\infty]{\text{a.s.}} 0, \quad \frac{\varphi_k}{k^c} \xrightarrow[k\to\infty]{\text{a.s.}} 0 \; \forall c > 0, \quad \text{and} \quad \frac{z_k}{k^{\frac{1}{3}}} \xrightarrow[k\to\infty]{\text{a.s.}} 0. \tag{4.3.111}$$

Proof. By noting

$$P[\frac{|u_k|}{k^c} > \varepsilon] = P[\frac{|u_k|^{2/c}}{k^2} > \varepsilon^{2/c}] < \frac{1}{\varepsilon^{2/c} k^2} E|u_k|^{2/c}$$

for any given $\varepsilon > 0$, it follows that

$$\sum_{k=1}^{\infty} P[\frac{|u_k|}{k^c} > \varepsilon] < \infty.$$

Hence, by the Borel–Cantelli lemma, we derive $\frac{u_k}{k^c} \xrightarrow[k\to\infty]{\text{a.s.}} 0$. By the growth rate restriction on $f(\cdot)$, the second assertion of the lemma can be proved in a similar way since v_k is Gaussian with variance $\sigma_k \xrightarrow[k\to\infty]{} \sigma$. Finally, taking notice of A4.3.4 and A4.3.5 we see $\limsup_{k\to\infty} E|z_k|^{\Delta} < \infty$. From this it follows that

$$\sum_{k=1}^{\infty} P[\frac{|z_k|}{k^{\frac{1}{3}}} > \varepsilon] = \sum_{k=1}^{\infty} P[\frac{|z_k|^{\Delta}}{k^{\frac{\Delta}{3}}} > \varepsilon^{\Delta}] \leq \sum_{k=1}^{\infty} \varepsilon^{-\Delta} k^{-\frac{\Delta}{3}} E|z_k|^{\Delta} < \infty,$$

which implies the last assertion of the lemma. □

Lemma 4.3.7 *Under Conditions A4.3.1, A4.3.2, and A4.3.4, for ω_k defined by (4.3.66) the following limits take place*

$$E[\omega_k] \xrightarrow[k\to\infty]{} \rho(y),\ E[\omega_k \varphi_k] \xrightarrow[k\to\infty]{} \rho(y)f(y), \tag{4.3.112}$$

$$E[\omega_k|\varphi_k|] \xrightarrow[k\to\infty]{} \rho(y)|f(y)|, \tag{4.3.113}$$

$$E[b_k^{\delta-1}\omega_k^{\delta}] \xrightarrow[k\to\infty]{} \frac{1}{(2\pi)^{\frac{\delta-1}{2}}\sqrt{\delta}}\rho(y), \tag{4.3.114}$$

$$E[b_k^{\delta-1}|\omega_k\varphi_k|^{\delta}] \xrightarrow[k\to\infty]{} \frac{1}{(2\pi)^{\frac{\delta-1}{2}}\sqrt{\delta}}\rho(y)|f(y)|^{\delta}, \tag{4.3.115}$$

where $\rho(y) = \frac{1}{\sqrt{2\pi}\sigma\vartheta}e^{-\frac{y^2}{2\sigma^2\vartheta^2}}$ and $\sigma^2 = \sum\limits_{i=0}^{\infty} l_i^2$.

Proof. We first prove (4.3.115). By (4.3.18), we have $v_k = \sum_{i=0}^{k-1} l_i u_{k-i-1}$, and hence $v_k \in \mathcal{N}(0, \sigma_k^2\vartheta^2)$ with $\sigma_k^2 = \sum\limits_{i=0}^{k-1} l_i^2$.

Consequently, we have

$$\begin{aligned}
Eb_k^{\delta-1}|\omega_k\varphi_k|^{\delta} &= \frac{1}{(2\pi)^{\frac{\delta+1}{2}}\sigma_k\vartheta b_k}\int_{\mathcal{R}} e^{-\frac{\delta(x-y)^2}{2b_k^2}-\frac{x^2}{2\sigma_k^2\vartheta^2}}|f(x)|^{\delta}\mathrm{d}x \\
&= \frac{1}{\sqrt{2\pi}\sigma_k\vartheta b_k}e^{-\frac{\delta t^2}{2b_k^2}-\frac{(y+t)^2}{2\sigma_k^2\vartheta^2}}|f(y+t)|^{\delta}\mathrm{d}t \\
&= \frac{1}{\sqrt{2\pi}\sigma_k\vartheta}e^{-\frac{\delta s^2}{2}-\frac{(y+b_k s)^2}{2\sigma_k^2\vartheta^2}}|f(y+b_k s)|^{\delta}\mathrm{d}s \\
&\xrightarrow[k\to\infty]{} \frac{1}{(2\pi)^{\frac{\delta-1}{2}}\sqrt{\delta}}\rho(y)|f(y)|^{\delta}.
\end{aligned}$$

The rest can be proved in a similar manner. □

Lemma 4.3.8 *Assume A4.3.1–A4.3.5 hold. There exists a constant $c > 0$ with $\frac{1}{6} - a - c > 0$ and $\frac{1}{2} - 3a - b - 3c > 0$ so that*

$$|v_k - \widehat{v}_k| = o\left(\frac{1}{k^{\frac{1}{2}-2c}}\right), \tag{4.3.116}$$

$$|\omega_k - \widehat{\omega}_k| = o\left(\frac{1}{k^{\frac{1}{2}-3a-2c}}\right), \tag{4.3.117}$$

$$|\mathrm{v}_k - \widehat{\mathrm{v}}_k| = o\left(\frac{1}{k^{\frac{1}{6}-c}}\right). \tag{4.3.118}$$

Proof. From (4.3.62)–(4.3.63), we have

$$\begin{aligned}\widehat{x}_{k+1}-x_{k+1}&=P_{k+1}\widehat{x}_k-Px_k+(Q_{k+1}-Q)u_k\\&=P_{k+1}(\widehat{x}_k-x_k)+(P_{k+1}-P)x_k\\&\quad+(Q_{k+1}-Q)u_k.\end{aligned}$$

Since P is stable and $P_k\to P$, there exists a $\lambda\in(0,1)$, such that

$$\|\widehat{x}_{k+1}-x_{k+1}\|\le N_1\lambda^{k+1}\|\widehat{x}_0-x_0\|+S(\lambda,k), \tag{4.3.119}$$

where

$$\begin{aligned}S(\lambda,k)=&N_2\sum_{j=1}^{k+1}\lambda^{k-j+1}(\|P_j-P\|\cdot\|x_{j-1}\|\\&+\|Q_j-Q\|\cdot\|u_{j-1}\|),\end{aligned}$$

and $N_1>0$ and $N_2>0$ are constants.

Since $\|P_k-P\|=o(\frac{1}{k^{\frac{1}{2}-c}})$ and $\|Q_k-Q\|=o(\frac{1}{k^{\frac{1}{2}-c}})$ by Theorem 4.3.1, and $\frac{u_k}{k^c}\xrightarrow[k\to\infty]{\text{a.s.}}0$ and $\frac{x_k}{k^c}\xrightarrow[k\to\infty]{\text{a.s.}}0$ by Lemma 4.3.6, we have

$$\begin{aligned}S(\lambda,k)&=\sum_{j=1}^{k+1}\lambda^{k-j+1}o(\frac{1}{j^{1/2-2c}})\\&=o(\frac{1}{k^{1/2-2c}})\sum_{j=1}^{k+1}\lambda^{k-j+1}(\frac{k}{j})^{1/2-2c}=o(\frac{1}{k^{1/2-2c}})\sum_{j=0}^{k}\lambda^{j}(\frac{k}{k-j+1})^{1/2-2c}\\&=o(\frac{1}{k^{1/2-2c}})(\sum_{j=0}^{[\frac{k}{2}]}+\sum_{j=[\frac{k}{2}]+1}^{k})\lambda^{j}(\frac{k}{k-j+1})^{1/2-2c}\\&=o(\frac{1}{k^{1/2-2c}})(\frac{2}{1-\lambda}+\lambda^{[\frac{k}{2}]+1}k^{\frac{3}{2}-2c})=o(\frac{1}{k^{\frac{1}{2}-2c}}).\end{aligned}$$

This implies that

$$\|\widehat{x}_k-x_k\|=o(\frac{1}{k^{1/2-2c}}),$$

because the first term at the right-hand side of (4.3.119) decays exponentially.

Since $|v_k-\widehat{v}_k|$ and $\|\widehat{x}_k-x_k\|$ are of the same order, we have

$$|\widehat{\omega}_k-\omega_k|=o(\frac{1}{b_k^3}|\widehat{v}_k-v_k|)=o(\frac{1}{b_k^3}\|\widehat{x}_k-x_k\|)=o(\frac{1}{k^{1/2-3a-2c}}).$$

Similar to (4.3.119), from (4.3.64) and (4.3.65) we have

$$\begin{aligned}\widehat{t}_{k+1}-t_{k+1}&=D_{k+1}\widehat{t}_k-Dx_k+(C_{k+1}-C)z_{k+1}\\&=D_{k+1}(\widehat{t}_k-t_k)+(D_{k+1}-D)t_k+(C_{k+1}-C)z_{k+1}.\end{aligned}$$

Since D is stable and $D_k \to D$, there exists a $\lambda \in (0,1)$ such that

$$\|\widehat{t}_{k+1} - t_{k+1}\| \le N_3 \lambda^{k+1} \|\widehat{t}_0 - t_0\| + S(\lambda, k), \tag{4.3.120}$$

where

$$S(\lambda,k) = N_4 \sum_{j=1}^{k+1} \lambda^{k-j+1} (\|D_j - D\| \cdot \|t_{j-1}\| + \|C_j - C\| \cdot \|z_j\|),$$

and $N_3 > 0$ and $N_4 > 0$ are constants.

Since $\|D_k - D\| = o(\frac{1}{k^{\frac{1}{2}-c}})$ and $\|C_k - C\| = o(\frac{1}{k^{\frac{1}{2}-c}})$ by Theorem 4.3.2 and $\frac{z_k}{k^{\frac{1}{3}}} \xrightarrow[k\to\infty]{\text{a.s.}} 0$ and $\frac{t_k}{k^{\frac{1}{3}}} \xrightarrow[k\to\infty]{\text{a.s.}} 0$ by Lemma 4.3.6, we have

$$\begin{aligned}
S(\lambda,k) &= \sum_{j=1}^{k+1} \lambda^{k-j+1} o(\frac{1}{j^{1/6-c}}) \\
&= o(\frac{1}{k^{1/6-c}}) \sum_{j=1}^{k+1} \lambda^{k-j+1} (\frac{k}{j})^{1/6-c} = o(\frac{1}{k^{1/6-c}}) \sum_{j=0}^{k} \lambda^{j} (\frac{k}{k-j+1})^{1/6-c} \\
&= o(\frac{1}{k^{1/6-c}}) (\sum_{j=0}^{[\frac{k}{2}]} + \sum_{j=[\frac{k}{2}]+1}^{k}) \lambda^{j} (\frac{k}{k-j+1})^{1/6-c} \\
&= o(\frac{1}{k^{1/6-c}}) (\frac{2}{1-\lambda} + \lambda^{[\frac{k}{2}]+1} k^{\frac{7}{6}-c}) = o(\frac{1}{k^{\frac{1}{6}-c}}).
\end{aligned}$$

The first term at the right side of (4.3.120) decays exponentially, and hence we have

$$\|\widehat{t}_k - t_k\| = o(\frac{1}{k^{1/6-c}}).$$

Since $|v_k - \widehat{v}_k|$ and $\|\widehat{t}_k - t_k\|$ are of the same order by (4.3.64) and (4.3.65), we have

$$|v_k - \widehat{v}_k| = o(\frac{1}{k^{1/6-c}}).$$

□

Lemma 4.3.9 *Assume A4.3.1–A4.3.5 hold. The following series converge* a.s.

$$\sum_{k=1}^{\infty} \frac{1}{k} (\omega_k - E\omega_k) < \infty, \tag{4.3.121}$$

$$\sum_{k=1}^{\infty} \frac{1}{k} (|\omega_k - E\omega_k| - E|\omega_k - E\omega_k|) < \infty, \tag{4.3.122}$$

$$\sum_{k=1}^{\infty} \frac{1}{k} (\omega_k |\varphi_k| - E[\omega_k |\varphi_k|]) < \infty, \tag{4.3.123}$$

$$\sum_{k=1}^{\infty} \frac{1}{k} (\omega_k \varphi_k - E\omega_k \varphi_k) < \infty, \tag{4.3.124}$$

$$\sum_{k=1}^{\infty} \frac{1}{k} \omega_k (v_k - \widehat{v}_k) < \infty, \tag{4.3.125}$$

$$\sum_{k=1}^{\infty} \frac{1}{k} \omega_k \eta_k < \infty, \ \sum_{k=1}^{\infty} \frac{1}{k} \omega_k \phi_k < \infty, \ \text{and} \ \sum_{k=1}^{\infty} \frac{1}{k} \omega_k \chi_k < \infty, \tag{4.3.126}$$

$$\sum_{k=1}^{\infty} \frac{1}{k} \omega_k (|\eta_k| - E|\eta_k|) < \infty, \ \sum_{k=1}^{\infty} \frac{1}{k} \omega_k (|\phi_k| - E|\phi_k|) < \infty,$$

and

$$\sum_{k=1}^{\infty} \frac{1}{k} \omega_k (|\chi_k| - E|\chi_k|) < \infty. \tag{4.3.127}$$

Proof. By Theorem 1.4.4 $z_k^{(2)} \triangleq \frac{1}{k}(\omega_k \varphi_k - E\omega_k \varphi_k)$ is a zero-mean α-mixing sequence with the mixing coefficient tending to zero exponentially fast. Noticing (4.3.115), we have $\sup_k E(k^{-a(1+\varepsilon)}|\omega_k \varphi_k|^{2+\varepsilon}) < a_1$, where a_1 is a constant and $0 < a < 1/6$, and hence by the C_r-inequality

$$\sum_{k=1}^{\infty} (E|z_k^{(2)}|^{2+\varepsilon})^{\frac{2}{2+\varepsilon}} \le \sum_{k=1}^{\infty} \frac{4}{k^2} (E|\omega_k \varphi_k|^{2+\varepsilon})^{\frac{2}{2+\varepsilon}}$$

$$= \sum_{k=1}^{\infty} \frac{4}{k^{2(1-\frac{a(1+\varepsilon)}{2+\varepsilon})}} (E(k^{-a(1+\varepsilon)}|\omega_k \varphi_k|^{2+\varepsilon}))^{\frac{2}{2+\varepsilon}} < \sum_{k=1}^{\infty} \frac{4a_1^{\frac{2}{2+\varepsilon}}}{k^{2(1-\frac{a(1+\varepsilon)}{2+\varepsilon})}} < \infty,$$

since $2(1 - \frac{a(1+\varepsilon)}{2+\varepsilon}) > 1$. Thus, by Theorem 1.4.2 we have proved (4.3.124), while (4.3.121)–(4.3.123) can be verified by a similar treatment.

By (4.3.66) there is a positive constant L such that $|\omega_k| \le Lk^a, a \in (0, \frac{1}{6})$, then by (4.3.118) we have

$$\left|\sum_{k=1}^{\infty} \frac{1}{k} \omega_k (v_k - \widehat{v}_k)\right| \le \sum_{k=1}^{\infty} o\left(\frac{1}{k^{\frac{7}{6}-c-a}}\right) < \infty \ \ a.s.$$

Defining $\mathcal{F}_k^1 \triangleq \sigma(u_i, \eta_i, 0 \le i \le k)$, we have $\omega_k \in \mathcal{F}_{k-1}^1$,

$$E(\omega_k \eta_k | \mathcal{F}_{k-1}^1) = \omega_k E(\eta_k | \mathcal{F}_{k-1}^1) = 0,$$

and

$$\sum_{k=1}^{\infty} \frac{1}{k^2} E(\omega_k^2 \eta_k^2 | \mathcal{F}_{k-1}^1) \le \sum_{k=1}^{\infty} \frac{L^2}{k^{2-2a}} E(\eta_k^2 | \mathcal{F}_{k-1}^1) < \infty \ \ a.s.$$

Then, convergence of the first series in (4.3.126) follows from Theorem 1.2.8.

Since ξ_k is with probability density, we have $\{\phi_k = D^{-1}(z)\xi_{k+1}\}_{k \ge 0}$ is a zero-mean α-mixing with mixing coefficients decaying to zero exponentially fast by Theorem 1.4.4. Noticing that $\{v_k\}$ and $\{\xi_k\}$ are mutually independent, by the heredity of the mixing process we see that $z_k^{(3)} \triangleq \frac{1}{k}\omega_k\phi_k$ is also a zero-mean α-mixing with mixing coefficients tending to zero exponentially fast.

By A4.3.5 and Lemma 4.2.1, we have

$$\sup_k E|\phi_k|^\Delta < \infty,$$

and hence by (4.3.114)

$$\sup_k E\left(k^{-a(\Delta-1)}\omega_k^\Delta|\phi_k|^\Delta\right) = \sup_k E\left(k^{-a(\Delta-1)}\omega_k^\Delta\right)E|\phi_k|^\Delta < \infty. \tag{4.3.128}$$

By (4.3.114), we have

$$\begin{aligned}\sum_{k=1}^{\infty}(E|z_k^{(3)}|^\Delta)^{\frac{2}{\Delta}} &= \sum_{k=1}^{\infty}\frac{1}{k^2}\left(E|w_k\phi_k|^\Delta\right)^{\frac{2}{\Delta}}\\ &= \sum_{k=1}^{\infty}\frac{1}{k^{2-\frac{2a(\Delta-1)}{\Delta}}}(E(k^{-a(\Delta-1)}|w_k\phi_k|^\Delta))^{\frac{2}{\Delta}} < \infty.\end{aligned}$$

Thus, by Theorem 1.4.2 we conclude $\sum_{k=1}^{\infty}\frac{1}{k}\omega_k\phi_k < \infty$.

Similarly, we have

$$\sum_{k=1}^{\infty}\frac{1}{k}\omega_k\chi_k < \infty \text{ a.s.},$$

where χ_k is given by (4.3.60). □

Lemma 4.3.10 *Assume A4.3.1–A4.3.5 hold. The following series are convergent:*

$$\sum_{k=1}^{\infty}\frac{1}{k}(\omega_k-\widehat{\omega}_k)(\widehat{v}_k-v_k) < \infty \text{ a.s.}, \tag{4.3.129}$$

$$\sum_{k=1}^{\infty}\frac{1}{k}(\omega_k-\widehat{\omega}_k)\eta_k < \infty \text{ a.s.}, \tag{4.3.130}$$

$$\sum_{k=1}^{\infty}\frac{1}{k}(\omega_k-\widehat{\omega}_k)\phi_k < \infty \text{ a.s.}, \tag{4.3.131}$$

$$\sum_{k=1}^{\infty}\frac{1}{k}(\omega_k-\widehat{\omega}_k)\chi_k < \infty \text{ a.s.} \tag{4.3.132}$$

Proof. By (4.3.117)–(4.3.118), we have

$$\left|\sum_{k=1}^{\infty}\frac{1}{k}(\omega_k-\widehat{\omega}_k)(\widehat{v}_k-v_k)\right| \le \sum_{k=1}^{\infty}o\left(\frac{1}{k^{(\frac{3}{2}-3a-2c)+(\frac{1}{6}-c)}}\right) < \infty \text{ a.s.}$$

By (4.3.117) and Theorem 1.2.8 we have

$$|\sum_{k=1}^{\infty}\frac{1}{k}(\omega_k-\widehat{\omega}_k)\eta_k|\le\sum_{k=1}^{\infty}\frac{1}{k}|\omega_k-\widehat{\omega}_k|\cdot|\eta_k|$$
$$=\sum_{k=1}^{\infty}o\Big(\frac{(|\eta_k|-E|\eta_k|)+E|\eta_k|}{k^{\frac{3}{2}-3a-2c}}\Big)<\infty \text{ a.s.}$$

By (4.3.117) we have

$$|\sum_{k=1}^{\infty}\frac{1}{k}(\omega_k-\widehat{\omega}_k)\phi_k|\le\sum_{k=1}^{\infty}o\Big(\frac{(|\phi_k|-E|\phi_k|)+E|\phi_k|}{k^{\frac{3}{2}-3a-2c}}\Big)<\infty \text{ a.s.}$$

Since $\frac{3}{2}-3a-2c>1$ by selection of c in Lemma 4.3.8, for proving $|\sum_{k=1}^{\infty}\frac{1}{k}(\omega_k-\widehat{\omega}_k)\phi_k|<\infty$ it suffices to show $\sum_{k=1}^{\infty}\frac{(|\phi_k|-E|\phi_k|)}{k^{\frac{3}{2}-3a-2c}}<\infty$.

By Theorem 1.4.4, we see that $\{|\phi_k|-E|\phi_k|\}$ is a zero-mean α-mixing with mixing coefficients exponentially decaying to zero. Noticing that

$$\sum_{k=1}^{\infty}\left[E\left(\frac{|\phi_k|-E|\phi_k|}{k^{3/2-3a-2c}}\right)^{\Delta}\right]^{\frac{2}{\Delta}}=\sum_{k=1}^{\infty}\frac{4E|\phi_k|^{\Delta}}{k^{3-6a-4c}}<\infty,$$

we conclude that $\sum_{k=1}^{\infty}\frac{|\phi_k|-E|\phi_k|}{k^{\frac{3}{2}-3a-2c}}<\infty$ by Theorem 1.4.2.

Similarly, we can show

$$\sum_{k=1}^{\infty}\frac{1}{k}(\omega_k-\widehat{\omega}_k)\chi_k<\infty \text{ a.s.}$$

□

Theorem 4.3.3 *Assume A4.3.1–A4.3.5 hold. Then* $\mu_k(y)$ *defined by (4.3.67) is strongly consistent*

$$\mu_k(y)\xrightarrow[k\to\infty]{}f(y) \text{ a.s., } y\in\mathbb{R}. \quad (4.3.133)$$

Proof. Since $f(y)$ is the unique root of $-\rho(y)(x-f(y))$, by Theorem 2.3.1, for (4.3.133) it suffices to prove

$$\lim_{T\to0}\limsup_{k\to\infty}\frac{1}{T}\Big|\sum_{j=n_k}^{m(n_k,T_k)}\frac{1}{j}\bar{\varepsilon}_{j+1}(y)\Big|=0\ \forall T_k\in[0,T] \quad (4.3.134)$$

for any convergent subsequence $\mu_{n_k}(y)$, where $\bar{\varepsilon}_{k+1}(y)=\widehat{\omega}_k(\mu_k(y)-\widehat{v}_k)-\rho(y)(\mu_k(y)-f(y))$.

Write $\bar{\varepsilon}_{k+1}(y)$ as

$$\bar{\varepsilon}_{k+1}(y)=\sum_{i-1}^{3}\bar{\varepsilon}_{k+1}^{(i)}(y),$$

where

$$\bar{\varepsilon}^{(1)}_{k+1}(y) = (\widehat{\omega}_k - \omega_k)(\mu_k(y) - \widehat{v}_k),$$
$$\bar{\varepsilon}^{(2)}_{k+1}(y) = (\omega_k - \rho(y))\mu_k(y),$$
$$\bar{\varepsilon}^{(3)}_{k+1}(y) = \rho(y)f(y) - \omega_k\widehat{v}_k.$$

We now prove (4.3.134) with $\bar{\varepsilon}_{j+1}(y)$, respectively, replaced by $\bar{\varepsilon}^{(i)}_{j+1}(y)$, $i = 1,2,3$.

For $i = 1$, by (4.3.61) we can write

$$\sum_{k=1}^{\infty}\frac{1}{k}(\widehat{\omega}_k - \omega_k)(\mu_k(y) - \widehat{v}_k) = \sum_{k=1}^{\infty}\frac{1}{k}(\widehat{\omega}_k - \omega_k)(\mu_k(y) - \varphi_k)$$
$$+\sum_{k=1}^{\infty}\frac{1}{k}(\omega_k - \widehat{\omega}_k)(\widehat{v}_k - v_k) + \sum_{k=1}^{\infty}\frac{1}{k}(\omega_k - \widehat{\omega}_k)(\eta_k + \phi_k + \chi_k). \quad (4.3.135)$$

From (4.3.67) we have $\mu_k(y) \le k^b$, which together with (4.3.117) implies

$$\sum_{k=1}^{\infty}\frac{1}{k}|(\widehat{\omega}_k - \omega_k)\mu_k(y)| = o\Big(\sum_{k=1}^{\infty}\frac{1}{k^{\frac{3}{2}-3a-b-2c}}\Big) < \infty. \quad (4.3.136)$$

By (4.3.117) and the second limit in (4.3.111) we have

$$\sum_{k=1}^{\infty}\frac{1}{k}|(\widehat{\omega}_k - \omega_k)\varphi_k| = \sum_{k=1}^{\infty}\frac{1}{k^{1-c}}|(\widehat{\omega}_k - \omega_k)\frac{\varphi_k}{k^c}| = o\Big(\sum_{k=1}^{\infty}\frac{1}{k^{\frac{3}{2}-3a-3c}}\Big) < \infty. \quad (4.3.137)$$

Combining (4.3.136) and (4.3.137), we see that the first term at the right-hand side of (4.3.135) converges, while its second and the third terms converge by (4.3.129) and (4.3.130)–(4.3.132), respectively. Thus, we have

$$\sum_{k=1}^{\infty}\frac{1}{k}(\widehat{\omega}_k - \omega_k)(\mu_k(y) - \widehat{v}_k) < \infty. \quad (4.3.138)$$

For $i = 3$, we have

$$\sum_{k=1}^{\infty}\frac{1}{k}(\rho(y)f(y) - \omega_k\widehat{v}_k) = \sum_{k=1}^{\infty}\frac{1}{k}(\rho(y)f(y) - \omega_k\varphi_k) + \sum_{k=1}^{\infty}\frac{1}{k}\omega_k(v_k - \widehat{v}_k)$$
$$-\sum_{k=1}^{\infty}\frac{1}{k}\omega_k(\eta_k + \phi_k + \chi_k). \quad (4.3.139)$$

At the right-hand side of (4.3.139) the first term converges by (4.3.124) and (4.3.112), the second term converges by (4.3.125), and the last term converges by (4.3.126).

Finally, we prove (4.3.134) for $i = 2$.

For this we first show that there exists an Ω_0 with $\mathbb{P}\Omega_0 = 1$ such that for any fixed sample path $\omega \in \Omega_0$ if $\mu_{n_k}(y)$ is a convergent subsequence of $\{\mu_k(y)\}$: $\mu_{n_k}(y) \xrightarrow[k\to\infty]{} \bar{\mu}(y)$, then for all large enough k and sufficiently small $T > 0$

$$\mu_{i+1}(y) = \mu_i(y) - \frac{1}{i}\widehat{\omega}_i(\mu_i(y) - \widehat{v}_i), \tag{4.3.140}$$

and

$$\|\mu_{i+1}(y) - \mu_{n_k}(y)\| \leq cT, \quad i = n_k, n_k + 1, \cdots, m(n_k, T), \tag{4.3.141}$$

where $c > 0$ is a constant which is independent of k but may depend on sample path ω.

Consider the recursion (4.3.140) starting from $\mu_{n_k}(y)$.

Let

$$\Phi_{i,j} \triangleq (1 - \frac{1}{i}\omega_i)\cdots(1 - \frac{1}{j}\omega_j),\ i \geq j,\ \Phi_{j,j+1} = 1.$$

Since $\lim_{k\to\infty} E[\omega_k] = \rho(y)$, by convergence of the series (4.3.121) it is clear that

$$\sum_{j=n_k}^{i} \frac{1}{j}\omega_j = O(T)\ \forall i \in [n_k, \cdots, m(n_k, T)], \tag{4.3.142}$$

which implies

$$\log\Phi_{i,n_k} = O\Big(\sum_{j=n_k}^{i} \frac{1}{j}\omega_j\Big)$$

and as $k \to \infty$ and $T \to 0$

$$\Phi_{i,j+1} = 1 + O(T)\ \ \forall j\ \text{ and }\ \forall i : n_k \leq j \leq i \in [n_k, \cdots, m(n_k, T)]. \tag{4.3.143}$$

By (4.3.113), (4.3.123), and (4.3.143) it is seen that $\forall i \in [n_k, \cdots, m(n_k, T)]$

$$\begin{aligned} &\Big|\sum_{j=n_k}^{i} \Phi_{i,j+1}\frac{1}{j}\omega_j\varphi_j\Big| \leq [1 + O(T)] \sum_{j=n_k}^{i} \frac{1}{j}\omega_j|\varphi_j| \\ &= [1 + O(T)] \cdot \sum_{j=n_k}^{i} \frac{1}{j}(\omega_j|\varphi_j| - E[\omega_j|\varphi_j|] + E[\omega_j|\varphi_j|]) = O(T). \end{aligned} \tag{4.3.144}$$

From (4.3.127), (4.3.142), and (4.3.143) it follows that

$$\begin{aligned} &\Big|\sum_{j=n_k}^{i} \Phi_{i,j+1}\frac{1}{j}\omega_j(\eta_j + \phi_j + \chi_j)\Big| \\ &\leq [1 + O(T)]\Big(\sum_{j=n_k}^{i} \frac{1}{j}\omega_j(|\eta_j| - E|\eta_j| + E|\eta_j|) \\ &\quad + \sum_{j=n_k}^{i} \frac{1}{j}\omega_j(|\phi_j| - E|\phi_j| + E|\phi_j|) + \sum_{j=n_k}^{i} \frac{1}{j}\omega_j(|\chi_j| - E|\chi_j| + E|\chi_j|)\Big) \\ &= O(T)\ \ \forall i \in [n_k, \cdots, m(n_k, T)]. \end{aligned} \tag{4.3.145}$$

By (4.3.125), (4.3.144), and (4.3.145) we have $\forall i \in [n_k, \cdots, m(n_k,T)]$

$$\Big|\sum_{j=n_k}^{i} \Phi_{i,j+1}\frac{1}{j}\omega_j\widehat{\nu}_j\Big| = \Big|\sum_{j=n_k}^{i} \Phi_{i,j+1}\frac{1}{j}\omega_j(\widehat{\nu}_j - \nu_j)\Big| + \Big|\sum_{j=n_k}^{i} \Phi_{i,j+1}\frac{1}{j}\omega_j\varphi_j\Big|$$
$$+\Big|\sum_{j=n_k}^{i} \Phi_{i,j+1}\frac{1}{j}\omega_j(\eta_j+\phi_j+\chi_j)\Big| = O(T) \tag{4.3.146}$$

as $k \to \infty$ and $T \to 0$.

From (4.3.138) and (4.3.143) it follows that $\forall i \in [n_k, \cdots, m(n_k,T)]$

$$\left|\sum_{j=n_k}^{i} \Phi_{i,j+1}\frac{1}{j}(\omega_j - \widehat{\omega}_j)(\mu_j(y) - \widehat{\nu}_j)\right| = O(T). \tag{4.3.147}$$

From the recursion (4.3.140) we have

$$\mu_{i+1}(y) = \mu_i(y) - \frac{1}{i}\omega_i(\mu_i(y) - \widehat{\nu}_i) + \frac{1}{i}(\omega_i - \widehat{\omega}_i)(\mu_i(y) - \widehat{\nu}_i)$$
$$= \Phi_{i,n_k}\mu_{n_k}(y) + \sum_{j=n_k}^{i} \Phi_{i,j+1}\frac{1}{j}\omega_j\widehat{\nu}_j + \sum_{j=n_k}^{i} \Phi_{i,j+1}\frac{1}{j}(\omega_j - \widehat{\omega}_j)(\mu_j(y) - \widehat{\nu}_j),$$

which incorporating with (4.3.143), (4.3.146), and (4.3.147) yields

$$\mu_{i+1}(y) = \mu_{n_k}(y) + O(T)\ \forall i \in [n_k, \cdots, m(n_k,T)]. \tag{4.3.148}$$

This means that the algorithm (4.3.67) has no truncation for $i \in [n_k, \cdots, m(n_k,T)]$, when k is large enough and $T > 0$ is sufficiently small. Since the possibly exceptional set is with probability zero, we may take Ω_0 with $\mathbb{P}\Omega_0 = 1$ such that for all $\omega \in \Omega_0$ (4.3.140) and (4.3.141) are true for large k and small $T > 0$.

Since $\mu_{n_k}(y) \xrightarrow[k\to\infty]{} \bar{\mu}(y)$, we have

$$\sum_{j=n_k}^{m(n_k,T_k)} \frac{1}{j}\bar{\varepsilon}^{(2)}_{k+1}(y) = \sum_{j=n_k}^{m(n_k,T_k)} \frac{1}{j}(\omega_j - \rho(y))\mu_j(y)$$
$$= \sum_{j=n_k}^{m(n_k,T_k)} \frac{1}{j}(\omega_j - E\omega_j)\mu_j(y) + \sum_{j=n_k}^{m(n_k,T_k)} \frac{1}{j}(E\omega_j - \rho(y)))\mu_j(y)$$
$$= \sum_{j=n_k}^{m(n_k,T_k)} \frac{1}{j}(\mu_j(y) - \bar{\mu}(y))(\omega_j - E\omega_j) + \bar{\mu}(y)\sum_{j=n_k}^{m(n_k,T_k)} \frac{1}{j}(\omega_j - E\omega_j)$$
$$+ \sum_{j=n_k}^{m(n_k,T_k)} \frac{1}{j}(E\omega_j - \rho(y)))\mu_j(y). \tag{4.3.149}$$

At the right-hand side of (4.3.149), as $k \to \infty$ the second term tends to zero by (4.3.121), the last term tends to zero by the first limit in (4.3.112) and (4.3.141), while the first term can be estimated as follows.

By (4.3.141) and $\mu_{n_k}(y) \xrightarrow[k\to\infty]{} \bar{\mu}(y)$, we have

$$\Big|\sum_{j=n_k}^{m(n_k,T_k)} \frac{1}{j}(\mu_j(y)-\bar{\mu}(y))(\omega_j - E\omega_j)\Big|$$
$$= O(T)\sum_{j=n_k}^{m(n_k,T_k)} \frac{1}{j}(|\omega_j - E\omega_j| - E|\omega_j - E\omega_j| + E|\omega_j - E\omega_j|). \qquad (4.3.150)$$

By (4.3.112) and (4.3.122) it follows that

$$\lim_{T\to 0}\limsup_{k\to\infty}\frac{1}{T}\sum_{j=n_k}^{m(n_k,T_k)} \frac{1}{j}(\mu_j(y)-\bar{\mu}(y))(\omega_j - E\omega_j)$$
$$= \lim_{T\to 0}\limsup_{k\to\infty}\frac{1}{T}O(T)\sum_{j=n_k}^{m(n_k,T_k)} \frac{1}{j}E|\omega_j - E\omega_j|$$
$$= \lim_{T\to 0}\limsup_{k\to\infty}O(1)\sum_{j=n_k}^{m(n_k,T_k)} \frac{1}{j}E\omega_j = 0.$$

Thus, (4.3.134) for $i = 2$ is proved, and meanwhile the proof of Theorem 4.3.3 is completed. □

Numerical Examples

We consider three examples with different nonlinearities being an odd function, an even function, and a general function, respectively. Let their linear parts be the same.

The first linear subsystem is given by

$$v_{k+1} + p_1 v_k + p_2 v_{k-2} = u_k + q_1 u_{k-1} + q_2 u_{k-2},$$

where $p_1 = 0.2$, $p_2 = 0.6$, $q_1 = -0.3$, and $q_2 = 1.2$, while the second linear subsystem is given by

$$y_{k+1} + c_1 y_k + c_2 y_{k-2} = \varphi_k^e + d_1 \varphi_{k-1}^e + d_2 \varphi_{k-2}^e + \xi_{k+1},$$

where $c_1 = -0.15$, $c_2 = 0.5$, $d_1 = 0.2$, and $d_2 = -0.4$.

Notice that $Q(z) = 1 + q_1 z + q_2 z^2$ is unstable. Let the input signal u_k be Gaussian: $u_k \in \mathcal{N}(0,1)$, and the mutually independent random variables $\{\eta_k\}$, $\{\xi_k\}$, and $\{\varepsilon_k\}$ be Gaussian: η_k, ξ_k, and $\varepsilon_k \in \mathcal{N}(0, 0.1^2)$.

The parameters used in the algorithms are as follows: $b_k = \frac{1}{k^{0.1}}$, $M_k = 2^k + 20$, and $m_k = k^{0.15} + 30$.

Example 1. $f(x) = 2\sin(x)$, which is an odd function. Figures 4.3.2 and 4.3.3

demonstrate the estimates for the coefficients of the first and the second linear subsystems, respectively, while Figure 4.3.4 gives the estimates at $k = 5000$ for the nonlinear function in the interval $[-2,2]$. The estimates for $f(-1.4)$, $f(-0.5)$, $f(0)$ and $f(1)$ are presented in Figure 4.3.5.

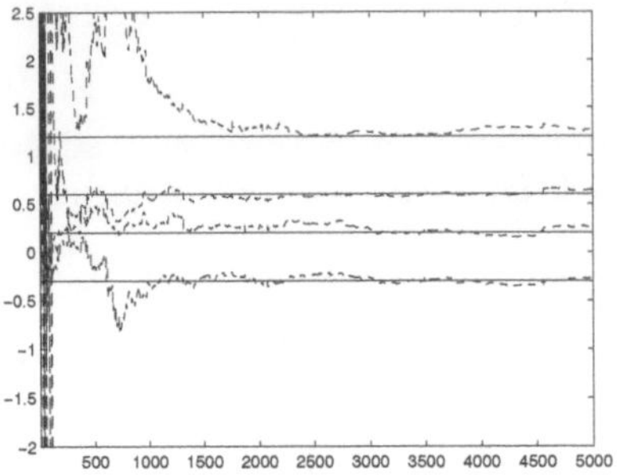

Figure 4.3.2: Estimates for p_1, p_2, c_1, c_2

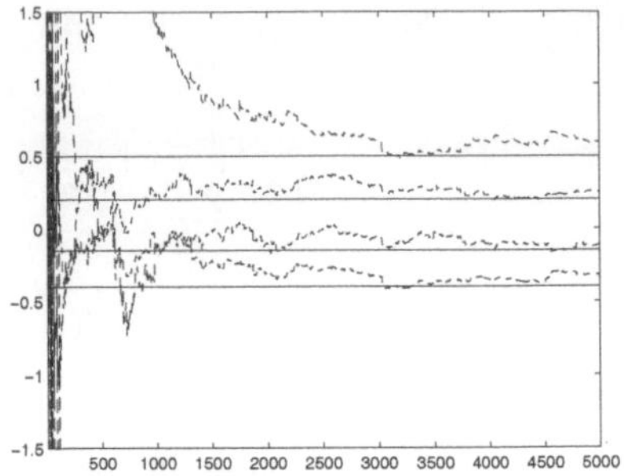

Figure 4.3.3: Estimates for c_1, c_2, d_1, d_2

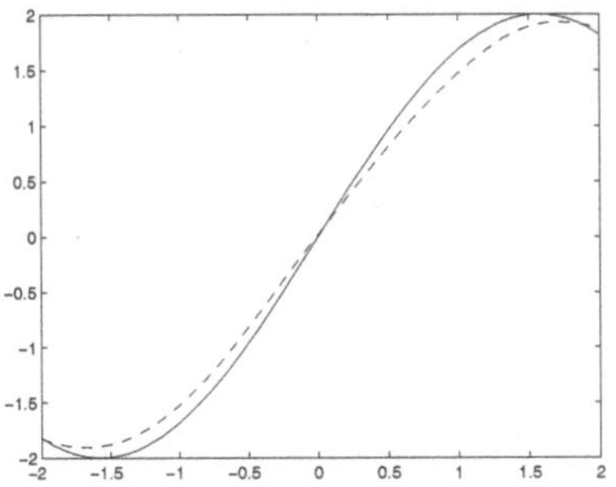

Figure 4.3.4: Estimates for $f(x) = 2\sin(x)$

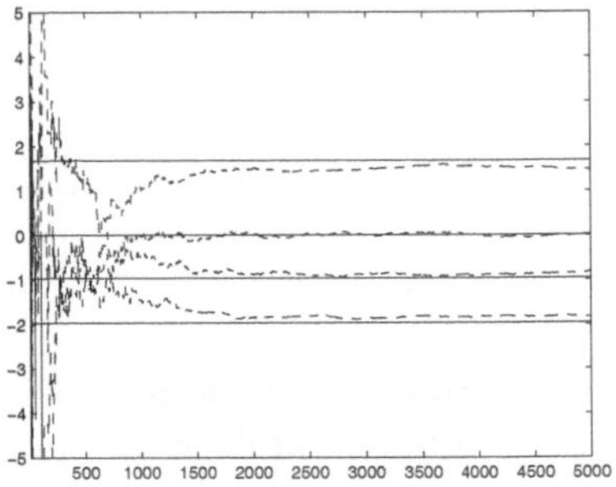

Figure 4.3.5: Estimates for $f(x)=2\sin(x)$ **at fixed points**

Example 2. The nonlinear function is even: $f(x) = x^2 - 1$. Figures 4.3.6 and 4.3.7 demonstrate the estimates for the coefficients of the first and the second linear subsystems, respectively, while Figure 4.3.8 gives the estimates at $k = 5000$ for the nonlinear function in the interval $[-2,2]$. The estimates for $f(-1.6)$, $f(-0.8)$, $f(0.2)$, and $f(1.2)$ are presented in Figure 4.3.9.

Example 3. The nonlinear function is neither odd nor even: $f(x) = 1.5x^2 + 2x + 1$. Figures 4.3.10 and 4.3.11 demonstrate the estimates for the coefficients of the first and the second linear subsystems, respectively, while Figure 4.3.12 gives the estimates at $k = 5000$ for the nonlinear function in the interval $[-2,2]$. The estimates for $f(-1)$, $f(0.4)$, $f(1.2)$, and $f(1.6)$ are presented in Figure 4.3.13.

It is worth noting that we cannot expect to have a high convergence rate of es-

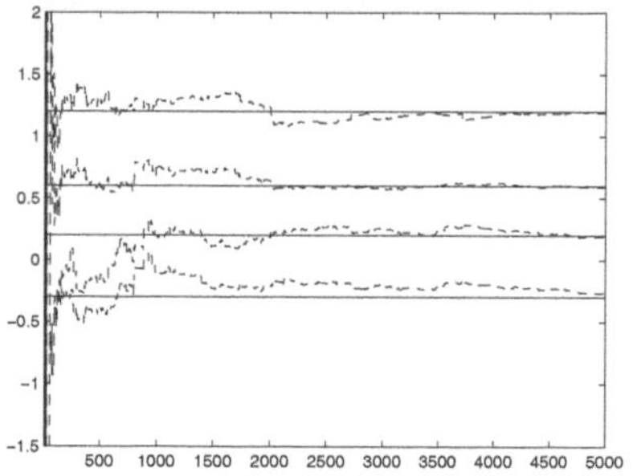

Figure 4.3.6: Estimates for p_1, p_2, c_1, c_2

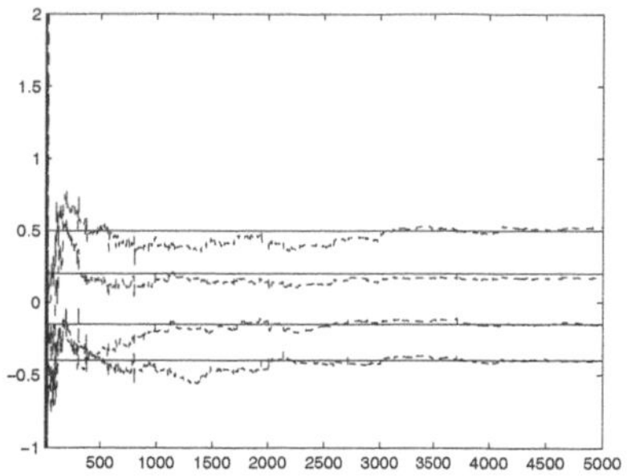

Figure 4.3.7: Estimates for c_1, c_2, d_1, d_2

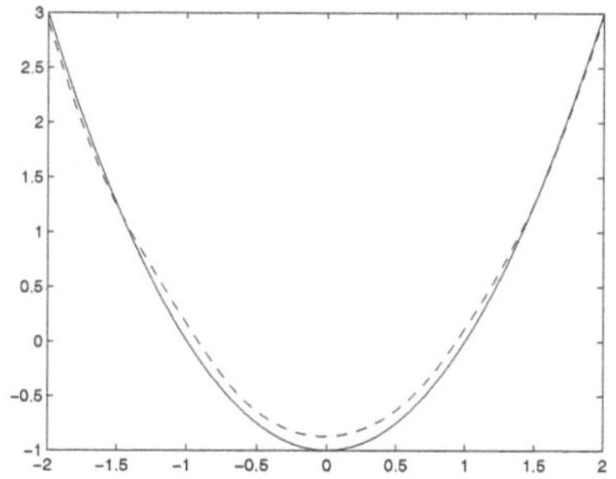

Figure 4.3.8: Estimates for $f(x) = x^2 - 1$

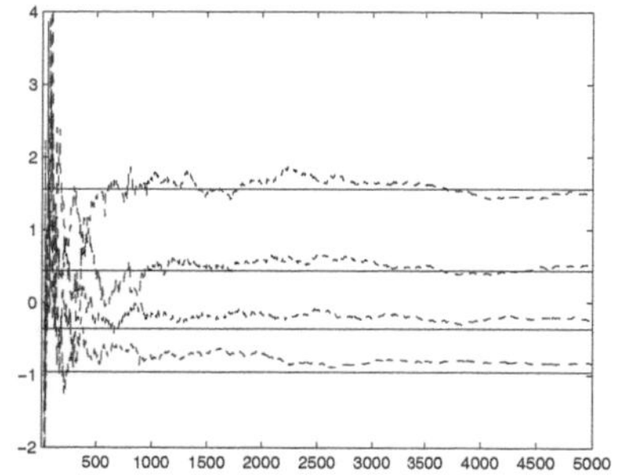

Figure 4.3.9: Estimates for $f(x)=x^2-1$ **at fixed points**

timates given for the system presented in Figure 4.3.1, because the system contains quite many uncertain sources: 1) For estimating the first linear subsystem instead of its output v_k the observation z_k of the whole system is used, and between v_k and z_k there are so many unknown factors and noises. So, it is conceivable that estimates for the first linear subsystem are with rather slow convergence rates; 2) The signal v_k, being the input to the Hammerstein system consisting of (4.3.2)–(4.3.4), is correlated and should be estimated with the help of "very rough" estimates of the first linear subsystem; 3) Even if v_k is exactly available, for identifying the Hammerstein system (4.3.2)–(4.3.4) one still has to overcome influence of not only observation noise ε_k, but also the internal noise η_k and the system noise ξ_k. Without these noises the SA-type algorithms for identifying Hammerstein systems may give estimates with much better rates of convergence as shown in Section 4.1.

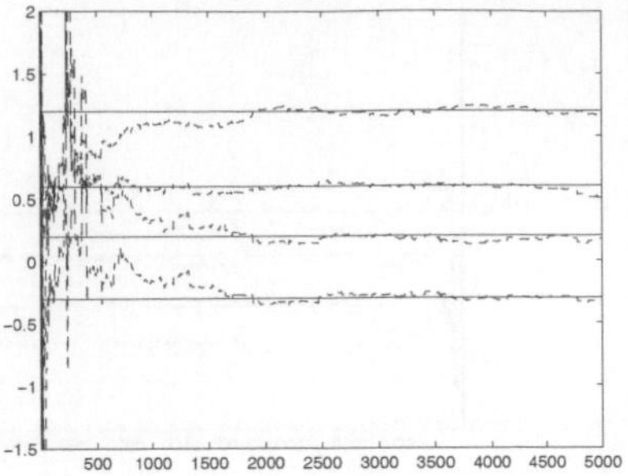

Figure 4.3.10: Estimates for p_1, p_2, c_1, c_2

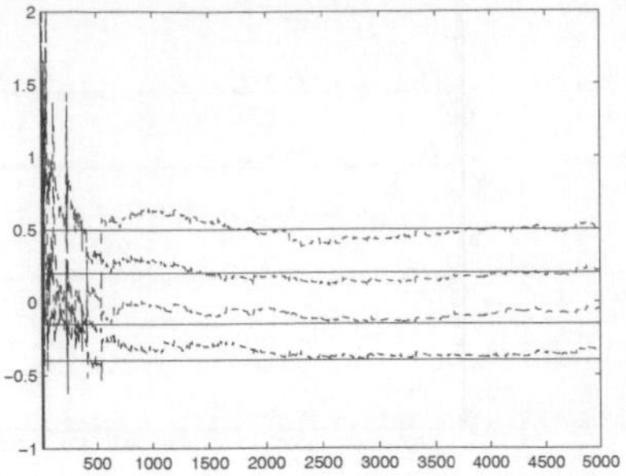

Figure 4.3.11: Estimates for c_1, c_2, d_1, d_2

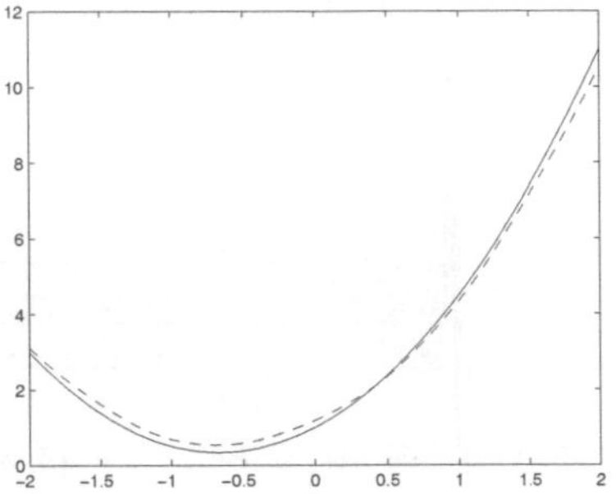

Figure 4.3.12: Estimates for $f(x) = 1.5x^2 + 2x + 1$

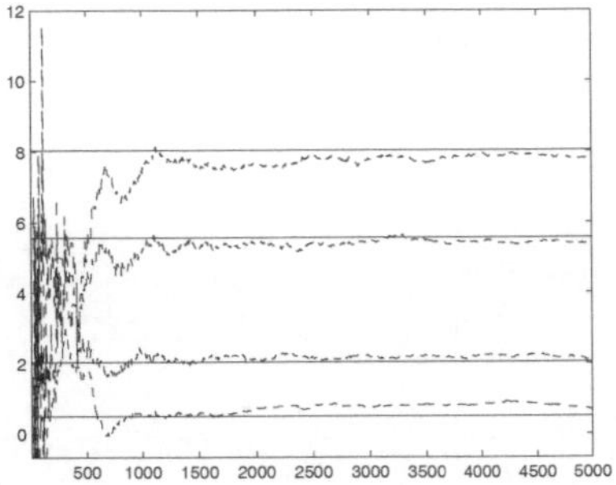

Figure 4.3.13: Estimates for $f(x)$ at fixed points

4.4 Recursive Identification of EIV Hammerstein Systems

In Section 4.1 we have discussed the recursive identification of Hammerstein systems. We now consider more general case: The observations are contaminated by noises not only for the system output but also for the input, and both the input noise $\{\eta_k\}$ and the output noise $\{\varepsilon_k\}$ are ARMA processes with unknown coefficients. To be precise, we consider the recursive identification of the MIMO EIV Hammerstein system described as follows:

$$v_k^0 = f(u_k^0) + w_k, \tag{4.4.1}$$

$$A(z)y_k^0 = B(z)v_k^0 + \xi_k, \tag{4.4.2}$$

$$A(z) = I + A_1 z + \cdots + A_p z^p, \tag{4.4.3}$$

$$B(z) = B_1 z + B_2 z^2 + \cdots + B_q z^q, \tag{4.4.4}$$

where $A(z)$ and $B(z)$ are the $m \times m$ and $m \times l$ matrix polynomials with unknown coefficients and known orders p, q, respectively. The noise-free system input and output are denoted by $u_k^0 \in \mathbb{R}^l$ and $y_k^0 \in \mathbb{R}^m$, respectively. Both $w_k \in \mathbb{R}^l$ and $\xi_k \in \mathbb{R}^m$ are the system noises. The nonlinearity $f(\cdot)$ is a vector-valued function: $f : \mathbb{R}^l \to \mathbb{R}^l$. The system input and output are observed with additive noises η_k and ε_k:

$$u_k = u_k^0 + \eta_k, \quad y_k = y_k^0 + \varepsilon_k. \tag{4.4.5}$$

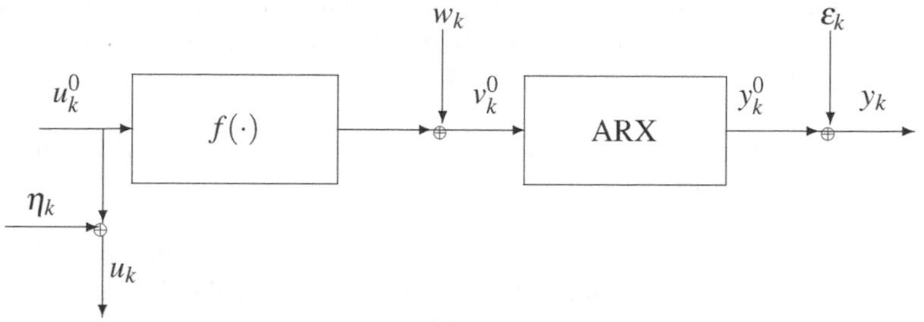

Figure 4.4.1: EIV Hammerstein System

The goal is to recursively estimate the unknown matrix coefficients $\{A_1, \cdots, A_p, B_1, \cdots, B_q\}$ of $A(z)$ and $B(z)$ and the value of $f(x)$ at any given x in its domain on the basis of the observed data $\{u_k, y_k\}$.

We now list the assumptions to be used.

A4.4.1 *The noise-free system input $\{u_k^0\}$ is a sequence of iid random vectors with zero-mean and known covariance matrix Λ. Furthermore, $\{u_k^0\}$ has the density function denoted by $p(\cdot)$, and $u_k^0 \in U$, where U is a bounded subset of $\mathbb{R}^l$.*

A4.4.2 *$A(z)$ and $B(z)$ have no common left factor, $[A_p, B_q]$ is of row-full-rank, and $A(z)$ is stable, namely, $\det A(z) \neq 0 \ \forall |z| \leq 1$.*

A4.4.3 *The measurement noises η_k and ε_k both are ARMA processes:*

$$P(z)\eta_k = Q(z)\zeta_k, \tag{4.4.6}$$
$$R(z)\varepsilon_k = S(z)\varsigma_k, \tag{4.4.7}$$

where

$$P(z) = I + P_1 z + P_2 z^2 + \cdots + P_{n_p} z^{n_p}, \tag{4.4.8}$$
$$Q(z) = I + Q_1 z + Q_2 z^2 + \cdots + Q_{n_q} z^{n_q}, \tag{4.4.9}$$
$$R(z) = I + R_1 z + R_2 z^2 + \cdots + R_{n_r} z^{n_r}, \tag{4.4.10}$$
$$S(z) = I + S_1 z + S_2 z^2 + \cdots + S_{n_s} z^{n_s}, \tag{4.4.11}$$

where $P(z)$ and $R(z)$ are stable. The driven noises $\{\zeta_k\}$, $\{\varsigma_k\}$, and the internal noises $\{\xi_k\}$, $\{w_k\}$ all are sequences of iid random vectors with zero-mean having probability densities. Moreover, ζ_k, ς_k, ξ_k, and w_k are independent of u_k^0, and $E(\|\zeta_k\|^{\Delta-1}) < \infty$, $E(\|\xi_k\|^{\Delta}) < \infty$, $E(\|w_k\|^{\Delta}) < \infty$, and $E(\|\varsigma_k\|^{\Delta}) < \infty$ for some $\Delta > 3$.

A4.4.4 *The function $f(\cdot)$ is measurable, locally bounded, and continuous at x where $f(x)$ is estimated. Furthermore, the correlation matrix $\Upsilon \triangleq E(f(u_k^0)u_k^{0T})$ is nonsingular.*

A4.4.5 *The dimension l of the system input is not greater than the dimension m of the system output, i.e., $l \le m$, B_1 is of column-full-rank, and $B_1^T z^{-1}B(z)$ is stable, namely, $\det(B_1^T z^{-1}B(z)) \neq 0 \ \forall |z| \le 1$.*

A4.4.6 *The driven noise $\{\zeta_k\}$ in (4.4.6) is a sequence of zero-mean iid Gaussian random vectors.*

Before proceeding further, let us explain these assumptions. To identify the linear subsystem it is sufficient to require A4.4.1–A4.4.4, while to estimate $f(\cdot)$ we have to additionally impose A4.4.5–A4.4.6. As can be seen from Lemma 4.4.1 presented below, Assumption A4.4.1 and the nonsingularity of $E(f(u_k^0)u_k^{0T})$ required in A4.4.4 are needed for defining the impulse responses of the linear subsystem from the observed inputs $\{u_k\}$ and outputs $\{y_k\}$. Assumption A4.4.2 is the necessary and sufficient condition for identifiability of the multivariable linear system $[A(z), B(z)]$.

In A4.4.3 the internal noises ξ_k and w_k are required to be iid, but they are allowed to be ARMA processes, because they enter the system after the nonlinearity $f(\cdot)$ and an ARMA process remains to be an ARMA if its MA part is driven by another ARMA process instead of an iid sequence.

It is natural to ask whether the system can uniquely be determined by the input-output data. Let us temporarily remove all the noises η_k, ξ_k, w_k, and ε_k. Then the system (4.4.1)–(4.4.2) can be written as

$$v_k = f(u_k), \quad A(z)y_k = B(z)v_k. \tag{4.4.12}$$

For any nonsingular matrix P, define $\tilde{v}_k \triangleq P^{-1}v_k$, $\tilde{B}(z) \triangleq B(z)P$. Then the following system

$$\tilde{v}_k = P^{-1}f(u_k), \quad A(z)y_k = \tilde{B}(z)\tilde{v}_k, \tag{4.4.13}$$

and the system (4.4.12) have the identical input and output, but they have the different structures. Namely, in (4.4.12) the nonlinearity is $f(\cdot)$ and the transfer function of the linear subsystem is $A^{-1}(z)B(z)$, while in (4.4.13) they are $P^{-1}f(\cdot)$ and $A^{-1}(z)B(z)P$, respectively. In order to uniquely define the system we have to fix a matrix P. As pointed out in Remark 4.1.1, if we take $P \triangleq \Upsilon$, then the corresponding input-output-correlation-matrix of the nonlinearity equals $E\Upsilon^{-1}f(u_k^0)u_k^{0T} = I$. So, without loss of generality, we may assume that the input-output-correlation-matrix of the nonlinearity is equal to I.

Estimation of $\{\mathbf{A_1}, \cdots, \mathbf{A_p}, \mathbf{B_1}, \cdots, \mathbf{B_q}\}$
Since $A(z)$ is stable by A4.4.2, we have

$$H(z) \triangleq A^{-1}(z)B(z) = \sum_{i=1}^{\infty} H_i z^i, \tag{4.4.14}$$

where $\|H_i\| = O(e^{-ri}), r > 0, i > 1$ and $H_1 = B_1$.

Defining $\bar{\xi}_k \triangleq A(z)^{-1}\xi_k$ and assuming $u_k^0 = 0 \; \forall k < 0$, we have

$$y_k^0 = \sum_{i=1}^{k} H_i v_{k-i}^0 + A(z)^{-1}\xi_k = \sum_{i=1}^{k} H_i f(u_{k-i}^0) + \sum_{i=1}^{k} H_i \omega_{k-i} + \bar{\xi}_k. \tag{4.4.15}$$

Lemma 4.4.1 *Assume A4.4.1–A4.4.4 hold but without need for availability of Λ in A4.4.1. The following formulas take place*

$$E(y_k u_{k-i}^T) = H_i, \quad for \;\; i \geq 1. \tag{4.4.16}$$

Proof. Since ξ_k, ζ_k, ς_k, w_k, and u_k^0 are zero-mean and mutually independent, it follows that

$$\begin{aligned} E(y_k u_{k-i}^T) &= E(y_k^0 + \varepsilon_k)(u_{k-i}^0 + \eta_{k-i})^T = E y_k^0 u_{k-i}^{0T} \\ &= E\left(\sum_{j=1}^{k} H_j v_{k-j}^0 + \bar{\xi}_k\right) u_{k-i}^{0T} = \sum_{j=1}^{k} H_j E(f(u_{k-j}^0) u_{k-i}^{0T}) \\ &= H_i E(f(u_{k-i}^0) u_{k-i}^{0T}) = H_i \Upsilon = H_i \;\; \forall \, i \geq 1. \end{aligned} \tag{4.4.17}$$

□

Remark 4.4.1 *If $f(\cdot)$ is an identity mapping, i.e., $f(u) = u$, then the system (4.4.1)–(4.4.2) turns to be an EIV linear system, and the results in this section remain true. However, from (4.4.17) it is seen that in this case the matrix Υ should be equal to $E(u_k^0 u_k^{0T})$. For identifiability, we may assume $E(u_k^0 u_k^{0T}) = I$, and (4.4.17) remains valid.*

For estimating the coefficients of the linear subsystem we first estimate the impulse responses $\{H_i\}$ and then obtain the estimates for the coefficients $\{A_1, \cdots, A_p, B_1, \cdots, B_q\}$ by using the linear algebraic equations connecting them with $\{H_i\}$.

We apply SAAWET to recursively estimate H_i motivated by (4.4.16):

$$\begin{aligned} H_{i,k+1} = &\left[H_{i,k} - \frac{1}{k}(H_{i,k} - y_{k+1} u_{k+1-i}^T)\right] \\ &\cdot I_{\left[\left\|H_{i,k} - \frac{1}{k}(H_{i,k} - y_{k+1} u_{k+1-i}^T)\right\| \leq M_{\delta_{i,k}}\right]}, \end{aligned} \tag{4.4.18}$$

$$\delta_{i,k} = \sum_{j=1}^{k-1} I_{\left[\left\|H_{i,j} - \frac{1}{j}(H_{i,j} - y_{j+1} u_{j+1-i}^T)\right\| > M_{\delta_{i,j}}\right]}, \tag{4.4.19}$$

where $\{M_k\}$ is an arbitrarily chosen sequence of positive real numbers increasingly diverging to infinity, $H_{i,0}$ is an arbitrary initial value, and I_A denotes the indicator function of a set A.

Once the estimates $H_{i,k+1}$ for the impulse responses H_i are obtained, the estimates for the parameters $\{A_1,\cdots,A_p,B_1,\cdots,B_q\}$ of the linear subsystem can be derived by the convolution relationship between the impulse response H_i and the parameters $\{A_1,\cdots,A_p,B_1,\cdots,B_q\}$.

Carrying out the operations similar to those done in Section 3.3, we still have (3.3.9) and (3.3.10), which are rewritten as follows:

$$[A_1,A_2,\cdots,A_p]L=-[H_{q+1},H_{q+2},\cdots,H_{q+mp}], \tag{4.4.20}$$

where

$$L\triangleq\begin{bmatrix} H_q & H_{q+1} & \cdots & H_{q+mp-1}\\ H_{q-1} & H_q & \cdots & H_{q+mp-2}\\ \vdots & \vdots & \ddots & \vdots\\ H_{q-p+1} & H_{q-p+2} & \cdots & H_{q+(m-1)p}\end{bmatrix}. \tag{4.4.21}$$

Noticing that the matrix L is of row-full-rank under A4.4.2 by Theorem 3.3.2 and that $H_{i,k}\xrightarrow[k\to\infty]{}H_i$ a.s. as to be shown by Theorem 4.4.1, we see that L_k is also of row-full-rank when k is sufficiently large, where

$$L_k\triangleq\begin{bmatrix} H_{q,k} & H_{q+1,k} & \cdots & H_{q+mp-1,k}\\ H_{q-1,k} & H_{q,k} & \cdots & H_{q+mp-2,k}\\ \vdots & \vdots & \ddots & \vdots\\ H_{q-p+1,k} & H_{q-p+2,k} & \cdots & H_{q+(m-1)p,k}\end{bmatrix}$$

serving as the kth estimate of L with $H_{i,k}=0$ for $i\le 0$.

The estimates for $\{A_1,\cdots,A_p,B_1,\cdots,B_q\}$ are naturally defined as follows:

$$[A_{1,k},A_{2,k},\cdots,A_{p,k}]=-[H_{q+1,k},H_{q+2,k},\cdots,H_{q+mp,k}]L_k^T(L_kL_k^T)^{-1}, \tag{4.4.22}$$

$$B_{i,k}=\sum_{j=0}^{i\wedge p}A_{j,k}H_{i-j,k}\ \ \forall\, 1\le i\le q. \tag{4.4.23}$$

In order to avoid taking the inverse in (4.4.22) for a large matrix at each time k, we provide the following recursive algorithm:

$$\theta_{k+1}=\left[\theta_k-\frac{1}{k}L_k(L_k^T\theta_k+W_k^T)\right]\cdot I_{\left[\|\theta_k-\frac{1}{k}L_k(L_k^T\theta_k+W_k^T)\|\le M_{\delta_k}\right]}, \tag{4.4.24}$$

$$\delta_k=\sum_{j=1}^{k-1}I_{[\|\theta_j-\frac{1}{j}L_j(L_j^T\theta_j+W_j^T)\|>M_{\delta_j}]}, \tag{4.4.25}$$

where

$$\theta_k\triangleq[A_{1,k},\cdots,A_{p,k}]^T$$

and

$$W_k\triangleq[H_{q+1,k},\cdots,H_{q+mp,k}].$$

Nonparametric Estimation of f(·)

We now recursively estimate $f(x)$, where x is an arbitrary point in the domain U. The idea consists in first estimating $f(u_k^0)$ by using the obtained estimates for coefficients of the linear subsystem, then applying SAAWET incorporated with a deconvolution kernel function to estimate $f(x)$, where the kernel function is imposed to concentrate the mass of u_k at x.

Let us start with estimating $f(u_k^0)$.

According to (4.4.1)–(4.4.3), we have

$$[z^{-1}B(z)](f(u_k^0)+w_k) = A(z)(y_{k+1}-\varepsilon_{k+1}) - \xi_{k+1}. \tag{4.4.26}$$

Multiplying the equality by B_1^T from left, we have

$$B_1^T[z^{-1}B(z)](f(u_k^0)+w_k) = B_1^T A(z)y_{k+1} - B_1^T A(z)\varepsilon_{k+1} - B_1^T\xi_{k+1}.$$

Set

$$\varphi_k \triangleq [B_1^T z^{-1}B(z)]^{-1}B_1^T A(z)y_{k+1}, \tag{4.4.27}$$
$$\psi_k \triangleq [B_1^T z^{-1}B(z)]^{-1}B_1^T A(z)\varepsilon_{k+1}, \tag{4.4.28}$$
$$\chi_k \triangleq [B_1^T z^{-1}B(z)]^{-1}B_1^T \xi_{k+1}. \tag{4.4.29}$$

Then $f(u_k^0)$ can be expressed by

$$\begin{aligned} f(u_k^0) &= [B_1^T z^{-1}B(z)]^{-1}B_1^T A(z)y_{k+1} - [B_1^T z^{-1}B(z)]^{-1}B_1^T A(z)\varepsilon_{k+1} \\ &\quad - [B_1^T z^{-1}B(z)]^{-1}B_1^T\xi_{k+1} - w_k = \varphi_k - \psi_k - \chi_k - w_k. \end{aligned} \tag{4.4.30}$$

Set

$$D \triangleq \begin{bmatrix} -(B_1^TB_1)^{-1}B_1^TB_2 & I & \cdots & 0 \\ \vdots & & \ddots & \vdots \\ \vdots & & & I \\ -(B_1^TB_1)^{-1}B_1^TB_{s+1} & 0 & \cdots & 0 \end{bmatrix}, C \triangleq \begin{bmatrix} (B_1^TB_1)^{-1}B_1^T \\ (B_1^TB_1)^{-1}B_1^TA_1 \\ \vdots \\ (B_1^TB_1)^{-1}B_1^TA_{s-1} \end{bmatrix},$$

and $H^T \triangleq [I\ 0\ \cdots\ 0]$, where $s \triangleq \max(p+1, q-1)$, $B_k \triangleq 0$ for $k > q$, and $A_k \triangleq 0$ for $k > p$.

Since (4.4.27) can be rewritten as

$$\begin{aligned} &B_1^TB_1\varphi_k + B_1^TB_2\varphi_{k-1} + \cdots + B_1^TB_q\varphi_{k-q+1} \\ &\qquad = B_1^Ty_{k+1} + B_1^TA_1y_k + \cdots + B_1^TA_py_{k-p+1}, \end{aligned}$$

φ_k can be presented in the state space form

$$\begin{cases} x_{k+1} = & Dx_k + Cy_{k+1} \\ \varphi_k = & H^Tx_{k+1}, \end{cases} \tag{4.4.31}$$

where D is an $ls \times ls$ matrix, C is an $ls \times m$ vector, and H is an $ls \times l$ vector.

Replacing B_i and A_j in D and C with $B_{i,k}$ and $A_{j,k}$ given by (4.4.22) and (4.4.23), respectively, $i = 1, \cdots, s,\ j = 1, \cdots, s$, we obtain the estimates D_k and C_k for D and C at time k, and, hence, the estimate $\widehat{\varphi}_k$ for φ_k is given as follows:

$$\widehat{x}_{k+1} = D_{k+1}\widehat{x}_k + C_{k+1}y_{k+1},\ \widehat{\varphi}_k = H^T\widehat{x}_{k+1} \tag{4.4.32}$$

with an arbitrary initial value $\widehat{x}_0$.

For estimating $f(u_k^0)$ we estimate φ_k, while the last three terms in (4.4.30) may be dealt with as noises. Since u_k^0 is corrupted by noise η_k, the exponential type kernel function would lead to a biased estimation. Instead, we use the deconvolution kernel function to achieve this.

We need the Sinc kernel function

$$K(x) = \prod_{j=1}^{l} \frac{\sin(x_j)}{\pi x_j}, \tag{4.4.33}$$

where $x = [x_1, \cdots, x_l]^T$.

Denote the Fourier transformation of the Sinc kernel function $K(x)$ by

$$\Phi_K(t) \triangleq \int_{\mathbb{R}^l} e^{it^T x} K(x)\mathrm{d}x = \prod_{j=1}^{l} I_{[|t_j| \le 1]},$$

where $t = [t_1, \cdots, t_l]^T$, i stands for the imaginary unit satisfying $i^2 = -1$, and $I_{[A]}$ represents the indicator function of a set A.

Under A4.4.6 the input measurement noise η_k is also Gaussian with zero-mean. Its characteristic function $\Phi_{\eta_k}(t)$ is given by

$$\Phi_{\eta_k}(t) \triangleq \int_{\mathbb{R}^l} e^{it^T x} \frac{1}{(2\pi)^{\frac{l}{2}} |\Sigma_k|^{\frac{1}{2}}} e^{-\frac{x^T \Sigma_k^{-1} x}{2}} \mathrm{d}t = e^{-\frac{t^T \Sigma_k t}{2}},$$

where $\Sigma_k \triangleq E(\eta_k \eta_k^T)$ is the covariance matrix of η_k. Denote by Σ the limit of Σ_k. It is clear that $\|\Sigma - \Sigma_k\| = O(e^{-r_1 k})$ for some fixed $r_1 > 0$.

We now introduce the deconvolution kernel function $\omega_k(x)$, but for this we first define

$$K_k(x) \triangleq \frac{1}{(2\pi)^l} \int_{\mathbb{R}^l} e^{-it^T x} \frac{\Phi_K(t)}{\Phi_{\eta_k}(t/b_k)} \mathrm{d}t = \frac{1}{(2\pi)^l} \int_{[-1,1]^l} e^{-it^T x} e^{\frac{t^T \Sigma_k t}{2b_k^2}} \mathrm{d}t. \tag{4.4.34}$$

The function $\omega_k(x)$ is defined by

$$\begin{aligned}
\omega_k(x) &\triangleq \frac{1}{b_k^l} K_k\Big(\frac{u_k - x}{b_k}\Big) = \frac{1}{(2\pi b_k)^l} \int_{\mathbb{R}^l} e^{\left[-it^T (u_k - x)/b_k\right]} \frac{\Phi_K(t)}{\Phi_{\eta_k}(t/b_k)} \mathrm{d}t \\
&= \frac{1}{(2\pi b_k)^l} \int_{[-1,1]^l} e^{\left[-it^T (u_k - x)/b_k\right]} e^{\frac{t^T \Sigma_k t}{2b_k^2}} \mathrm{d}t \\
&= \frac{1}{(2\pi b_k)^l} \int_{[-1,1]^l} \cos\big(t^T (u_k - x)/b_k\big) e^{\frac{t^T \Sigma_k t}{2b_k^2}} \mathrm{d}t,
\end{aligned} \tag{4.4.35}$$

where $b_k = \left(\frac{b\lambda_{\max}(\Sigma_k)}{\log k}\right)^{\frac{1}{2}}$ is the bandwidth with $b > 3l$ being a constant, and $\lambda_{\max}(\Sigma_k)$ denotes the maximal eigenvalue of Σ_k.

As a matter of fact, as b increases, more samples around x will be used for estimating and this may reduce the variance but enlarge the bias of the estimate. So, b is selected on the basis of tradeoff between the variance and bias of the estimate. The unknown variance matrix Σ_k is recursively estimated by SAAWET:

$$\widehat{\Sigma}_{k+1} = \Big[\widehat{\Sigma}_k - \frac{1}{k+1}(\widehat{\Sigma}_k + \Lambda - u_{k+1}u_{k+1}^T)\Big] \cdot I_{\left[|\widehat{\Sigma}_k - \frac{1}{k+1}(\widehat{\Sigma}+\Lambda - u_{k+1}u_{k+1}^T)| \le M_{\phi_k}\right]}, \tag{4.4.36}$$

$$\phi_k = \sum_{j=1}^{k-1} I_{\left[|\widehat{\Sigma}_j - \frac{1}{j+1}(\widehat{\Sigma}_j+\Lambda - u_{j+1}u_{j+1}^T)| > M_{\phi_j}\right]}, \tag{4.4.37}$$

where $\widehat{\Sigma}_k$ represents the estimate for Σ_k at time k.

Therefore, $\omega_k(x)$ can be estimated at time k by

$$\widehat{\omega}_k(x) \triangleq \frac{1}{(2\pi\widehat{b}_k)^l} \int_{[-1,1]^l} \cos\left(t^T(u_k - x)/\widehat{b}_k\right) e^{\frac{t^T\widehat{\Sigma}_k t}{2\widehat{b}_k^2}} \, \mathrm{d}t, \tag{4.4.38}$$

where $\widehat{b}_k = \left(\frac{b\lambda_{\max}(\widehat{\Sigma}_k)}{\log k}\right)^{\frac{1}{2}}$. For calculating (4.4.38), the numerical integration is needed.

We now give the algorithms to estimate $f(x)$:

$$\tau_{k+1}(x) = \Big[\tau_k(x) - \frac{1}{k}(\tau_k(x) - \widehat{\omega}_k(x))\Big] \cdot I_{\left[\left|\tau_k(x) - \frac{1}{k}(\tau_k(x) - \widehat{\omega}_k(x))\right| \le M_{\Delta_k^\tau(x)}\right]}, \tag{4.4.39}$$

$$\Delta_k^\tau(x) = \sum_{j=1}^{k-1} I_{\left[\left|\tau_j(x) - \frac{1}{j}(\tau_j(x) - \widehat{\omega}_j(x))\right| > M_{\Delta_j^\tau(x)}\right]}, \tag{4.4.40}$$

$$\beta_{k+1}(x) = \Big[\beta_k(x) - \frac{1}{k}(\beta_k(x) - \widehat{\omega}_k(x)\widehat{\varphi}_k)\Big] \cdot I_{\left[\left\|\beta_k(x) - \frac{1}{k}(\beta_k(x) - \widehat{\omega}_k(x)\widehat{\varphi}_k)\right\| \le M_{\Delta_k^\beta(x)}\right]}, \tag{4.4.41}$$

$$\Delta_k^\beta(x) = \sum_{j=1}^{k-1} I_{\left[\left\|\beta_j(x) - \frac{1}{j}(\beta_j(x) - \widehat{\omega}_j(x)\widehat{\varphi}_j)\right\| > M_{\Delta_j^\beta(x)}\right]}. \tag{4.4.42}$$

It is noticed that $\tau_{k+1}(x)$ defined by (4.4.39)–(4.4.40) and $\beta_{k+1}(x)$ defined by (4.4.41)–(4.4.42) are applied to estimate $p(x)$ and $p(x)f(x)$, respectively. The estimate for $f(x)$ is defined naturally as:

$$f_{k+1}(x) \triangleq \begin{cases} \frac{\beta_{k+1}(x)}{\tau_{k+1}(x)}, & \text{if } \tau_{k+1}(x) \neq 0 \\ 0, & \text{if } \tau_{k+1}(x) = 0. \end{cases} \tag{4.4.43}$$

Strong Consistency of Estimates for Linear Subsystem

Lemma 4.4.2 *Assume A4.4.1–A4.4.4 hold but without need for availability of Λ in A4.4.1. Then, for any $0 \le \nu < 1/2$, the following series converge.*

$$\sum_{k=1}^{\infty} \frac{1}{k^{1-\nu}} (E(y_k u_{k-i}^T) - y_k u_{k-i}^T) < \infty \text{ a.s. for } i \ge 1. \tag{4.4.44}$$

Proof. It is noticed that u_{k-i}^0, η_{k-i}, $\bar{\xi}_k$, w_k, and ε_k are mutually independent and by (4.4.16), we have

$$\begin{aligned} E(y_k u_{k-i}^T) - y_k u_{k-i}^T &= \left[E(y_k^0 u_{k-i}^{0T}) - y_k^0 u_{k-i}^{0T}\right] - y_k^0 \eta_{k-i}^T - \varepsilon_k u_{k-i}^{0T} - \varepsilon_k \eta_{k-i}^T \\ &= H_i\left[Ef(u_{k-i}^0)u_{k-i}^{0T} - f(u_{k-i}^0)u_{k-i}^{0T}\right] - \sum_{j=1, j\ne i}^{k} H_j f(u_{k-j}^0)u_{k-i}^{0T} - \sum_{j=1}^{k} H_j f(u_{k-j}^0)\eta_{k-i}^T \\ &\quad - \sum_{j=1}^{k} H_j w_{k-j} u_{k-i}^{0T} - \sum_{j=1}^{k} H_j w_{k-j} \eta_{k-i}^T - \bar{\xi}_k u_{k-i}^{0T} - \bar{\xi}_k \eta_{k-i}^T - \varepsilon_k u_{k-i}^{0T} - \varepsilon_k \eta_{k-i}^T. \end{aligned} \tag{4.4.45}$$

From (4.4.45) it follows that

$$\begin{aligned} \sum_{k=1}^{\infty} \frac{1}{k^{1-\nu}} \left[E(y_k u_{k-i}^T) - y_k u_{k-i}^T\right] &= H_i \sum_{k=1}^{\infty} \frac{1}{k^{1-\nu}} \left[Ef(u_{k-i}^0)u_{k-i}^{0T} - f(u_{k-i}^0)u_{k-i}^{0T}\right] \\ &- \sum_{k=1}^{\infty} \frac{1}{k^{1-\nu}} \left[\sum_{j=1, j\ne i}^{k} H_j f(u_{k-j}^0)u_{k-i}^{0T}\right] - \sum_{k=1}^{\infty} \frac{1}{k^{1-\nu}} \left[\sum_{j=1}^{k} H_j f(u_{k-j}^0)\eta_{k-i}^T\right] \\ &- \sum_{k=1}^{\infty} \frac{1}{k^{1-\nu}} \left[\sum_{j=1}^{k} H_j w_{k-j} u_{k-i}^{0T}\right] - \sum_{k=1}^{\infty} \frac{1}{k^{1-\nu}} \left[\sum_{j=1}^{k} H_j w_{k-j} \eta_{k-i}^T\right] \\ &- \sum_{k=1}^{\infty} \frac{1}{k^{1-\nu}} \left[\bar{\xi}_k u_{k-i}^{0T}\right] - \sum_{k=1}^{\infty} \frac{1}{k^{1-\nu}} \left[\bar{\xi}_k \eta_{k-i}^T\right] - \sum_{k=1}^{\infty} \frac{1}{k^{1-\nu}} \left[\varepsilon_k u_{k-i}^{0T}\right] - \sum_{k=1}^{\infty} \frac{1}{k^{1-\nu}} \left[\varepsilon_k \eta_{k-i}^T\right]. \end{aligned} \tag{4.4.46}$$

Define $z_k^{(1)} \triangleq \frac{1}{k^{1-\nu}} \left[E(f(u_{k-i}^0)u_{k-i}^{0T}) - f(u_{k-i}^0)u_{k-i}^{0T}\right]$. It is clear that $z_k^{(1)}$ is a sequence of independent random matrices with zero-mean. Noticing that both u_k^0 and $f(u_k^0)$ are bounded, then by the C_r-inequality we have

$$\begin{aligned} \sum_{k=1}^{\infty} E\|z_k^{(1)}\|^2 &= \sum_{k=1}^{\infty} E\left\| \frac{1}{k^{1-\nu}} (E(f(u_{k-i}^0)u_{k-i}^{0T}) - f(u_{k-i}^0)u_{k-i}^{0T}) \right\|^2 \\ &\le \sum_{k=1}^{\infty} \frac{2}{k^{2(1-\nu)}} E\left\|f(u_{k-i}^0)u_{k-i}^{0T}\right\|^2 < \infty, \end{aligned}$$

which implies that the first term at the right-hand side of (4.4.46) converges a.s. by Theorem 1.2.8.

For the second term at the right-hand side of (4.4.46), we have

$$\begin{aligned}
&\sum_{k=1}^{\infty}\frac{1}{k^{1-\nu}}\Big[\sum_{j=1,j\neq i}^{k} H_j f(u_{k-j}^0)u_{k-i}^{0T}\Big] \\
&=\sum_{k=1}^{\infty}\frac{1}{k^{1-\nu}}\Big[\sum_{j=1}^{i-1} H_j f(u_{k-j}^0)u_{k-i}^{0T}\Big]+\sum_{k=1}^{\infty}\frac{1}{k^{1-\nu}}\Big[\sum_{j=i+1}^{k} H_j f(u_{k-j}^0)u_{k-i}^{0T}\Big] \\
&=\sum_{j=1}^{i-1}H_j\sum_{l=0}^{i-j}\sum_{k=1}^{\infty}\frac{1}{((i-j+1)k+l)^{1-\nu}}\Big[f(u^0_{(i-j+1)k+l-j})u^{0T}_{(i-j+1)k+l-i}\Big] \\
&\quad+\sum_{k=1}^{\infty}\frac{1}{k^{1-\nu}}\Big[\sum_{j=i+1}^{k} H_j f(u_{k-j}^0)u_{k-i}^{0T}\Big].
\end{aligned} \tag{4.4.47}$$

Defining $z_k^{(2)} \triangleq \frac{1}{((i-j+1)k+l)^{1-\nu}}\Big[f(u^0_{(i-j+1)k+l-j})u^{0T}_{(i-j+1)k+l-i}\Big]$, we see that $z_k^{(2)}$ is a sequence of independent random matrices with zero-mean. Noticing that u_k^0 and $f(u_k^0)$ are bounded, we have

$$\begin{aligned}
\sum_{k=1}^{\infty}E\|z_k^{(2)}\|^2 &= \sum_{k=1}^{\infty}E\Big\|\frac{1}{((i-j+1)k+l)^{1-\nu}}\Big[f(u^0_{(i-j+1)k+l-j})u^{0T}_{(i-j+1)k+l-i}\Big]\Big\|^2 \\
&=\sum_{k=1}^{\infty}\frac{1}{((i-j+1)k+l)^{2(1-\nu)}}E\Big\|f(u^0_{(i-j+1)k+l-j})u^{0T}_{(i-j+1)k+l-i}\Big\|^2<\infty,
\end{aligned}$$

which implies the first term at the right-hand side of (4.4.47) converges a.s. by Theorem 1.2.8.

Define $z_k^{(3)} \triangleq \frac{1}{k^{1-\nu}}\Big[\sum_{j=i+1}^{k} H_j f(u_{k-j}^0)u_{k-i}^{0T}\Big]$ and $\mathcal{F}_k \triangleq \{u_{j-i}^0, i\le j\le k\}$. We have $E\big[z_k^{(3)}|\mathcal{F}_{k-1}\big]=0$, i.e., $\{z_k^{(3)},\mathcal{F}_k\}$ is an mds.

By Lemma 4.2.1, we have $E\big\|\sum_{j=i+1}^{k}H_jf(u_{k-j}^0)\big\|^2<\infty$ for any fixed i, and hence

$$\begin{aligned}
\sum_{k=1}^{\infty}E\|z_k^{(3)}\|^2 &= \sum_{k=1}^{\infty}E\Big\|\frac{1}{k^{1-\nu}}\Big[\sum_{j=i+1}^{k} H_j f(u_{k-j}^0)u_{k-i}^{0T}\Big]\Big\|^2 \\
&=\sum_{k=1}^{\infty}\frac{1}{k^{2(1-\nu)}}E\Big\|\sum_{j=i+1}^{k} H_j f(u_{k-j}^0)u_{k-i}^{0T}\Big\|^2 \\
&=\sum_{k=1}^{\infty}\frac{1}{k^{2(1-\nu)}}E\Big\|\sum_{j=i+1}^{k} H_j f(u_{k-j}^0)\Big\|^2E\big\|u_{k-i}^0\big\|^2<\infty,
\end{aligned}$$

which implies

$$\sum_{k=1}^{\infty} E\left[\|z_k^{(3)}\|^2 | \mathcal{F}_{k-1}\right] < \infty \ a.s.$$

Therefore, the second term at the right-hand side of (4.4.47) converges a.s. by Theorem 1.2.8, and hence the second term at the right-hand side of (4.4.46) also converges a.s.

Define $z_k^{(4)} \triangleq \frac{1}{k^{1-\nu}}\left[\bar{\xi}_k u_{k-i}^{0T}\right]$ and $\mathcal{F}_k \triangleq \{u_{j-i}^0, \bar{\xi}_{j+1}, i \leq j \leq k\}$. Then we have $\frac{1}{k^{1-\nu}}\bar{\xi}_k \in \mathcal{F}_{k-1}$ and $E\left[z_k^{(4)}|\mathcal{F}_{k-1}\right] = \frac{1}{k^{1-\nu}}\bar{\xi}_k E\left[u_{k-i}^{0T}|\mathcal{F}_{k-1}\right] = 0$, so $\{z_k^{(4)}, \mathcal{F}_k\}$ is an mds.

Noticing that $\bar{\xi}_k$ and u_k^0 are mutually independent and $E\|\bar{\xi}_k\|^2 < \infty$ by Lemma 4.2.1, we have

$$\sum_{k=1}^{\infty} E\|z_k^{(4)}\|^2 = \sum_{k=1}^{\infty} E\left\|\frac{1}{k^{1-\nu}}(\bar{\xi}_k u_{k-i}^{0T})\right\|^2$$
$$= \sum_{k=1}^{\infty} \frac{1}{k^{2(1-\nu)}} E\|\bar{\xi}_k\|^2 E\|u_{k-i}^0\|^2 < \infty,$$

which implies

$$\sum_{k=1}^{\infty} E\left[\|z_k^{(4)}\|^2 | \mathcal{F}_{k-1}\right] < \infty \ a.s.$$

Therefore, the sixth term at the right-hand side of (4.4.46) converges a.s. by Theorem 1.2.8. The eighth term at the right-hand side of (4.4.46) can be proved in a similar way.

According to Theorem 1.4.4, both ε_k and η_k are zero-mean α-mixing sequences with mixing coefficients decaying exponentially to zero. Since ε_k and η_k are mutually independent, $z_k^{(5)} \triangleq \frac{1}{k^{1-\nu}}\left[\varepsilon_k \eta_{k-i}^T\right]$ is a zero-mean α-mixing sequence with mixing coefficients decaying exponentially to zero. Moreover, $E\|\varepsilon_k\|^{2+\varepsilon} < \infty$ and $E\|\eta_k\|^{2+\varepsilon} < \infty$ by Lemma 4.2.1, then for any $\varepsilon > 0$ we have

$$E\|z_k^{(5)}\|^{2+\varepsilon} = E\left\|\frac{1}{k^{1-\nu}}\varepsilon_k \eta_{k-i}^T\right\|^{2+\varepsilon}$$
$$= \frac{1}{k^{(1-\nu)(2+\varepsilon)}}(E\|\varepsilon_k\|^{2+\varepsilon}) \cdot (E\|\eta_{k-i}\|^{2+\varepsilon}) = O\Big(\frac{1}{k^{(1-\nu)(2+\varepsilon)}}\Big),$$

which leads to

$$\sum_{k=1}^{\infty} (E\|z_k^{(5)}\|^{2+\varepsilon})^{\frac{2}{2+\varepsilon}} = O\Big(\sum_{k=1}^{\infty} \frac{1}{k^{2(1-\nu)}}\Big) < \infty.$$

Therefore, by Theorem 1.4.2 the last term at the right-hand side of (4.4.46) converges a.s. The seventh term at the right-hand side of (4.4.46) can be proved in a similar way as that used for the last term at the right-hand side of (4.4.46).

Finally, we prove the third term at the right-hand side of (4.4.46) converges a.s. Since $\{f(u_k^0)\}$ is a sequence of iid random vectors with density function, $\{\sum_{j=1}^k H_j f(u_{k-j}^0)\}$ is an α-mixing sequence with mixing coefficients decaying exponentially to zero by Theorem 1.4.4. Similarly, by Theorem 1.4.4, $\{\eta_k\}$ is a zero-mean α-mixing sequence with mixing coefficients decaying exponentially to zero. Since u_k^0 and η_k are mutually independent, $z_k^{(6)} \triangleq \frac{1}{k^{1-\nu}}\{\sum_{j=1}^k H_j f(u_{k-j}^0)\eta_{k-i}^T\}$ is also a zero-mean α-mixing sequence with mixing coefficients decaying exponentially to zero. Moreover, $E\|\sum_{j=1}^k H_j f(u_{k-j}^0)\|^{2+\varepsilon} < \infty$ and $E\|\eta_k\|^{2+\varepsilon} < \infty$ by Lemma 4.2.1, hence for any $\varepsilon > 0$ we have

$$E\|z_k^{(6)}\|^{2+\varepsilon} = E\Big\|\frac{1}{k^{1-\nu}}\sum_{j=1}^k H_j f(u_{k-j}^0)\eta_{k-i}^T\Big\|^{2+\varepsilon}$$

$$= \frac{1}{k^{(1-\nu)(2+\varepsilon)}}E\Big\|\sum_{j=1}^k H_j f(u_{k-j}^0)\Big\|^{2+\varepsilon}\cdot E\|\eta_{k-i}\|^{2+\varepsilon} = O\Big(\frac{1}{k^{(1-\nu)(2+\varepsilon)}}\Big),$$

which leads to

$$\sum_{k=1}^{\infty}\big(E\|z_k^{(6)}\|^{2+\varepsilon}\big)^{\frac{2}{2+\varepsilon}} = O\Big(\sum_{k=1}^{\infty}\frac{1}{k^{2(1-\nu)}}\Big) < \infty.$$

Therefore, by Theorem 1.4.2 the third term at the right-hand side of (4.4.46) converges a.s. The fourth and fifth terms at the right-hand side of (4.4.46) can be proved in a similar way as that used for the third term at the right-hand side of (4.4.46).

The proof is completed. □

Theorem 4.4.1 *Assume A4.4.1–A4.4.4 hold but without need for availability of Λ in A4.4.1. Then, $H_{i,k}$, $i \ge 1$, defined by (4.4.18)–(4.4.19) have the convergence rates*

$$\|H_{i,k} - H_i\| = o(k^{-\nu}) \text{ a.s., } i \ge 1\ \forall\, \nu \in (0, 1/2). \tag{4.4.48}$$

As a consequence, $A_{i,k}$, $i = 1, \cdots, p$, and $B_{j,k}$, $j = 1, \cdots, q$ converge and have the following convergence rate:

$$\|A_{i,k} - A_i\| = o(k^{-\nu}) \text{ a.s. } \forall \nu \in (0, 1/2),\ i = 1, \cdots, p, \tag{4.4.49}$$

$$\|B_{j,k} - B_j\| = o(k^{-\nu}) \text{ a.s. } \forall \nu \in (0, 1/2),\ j = 1, \cdots, q. \tag{4.4.50}$$

Proof. We rewrite (4.4.18) as

$$H_{i,k+1} = \big[H_{i,k} - \frac{1}{k}(H_{i,k} - H_i) - \frac{1}{k}e_{k+1}(i)\big]\cdot I_{\left[\left\|H_{i,k} - \frac{1}{k}(H_{i,k} - H_i) - \frac{1}{k}e_{k+1}(i)\right\| \le M_{\delta_{i,k}}\right]},$$

where $e_{k+1}(i) = H_i - y_{k+1}u_{k+1-i}^T = [Ey_{k+1}u_{k+1-i}^T - y_{k+1}u_{k+1-i}^T]$.

Since H_i is the single root of the linear function $-(y - H_i)$, by Theorem 2.6.1, it suffices to prove

$$\sum_{k=1}^{\infty}\frac{1}{k^{1-\nu}}e_k^{(i)} < \infty \text{ a.s., } i \ge 1\ \forall \nu \in (0, 1/2). \tag{4.4.51}$$

By Lemma 4.4.2, we find that (4.4.51) is true, and hence (4.4.48) holds, while the assertions (4.4.49)–(4.4.50) directly follow from (4.4.48). □

Strong Consistency of Estimates for f(·)

Lemma 4.4.3 *Assume A4.4.1–A4.4.4 hold. Then the following limit takes place*

$$\frac{y_k}{k^{\frac{1}{3}}} \xrightarrow[k\to\infty]{\text{a.s.}} 0. \tag{4.4.52}$$

Proof. By Lemma 4.2.1, we have $E\left\|\sum_{j=1}^{k} H_j v_{k-j}^0\right\|^{\Delta} < \infty$, $E\left\|\bar{\xi}_k\right\|^{\Delta} < \infty$, and $E\left\|\varepsilon_k\right\|^{\Delta} < \infty$ for some $\Delta > 3$. Since $y_k = \sum_{j=1}^{k} H_j v_{k-j}^0 + \bar{\xi}_k + \varepsilon_k$, by the C_r-inequality we have

$$E|y_k|^{\Delta} \le 3^{\Delta-1}\Big(E\Big\|\sum\nolimits_{j=1}^{k} H_j v_{k-j}^0\Big\|^{\Delta} + E\left\|\bar{\xi}_k\right\|^{\Delta} + E\left\|\varepsilon_k\right\|^{\Delta}\Big) < \infty.$$

Noticing that

$$\sum_{k=1}^{\infty} P\Big[\frac{\|y_k\|}{k^{\frac{1}{3}}} > \varepsilon\Big] = \sum_{k=1}^{\infty} P\Big[\frac{\|y_k\|^{\Delta}}{k^{\frac{\Delta}{3}}} > \varepsilon^{\Delta}\Big] \le \sum_{k=1}^{\infty} \frac{E\|y_k\|^{\Delta}}{\varepsilon^{\Delta} k^{\frac{\Delta}{3}}} < \infty$$

for any given $\varepsilon > 0$, by the Borel–Cantelli lemma, we derive $\frac{y_k}{k^{\frac{1}{3}}} \xrightarrow[k\to\infty]{a.s.} 0$. □

Lemma 4.4.4 *Under Conditions A4.4.1, A4.4.3, and A4.4.6, for $\omega_k(x)$ defined by (4.4.35) the following limits take place*

$$E[\omega_k(x)] \xrightarrow[k\to\infty]{} p(x),\ E[\omega_k(x) f(u_k^0)] \xrightarrow[k\to\infty]{} p(x)f(x), \tag{4.4.53}$$

$$|\omega_k(x)|^{\delta} = O\big(k^{\frac{l\delta}{2b}} (\log k)^{\frac{l\delta}{2}}\big)\ \ \forall \delta \ge 1, \tag{4.4.54}$$

$$|\omega_k(x) f(u_k^0)|^{\delta} = O\big(k^{\frac{l\delta}{2b}} (\log k)^{\frac{l\delta}{2}}\big)\ \ \forall \delta \ge 1. \tag{4.4.55}$$

Proof. By the Fubini theorem changing the order of taking expectation and integral, and noticing that the density function of η_k is even, we have

$$\begin{aligned}
E\omega_k(x) =& \frac{1}{(2\pi b_k)^l} \int_{\mathbb{R}^l} E e^{\left[-it^T (u_k - x)/b_k\right]} \frac{\Phi_K(t)}{\Phi_{\eta_k}(t/b_k)} \mathrm{d}t \\
=& \frac{1}{(2\pi b_k)^l} \int_{\mathbb{R}^l} E e^{\left[-it^T (u_k^0 - x)/b_k\right]} E e^{\left[-it^T \eta_k / b_k\right]} \frac{\Phi_K(t)}{\Phi_{\eta_k}(t/b_k)} \mathrm{d}t \\
=& \frac{1}{(2\pi b_k)^l} \int_{\mathbb{R}^l} E e^{\left[-it^T (u_k^0 - x)/b_k\right]} \Phi_K(t) \mathrm{d}t \\
=& \frac{1}{(2\pi b_k)^l} \int_{\mathbb{R}^l} \int_U e^{\left[-it^T (y - x)/b_k\right]} p(y) \mathrm{d}y \Phi_K(t) \mathrm{d}t \\
=& \frac{1}{b_k^l} \int_U \Big(\frac{1}{(2\pi)^l} \int_{\mathbb{R}^l} e^{\left[-it^T (y - x)/b_k\right]} \Phi_K(t) \mathrm{d}t\Big) p(y) \mathrm{d}y
\end{aligned}$$

$$
\begin{aligned}
&= \frac{1}{b_k^l}\int_U K\Big(\frac{y-x}{b_k}\Big)p(y)\mathrm{d}y \\
&= \int_{U_k} K(t)p(x+b_k t)\mathrm{d}t \xrightarrow[k\to\infty]{} p(x),
\end{aligned}
$$

where U_k tends to $\mathbb{R}^l$ as $k \to \infty$.

The second limit of (4.4.53) can be proved by a similar treatment.

By (4.4.35) and $t^T t \le l$ in the domain $[-1,1]^l$, we have

$$
\begin{aligned}
|\omega_k(x)|^\delta &= \Big|\frac{1}{(2\pi b_k)^l}\int_{[-1,1]^l} e^{\left[-it^T(u_k-x)/b_k\right]} e^{\frac{t^T\Sigma_k t}{2b_k^2}}\mathrm{d}t\Big|^\delta \\
&\le \frac{1}{(2\pi b_k)^{l\delta}}\Big|\int_{[-1,1]^l} e^{\frac{t^T\Sigma_k t}{2b_k^2}}\mathrm{d}t\Big|^\delta \\
&\le \frac{1}{(2\pi b_k)^{l\delta}}\Big|\int_{[-1,1]^l} e^{\frac{t^T t\lambda_{\max}(\Sigma_k)}{2b_k^2}}\mathrm{d}t\Big|^\delta \\
&= \frac{1}{(2\pi b_k)^{l\delta}}\Big|\int_{[-1,1]^l} e^{\frac{t^T t\lambda_{\max}(\Sigma_k)}{2}\frac{\log k}{b\lambda_{\max}(\Sigma_k)}}\mathrm{d}t\Big|^\delta \\
&\le \frac{1}{(2\pi b_k)^{l\delta}}\Big|\int_{[-1,1]^l} k^{\frac{t^T t}{2b}}\mathrm{d}t\Big|^\delta = O\big(k^{\frac{l\delta}{2b}}(\log k)^{\frac{l\delta}{2}}\big).
\end{aligned}
\tag{4.4.56}
$$

Finally, (4.4.55) can be similarly proved as $f(u_k^0)$ is bounded. □

Lemma 4.4.5 *Assume that A4.4.1, A4.4.3, and A4.4.6 hold. Then $\widehat{\Sigma}_k$ defined by (4.4.36) and (4.4.37) has the convergence rate*

$$
\|\widehat{\Sigma}_k - \Sigma_k\| = o\Big(\frac{1}{k^{1/2-c}}\Big)\ \ \forall c > 0.
$$

Proof. The algorithm (4.4.36) can be rewritten as

$$
\widehat{\Sigma}_{k+1} = \Big[\widehat{\Sigma}_k - \frac{1}{k+1}(\widehat{\Sigma}_k - \Sigma) - \frac{1}{k+1}\gamma_{k+1}\Big]\cdot I_{\left[\left\|\widehat{\Sigma}_k - \frac{1}{k+1}(\widehat{\Sigma}_k - \Sigma) - \frac{1}{k+1}\gamma_{k+1}\right\| \le M_{\phi_k}\right]},
$$

where

$$
\begin{aligned}
\gamma_{k+1} = {} & (\Sigma - \Sigma_{k+1}) + (\Sigma_{k+1} - \eta_{k+1}\eta_{k+1}^T) + (\Lambda - u_{k+1}^0 u_{k+1}^{0T}) \\
& - u_{k+1}^0\eta_{k+1}^T - \eta_{k+1}u_{k+1}^{0T}.
\end{aligned}
\tag{4.4.57}
$$

Since Σ is the single root of the linear function $-(y-\Sigma)$, by Theorem 2.6.1, for proving that $\|\widehat{\Sigma}_k - \Sigma\| = o\left(\frac{1}{k^{1/2-c}}\right)$ it suffices to show

$$
\sum_{k=1}^{\infty}\frac{1}{k^{1-\nu}}\gamma_k < \infty \ \text{ a.s. } \forall\ 0 < \nu < 1/2. \tag{4.4.58}
$$

Since $\|\Sigma-\Sigma_k\| = O(e^{-r_1 k})$ for some fixed r_1, (4.4.58) holds with γ_k replaced by $\Sigma-\Sigma_k$.

By Theorem 1.4.4 $\{\Sigma_k - \eta_k\eta_k^T\}$ is a zero-mean α-mixing sequence with mixing coefficients decaying exponentially to zero. Further, $E\|\eta_k\|^{2(2+\varepsilon)} < \infty$ by A4.4.6 and Lemma 4.2.1, for any $\varepsilon > 0$ we have

$$\sum_{k=1}^{\infty}\Big[E\Big\|\frac{1}{k^{1-v}}(\Sigma_k - \eta_k\eta_k^T)\Big\|^{2+\varepsilon}\Big]^{\frac{2}{2+\varepsilon}}$$
$$\le \sum_{k=1}^{\infty}\frac{2^{1+\varepsilon}}{k^{2(1-v)}}(\|\Sigma_k\|^{2+\varepsilon} + E\|\eta_k\eta_k^T\|^{2+\varepsilon})^{\frac{2}{2+\varepsilon}} < \infty.$$

Therefore, (4.4.58) also holds with γ_k replaced by the second term at the right-hand side of (4.4.57) by Theorem 1.4.2.

Since $\{\Lambda - u_k^0 u_k^{0T}\}$ is a sequence of zero-mean iid random matrices and $\sum_{k=1}^{\infty}\frac{1}{k^{2(1-v)}}E\|\Lambda - u_k^0 u_k^{0T}\|^2 < \infty$, it follows that (4.4.58) holds with γ_k replaced by the third term at the right-hand side of (4.4.57) by Theorem 1.2.8.

Define $\mathcal{F}_k \triangleq \{u_j^0, \eta_{j+1}, 1 \le j \le k\}$. Then $E(u_k^0\eta_k^T|\mathcal{F}_{k-1}) = 0$, which indicates that $\{u_k^0\eta_k^T, \mathcal{F}_k\}$ is an mds. Moreover, $E\|\eta_k\|^2 < \infty$ by Lemma 4.2.1, we have

$$\sum_{k=1}^{\infty}E\Big\|\frac{1}{k^{1-v}}u_k^0\eta_k^T\Big\|^2 = \sum_{k=1}^{\infty}\frac{1}{k^{2(1-v)}}E\|u_k^0\|^2 E\|\eta_k\|^2 < \infty,$$

which implies that

$$\sum_{k=1}^{\infty}E\Big[\Big\|\frac{1}{k^{1-v}}u_k^0\eta_k^T\Big\|^2|\mathcal{F}_{k-1}\Big] < \infty \text{ a.s.}$$

It follows that (4.4.58) holds with γ_k replaced by the fourth term of (4.4.57) by Theorem 1.2.8. Similarly, (4.4.58) also holds with γ_k replaced by the last term of (4.4.57).

Since $\|\widehat{\Sigma}_k - \Sigma_k\| \le \|\widehat{\Sigma}_k - \Sigma\| + \|\Sigma - \Sigma_k\|$ and $\|\Sigma-\Sigma_k\| = O(e^{-r_1 k})$ for some fixed r_1, we have

$$\|\widehat{\Sigma}_k - \Sigma_k\| = o\Big(\frac{1}{k^{1/2-c}}\Big)\ \forall c > 0. \qquad \square$$

Lemma 4.4.6 *Assume A4.4.1–A4.4.6 hold. Then there is a constant* $c > 0$ *with* $\frac{1}{6} - \frac{l}{2b} - c > 0$ *such that*

$$|\omega_k(x) - \widehat{\omega}_k(x)| = o\Big(\frac{(\log k)^{\frac{l}{2}+1}}{k^{\frac{1}{2}-\frac{l}{2b}-c}}\Big), \tag{4.4.59}$$

$$\|\varphi_k - \widehat{\varphi}_k\| = o\Big(\frac{1}{k^{1/6-c}}\Big). \tag{4.4.60}$$

Proof. According to (4.4.35)–(4.4.38), we have

$$\omega_k(x) - \widehat{\omega}_k(x) = I_1 + I_2 + I_3,$$

where

$$\begin{aligned}
I_1 =& \frac{1}{(2\pi b_k)^l} \int_{[-1,1]^l} \cos\big(t^T(u_k - x)/b_k\big) e^{\frac{t^T \Sigma_k t}{2b_k^2}} \,\mathrm{d}t \\
&- \frac{1}{(2\pi \widehat{b}_k)^l} \int_{[-1,1]^l} \cos\big(t^T(u_k - x)/b_k\big) e^{\frac{t^T \Sigma_k t}{2b_k^2}} \,\mathrm{d}t \\
=& \frac{1}{(2\pi)^l} \Big(\frac{1}{b_k^l} - \frac{1}{\widehat{b}_k^l}\Big) \int_{[-1,1]^l} \cos\big(t^T(u_k - x)/b_k\big) e^{\frac{t^T \Sigma_k t}{2b_k^2}} \,\mathrm{d}t, \\
I_2 =& \frac{1}{(2\pi \widehat{b}_k)^l} \int_{[-1,1]^l} \cos\big(t^T(u_k - x)/b_k\big) e^{\frac{t^T \Sigma_k t}{2b_k^2}} \,\mathrm{d}t \\
&- \frac{1}{(2\pi \widehat{b}_k)^l} \int_{[-1,1]^l} \cos\big(t^T(u_k - x)/b_k\big) e^{\frac{t^T \widehat{\Sigma}_k t}{2b_k^2}} \,\mathrm{d}t \\
=& \frac{1}{(2\pi \widehat{b}_k)^l} \int_{[-1,1]^l} \cos\big(t^T(u_k - x)/b_k\big) \Big(e^{\frac{t^T \Sigma_k t}{2b_k^2}} - e^{\frac{t^T \widehat{\Sigma}_k t}{2b_k^2}}\Big) \mathrm{d}t, \\
I_3 =& \frac{1}{(2\pi \widehat{b}_k)^l} \int_{[-1,1]^l} \cos\big(t^T(u_k - x)/b_k\big) e^{\frac{t^T \widehat{\Sigma}_k t}{2\widehat{b}_k^2}} \,\mathrm{d}t \\
&- \frac{1}{(2\pi \widehat{b}_k)^l} \int_{[-1,1]^l} \cos\big(t^T(u_k - x)/\widehat{b}_k\big) e^{\frac{t^T \widehat{\Sigma}_k t}{2\widehat{b}_k^2}} \,\mathrm{d}t \\
=& \frac{1}{(2\pi \widehat{b}_k)^l} \int_{[-1,1]^l} \Big(\cos\big(t^T(u_k - x)/b_k\big) - \cos\big(t^T(u_k - x)/\widehat{b}_k\big)\Big) e^{\frac{t^T \widehat{\Sigma}_k t}{2\widehat{b}_k^2}} \,\mathrm{d}t.
\end{aligned}$$

Since $\|\widehat{\Sigma}_k - \Sigma_k\| = o\big(\frac{1}{k^{1/2-c}}\big)$ for any $c > 0$, we have

$$\Big|\frac{1}{b_k^l} - \frac{1}{\widehat{b}_k^l}\Big| = \Big(\frac{\log k}{b}\Big)^{\frac{l}{2}} \frac{\big|(\lambda_{\max}(\Sigma_k))^{\frac{l}{2}} - (\lambda_{\max}(\widehat{\Sigma}_k))^{\frac{l}{2}}\big|}{(\lambda_{\max}(\widehat{\Sigma}_k))^{\frac{l}{2}} (\lambda_{\max}(\Sigma_k))^{\frac{l}{2}}} = o\Big(\frac{(\log k)^{\frac{l}{2}}}{k^{\frac{1}{2}-c}}\Big). \tag{4.4.61}$$

Consequently, it follows that

$$\begin{aligned}
|I_1| \le& \frac{1}{(2\pi)^l} \Big|\frac{1}{b_k^l} - \frac{1}{\widehat{b}_k^l}\Big| \Big| \int_{[-1,1]^l} \cos\big(t^T(u_k - x)/b_k\big) e^{\frac{t^T \Sigma_k t}{2b_k^2}} \,\mathrm{d}t \Big| \\
\le& o\Big(\frac{(\log k)^{\frac{l}{2}}}{k^{\frac{1}{2}-c}}\Big) \Big| \int_{[-1,1]^l} e^{\frac{t^T \Sigma_k t}{2b_k^2}} \,\mathrm{d}t \Big| = o\Big(\frac{(\log k)^{\frac{l}{2}}}{k^{\frac{1}{2}-c}}\Big) O\big(k^{\frac{l}{2b}}\big) \\
=& o\Big(\frac{(\log k)^{\frac{l}{2}}}{k^{\frac{1}{2}-\frac{l}{2b}-c}}\Big).
\end{aligned} \tag{4.4.62}$$

Again by $\|\widehat{\Sigma}_k - \Sigma_k\| = o\big(\frac{1}{k^{1/2-c}}\big)$ for any $c > 0$ and (4.4.61), we have

$$\Big| e^{\frac{t^T\Sigma_k t}{2b_k^2}} - e^{\frac{t^T\widehat{\Sigma}_k t}{2\widehat{b}_k^2}} \Big| = e^{\frac{t^T\Sigma_k t}{2b_k^2}} \Big| 1 - e^{t^T\big(\frac{\widehat{\Sigma}_k}{2\widehat{b}_k^2} - \frac{\Sigma_k}{2b_k^2}\big)t} \Big| \le e^{\frac{t^T\Sigma_k t}{2b_k^2}} o\Big(\frac{\log k}{k^{\frac{1}{2}-c}}\Big).$$

As a consequence, it follows that

$$\begin{aligned}
|I_2| &\le \frac{1}{(2\pi\widehat{b}_k)^l}\Big|\int_{[-1,1]^l} \cos\big(t^T(u_k - x)/b_k\big)\big(e^{\frac{t^T\Sigma_k t}{2b_k^2}} - e^{\frac{t^T\widehat{\Sigma}_k t}{2\widehat{b}_k^2}}\big)\mathrm{d}t\Big| \\
&\le \frac{1}{(2\pi\widehat{b}_k)^l}\int_{[-1,1]^l} \Big| e^{\frac{t^T\Sigma_k t}{2b_k^2}} - e^{\frac{t^T\widehat{\Sigma}_k t}{2\widehat{b}_k^2}} \Big|\mathrm{d}t \\
&\le \frac{1}{(2\pi\widehat{b}_k)^l} o\Big(\frac{\log k}{k^{\frac{1}{2}-c}}\Big)\int_{[-1,1]^l} e^{\frac{t^T\Sigma_k t}{2b_k^2}}\mathrm{d}t = o\Big(\frac{(\log k)^{\frac{l}{2}+1}}{k^{\frac{1}{2}-\frac{l}{2b}-c}}\Big). \qquad (4.4.63)
\end{aligned}$$

Finally, by (4.4.61) we have

$$\begin{aligned}
|I_3| &= \frac{1}{(2\pi\widehat{b}_k)^l}\Big|\int_{[-1,1]^l} \Big(\cos\big(t^T(u_k - x)/b_k\big) \\
&\quad - \cos\big(t^T(u_k - x)/\widehat{b}_k\big)\Big) e^{\frac{t^T\widehat{\Sigma}_k t}{2\widehat{b}_k^2}}\mathrm{d}t\Big| \\
&\le \frac{1}{(2\pi\widehat{b}_k)^l}\int_{[-1,1]^l} \Big| -2\sin\Big(t^T(u_k - x)\frac{1}{2}\big(\frac{1}{b_k} + \frac{1}{\widehat{b}_k}\big)\Big) \\
&\quad \cdot \sin\Big(t^T(u_k - x)\frac{1}{2}\big(\frac{1}{b_k} - \frac{1}{\widehat{b}_k}\big)\Big)\Big| e^{\frac{t^T\widehat{\Sigma}_k t}{2\widehat{b}_k^2}}\mathrm{d}t \\
&\le \frac{1}{(2\pi\widehat{b}_k)^l}\big(\frac{1}{b_k} - \frac{1}{\widehat{b}_k}\big)\int_{[-1,1]^l} e^{\frac{t^T\widehat{\Sigma}_k t}{2\widehat{b}_k^2}}\mathrm{d}t = o\Big(\frac{(\log k)^{\frac{l+1}{2}}}{k^{\frac{1}{2}-\frac{l}{2b}-c}}\Big). \qquad (4.4.64)
\end{aligned}$$

Therefore, by (4.4.62)–(4.4.64), we have

$$|\omega_k(x) - \widehat{\omega}_k(x)| = o\Big(\frac{(\log k)^{\frac{l}{2}+1}}{k^{\frac{1}{2}-\frac{l}{2b}-c}}\Big).$$

From (4.4.31) and (4.4.32) we have

$$\begin{aligned}
\widehat{x}_{k+1} - x_{k+1} &= D_{k+1}\widehat{x}_k - Dx_k + (C_{k+1} - C)y_{k+1} \\
&= D_{k+1}(\widehat{x}_k - x_k) + (D_{k+1} - D)x_k + (C_{k+1} - C)y_{k+1}.
\end{aligned}$$

Since D is stable and $D_k \to D$, there exists a $\lambda \in (0,1)$ such that

$$\|\widehat{x}_{k+1} - x_{k+1}\| \le N_1\lambda^{k+1}\|\widehat{x}_0 - x_0\| + S(\lambda, k), \qquad (4.4.65)$$

where

$$S(\lambda,k) = N_2\sum_{j=1}^{k+1}\lambda^{k-j+1}(\|D_j - D\|\cdot\|x_{j-1}\| + \|C_j - C\|\cdot\|y_j\|),$$

and $N_1 > 0$ and $N_2 > 0$ are constants.

Since $\|D_k - D\| = o(\frac{1}{k^{\frac{1}{2}-c}})$ and $\|C_k - C\| = o(\frac{1}{k^{\frac{1}{2}-c}})$ by Theorem 4.4.1 and $\frac{y_k}{k^{\frac{1}{3}}} \xrightarrow[k\to\infty]{\text{a.s.}} 0$ and $\frac{x_k}{k^{\frac{1}{3}}} \xrightarrow[k\to\infty]{\text{a.s.}} 0$ by Lemma 4.4.3, we have

$$\begin{aligned}
S(\lambda,k) &= \sum_{j=1}^{k+1}\lambda^{k-j+1}o\Big(\frac{1}{j^{1/6-c}}\Big)\\
&= o\Big(\frac{1}{k^{1/6-c}}\Big)\sum_{j=1}^{k+1}\lambda^{k-j+1}\Big(\frac{k}{j}\Big)^{1/6-c} = o\Big(\frac{1}{k^{1/6-c}}\Big)\sum_{j=0}^{k}\lambda^{j}\Big(\frac{k}{k-j+1}\Big)^{1/6-c}\\
&= o\Big(\frac{1}{k^{1/6-c}}\Big)\Big(\sum_{j=0}^{[\frac{k}{2}]}+\sum_{j=[\frac{k}{2}]+1}^{k}\Big)\lambda^{j}\Big(\frac{k}{k-j+1}\Big)^{1/6-c}\\
&= o\Big(\frac{1}{k^{1/6-c}}\Big)\Big(\frac{2}{1-\lambda}+\lambda^{[\frac{k}{2}]+1}k^{\frac{7}{6}-c}\Big) = o\Big(\frac{1}{k^{\frac{1}{6}-c}}\Big).
\end{aligned}$$

The first term at the right-hand side of (4.4.65) decays exponentially, and hence we have

$$\|\widehat{x}_k - x_k\| = o\Big(\frac{1}{k^{1/6-c}}\Big).$$

Since $\|\varphi_k - \widehat{\varphi}_k\|$ and $\|\widehat{x}_k - x_k\|$ are of the same order by (4.4.31) and (4.4.32), we have

$$\|\varphi_k - \widehat{\varphi}_k\| = o\Big(\frac{1}{k^{1/6-c}}\Big). \qquad \square$$

Lemma 4.4.7 *Assume A4.4.1–A4.4.6 hold. The following series converge* a.s.

$$\sum_{k=1}^{\infty}\frac{1}{k}(E\omega_k(x) - \omega_k(x)) < \infty, \tag{4.4.66}$$

$$\sum_{k=1}^{\infty}\frac{1}{k}(E\omega_k(x)f(u_k^0) - \omega_k(x)f(u_k^0)) < \infty, \tag{4.4.67}$$

$$\sum_{k=1}^{\infty}\frac{1}{k}\omega_k(x)(\varphi_k - \widehat{\varphi}_k) < \infty, \tag{4.4.68}$$

$$\sum_{k=1}^{\infty}\frac{1}{k}\omega_k(x)\psi_k < \infty, \quad \sum_{k=1}^{\infty}\frac{1}{k}\omega_k(x)\chi_k < \infty, \quad \sum_{k=1}^{\infty}\frac{1}{k}\omega_k(x)\omega_k(x) < \infty. \tag{4.4.69}$$

Proof. By Theorem 1.4.4, $\{\eta_k\}$ is a zero-mean α-mixing sequence with mixing coefficients decaying exponentially to zero, and $\{u_k^0\}$ is a sequence of iid random vectors with zero-mean. Since u_k^0 and η_k are mutually independent, $\{u_k\}$ is also a zero-mean α-mixing sequence with mixing coefficients decaying exponentially to zero. Since the mixing property is hereditary, $\omega_k(x)$ possesses the same mixing property as u_k does.

Define $z_k^{(7)} \triangleq \frac{1}{k}(E\omega_k(x) - \omega_k(x))$. Then $z_k^{(7)}$ is a zero-mean α-mixing sequence with mixing coefficients decaying exponentially to zero. Noticing $E|\omega_k(x)|^{2+\varepsilon} = O(k^{\frac{l(2+\varepsilon)}{2b}}(\log k)^{\frac{l(2+\varepsilon)}{2}})$ by (4.4.54), by the C_r-inequality for any $\varepsilon > 0$ we have

$$
\begin{aligned}
E\big|z_k^{(7)}\big|^{2+\varepsilon} &= E\Big|\frac{1}{k}(E\omega_k(x) - \omega_k(x))\Big|^{2+\varepsilon} \le \frac{2^{(2+\varepsilon)}}{k^{(2+\varepsilon)}}E\big|\omega_k(x)\big|^{2+\varepsilon} \\
&= O\Big(\frac{1}{k^{(2+\varepsilon)}}k^{\frac{l(2+\varepsilon)}{2b}}(\log k)^{\frac{l(2+\varepsilon)}{2}}\Big),
\end{aligned}
$$

which leads to

$$
\sum_{k=1}^{\infty}\big(E|z_k^{(7)}|^{2+\varepsilon}\big)^{\frac{2}{2+\varepsilon}} = O\Big(\sum_{k=1}^{\infty}\frac{(\log k)^l}{k^{2-\frac{l}{b}}}\Big) < \infty.
$$

Therefore, by Theorem 1.4.2 we have proved (4.4.66), while (4.4.67) can be verified by the similar treatment.

By (4.4.54), there is a positive constant L such that $|\omega_k(x)| \le Lk^{\frac{l}{2b}}(\log k)^{\frac{l}{2}}$, then by (4.4.60) we have

$$
\Big|\sum_{k=1}^{\infty}\frac{1}{k}\omega_k(x)(\varphi_k - \widehat{\varphi}_k)\Big| \le \sum_{k=1}^{\infty} o\Big(\frac{(\log k)^{\frac{l}{2}}}{k^{\frac{7}{6}-\frac{l}{2b}-c}}\Big) < \infty \text{ a.s.}
$$

By Theorem 1.4.4 $\{\psi_k\}$ is a zero-mean α-mixing sequence with mixing coefficients decaying exponentially to zero. Since u_k and ψ_k are mutually independent, $z_k^{(8)} \triangleq \omega_k(x)\psi_k$ is also a zero-mean α-mixing sequence with mixing coefficients decaying exponentially to zero. Noticing $E|\omega_k(x)|^{2+\varepsilon} = O(k^{\frac{l(2+\varepsilon)}{2b}}(\log k)^{\frac{l(2+\varepsilon)}{2}})$ by (4.4.54) and $E\big\|\psi_k\big\|^{2+\varepsilon} < \infty$ by Lemma 4.2.1, for any $\varepsilon > 0$ we have

$$
\begin{aligned}
E\big\|z_k^{(8)}\big\|^{2+\varepsilon} &= E\Big\|\frac{1}{k}\omega_k(x)\psi_k\Big\|^{2+\varepsilon} \\
&= \frac{1}{k^{(2+\varepsilon)}}E|\omega_k(x)|^{2+\varepsilon}E\|\psi_k\|^{2+\varepsilon} = O\Big(\frac{(\log k)^{\frac{l(2+\varepsilon)}{2}}}{k^{(2+\varepsilon)\left(1-\frac{l}{2b}\right)}}\Big),
\end{aligned}
$$

which leads to

$$
\sum_{k=1}^{\infty}\big(E\|z_k^{(8)}\|^{2+\varepsilon}\big)^{\frac{2}{2+\varepsilon}} = O\Big(\sum_{k=1}^{\infty}\frac{(\log k)^l}{k^{2-\frac{l}{b}}}\Big) < \infty.
$$

Therefore, by Theorem 1.4.2 we have proved the first assertion in (4.4.69), while the rest in (4.4.69) can be verified by the similar treatment. □

Lemma 4.4.8 *Assume A4.4.1–A4.4.6 hold. The following series are convergent:*

$$\sum_{k=1}^{\infty}\frac{1}{k}(\omega_k(x)-\widehat{\omega}_k(x))(\widehat{\varphi}_k-\varphi_k)<\infty \text{ a.s.,} \tag{4.4.70}$$

$$\sum_{k=1}^{\infty}\frac{1}{k}(\omega_k(x)-\widehat{\omega}_k(x))\psi_k<\infty \text{ a.s.,} \tag{4.4.71}$$

$$\sum_{k=1}^{\infty}\frac{1}{k}(\omega_k(x)-\widehat{\omega}_k(x))\chi_k<\infty \text{ a.s.,} \tag{4.4.72}$$

$$\sum_{k=1}^{\infty}\frac{1}{k}(\omega_k(x)-\widehat{\omega}_k(x))\omega_k<\infty \text{ a.s.} \tag{4.4.73}$$

Proof. By (4.4.59) and (4.4.60), we have

$$\Big\|\sum_{k=1}^{\infty}\frac{1}{k}(\omega_k(x)-\widehat{\omega}_k(x))(\widehat{\varphi}_k-\varphi_k)\Big\|\le o\Big(\sum_{k=1}^{\infty}\frac{(\log k)^{\frac{l}{2}+1}}{k^{\frac{5}{3}-\frac{l}{2b}-2c}}\Big)<\infty \text{ a.s.}$$

By (4.4.59), we have

$$\Big\|\sum_{k=1}^{\infty}\frac{1}{k}(\omega_k(x)-\widehat{\omega}_k(x))\psi_k\Big\|$$
$$\le o\Big(\sum_{k=1}^{\infty}\frac{(\log k)^{\frac{l}{2}+1}[(\|\psi_k\|-E\|\psi_k\|)+E\|\psi_k\|]}{k^{\frac{3}{2}-\frac{l}{2b}-c}}\Big)<\infty \text{ a.s.}$$

Since $\sum_{k=1}^{\infty}\frac{(\log k)^{\frac{l}{2}+1}E\|\psi_k\|}{k^{\frac{3}{2}-\frac{l}{2b}-c}}<\infty$, for proving $\sum_{k=1}^{\infty}\frac{1}{k}(\omega_k(x)-\widehat{\omega}_k(x))\psi_k<\infty$ it suffices to show $\sum_{k=1}^{\infty}\frac{(\log k)^{\frac{l}{2}+1}(\|\psi_k\|-E\|\psi_k\|)}{k^{\frac{3}{2}-\frac{l}{2b}-c}}<\infty$.

By Theorem 1.4.4, $\Big\{\frac{(\log k)^{\frac{l}{2}+1}(\|\psi_k\|-E\|\psi_k\|)}{k^{\frac{3}{2}-\frac{l}{2b}-c}}\Big\}$ is a zero-mean α-mixing sequence with mixing coefficients decaying exponentially to zero. We have $E\big\|\psi_k\big\|^{2+\varepsilon}<\infty$ by Lemma 4.2.1. Then by the C_r-inequality for any $\varepsilon>0$ it follows that

$$E\Big|\frac{(\log k)^{\frac{l}{2}+1}}{k^{\frac{3}{2}-\frac{l}{2b}-c}}(\|\psi_k\|-E\|\psi_k\|)\Big|^{2+\varepsilon}\le\frac{2^{(2+\varepsilon)}(\log k)^{(\frac{l}{2}+1)(2+\varepsilon)}}{k^{(\frac{3}{2}-\frac{l}{2b}-c)(2+\varepsilon)}}E\|\psi_k\|^{2+\varepsilon}$$
$$=O\Big(\frac{(\log k)^{(\frac{l}{2}+1)(2+\varepsilon)}}{k^{(\frac{3}{2}-\frac{l}{2b}-c)(2+\varepsilon)}}\Big),$$

which leads to

$$\sum_{k=1}^{\infty}\Big(E\Big|\frac{(\log k)^{\frac{l}{2}+1}}{k^{\frac{3}{2}-\frac{l}{2b}-c}}(\|\psi_k\|-E\|\psi_k\|)\Big|^{2+\varepsilon}\Big)^{\frac{2}{2+\varepsilon}}=O\Big(\frac{(\log k)^{l+2}}{k^{(3-\frac{l}{b}-2c)}}\Big)<\infty.$$

Therefore, by Theorem 1.4.2 we have proved (4.4.71), while (4.4.72) and (4.4.73) can be proved in a similar manner. □

Theorem 4.4.2 *Assume A4.4.1–A4.4.6 hold. Then $\tau_k(x)$ defined by (4.4.39) and (4.4.40) and $\beta_k(x)$ defined by (4.4.41) and (4.4.42) are convergent:*

$$\tau_k(x) \xrightarrow[k\to\infty]{} p(x) \text{ a.s.}, \tag{4.4.74}$$

$$\beta_k(x) \xrightarrow[k\to\infty]{} p(x)f(x) \text{ a.s.} \tag{4.4.75}$$

As a consequence, $f_k(x)$ defined by (4.4.43) is strongly consistent

$$f_k(x) \xrightarrow[k\to\infty]{} f(x) \text{ a.s.} \tag{4.4.76}$$

Proof. The algorithm (4.4.39) can be rewritten as

$$\begin{aligned}\tau_{k+1}(x) =& [\tau_k(x) - \frac{1}{k}(\tau_k(x) - p(x)) - \frac{1}{k}\bar{e}_{k+1}(x)] \\ & \cdot I_{[|\tau_k(x) - \frac{1}{k}(\tau_k(x)-p(x)) - \frac{1}{k}\bar{e}_{k+1}(x)| \le M_{\Delta_k^{\tau}(x)}]},\end{aligned}$$

where

$$\begin{aligned}\bar{e}_{k+1}(x) &= p(x) - \widehat{\omega}_k(x) \\ &= \big(p(x) - E\omega_k(x)\big) + \big(E\omega_k(x) - \omega_k(x)\big) + \big(\omega_k(x) - \widehat{\omega}_k(x)\big).\end{aligned} \tag{4.4.77}$$

Since $p(x)$ is the single root of the linear function $-(y-p(x))$, by Theorem 2.3.1, for convergence of $\tau_k(x)$ it suffices to show for any convergent subsequence $\tau_{n_k}(x)$,

$$\lim_{T\to 0} \limsup_{k\to\infty} \frac{1}{T} \Big| \sum_{j=n_k}^{m(n_k,T_k)} \frac{1}{j} \bar{e}_j(x) \Big| = 0 \ \ \forall T_k \in [0,T] \tag{4.4.78}$$

where $m(k,T) \triangleq \max\left\{m : \sum_{j=k}^{m} \frac{1}{j} \le T\right\}$. By the first assertion in (4.4.53) together with (4.4.59) and (4.4.66), it follows that (4.4.78) holds for $\{\bar{e}_{k+1}(x)\}$.

The proof of (4.4.75) can similarly be carried out, if we rewrite the algorithm (4.4.41) as follows:

$$\begin{aligned}\beta_{k+1}(x) =& [\beta_k(x) - \frac{1}{k}(\beta_k(x) - p(x)f(x)) - \frac{1}{k}\tilde{e}_{k+1}(x)] \\ & \cdot I_{[\|\beta_k(x) - \frac{1}{k}(\beta_k(x)-p(x)f(x)) - \frac{1}{k}\tilde{e}_{k+1}(x)\| \le M_{\Delta_k^{\beta}(x)}]},\end{aligned}$$

where

$$\begin{aligned}\tilde{e}_{k+1}(x) =& p(x)f(x) - \widehat{\omega}_k(x)\widehat{\varphi}_k = \Big(p(x)f(x) - E\omega_k(x)f(u_k^0)\Big) \\ & + \Big(E\omega_k(x)f(u_k^0) - \omega_k(x)f(u_k^0)\Big) - (\widehat{\omega}_k(x) - \omega_k(x))(\widehat{\varphi}_k - \varphi_k) \\ & - (\widehat{\omega}_k(x) - \omega_k(x))(f(u_k^0) + \psi_k + \chi_k + \omega_k) \\ & - \omega_k(x)(\widehat{\varphi}_k - \varphi_k) - \omega_k(x)(\psi_k + \chi_k + \omega_k).\end{aligned} \tag{4.4.79}$$

Noticing the second part of (4.4.53) and also (4.4.67), (4.4.70), (4.4.59), (4.4.71), (4.4.72), (4.4.73), (4.4.68), and (4.4.69) we see that each term at the right-hand side of (4.4.79) satisfies the convergence condition (4.4.78) with $\overline{e}_j(x)$ replaced by it.

Therefore, the estimate (4.4.43) is strongly consistent. □

Numerical Example

Let the nonlinear function be

$$f(x) = \arctan(x).$$

Let the linear subsystem be

$$y_k^0 + A_1 y_{k-1}^0 + A_2 y_{k-2}^0 + A_3 y_{k-3}^0 = B_1 v_{k-1}^0 + B_2 v_{k-2}^0 + B_3 v_{k-3}^0 + \xi_k,$$

where

$$A_1 = \begin{bmatrix} 0 & 0.5 \\ 1 & 0 \end{bmatrix}, A_2 = \begin{bmatrix} 1 & 0 \\ 0 & 0.6 \end{bmatrix},$$
$$A_3 = \begin{bmatrix} 0 & 0.4 \\ -0.6 & 0 \end{bmatrix}, B_1 = \begin{bmatrix} 0.6 \\ -0.8 \end{bmatrix},$$
$$B_2 = \begin{bmatrix} 0.3 \\ -0.5 \end{bmatrix}, B_3 = \begin{bmatrix} -0.2 \\ 0.9 \end{bmatrix}.$$

Let the input signal $\{u_k^0\}$ be a sequence of random variables uniformly distributed over the domain $[-U, U]$, where $U \approx 2.1727$. All the noises $\{\zeta_k\}$, $\{\omega_k\}$, $\{\xi_k\}$, and $\{\varsigma_k\}$ are mutually independent Gaussian random variables, ζ_k and $\omega_k \in \mathcal{N}(0, 0.4^2)$, ξ_k and $\varsigma_k \in \mathcal{N}(0, 0.4^2 I_2)$. The measurement noises η_k and ε_k are ARMA processes:

$$\eta_k + 0.3\eta_{k-1} = \zeta_k + 0.5\zeta_{k-1},$$
$$\varepsilon_k + \begin{bmatrix} 0.7 & -0.3 \\ 0.2 & 0.1 \end{bmatrix} \varepsilon_{k-1}$$
$$= \varsigma_k + \begin{bmatrix} 0.25 & 0.3 \\ -0.2 & 0.2 \end{bmatrix} \varsigma_{k-1}.$$

The parameters used in the algorithms are as follows: $b = 6$ and $M_k = 2^k + 10$.

A straightforward calculation shows that $\Upsilon \triangleq E(f(u_k^0) u_k^{0T}) = 1$.

The estimates for A_1, A_2, A_3, B_1, B_2, and B_3 are, respectively, given in Fig. 4.4.2–Fig. 4.4.7 with the convergence rate faster than that for the estimate of $f(\cdot)$ as shown in Table 4.4.1. Though the estimates for $f(\cdot)$ in the interval $[-1.5, 1.5]$ at $k = 5000$ are acceptably accurate as presented in Figure 4.4.8, Table 4.4.1 shows that the errors at some points are still not negligible if k is not large enough.

In all figures, the solid lines and the dash lines denote the true value and the corresponding estimates, respectively.

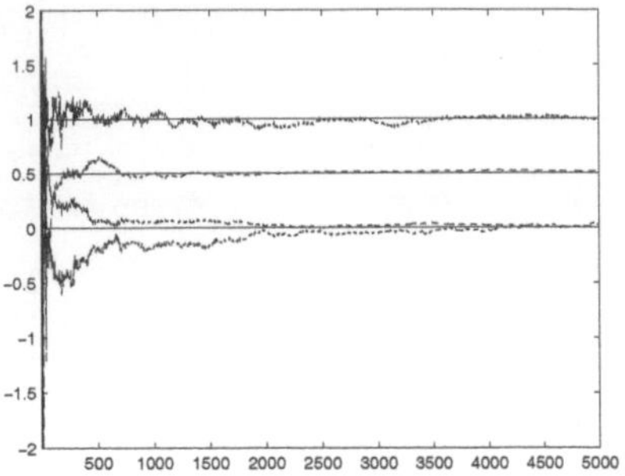

Figure 4.4.2: Estimate for A_1

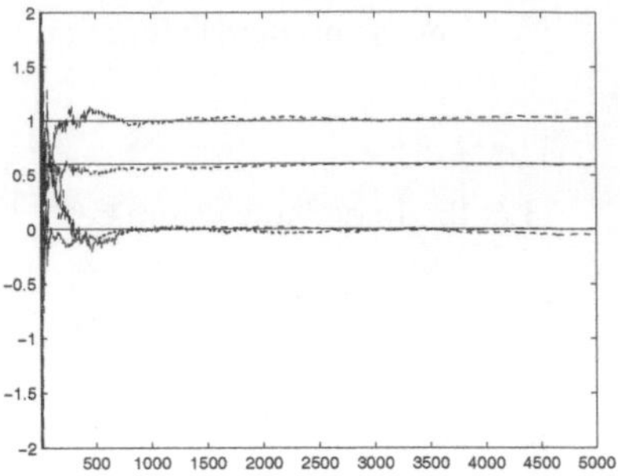

Figure 4.4.3: Estimates for A_2

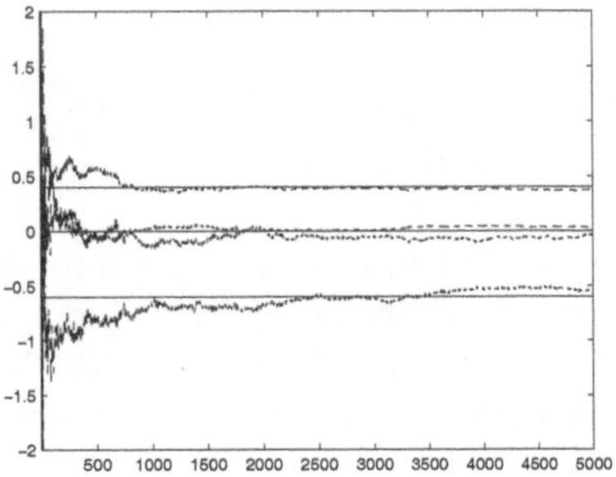

Figure 4.4.4: Estimate for A_3

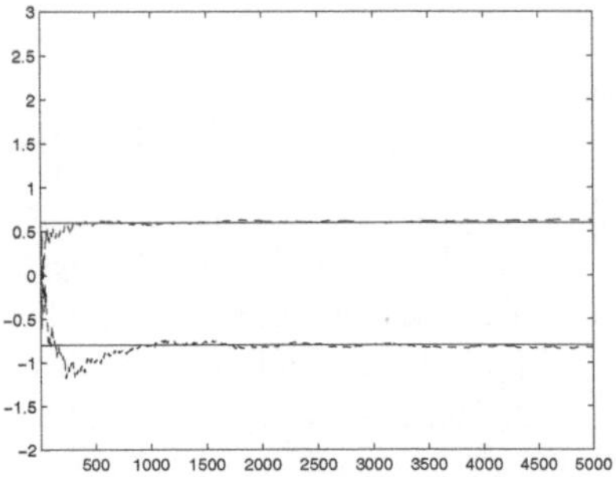

Figure 4.4.5: Estimates for B_1

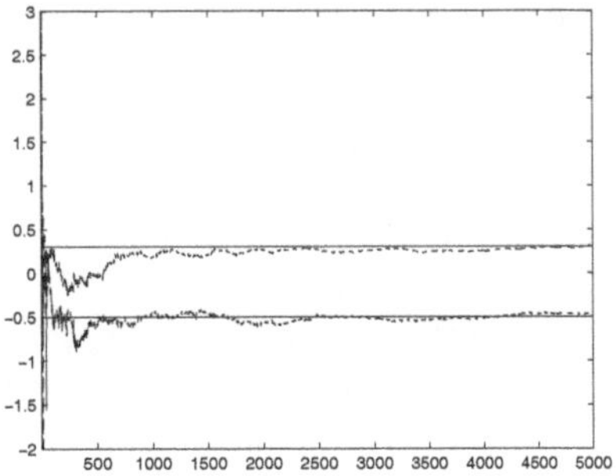

Figure 4.4.6: Estimate for B_2

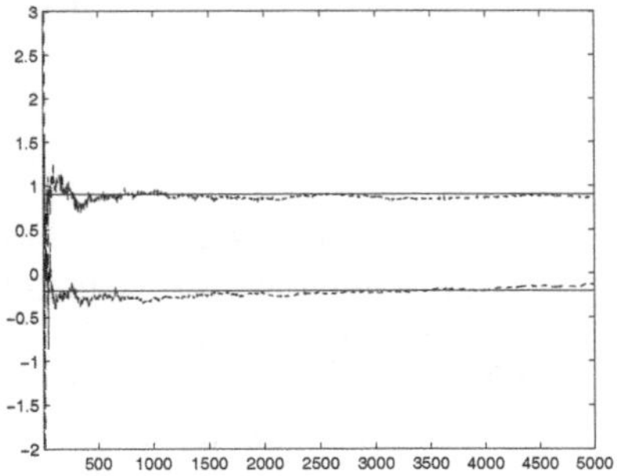

Figure 4.4.7: Estimates for B_3

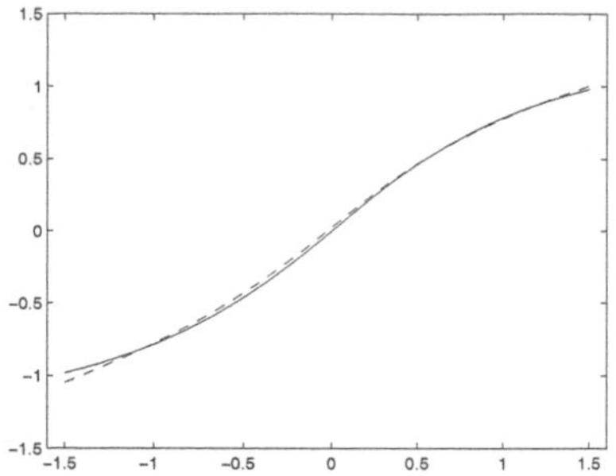

Figure 4.4.8: Estimate for $f(x) = \arctan(x)$

Table 4.4.1: Estimates and estimation errors for $f(x)$ at fixed points

		$k=1000$		$k=2000$		$k=3000$		$k=4000$		$k=5000$	
x	$f(x)$	Est.	Errors	Est.	Errors	Est.	Errors	Est.	Errors	Est.	Errors
-1.5	-0.983	-0.929	0.054	-0.951	0.032	-1.005	-0.022	-1.011	-0.028	-1.049	-0.066
-1	-0.785	-0.893	-0.107	-0.700	0.085	-0.711	0.074	-0.725	0.061	-0.779	0.007
-0.5	-0.464	-0.665	-0.201	-0.448	0.015	-0.403	0.060	-0.401	0.063	-0.432	0.032
0	0.000	-0.170	-0.170	-0.117	-0.117	-0.039	-0.039	0.006	0.006	0.025	0.025
0.5	0.464	0.429	-0.035	0.312	-0.152	0.362	-0.102	0.427	-0.036	0.466	0.002
1	0.785	0.867	0.082	0.727	-0.059	0.725	-0.060	0.765	-0.020	0.780	-0.006
1.5	0.983	1.041	0.058	1.042	0.059	1.025	0.043	1.024	0.042	1.005	0.022

4.5 Recursive Identification of EIV Wiener Systems

In Section 4.2 we have discussed the recursive identification for the Weiner systems. We now consider the recursive identification for the SISO EIV Wiener system presented by Fig. 4.5.1.

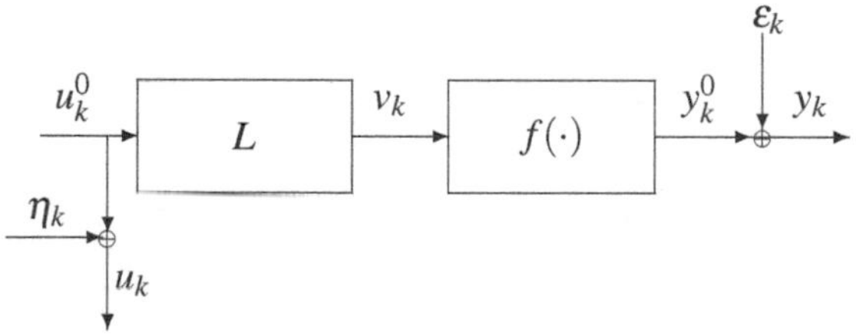

Figure 4.5.1: EIV Wiener system

Mathematically, the system is described as follows:

$$C(z)v_k = D(z)u_k^0, \tag{4.5.1}$$

$$y_k^0 = f(v_k), \tag{4.5.2}$$

$$C(z) = 1 + c_1 z + \cdots + c_p z^p, \tag{4.5.3}$$

$$D(z) = z + d_2 z^2 + \cdots + d_q z^q, \tag{4.5.4}$$

where $C(z)$ and $D(z)$ are polynomials with unknown coefficients but known orders p, q, respectively. The intermediate signal v_k is not directly observed. The noise-free input and output are denoted by u_k^0 and y_k^0, respectively, but they are observed with additive noises η_k and ε_k:

$$u_k = u_k^0 + \eta_k, \quad y_k = y_k^0 + \varepsilon_k. \tag{4.5.5}$$

The target is to recursively estimate the unknown parameters $\{c_1, \cdots, c_p, d_2, \cdots, d_q\}$ of the polynomials $C(z)$ and $D(z)$ and the value of $f(x)$ at any given x in its domain on the basis of the observed data $\{u_k, y_k\}$.

Recursive algorithms for estimating $\{\mathbf{c_1}, \cdots, \mathbf{c_p}, \mathbf{d_2}, \cdots, \mathbf{d_q}\}$

We first give the conditions for identifying the linear subsystem.

A4.5.1 *The noise-free input $\{u_k^0\}$ is a sequence of iid Gaussian random variables: $u_k^0 \in \mathcal{N}(0, \vartheta^2)$ with unknown $\vartheta > 0$ and is independent of $\{\eta_k\}$ and $\{\varepsilon_k\}$.*

A4.5.2 *$C(z)$ and $D(z)$ are coprime and $C(z)$ is stable: $C(z) \neq 0 \ \forall |z| \leq 1$.*

Since $C(z)$ is stable, from A4.5.2 we have

$$H(z) \triangleq \frac{D(z)}{C(z)} = \sum_{i=1}^{\infty} h_i z^i, \tag{4.5.6}$$

where $|h_i| = O(e^{-ri}), r > 0, i \geq 2$, and $h_1 = 1$.

A4.5.3 *The measurement noises η_k and ε_k both are ARMA processes:*

$$P(z)\eta_k = Q(z)\zeta_k, \tag{4.5.7}$$
$$F(z)\varepsilon_k = G(z)\varsigma_k, \tag{4.5.8}$$

where

$$P(z) = 1 + p_1 z + p_2 z^2 + \cdots + p_{n_p} z^{n_p}, \tag{4.5.9}$$
$$Q(z) = 1 + q_1 z + q_2 z^2 + \cdots + q_{n_q} z^{n_q}, \tag{4.5.10}$$
$$F(z) = 1 + f_1 z + f_2 z^2 + \cdots + f_{n_f} z^{n_f}, \tag{4.5.11}$$
$$G(z) = 1 + g_1 z + g_2 z^2 + \cdots + g_{n_g} z^{n_g}. \tag{4.5.12}$$

The polynomial $P(z)$ has no common roots with both $Q(z)$ and $Q(z^{-1})z^{n_q}$, and $P(z)$ and $F(z)$ are stable. The driven noises $\{\zeta_k\}$ and $\{\varsigma_k\}$ are mutually independent, and each of them is a sequence of iid zero-mean random variables with probability density. Moreover, $E(|\zeta_k|^{\Delta+2}) < \infty$ and $E(|\varsigma_k|^{\Delta}) < \infty$ for some $\Delta > 2$.

A4.5.4 *$f(\cdot)$ is a measurable function and has both the left limit $f(x-)$ and the right limit $f(x+)$ at any x. The growth rate of $f(x)$ as $|x| \to \infty$ is not faster than a polynomial. Further, at least one of the constants τ and ρ is nonzero, where*

$$\tau \triangleq \frac{1}{\sqrt{2\pi}\sigma^3\vartheta} \int_{\mathbb{R}} x f(x) e^{-\frac{x^2}{2\sigma^2\vartheta^2}} \mathrm{d}x, \tag{4.5.13}$$

$$\rho \triangleq \frac{1}{\sqrt{2\pi}\sigma^5\vartheta} \int_{\mathbb{R}} (x^2 - \sigma^2\vartheta^2) f(x) e^{-\frac{x^2}{2\sigma^2\vartheta^2}} \mathrm{d}x, \tag{4.5.14}$$

where $\sigma^2 \triangleq \sum_{i=1}^{\infty} h_i^2$.

Remark 4.5.1 *The growth rate restriction required in A4.5.4 implies that there are constants $\alpha > 0$ and $\beta \geq 1$ such that*

$$|f(x)| \leq \alpha(1 + |x|^{\beta}) \ \ \forall x \in \mathbb{R}. \tag{4.5.15}$$

Therefore, under A4.5.4 the finiteness of integrals (4.5.13) and (4.5.14) is guaranteed.

We now show that for polynomial functions either τ or ρ or even both are nonzero indeed.

Let $f(\cdot)$ be a monic polynomial with arbitrary coefficients.

If $f(x) = x^2 + ax + b$, then $\tau = a\vartheta^2$ and $\rho = 2\vartheta^4 > 0$. So, A4.5.4 is satisfied, and $f(\cdot)$ is even when $a = 0$.

If $f(x) = x^3 + ax^2 + bx + c$, then $\tau = (3\sigma^2\vartheta^2 + b)\vartheta^2$ and $\rho = 2a\vartheta^4$.

Both τ and ρ are zero only in the special case where $3\sigma^2\vartheta^2 + b = 0$ and $a = 0$. It is worth noting that the equality $3\sigma^2\vartheta^2 + b = 0$ can be avoided by slightly changing the input variance ϑ^2.

If $f(x) = x^4 + ax^3 + bx^2 + cx + d$, then $\tau = (3\sigma^2\vartheta^2 a + c)\vartheta^2$ and $\rho = 2\vartheta^4(6\sigma^2\vartheta^2 - b)$. Similarly, both τ and ρ are zero only in the special case where $3\sigma^2\vartheta^2 a = -c$ and $6\sigma^2\vartheta^2 = b$, and this can also be avoided by slightly changing the input variance ϑ^2.

The higher order polynomials can be discussed in a similar manner.

From the above discussion we see that A4.5.4 is not restrictive.

Assuming $u_k^0 = 0 \ \forall k < 0$, we have

$$v_k = \sum_{i=1}^{k} h_i u_{k-i}^0, \tag{4.5.16}$$

and hence $v_k \in \mathcal{N}(0, \sigma_k^2)$ where $\sigma_k^2 \triangleq \sum_{i=1}^{k} h_i^2 \xrightarrow[k\to\infty]{} \sigma^2$.

The following lemma is important to identify the linear subsystem.

Lemma 4.5.1 *Assume A4.5.1–A4.5.4 hold. Then the following limits take place:*

$$Ey_k u_{k-i} \xrightarrow[k\to\infty]{} \tau h_i \ \ \forall i \geq 1, \tag{4.5.17}$$

$$E(y_k - Ey_k)u_{k-1}^2 \xrightarrow[k\to\infty]{} \rho, \tag{4.5.18}$$

$$E(y_k - Ey_k)u_{k-1}u_{k-i} \xrightarrow[k\to\infty]{} \rho h_i \ \ \forall i \geq 2. \tag{4.5.19}$$

Proof. Similar to Lemma 4.3.2 we have

$$E(f(v_k)u_{k-i}^0) = \tau_k h_i \xrightarrow[k\to\infty]{} \tau h_i,$$

where $\tau_k \triangleq \frac{1}{\sigma_k^2}E\big(f(v_k)v_k\big)$.

Since u_k^0 is independent of η_k and ε_k, we have

$$Ey_k u_{k-i} = E(y_k^0+\varepsilon_k)(u_{k-i}^0+\eta_{k-i}) = E(f(v_k)u_{k-i}^0) \xrightarrow[k\to\infty]{} \tau h_i.$$

By Lemma 4.3.1 we obtain

$$E\left(f(v_k)((u_{k-1}^0)^2-\vartheta^2)\right) = \frac{1}{\sigma_k^4}E(f(v_k)(v_k)^2) - \frac{\vartheta^2}{\sigma_k^2}Ef(v_k) \triangleq \rho_k,$$

and

$$E(f(v_k)u_{k-1}^0 u_{k-j}^0) = \Big(\frac{1}{\sigma_k^4}E(f(v_k)(v_k)^2) - \frac{\vartheta^2}{\sigma_k^2}Ef(v_k)\Big)h_j = \rho_k h_j,\ j\geq 2.$$

It is noticed that $\sigma_k^2 \xrightarrow[k\to\infty]{} \sigma^2$, by (4.5.14) we see that

$$\begin{aligned}
E(y_k - Ey_k)u_{k-1}^2 &= E(y_k^0+\varepsilon_k - Ey_k^0)((u_{k-1}^0)^2+\eta_{k-1}^2+2u_{k-1}^0\eta_{k-1})\\
&= E(y_k^0-Ey_k^0)((u_{k-1}^0)^2+\eta_{k-1}^2+2u_{k-1}^0\eta_{k-1}) = E[(y_k^0-Ey_k^0)(u_{k-1}^0)^2]\\
&= Ey_k^0((u_{k-1}^0)^2-\vartheta^2) = \rho_k \xrightarrow[k\to\infty]{} \rho
\end{aligned}$$

and

$$\begin{aligned}
E(y_k - Ey_k)u_{k-1}u_{k-i} &= E(y_k^0+\varepsilon_k - Ey_k^0)(u_{k-1}^0+\eta_{k-1})(u_{k-i}^0+\eta_{k-i})\\
&= E(y_k^0-Ey_k^0)(u_{k-1}^0u_{k-i}^0+u_{k-1}^0\eta_{k-i}+\eta_{k-1}u_{k-i}^0+\eta_{k-1}\eta_{k-i})\\
&= E(y_k^0-Ey_k^0)(u_{k-1}^0u_{k-i}^0) = Ey_k^0u_{k-1}^0u_{k-i}^0 = \rho_k h_i \xrightarrow[k\to\infty]{} \rho h_i,\ i\geq 2.
\end{aligned}$$

The proof of the lemma is completed. □

For estimating the coefficients of the linear subsystem we first estimate the impulse responses $\{h_i\}$ and then obtain the estimates for the coefficients $\{c_1,\cdots,c_p,d_2,\cdots,d_q\}$ by using the linear algebraic equations connecting them with $\{h_i\}$.

Since Ey_k in (4.5.18) and (4.5.19) is unknown, we first use SAAWET to recursively estimate Ey_k:

$$\lambda_k = \Big[\lambda_{k-1}-\frac{1}{k}(\lambda_{k-1}-y_k)\Big]\cdot I_{\left[\left|\lambda_{k-1}-\frac{1}{k}\left(\lambda_{k-1}-y_k\right)\right|\leq M_{\delta_k^{(\lambda)}}\right]}, \tag{4.5.20}$$

$$\delta_k^{(\lambda)} = \sum_{j=1}^{k-1} I_{\left[\left|\lambda_{j-1}-\frac{1}{j}\left(\lambda_{j-1}-y_j\right)\right|> M_{\delta_j^{(\lambda)}}\right]}, \tag{4.5.21}$$

where $\{M_k\}$ is an arbitrarily chosen sequence of positive real numbers increasingly diverging to infinity, λ_0 is an arbitrary initial value, and $I_{[A]}$ denotes the indicator function of a set A.

Before giving the estimates for h_i, the constants τ and ρ are to be estimated on the basis of (4.5.17) and (4.5.18), respectively. Their estimates are given as follows:

$$\theta_k^{(1,\tau)} = \Big[\theta_{k-1}^{(1,\tau)} - \frac{1}{k}(\theta_{k-1}^{(1,\tau)} - y_k u_{k-1})\Big] \cdot I_{\Big[\big|\theta_{k-1}^{(1,\tau)} - \frac{1}{k}\big(\theta_{k-1}^{(1,\tau)} - y_k u_{k-1}\big)\big| \le M_{\delta_k^{(1,\tau)}}\Big]}, \tag{4.5.22}$$

$$\delta_k^{(1,\tau)} = \sum_{j=1}^{k-1} I_{\Big[\big|\theta_{j-1}^{(1,\tau)} - \frac{1}{j}\big(\theta_{j-1}^{(1,\tau)} - y_j u_{j-1}\big)\big| > M_{\delta_j^{(1,\tau)}}\Big]}, \tag{4.5.23}$$

$$\theta_k^{(1,\rho)} = \Big[\theta_{k-1}^{(1,\rho)} - \frac{1}{k}(\theta_{k-1}^{(1,\rho)} - (y_k - \lambda_k) u_{k-1}^2)\Big] \cdot I_{\Big[\big|\theta_{k-1}^{(1,\rho)} - \frac{1}{k}\big(\theta_{k-1}^{(1,\rho)} - (y_k - \lambda_k) u_{k-1}^2\big)\big| \le M_{\delta_k^{(1,\rho)}}\Big]}, \tag{4.5.24}$$

$$\delta_k^{(1,\rho)} = \sum_{j=1}^{k-1} I_{\Big[\big|\theta_{j-1}^{(1,\rho)} - \frac{1}{j}\big(\theta_{j-1}^{(1,\rho)} - (y_j - \lambda_j) u_{j-1}^2\big)\big| > M_{\delta_j^{(1,\rho)}}\Big]}. \tag{4.5.25}$$

If $|\theta_k^{(1,\tau)}| \ge |\theta_k^{(1,\rho)}|$, then the following algorithm based on (4.5.17) is used to estimate τh_i:

$$\theta_k^{(i,\tau)} = \Big[\theta_{k-1}^{(i,\tau)} - \frac{1}{k}(\theta_{k-1}^{(i,\tau)} - y_k u_{k-i})\Big] \cdot I_{\Big[\big|\theta_{k-1}^{(i,\tau)} - \frac{1}{k}\big(\theta_{k-1}^{(i,\tau)} - y_k u_{k-i}\big)\big| \le M_{\delta_k^{(i,\tau)}}\Big]}, \tag{4.5.26}$$

$$\delta_k^{(i,\tau)} = \sum_{j=1}^{k-1} I_{\Big[\big|\theta_{j-1}^{(i,\tau)} - \frac{1}{j}\big(\theta_{j-1}^{(i,\tau)} - y_j u_{j-i}\big)\big| > M_{\delta_j^{(i,\tau)}}\Big]}, \quad i \ge 2. \tag{4.5.27}$$

Here $\theta_{k-1}^{(i,\tau)}$ is obtained from the previous step of the recursion if $|\theta_{k-1}^{(1,\tau)}| \ge |\theta_{k-1}^{(1,\rho)}|$; otherwise, i.e., if $|\theta_{k-1}^{(1,\tau)}| < |\theta_{k-1}^{(1,\rho)}|$, then $\theta_{k-1}^{(i,\tau)}$ has not been computed in accordance with (4.5.26)–(4.5.27). In this case $\theta_{k-1}^{(i,\tau)}$ in (4.5.26) is set to equal $\theta_{k-1}^{(1,\tau)} h_{i,k-1}$. After having the estimates for τ and τh_i, the estimates for the impulse responses $h_i, i \ge 2$ at time k are given by

$$h_{i,k} \triangleq \begin{cases} \dfrac{\theta_k^{(i,\tau)}}{\theta_k^{(1,\tau)}}, & \text{if } \theta_k^{(1,\tau)} \ne 0, \\ 0, & \text{if } \theta_k^{(1,\tau)} = 0. \end{cases} \tag{4.5.28}$$

Conversely, if $|\theta_k^{(1,\rho)}| > |\theta_k^{(1,\tau)}|$, then based on (4.5.19), ρh_i is estimated by the following algorithm:

$$\theta_k^{(i,\rho)} = \Big[\theta_{k-1}^{(i,\rho)} - \frac{1}{k}\big(\theta_{k-1}^{(i,\rho)} - (y_k - \lambda_k)u_{k-1}u_{k-i}\big)\Big] \cdot I_{\left[\left|\theta_{k-1}^{(i,\rho)} - \frac{1}{k}\left(\theta_{k-1}^{(i,\rho)} - (y_k-\lambda_k)u_{k-1}u_{k-i}\right)\right| \le M_{\delta_k^{(i,\rho)}}\right]}, \tag{4.5.29}$$

$$\delta_k^{(i,\rho)} = \sum_{j=1}^{k-1} I_{\left[\left|\theta_{j-1}^{(i,\rho)} - \frac{1}{j}\left(\theta_{j-1}^{(i,\rho)} - (y_j-\lambda_j)u_{j-1}u_{j-i}\right)\right| > M_{\delta_j^{(i,\rho)}}\right]}, \quad i \ge 2. \tag{4.5.30}$$

Similar to the previous case, $\theta_{k-1}^{(i,\rho)}$ is derived from the previous step of the recursion if $|\theta_{k-1}^{(1,\tau)}| < |\theta_{k-1}^{(1,\rho)}|$; otherwise, i.e., if $|\theta_{k-1}^{(1,\tau)}| \ge |\theta_{k-1}^{(1,\rho)}|$, then $\theta_{k-1}^{(i,\rho)}$ has not been computed in accordance with (4.5.29)–(4.5.30). In this case $\theta_{k-1}^{(i,\rho)}$ in (4.5.29) is set to equal $\theta_{k-1}^{(1,\rho)} h_{i,k-1}$. After having the estimates for ρ and ρh_i, the estimates for the impulse responses $h_i, i \ge 2$ at time k are given by

$$h_{i,k} \triangleq \begin{cases} \frac{\theta_k^{(i,\rho)}}{\theta_k^{(1,\rho)}}, & \text{if } \theta_k^{(1,\rho)} \neq 0, \\ 0, & \text{if } \theta_k^{(1,\rho)} = 0. \end{cases} \tag{4.5.31}$$

It is important to note that after establishing the strong consistency of $\theta_k^{(1,\tau)}$ and $\theta_k^{(1,\rho)}$, switching between the algorithms (4.5.29)–(4.5.31) and (4.5.26)–(4.5.28) ceases in a finite number of steps, because by A4.5.4 at least one of τ and ρ is nonzero and hence either $\theta_k^{(1,\tau)} \ge \theta_k^{(1,\rho)}$ or $\theta_k^{(1,\tau)} < \theta_k^{(1,\rho)}$ takes place for all sufficiently large k.

Once the estimates $h_{i,k}$ for the impulse response h_i are obtained, the parameters $\{c_1, \cdots, c_p,\ d_2, \cdots, d_q\}$ of the linear subsystem can be derived by the convolution relationship between the impulse responses $\{h_i\}$ and the parameters $\{c_1, \cdots, c_p, d_2, \cdots, d_q\}$.

Carrying out the treatment similar to that done in Section 3.3, we have the following linear algebraic equation:

$$L[c_1, c_2, \cdots, c_p]^T = -[h_{q+1}, h_{q+2}, \cdots, h_{q+p}]^T, \tag{4.5.32}$$

where

$$L \triangleq \begin{bmatrix} h_q & h_{q-1} & \cdots & h_{q-p+1} \\ h_{q+1} & h_q & \cdots & h_{q-p+2} \\ \vdots & \vdots & \ddots & \vdots \\ h_{q+p-1} & h_{q+p-2} & \cdots & h_q \end{bmatrix}, \tag{4.5.33}$$

where $h_i \triangleq 0$ for $i \le 0$.

Noticing that by Theorem 3.3.2 the matrix L is nonsingular under A4.5.2 and that $h_{i,k} \xrightarrow[k\to\infty]{} h_i$ a.s. as to be shown by Theorem 4.5.1, we see that L_k is nonsingular when k is sufficiently large, where

$$L_k \triangleq \begin{bmatrix} h_{q,k} & h_{q-1,k} & \cdots & h_{q-p+1,k} \\ h_{q+1,k} & h_{q,k} & \cdots & h_{q-p+2,k} \\ \vdots & \vdots & \ddots & \vdots \\ h_{q+p-1,k} & h_{q+p-2,k} & \cdots & h_{q,k} \end{bmatrix} \tag{4.5.34}$$

serving as the kth estimate for L with $h_{i,k} = 0$ for $i \leq 0$.

The estimates for $\{c_1, \cdots, c_p, d_2, \cdots, d_q\}$ are naturally defined as:

$$[c_{1,k}, c_{2,k}, \cdots, c_{p,k}]^T \triangleq -L_k^{-1}[h_{q+1,k}, h_{q+2,k}, \cdots, h_{q+p,k}]^T, \tag{4.5.35}$$

$$d_{i,k} \triangleq h_{i,k} + \sum_{j=1}^{(i-1)\wedge p} c_{j,k} h_{i-j,k}, \quad i = 2, \cdots, q. \tag{4.5.36}$$

Estimation of f(·)

We now recursively estimate $f(x)$, where x is an arbitrary point on the real axis. For this the useful information is the sequences $\{v_k\}$, the inputs to the nonlinear function, and the corresponding outputs $\{f(v_k)\}$. Since v_k is not directly available, the conventional kernel estimation method can not be used. We apply the SAAWET algorithm incorporated with the deconvolution kernel functions to estimate $f(x)$.

Instead of directly estimating v_k let us estimate the signal ψ_k defined below, which, in fact, is a noisy v_k.

Define

$$\psi_k \triangleq C^{-1}(z)D(z)u_k, \tag{4.5.37}$$

$$e_k \triangleq C^{-1}(z)D(z)\eta_k = (C(z)P(z))^{-1}(D(z)Q(z))\zeta_k. \tag{4.5.38}$$

According to (4.5.1), (4.5.5), and (4.5.7), we have

$$\psi_k = C^{-1}(z)D(z)u_k^0 + [C(z)P(z)]^{-1}D(z)Q(z)\zeta_k = v_k + e_k. \tag{4.5.39}$$

Define

$$C \triangleq \begin{bmatrix} -c_1 & 1 & \cdots & 0 \\ \vdots & & \ddots & \vdots \\ \vdots & & & 1 \\ -c_s & 0 & \cdots & 0 \end{bmatrix}, D \triangleq \begin{bmatrix} 1 \\ d_2 \\ \vdots \\ d_s \end{bmatrix}, \text{ and } H \triangleq \begin{bmatrix} 1 \\ 0 \\ \vdots \\ 0 \end{bmatrix},$$

where $s \triangleq \max(p,q)$, $c_i \triangleq 0$ for $i > p$, and $d_j \triangleq 0$ for $j > q$.

Then, the equation (4.5.37) connecting ψ_k and u_k can be written as

$$\psi_k + c_1\psi_{k-1} + \cdots + c_p\psi_{k-p} = u_{k-1} + d_2 u_{k-2} + \cdots + d_q u_{k-q},$$

or in the state space form

$$\begin{cases} x_{k+1} = & Cx_k + Du_k \\ \psi_{k+1} = & H^T x_{k+1}, \end{cases} \tag{4.5.40}$$

where C is an $s \times s$ matrix, D is an $s \times 1$ vector, and H is an $s \times 1$ vector.

Replacing c_i and d_j in C and D with $c_{i,k}$ and $d_{j,k}$ given by (4.5.35) and (4.5.36), respectively, $i = 1, \cdots, s,\ j = 1, \cdots, s$, we obtain the estimates C_k and D_k for C and D at time k, and, hence, the estimate $\widehat{\psi}_k$ for ψ_k is given as follows:

$$\widehat{x}_{k+1} = C_{k+1}\widehat{x}_k + D_{k+1}u_k, \quad \widehat{\psi}_{k+1} = H^T \widehat{x}_{k+1} \tag{4.5.41}$$

with an arbitrary initial value $\widehat{x}_0$.

In order to eliminate the influence of e_k involved in ψ_k, the deconvolution kernel functions are applied, but for this the additional assumptions are needed.

A4.5.5 *The variance ϑ^2 of the noise-free input u_k^0 is known.*

A4.5.6 *The driven noise $\{\zeta_k\}$ in (4.5.7) is a sequence of zero-mean iid Gaussian random variables.*

Let us introduce the Sinc kernel function

$$K(x) = \frac{\sin(x)}{\pi x}. \tag{4.5.42}$$

Its Fourier transformation is equal to

$$\Phi_K(t) \triangleq \int_{\mathbb{R}} e^{\imath tx} K(x) \mathrm{d}x = I_{[|t| \le 1]},$$

where $\imath$ stands for the imaginary unit $\imath^2 = -1$ and $I_{[A]}$ represents the indicator function of a set A.

Under A4.5.6 $\{e_k\}$ is also a sequence of zero-mean Gaussian random variables, and the characteristic function $\Phi_{e_k}(t)$ of e_k is

$$\Phi_{e_k}(t) \triangleq \int_{\mathbb{R}} e^{\imath tx} \frac{1}{\sqrt{2\pi}\sigma_k(e)} e^{-\frac{x^2}{2\sigma_k^2(e)}} \mathrm{d}t = e^{-\frac{\sigma_k^2(e)t^2}{2}},$$

where $\sigma_k^2(e) \triangleq E(e_k)^2$ is the variance of e_k. Denote by $\sigma^2(e)$ the limit of $\sigma_k^2(e)$. It is clear that $|\sigma^2(e) - \sigma_k^2(e)| = O(e^{-r_e k})$ for some $r_e > 0$.

We now introduce the deconvolution kernel $\omega_k(x)$, but for this we first define

$$K_k(x) \triangleq \frac{1}{2\pi} \int_{\mathbb{R}} e^{-\imath tx} \frac{\Phi_K(t)}{\Phi_{e_k}(t/b_k)} \mathrm{d}t = \frac{1}{2\pi} \int_{-1}^{1} e^{-\imath tx} e^{\frac{\sigma_k^2(e)t^2}{2b_k^2}} \mathrm{d}t, \tag{4.5.43}$$

where $b_k = \left(\frac{b\sigma_k^2(e)}{\log k}\right)^{\frac{1}{2}}$ is the bandwidth with $b > 3$ being a constant chosen in advance.

The function $\omega_k(x)$ is defined by

$$\omega_k(x) \triangleq \frac{1}{b_k} K_k\Big(\frac{\psi_k - x}{b_k}\Big) = \frac{1}{2\pi b_k}\int_{-1}^{1} e^{\left[-\iota t(\psi_k - x)/b_k\right]} e^{\frac{\sigma_k^2(e)t^2}{2b_k^2}}\,\mathrm{d}t$$

$$= \frac{1}{2\pi b_k}\int_{-1}^{1} \cos[(\psi_k - x)t/b_k] e^{\frac{\sigma_k^2(e)t^2}{2b_k^2}}\,\mathrm{d}t$$

$$= \frac{1}{\pi}\int_{0}^{\frac{1}{b_k}} \cos[(\psi_k - x)t] e^{\frac{\sigma_k^2(e)t^2}{2}}\,\mathrm{d}t. \tag{4.5.44}$$

Notice that $\sigma_k^2(e)$ in (4.5.43) and (4.5.44) is unknown. To obtain its estimate $\widehat{\sigma}_k^2(e)$ we first estimate the spectral density of η_k, and then derive the estimate for the spectral density of e_k with the help of the estimates for the linear subsystem. Finally, the estimate $\widehat{\sigma}_k^2(e)$ for $\sigma_k^2(e)$ can be derived by the inverse Fourier transformation of the spectral density estimate for e_k.

For simplicity, we assume that the orders n_p and n_q in (4.5.9) and (4.5.10) are known in the procedure of estimating the spectral density of η_k. When the orders n_p and n_q are unknown, their strongly consistent estimates can be derived by the method provided in Section 3.5.

The correlation function $a_i(\eta) \triangleq E(\eta_k \eta_{k-i}), i \geq 0$ of η_k can recursively be estimated by SAAWET:

$$a_{0,k}(\eta) = \Big[a_{0,k-1}(\eta) - \frac{1}{k}(a_{0,k-1}(\eta) + \vartheta^2 - u_k^2)\Big] \cdot I_{\left[\left|a_{0,k-1}(\eta) - \frac{1}{k}(a_{0,k-1}(\eta)+\vartheta^2 - u_k^2)\right| \leq M_{\delta_k^{(0,\eta)}}\right]}, \tag{4.5.45}$$

$$\delta_k^{(0,\eta)} = \sum_{j=1}^{k-1} I_{\left[\left|a_{0,j-1} - \frac{1}{j}(a_{0,j-1}+\vartheta^2 - u_j^2)\right| > M_{\delta_j^{(0,\eta)}}\right]}, \tag{4.5.46}$$

$$a_{i,k}(\eta) = \Big[a_{i,k-1}(\eta) - \frac{1}{k}(a_{i,k-1}(\eta) - u_k u_{k-i})\Big] \cdot I_{\left[\left|a_{i,k-1}(\eta) - \frac{1}{k}(a_{i,k-1}(\eta) - u_k u_{k-i})\right| \leq M_{\delta_k^{(i,\eta)}}\right]}, \tag{4.5.47}$$

$$\delta_k^{(i,\eta)} = \sum_{j=1}^{k-1} I_{\left[\left|a_{i,j-1}(\eta) - \frac{1}{j}(a_{i,j-1}(\eta) - u_j u_{j-i})\right| > M_{\delta_j^{(i,\eta)}}\right]}, \quad i \geq 1. \tag{4.5.48}$$

Define the Hankel matrix

$$\Gamma_k(\eta) \triangleq \begin{bmatrix} a_{n_q,k}(\eta) & a_{n_q-1,k}(\eta) & \cdots & a_{n_q-n_p+1,k}(\eta) \\ a_{n_q+1,k}(\eta) & a_{n_q,k}(\eta) & \cdots & a_{n_q-n_p+2,k}(\eta) \\ \vdots & \vdots & \ddots & \vdots \\ a_{n_q+n_p-1,k}(\eta) & a_{n_q+n_p-2,k}(\eta) & \cdots & a_{n_q,k}(\eta) \end{bmatrix}$$

where $a_{i,k}(\eta) \triangleq a_{-i,k}(\eta)$ for $i<0$. Since $a_{i,k}(\eta) \xrightarrow[k\to\infty]{} a_i(\eta), i \geq 0$ as to be shown in Lemma 4.5.7 and by Theorem 3.3.3 the limit of $\Gamma_k(\eta)$ is nonsingular under A4.5.3, the matrix $\Gamma_k(\eta)$ is nonsingular for sufficiently large k. Therefore, at time k, the parameters $\{p_1, \cdots, p_{n_p}\}$ can be estimated by the Yule–Walker equation

$$[p_{1,k}, \cdots, p_{n_p,k}]^T = -\Gamma_k^{-1}(\eta)[a_{n_q+1,k}(\eta), a_{n_q+2,k}(\eta), \cdots, a_{n_q+n_p,k}(\eta)]^T.$$

The spectral density $S_{\eta_k}(z)$ of η_k is equal to

$$S_{\eta_k}(z) \triangleq \sum_{l=-\infty}^{\infty} a_l(\eta) z^l = \frac{Q(z)Q(z^{-1})\sigma_\zeta^2}{P(z)P(z^{-1})}, \tag{4.5.49}$$

where σ_ζ^2 denotes the variance of ζ_k.

Identifying coefficients of the same order of z at both sides of the equation

$$P(z)P(z^{-1}) \sum_{l=-\infty}^{\infty} a_l(\eta) z^l = Q(z)Q(z^{-1})\sigma_\zeta^2, \tag{4.5.50}$$

we derive

$$Q(z)Q(z^{-1})\sigma_\zeta^2 = \sum_{l=-n_q}^{n_q} \Big(\sum_{i=0}^{n_p} \sum_{j=0}^{n_p} a_{l+j-i}(\eta) p_i p_j \Big) z^l, \tag{4.5.51}$$

where only a finite number of correlation functions $a_l(\eta), -n_p - n_q \leq l \leq n_p + n_q$ are involved. As a consequence, the estimate for $S_{\eta_k}(z)$ can be obtained as follows:

$$\widehat{S}_{\eta_k}(z) = \frac{\sum_{l=-n_q}^{n_q} \Big(\sum_{i=0}^{n_p} \sum_{j=0}^{n_p} a_{l+j-i,k}(\eta) p_{i,k} p_{j,k} \Big) z^l}{(\sum_{i=0}^{n_p} p_{i,k} z^i)(\sum_{j=0}^{n_p} p_{j,k} z^{-j})},$$

and by (4.5.38) the spectral density $S_{e_k}(z)$ of e_k can be estimated by

$$\widehat{S}_{e_k}(z) = \frac{(\sum_{i=1}^{q} d_{i,k} z^i)(\sum_{j=1}^{q} d_{j,k} z^{-j})}{(\sum_{i=0}^{p} c_{i,k} z^i)(\sum_{j=0}^{p} c_{j,k} z^{-j})} \widehat{S}_{\eta_k}(z). \tag{4.5.52}$$

Finally, the variance $\sigma_k^2(e)$ of e_k can be approximated by the inverse Fourier transformation:

$$\widehat{\sigma}_k^2(e) = \frac{1}{2\pi} \int_{-\pi}^{\pi} \widehat{S}_{e_k}(e^{\iota\omega}) d\omega. \tag{4.5.53}$$

Therefore, $\omega_k(x)$ can be estimated at time k by

$$\widehat{\omega}_k(x) \triangleq \frac{1}{\pi} \int_0^{\frac{1}{b_k}} \cos[(\widehat{\psi}_k - x)t] e^{\frac{\widehat{\sigma}_k^2(e)t^2}{2}} dt, \tag{4.5.54}$$

where $\widehat{b}_k = \left(\frac{b\widehat{\sigma}_k^2(e)}{\log k}\right)^{\frac{1}{2}}$.

We now give the algorithms to estimate $f(x)$:

$$\mu_k(x) = \left[\mu_{k-1}(x) - \frac{1}{k}(\mu_{k-1}(x) - \widehat{\omega}_k(x))\right] \cdot I_{\left[\left|\mu_{k-1}(x) - \frac{1}{k}(\mu_{k-1}(x) - \widehat{\omega}_k(x))\right| \le M_{\delta_k^{(\mu)}(x)}\right]}, \tag{4.5.55}$$

$$\delta_k^{(\mu)}(x) = \sum_{j=1}^{k-1} I_{\left[\left|\mu_{j-1}(x) - \frac{1}{j}(\mu_{j-1}(x) - \widehat{\omega}_j(x))\right| > M_{\delta_j^{(\mu)}(x)}\right]}, \tag{4.5.56}$$

$$\beta_k(x) = \left[\beta_{k-1}(x) - \frac{1}{k}(\beta_{k-1}(x) - \widehat{\omega}_k(x)y_k)\right] \cdot I_{\left[\left|\beta_{k-1}(x) - \frac{1}{k}(\beta_{k-1}(x) - \widehat{\omega}_k(x)y_k)\right| \le M_{\delta_k^{(\beta)}(x)}\right]}, \tag{4.5.57}$$

$$\delta_k^{(\beta)}(x) = \sum_{j=1}^{k-1} I_{\left[\left|\beta_{j-1}(x) - \frac{1}{j}(\beta_{j-1}(x) - \widehat{\omega}_j(x)y_j)\right| > M_{\delta_j^{(\beta)}(x)}\right]}. \tag{4.5.58}$$

As a matter of fact, $\mu_k(x)$ defined by (4.5.55)–(4.5.56) and $\beta_k(x)$ defined by (4.5.57)–(4.5.58) are applied to estimate $p(x)$ and $p(x)\widetilde{f}(x)$ (see (4.5.84) and (4.5.85)), respectively, where $p(x) = \frac{1}{\sqrt{2\pi}\sigma\vartheta}e^{-\frac{x^2}{2\sigma^2\vartheta^2}}$ is the limit of the density function of v_k.

The estimate for $f(x)$ is naturally defined as:

$$f_k(x) \triangleq \begin{cases} \frac{\beta_k(x)}{\mu_k(x)}, & \text{if } \mu_k(x) \neq 0, \\ 0, & \text{if } \mu_k(x) = 0. \end{cases} \tag{4.5.59}$$

Strong Consistency for Linear Subsystem

Lemma 4.5.2 *Assume that A4.5.1–A4.5.4 hold. Then, for any* $0 \le \nu < 1/2$, *the following series converge.*

$$\sum_{k=1}^{\infty} \frac{1}{k^{1-\nu}}(\tau h_i - Ey_k u_{k-i}) < \infty \ \forall i \ge 1, \tag{4.5.60}$$

$$\sum_{k=1}^{\infty} \frac{1}{k^{1-\nu}}(\rho - E(y_k - Ey_k)u_{k-1}^2) < \infty, \tag{4.5.61}$$

$$\sum_{k=1}^{\infty} \frac{1}{k^{1-\nu}}(\rho h_i - E(y_k - Ey_k)u_{k-1}u_{k-i}) < \infty \ \forall i \ge 2. \tag{4.5.62}$$

Proof. The proof is based on the fact $|\sigma^2 - \sigma_k^2| = O(e^{-rk})$ for some $r > 0$ by the same method as that used for Lemma 4.3.3. □

Lemma 4.5.3 *Assume A4.5.1–A4.5.4 hold. Then, λ_k defined by (4.5.20)–(4.5.21) has the following convergence rate:*

$$|\lambda_k - Ey_k| = o\Big(\frac{1}{k^{1/2-c}}\Big) \ \ \forall c > 0. \tag{4.5.63}$$

Proof. By (4.5.5) and A4.5.3 we see

$$Ey_k = Ey_k^0 = Ef(v_k) \xrightarrow[k\to\infty]{} \frac{1}{\sqrt{2\pi}\sigma\vartheta}\int_{\mathbb{R}} f(x)e^{-\frac{x^2}{2\sigma^2\vartheta^2}}\mathrm{d}x \triangleq \bar{\lambda},$$

where $\sigma^2 = \sum_{i=1}^{\infty} h_i^2$.

The algorithm (4.5.20) can be written as

$$\begin{aligned}\lambda_k = &\big[\lambda_{k-1} - \frac{1}{k}(\lambda_{k-1} - \bar{\lambda}) - \frac{1}{k}e_k^{(\lambda)}\big] \\ &\cdot I_{\left[\left|\lambda_{k-1} - \frac{1}{k}(\lambda_{k-1}-\bar{\lambda}) - \frac{1}{k}e_k^{(\lambda)}\right| \leq M_{\delta_k^{(\lambda)}}\right]},\end{aligned} \tag{4.5.64}$$

where

$$e_k^{(\lambda)} = \bar{\lambda} - y_k = (\bar{\lambda} - Ey_k^0) + (Ey_k^0 - y_k^0) - \varepsilon_k. \tag{4.5.65}$$

Since $\bar{\lambda}$ is the single root of the linear function $-(y-\bar{\lambda})$, by Theorem 2.6.1 for proving $|\lambda_k - \bar{\lambda}| = o\left(\frac{1}{k^{1/2-c}}\right)$, it suffices to show

$$\sum_{k=1}^{\infty} \frac{1}{k^{1-\nu}} e_k^{(\lambda)} < \infty \ \text{a.s.} \ \forall \ 0 < \nu < 1/2. \tag{4.5.66}$$

Since $|\sigma^2 - \sigma_k^2| = O(e^{-rk})$ for some $r > 0$, we have $|\bar{\lambda} - Ey_k^0| = O(e^{-r_\lambda k})$ for some $r_\lambda > 0$. Thus, (4.5.66) holds for the first term at the right-hand side of (4.5.65).

By Theorem 1.4.4, both $\{Ey_k^0 - y_k^0\}$ and $\{\varepsilon_k\}$ are the zero-mean α-mixing sequences with mixing coefficients decaying exponentially to zero. Further, by Lemma 4.2.1, we have $E|y_k^0|^{2+\varepsilon} < \infty$ and $E|\varepsilon_k|^{2+\varepsilon} < \infty$ for some $\varepsilon > 0$. Thus, by Theorem 1.4.2, (4.5.66) holds for the last two terms at the right-hand side of (4.5.65).

Since $|\lambda_k - Ey_k| \leq |\lambda_k - \bar{\lambda}| + |\bar{\lambda} - Ey_k|$ and $|\bar{\lambda} - Ey_k| = O(e^{-r_\lambda k})$ for some $r_\lambda > 0$, we have

$$|\lambda_k - Ey_k| = o\Big(\frac{1}{k^{1/2-c}}\Big) \ \ \forall c > 0.$$

□

Lemma 4.5.4 *Assume that A4.5.1–A4.5.4 hold. Then, for any $0 \leq \nu < 1/2$, the following series converge.*

$$\sum_{k=1}^{\infty} \frac{1}{k^{1-\nu}}(Ey_k u_{k-i} - y_k u_{k-i}) < \infty \ \text{a.s.} \ \forall i \geq 1, \tag{4.5.67}$$

$$\sum_{k=1}^{\infty} \frac{1}{k^{1-\nu}}(E(y_k - Ey_k)u_{k-1}^2 - (y_k - Ey_k)u_{k-1}^2) < \infty \ \text{a.s.}, \tag{4.5.68}$$

$$\sum_{k=1}^{\infty}\frac{1}{k^{1-\nu}}(E(y_k-Ey_k)u_{k-1}u_{k-i}$$
$$-(y_k-Ey_k)u_{k-1}u_{k-i})<\infty \text{ a.s. } \forall i\geq 2, \tag{4.5.69}$$
$$\sum_{k=1}^{\infty}\frac{1}{k^{1-\nu}}((\lambda_k-Ey_k)u_{k-1}^2)<\infty \text{ a.s.}, \tag{4.5.70}$$
$$\sum_{k=1}^{\infty}\frac{1}{k^{1-\nu}}((\lambda_k-Ey_k)u_{k-1}u_{k-i})<\infty \text{ a.s. } \forall i\geq 2. \tag{4.5.71}$$

Proof. Since u_k^0, η_k, and ε_k are mutually independent, we have

$$\begin{aligned}&E(y_k-Ey_k)u_{k-1}^2-(y_k-Ey_k)u_{k-1}^2\\&=\left[E(y_k^0-Ey_k^0)(u_{k-1}^0)^2-(y_k^0-Ey_k^0)(u_{k-1}^0)^2\right]\\&\quad-(y_k^0-Ey_k^0)\eta_{k-1}^2-2(y_k^0-Ey_k^0)u_{k-1}^0\eta_{k-1}-\varepsilon_k u_{k-1}^2.\end{aligned} \tag{4.5.72}$$

From (4.5.72) it follows that

$$\begin{aligned}&\sum_{k=1}^{\infty}\frac{1}{k^{1-\nu}}\left[E(y_k-Ey_k)u_{k-1}^2-(y_k-Ey_k)u_{k-1}^2\right]\\&=\sum_{k=1}^{\infty}\frac{1}{k^{1-\nu}}\left[E(y_k^0-Ey_k^0)(u_{k-1}^0)^2-(y_k^0-Ey_k^0)(u_{k-1}^0)^2\right]-\sum_{k=1}^{\infty}\frac{1}{k^{1-\nu}}\left[(y_k^0-Ey_k^0)\eta_{k-1}^2\right]\\&\quad-2\sum_{k=1}^{\infty}\frac{1}{k^{1-\nu}}\left[(y_k^0-Ey_k^0)u_{k-1}^0\eta_{k-1}\right]-\sum_{k=1}^{\infty}\frac{1}{k^{1-\nu}}\left[\varepsilon_k u_{k-1}^2\right].\end{aligned} \tag{4.5.73}$$

Define $z_k^{(1)}\triangleq\frac{1}{k^{1-\nu}}E(y_k^0-Ey_k^0)(u_{k-1}^0)^2-(y_k^0-Ey_k^0)(u_{k-1}^0)^2$. Thus, by Theorem 1.4.4, $z_k^{(1)}$ is a zero-mean α-mixing sequence with the mixing coefficient decaying exponentially to zero. By Lemma 4.2.1, Hölder, and C_r-inequalities, for some $\varepsilon>0$ we have

$$\begin{aligned}\sum_{k=1}^{\infty}\left(E|z_k^{(1)}|^{2+\varepsilon}\right)^{\frac{2}{2+\varepsilon}}&\leq\sum_{k=1}^{\infty}\frac{4}{k^{2(1-\nu)}}\left(E|(y_k^0-Ey_k^0)(u_{k-1}^0)^2|^{2+\varepsilon}\right)^{\frac{2}{2+\varepsilon}}\\&\leq O\left(\sum_{k=1}^{\infty}\frac{1}{k^{2(1-\nu)}}\right)<\infty.\end{aligned}$$

Therefore, by Theorem 1.4.2, the first term at the right-hand side of (4.5.73) converges a.s. The convergence of the remaining terms at the right-hand side of (4.5.73) can be proved in a similar way, and hence (4.5.68) holds. Similarly, the assertions (4.5.67) and (4.5.69) also hold.

According to (4.5.63), we have

$$\left|\sum_{k=1}^{\infty}\frac{1}{k^{1-\nu}}((\lambda_k-Ey_k)u_{k-1}^2)\right|\leq\sum_{k=1}^{\infty}\frac{1}{k^{\frac{3}{2}-\nu-c}}(u_{k-1}^2-Eu_{k-1}^2)+\sum_{k=1}^{\infty}\frac{1}{k^{\frac{3}{2}-\nu-c}}Eu_{k-1}^2. \tag{4.5.74}$$

Since $\sum_{k=1}^{\infty} \frac{1}{k^{\frac{3}{2}-\nu-c}} Eu_{k-1}^2 < \infty$, for proving $\sum_{k=1}^{\infty} \frac{1}{k^{1-\nu}}((\lambda_k - Ey_k)u_{k-1}^2) < \infty$ it suffices to show the first term at the right-hand side of (4.5.74) converges a.s.

Define $z_k^{(2)} \triangleq \frac{1}{k^{\frac{3}{2}-\nu-c}}(u_{k-1}^2 - Eu_{k-1}^2)$. By Theorem 1.4.4, $z_k^{(2)}$ is a zero-mean α-mixing sequence with the mixing coefficient decaying exponentially to zero. Noticing that $E|u_{k-1}^2|^{2+\varepsilon} < \infty$, by C_r-inequality for some $\varepsilon > 0$ we have

$$\sum_{k=1}^{\infty} \left(E|z_k^{(2)}|^{2+\varepsilon}\right)^{\frac{2}{2+\varepsilon}} \le \sum_{k=1}^{\infty} \frac{4}{k^{3-2\nu-2c}} (E|u_{k-1}^2|^{2+\varepsilon})^{\frac{2}{2+\varepsilon}} = O(\sum_{k=1}^{\infty} \frac{1}{k^{3-2\nu-2c}}) < \infty.$$

Therefore, by Theorem 1.4.2 the assertion (4.5.70) holds. Similarly, (4.5.71) is also true.

The proof is finished. □

Theorem 4.5.1 *Assume that A4.5.1–A4.5.4 hold. Then, $h_{i,k}$ defined by (4.5.28) and (4.5.31) converges to $\{h_i \; \forall i \ge 2\}$ with the rate of convergence*

$$|h_{i,k} - h_i| = o(k^{-\nu}) \text{ a.s. } \forall \; \nu \in (0, 1/2), \;\; i \ge 2. \tag{4.5.75}$$

As consequences, from (4.5.33)–(4.5.36) the following convergence rates also take place: $\forall \; \nu \in (0, 1/2)$,

$$|c_{i,k} - c_i| = o(k^{-\nu}) \text{ a.s., } \; 1 \le i \le p, \tag{4.5.76}$$

$$|d_{j,k} - d_j| = o(k^{-\nu}) \text{ a.s., } \; 2 \le j \le q. \tag{4.5.77}$$

Proof. As pointed out before, by A4.5.4 at least one of τ and ρ is nonzero, so switching between (4.5.26)–(4.5.28) and (4.5.29)–(4.5.31) may happen only a finite number of times. Therefore, for proving (4.5.75) it suffices to show

$$\left|\theta_k^{(i,\tau)} - \tau h_i\right| = o(k^{-\nu}) \text{ a.s. } \forall \nu \in (0, 1/2), \;\; i \ge 1, \tag{4.5.78}$$

$$\left|\theta_k^{(1,\rho)} - \rho\right| = o(k^{-\nu}) \text{ a.s. } \forall \nu \in (0, 1/2), \tag{4.5.79}$$

$$\left|\theta_k^{(i,\rho)} - \rho h_i\right| = o(k^{-\nu}) \text{ a.s. } \forall \nu \in (0, 1/2), \;\; i \ge 2. \tag{4.5.80}$$

We rewrite (4.5.24) as

$$\theta_k^{(1,\rho)} = \left[\theta_{k-1}^{(1,\rho)} - \frac{1}{k}(\theta_{k-1}^{(1,\rho)} - \rho) - \frac{1}{k} e_k^{(1,\rho)}\right] \cdot I_{\left[\left|\theta_{k-1}^{(1,\rho)} - \frac{1}{k}(\theta_{k-1}^{(1,\rho)} - \rho) - \frac{1}{k} e_k^{(1,\rho)}\right| \le M_{\delta_k^{(1,\rho)}}\right]},$$

where $e_k^{(1,\rho)} = \rho - (y_k - \lambda_k)u_{k-1}^2$.

Since ρ is the single root of the linear function $-(y - \rho)$, by Theorem 2.6.1 it suffices to prove

$$\sum_{k=1}^{\infty} \frac{1}{k^{1-\nu}} e_k^{(1,\rho)} < \infty \text{ a.s. } \forall \nu \in (0, 1/2), \;\; i \ge 1. \tag{4.5.81}$$

Write $e_k^{(1,\rho)}$ as

$$\begin{aligned} e_k^{(1,\rho)} =& \rho - \rho_k + (E(y_k - Ey_k)u_{k-1}^2 - (y_k - Ey_k)u_{k-1}^2) \\ &+ (\lambda_k - Ey_k)u_{k-1}^2. \end{aligned} \tag{4.5.82}$$

By (4.5.61), (4.5.68), and (4.5.70) we find that (4.5.81) is true for (4.5.82), and hence (4.5.79) holds. Similarly, (4.5.78) and (4.5.80) can be proved by Lemmas 4.5.2 and 4.5.4, while the assertions (4.5.76)–(4.5.77) directly follow from (4.5.75). □

Strong Consistency of Estimation of f$(\cdot)$

Lemma 4.5.5 *Assume that A4.5.1–A4.5.4 and A4.5.6 hold. Then the following limits take place*

$$\frac{u_k}{k^c} \xrightarrow[k\to\infty]{\text{a.s.}} 0, \quad \frac{f(v_k)}{k^c} \xrightarrow[k\to\infty]{\text{a.s.}} 0 \ \ \forall c > 0. \tag{4.5.83}$$

Proof. The lemma can be proved by the same treatment as that used in the proof of Lemma 4.3.7. □

Lemma 4.5.6 *Under Conditions H1–H6, the following assertions for $\omega_k(x)$ defined by (4.5.44) take place*

$$E[\omega_k(x)] \xrightarrow[k\to\infty]{} p(x), \tag{4.5.84}$$

$$E[\omega_k(x)f(v_k)] \xrightarrow[k\to\infty]{} p(x)\widetilde{f}(x), \tag{4.5.85}$$

$$|\omega_k(x)|^\delta = O\big(k^{\frac{\delta}{2b}}(\log k)^{\frac{\delta}{2}}\big) \ \ \forall \delta \geq 1, \tag{4.5.86}$$

$$|\omega_k(x)f(v_k)|^\delta = O\big(k^{\frac{\delta}{2b}+c}(\log k)^{\frac{\delta}{2}}\big) \ \forall \delta \geq 1, \ c > 0, \tag{4.5.87}$$

where $p(x) = \frac{1}{\sqrt{2\pi}\sigma\vartheta}e^{-\frac{x^2}{2\sigma^2\vartheta^2}}$, $\sigma^2 = \sum_{i=1}^\infty h_i^2$, *and*

$$\widetilde{f}(x) = f(x-)\int_{-\infty}^{x} K(t)\mathrm{d}t + f(x+)\int_{x}^{\infty} K(t)\mathrm{d}t,$$

which equals $f(x)$ for any x where $f(\cdot)$ is continuous.

Proof. By the Fubini theorem changing the order of taking expectation and integral, and noticing that the density function of e_k is even, we have

$$\begin{aligned} E[\omega_k(x)f(v_k)] =& \frac{1}{2\pi b_k}\int_{\mathbb{R}} E\big(e^{[-\iota t(\psi_k - x)/b_k]}f(v_k)\big)\frac{\Phi_K(t)}{\Phi_{e_k}(t/b_k)}\mathrm{d}t \\ =& \frac{1}{2\pi b_k}\int_{\mathbb{R}} E\big(e^{[-\iota t(v_k - x)/b_k]}f(v_k)\big)Ee^{[-\iota te_k/b_k]}\frac{\Phi_K(t)}{\Phi_{e_k}(t/b_k)}\mathrm{d}t \\ =& \frac{1}{2\pi b_k}\int_{\mathbb{R}} E\big(e^{[-\iota t(v_k - x)/b_k]}f(v_k)\big)\Phi_K(t)\mathrm{d}t \\ =& \frac{1}{2\pi b_k}\int_{\mathbb{R}}\int_{\mathbb{R}} \big(e^{[-\iota t(y - x)/b_k]}f(y)\frac{1}{\sqrt{2\pi}\sigma_k\vartheta}e^{-\frac{y^2}{2\sigma_k^2\vartheta^2}}\big)\mathrm{d}y\Phi_K(t)\mathrm{d}t \end{aligned}$$

$$
\begin{aligned}
&= \frac{1}{b_k}\int_{\mathbb{R}} \Big(\frac{1}{2\pi}\int_{\mathbb{R}} e^{[-\iota t(y-x)/b_k]}\Phi_K(t)\mathrm{d}t\Big) f(y)\frac{1}{\sqrt{2\pi}\sigma_k\vartheta} e^{-\frac{y^2}{2\sigma_k^2\vartheta^2}}\mathrm{d}y\\
&= \frac{1}{b_k}\int_{\mathbb{R}} K\Big(\frac{y-x}{b_k}\Big) f(y)\frac{1}{\sqrt{2\pi}\sigma_k\vartheta} e^{-\frac{y^2}{2\sigma_k^2\vartheta^2}}\mathrm{d}y\\
&= \frac{1}{\sqrt{2\pi}\sigma_k\vartheta} e^{-\frac{x^2}{2\sigma_k^2\vartheta^2}} \int_{\mathbb{R}} K(t) f(x+b_k t) e^{-\frac{2xb_k t + b_k^2 t^2}{2\sigma_k^2\vartheta^2}}\mathrm{d}t\\
&= \frac{1}{\sqrt{2\pi}\sigma_k\vartheta} e^{-\frac{x^2}{2\sigma_k^2\vartheta^2}} \int_{-\infty}^{x} K(t) f(x+b_k t) e^{-\frac{2xb_k t + b_k^2 t^2}{2\sigma_k^2\vartheta^2}}\mathrm{d}t\\
&+ \frac{1}{\sqrt{2\pi}\sigma_k\vartheta} e^{-\frac{x^2}{2\sigma_k^2\vartheta^2}} \int_{x}^{\infty} K(t) f(x+b_k t) e^{-\frac{2xb_k t + b_k^2 t^2}{2\sigma_k^2\vartheta^2}}\mathrm{d}t\\
&\xrightarrow[k\to\infty]{} p(x)\Big(f(x-)\int_{-\infty}^{x} K(t)\mathrm{d}t + f(x+)\int_{x}^{\infty} K(t)\mathrm{d}t\Big),
\end{aligned}
$$

while the limit (4.5.84) can be proved by a similar treatment.

By (4.5.44), we have

$$
\begin{aligned}
|\omega_k(x)|^\delta &= \Big|\frac{1}{\pi}\int_0^{\frac{1}{b_k}} \cos[(\psi_k - x)t] e^{\frac{\sigma_k^2(e)t^2}{2}}\mathrm{d}t\Big|^\delta \le \frac{1}{\pi^\delta}\Big|\int_0^{\frac{1}{b_k}} e^{\frac{\sigma_k^2(e)t^2}{2}}\mathrm{d}t\Big|^\delta\\
&\le \frac{1}{\pi^\delta}\Big[\frac{1}{2}\Big(1+e^{\frac{\sigma_k^2(e)}{2}\frac{\log k}{b\sigma_k^2(e)}}\Big)\Big(\frac{\log k}{b\sigma_k^2(e)}\Big)^{\frac{1}{2}}\Big]^\delta \le \frac{1}{(2\pi)^\delta}\frac{(\log k)^{\frac{\delta}{2}}}{(b\sigma_k^2(e))^{\delta/2}}\big(1+k^{\frac{1}{2b}}\big)^\delta\\
&\le \frac{(\log k)^{\frac{\delta}{2}}}{(2\pi\sigma_k(e))^\delta b^{\delta/2}}\big(1+k^{\frac{\delta}{2b}}\big) = O\big(k^{\frac{\delta}{2b}}(\log k)^{\frac{\delta}{2}}\big). \qquad (4.5.88)
\end{aligned}
$$

Similarly, the assertion (4.5.87) can be proved by noticing the second limit of (4.5.83). □

Lemma 4.5.7 *Assume that A4.5.1, A4.5.3, A4.5.5, and A4.5.6 hold. Then both $a_{0,k}(\eta)$ defined by (4.5.45) and (4.5.46) and $a_{i,k}(\eta), i \ge 1$ defined by (4.5.47) and (4.5.48) have the convergence rate*

$$
|a_{i,k}(\eta) - a_i(\eta)| = o\Big(\frac{1}{k^{1/2-c}}\Big) \ \forall c > 0,\ i \ge 0. \qquad (4.5.89)
$$

Proof. The proof of the lemma is similar to that for Lemma 4.5.3. □

By Lemma 4.5.7, from (4.5.49)–(4.5.53) we have the following convergence rate of $\widehat{\sigma}_k^2(e)$, the estimate for the variance of e_k.

Corollary 4.5.1 *Assume that A4.5.1–A4.5.6 hold. Then $\widehat{\sigma}_k^2(e)$ defined by (4.5.53) has the following convergence rate*

$$
|\widehat{\sigma}_k^2(e) - \sigma_k^2(e)| = o\Big(\frac{1}{k^{1/2-c}}\Big) \ \forall c > 0.
$$

Lemma 4.5.8 *Assume A4.5.1–A4.5.6 hold. Then there is a constant* $c > 0$ *with* $\frac{1}{2} - \frac{1}{2b} - 2c > 0$ *such that*

$$|\psi_k - \widehat{\psi}_k| = o\Big(\frac{1}{k^{\frac{1}{2}-2c}}\Big), \tag{4.5.90}$$

$$|\omega_k(x) - \widehat{\omega}_k(x)| = o\Bigg(\frac{(\log k)^{\frac{3}{2}}}{k^{\frac{1}{2}(1-\frac{1}{b})-2c}}\Bigg). \tag{4.5.91}$$

Proof. The convergence rate (4.5.90) is established with the help of Lemma 4.3.8.

According to (4.5.44) and (4.5.54), we have

$$\omega_k(x) - \widehat{\omega}_k(x) = I_1 + I_2 + I_3,$$

where

$$\begin{aligned}
I_1 =& \frac{1}{\pi}\int_0^{\frac{1}{b_k}} \cos[(\psi_k - x)t] e^{\frac{\sigma_k^2(e)t^2}{2}} \mathrm{d}t - \frac{1}{\pi}\int_0^{\frac{1}{\widehat{b}_k}} \cos[(\psi_k - x)t] e^{\frac{\sigma_k^2(e)t^2}{2}} \mathrm{d}t \\
=& \frac{1}{\pi}\int_{\frac{1}{\widehat{b}_k}}^{\frac{1}{b_k}} \cos[(\psi_k - x)t] e^{\frac{\sigma_k^2(e)t^2}{2}} \mathrm{d}t, \\
I_2 =& \frac{1}{\pi}\int_0^{\frac{1}{\widehat{b}_k}} \cos[(\psi_k - x)t] e^{\frac{\sigma_k^2(e)t^2}{2}} \mathrm{d}t - \frac{1}{\pi}\int_0^{\frac{1}{\widehat{b}_k}} \cos[(\psi_k - x)t] e^{\frac{\widehat{\sigma}_k^2(e)t^2}{2}} \mathrm{d}t \\
=& \frac{1}{\pi}\int_0^{\frac{1}{\widehat{b}_k}} \cos[(\psi_k - x)t] \big(e^{\frac{\sigma_k^2(e)t^2}{2}} - e^{\frac{\widehat{\sigma}_k^2(e)t^2}{2}}\big) \mathrm{d}t, \\
I_3 =& \frac{1}{\pi}\int_0^{\frac{1}{\widehat{b}_k}} \cos[(\psi_k - x)t] e^{\frac{\widehat{\sigma}_k^2(e)t^2}{2}} \mathrm{d}t - \frac{1}{\pi}\int_0^{\frac{1}{\widehat{b}_k}} \cos[(\widehat{\psi}_k - x)t] e^{\frac{\widehat{\sigma}_k^2(e)t^2}{2}} \mathrm{d}t \\
=& \frac{1}{\pi}\int_0^{\frac{1}{\widehat{b}_k}} \big(\cos[(\psi_k - x)t] - \cos[(\widehat{\psi}_k - x)t]\big) e^{\frac{\widehat{\sigma}_k^2(e)t^2}{2}} \mathrm{d}t.
\end{aligned}$$

Since $|\widehat{\sigma}_k^2(e) - \sigma_k^2(e)| = o\big(\frac{1}{k^{1/2-c}}\big)$ and $k^{\frac{\sigma_k^2(e)}{2b\widehat{\sigma}_k^2(e)}} = o(k^{\frac{1}{2b}+c})$ for any $c > 0$, we have

$$\begin{aligned}
|I_1| \le& \frac{1}{\pi}\int_{\frac{1}{\widehat{b}_k}}^{\frac{1}{b_k}} e^{\frac{\sigma_k^2(e)t^2}{2}} \mathrm{d}t \le \frac{1}{2\pi}\big(e^{\frac{\sigma_k^2(e)}{2}\frac{1}{b_k^2}} + e^{\frac{\sigma_k^2(e)}{2}\frac{1}{\widehat{b}_k^2}}\big)\Big|\frac{1}{b_k} - \frac{1}{\widehat{b}_k}\Big| \\
\le& \frac{1}{2\pi}\big(e^{\frac{\sigma_k^2(e)}{2}\frac{\log k}{b\sigma_k^2(e)}} + e^{\frac{\sigma_k^2(e)}{2}\frac{\log k}{b\widehat{\sigma}_k^2(e)}}\big)\Big(\frac{\log k}{b}\Big)^{\frac{1}{2}} \cdot \frac{|\widehat{\sigma}_k^2(e) - \sigma_k^2(e)|}{\widehat{\sigma}_k(e)\sigma_k(e)(\widehat{\sigma}_k(e) + \sigma_k(e))} \\
\le& \frac{1}{2\pi}\big(k^{\frac{1}{2b}} + k^{\frac{\sigma_k^2(e)}{2b\widehat{\sigma}_k^2(e)}}\big)\Big(\frac{\log k}{b}\Big)^{\frac{1}{2}} \cdot \frac{|\widehat{\sigma}_k^2(e) - \sigma_k^2(e)|}{\widehat{\sigma}_k(e)\sigma_k(e)(\widehat{\sigma}_k(e) + \sigma_k(e))} \\
=& o\Bigg(\frac{(\log k)^{\frac{1}{2}}}{k^{\frac{1}{2}(1-\frac{1}{b})-2c}}\Bigg).
\end{aligned} \tag{4.5.92}$$

By the mean value theorem, there is an $\bar{s} \in (\widehat{\sigma}_k^2(e), \sigma_k^2(e))$ or $\bar{s} \in (\sigma_k^2(e), \widehat{\sigma}_k^2(e))$ such that

$$\left|e^{\frac{\sigma_k^2(e)t^2}{2}} - e^{\frac{\widehat{\sigma}_k^2(e)t^2}{2}}\right| = \frac{t^2}{2} e^{\frac{\bar{s}t^2}{2}} |\sigma_k^2(e) - \widehat{\sigma}_k^2(e)|.$$

Again by $|\widehat{\sigma}_k^2(e) - \sigma_k^2(e)| = o\big(\frac{1}{k^{1/2-c}}\big)$ and $k^{\frac{\bar{s}}{2b\widehat{\sigma}_k^2(e)}} = o(k^{\frac{1}{2b}+c})$ for any $c > 0$, we have

$$\begin{aligned}
|I_2| \leq & \frac{1}{\pi} \int_0^{\frac{1}{b_k}} \left|e^{\frac{\sigma_k^2(e)t^2}{2}} - e^{\frac{\widehat{\sigma}_k^2(e)t^2}{2}}\right| dt = \frac{1}{\pi} \int_0^{\frac{1}{b_k}} \frac{t^2}{2} e^{\frac{\bar{s}t^2}{2}} dt |\sigma_k^2(e) - \widehat{\sigma}_k^2(e)| \\
\leq & \frac{1}{2\pi} \Big(\frac{\log k}{2b\widehat{\sigma}_k^2(e)} e^{\frac{\bar{s}}{2} \frac{\log k}{b\widehat{\sigma}_k^2(e)}}\Big) \Big(\frac{\log k}{b\widehat{\sigma}_k^2(e)}\Big)^{\frac{1}{2}} |\sigma_k^2(e) - \widehat{\sigma}_k^2(e)| \\
= & \frac{1}{4\pi} \Big(\frac{\log k}{b\widehat{\sigma}_k^2(e)}\Big)^{\frac{3}{2}} k^{\frac{\bar{s}}{2b\widehat{\sigma}_k^2(e)}} |\sigma_k^2(e) - \widehat{\sigma}_k^2(e)| = o\Big(\frac{(\log k)^{\frac{3}{2}}}{k^{\frac{1}{2}(1-\frac{1}{b})-2c}}\Big).
\end{aligned} \tag{4.5.93}$$

From (4.5.90) it follows that

$$\begin{aligned}
|I_3| \leq & \frac{1}{\pi} \int_0^{\frac{1}{b_k}} \big|\cos[(\psi_k - x)t] - \cos[(\widehat{\psi}_k - x)t]\big| e^{\frac{\widehat{\sigma}_k^2(e)t^2}{2}} dt \\
\leq & \frac{1}{\pi} \int_0^{\frac{1}{b_k}} \Big| -2\sin\Big(\frac{(\psi_k + \widehat{\psi}_k)t - 2xt}{2}\Big) \sin\Big(\frac{(\psi_k - \widehat{\psi}_k)t}{2}\Big)\Big| e^{\frac{\widehat{\sigma}_k^2(e)t^2}{2}} dt \\
\leq & \frac{1}{\pi} \int_0^{\frac{1}{b_k}} t e^{\frac{\widehat{\sigma}_k^2(e)t^2}{2}} dt \big|\psi_k - \widehat{\psi}_k\big| = \frac{1}{\pi\widehat{\sigma}_k^2(e)} (k^{\frac{1}{2b}} - 1) \big|\psi_k - \widehat{\psi}_k\big| \\
= & o\Big(\frac{1}{k^{\frac{1}{2}(1-\frac{1}{b})-2c}}\Big).
\end{aligned} \tag{4.5.94}$$

By (4.5.92)–(4.5.94), we have

$$|\omega_k(x) - \widehat{\omega}_k(x)| = o\Big(\frac{(\log k)^{\frac{3}{2}}}{k^{\frac{1}{2}(1-\frac{1}{b})-2c}}\Big).$$

□

Lemma 4.5.9 *Assume A4.5.1–A4.5.6 hold. The following series converge* a.s.

$$\sum_{k=1}^{\infty} \frac{1}{k} (E\omega_k(x) - \omega_k(x)) < \infty, \tag{4.5.95}$$

$$\sum_{k=1}^{\infty} \frac{1}{k} (E\omega_k(x) f(v_k) - \omega_k(x) f(v_k)) < \infty, \tag{4.5.96}$$

$$\sum_{k=1}^{\infty} \frac{1}{k} (\omega_k(x) - \widehat{\omega}_k(x)) \varepsilon_k < \infty, \; \sum_{k=1}^{\infty} \frac{1}{k} \omega_k(x) \varepsilon_k < \infty. \tag{4.5.97}$$

Proof. Define $z_k^{(3)} \triangleq \frac{1}{k}(E\omega_k(x) - \omega_k(x))$. Then, by Theorem 1.4.4, $z_k^{(3)}$ is a zero-mean α-mixing sequence with mixing coefficients decaying exponentially to zero, because $\omega_k(x)$ is a measurable function of ψ_k. Noticing $E|\omega_k(x)|^{2+\varepsilon} = O(k^{\frac{2+\varepsilon}{2b}}(\log k)^{\frac{2+\varepsilon}{2}})$ by (4.5.86), by the C_r-inequality for some $\varepsilon > 0$ we have

$$\sum_{k=1}^{\infty}\left(E|z_k^{(3)}|^{2+\varepsilon}\right)^{\frac{2}{2+\varepsilon}} \leq \sum_{k=1}^{\infty}\frac{4}{k^2}\left(E|\omega_k(x)|^{2+\varepsilon}\right)^{\frac{2}{2+\varepsilon}} = O\Big(\sum_{k=1}^{\infty}\frac{\log k}{k^{2-\frac{1}{b}}}\Big) < \infty.$$

Therefore, by Theorem 1.4.2 we have proved (4.5.95), while (4.5.96) can be verified in a similar way.

The convergence of the first series in (4.5.97) can be proved by the treatment similar to that used for proving (4.5.70).

Define $z_k^{(4)} \triangleq \frac{1}{k}\omega_k(x)\varepsilon_k$. By Theorem 1.4.4, $z_k^{(4)}$ is a zero-mean α-mixing sequence with mixing coefficients decaying exponentially to zero. Noticing $E|\omega_k(x)|^{2+\varepsilon} = O(k^{\frac{2+\varepsilon}{2b}}(\log k)^{\frac{2+\varepsilon}{2}})$ by (4.5.86) and $E\left|\varepsilon_k\right|^{2+\varepsilon} < \infty$ by Lemma 4.2.1, for some $\varepsilon > 0$ we have

$$\begin{aligned}\sum_{k=1}^{\infty}\left(E|z_k^{(4)}|^{2+\varepsilon}\right)^{\frac{2}{2+\varepsilon}} \leq &\sum_{k=1}^{\infty}\frac{1}{k^2}\left(E|\omega_k(x)|^{2+\varepsilon}\right)^{\frac{2}{2+\varepsilon}}\left(E\left|\varepsilon_k\right|^{2+\varepsilon}\right)^{\frac{2}{2+\varepsilon}}\\ =&O\Big(\sum_{k=1}^{\infty}\frac{\log k}{k^{2-\frac{1}{b}}}\Big) < \infty.\end{aligned}$$

Therefore, by Theorem 1.4.2 we have proved the convergence of the last series in (4.5.97). □

Theorem 4.5.2 *Assume A4.5.1–A4.5.6 hold. Then $\mu_k(x)$ defined by (4.5.55) and (4.5.56) and $\beta_k(x)$ defined by (4.5.57) and (4.5.58) are convergent:*

$$\mu_k(x) \xrightarrow[k\to\infty]{} p(x) \text{ a.s.,} \tag{4.5.98}$$

$$\beta_k(x) \xrightarrow[k\to\infty]{} p(x)\widetilde{f}(x) \text{ a.s.} \tag{4.5.99}$$

As a consequence, $f_k(x)$ defined by (4.5.59) is strongly consistent

$$f_k(x) \xrightarrow[k\to\infty]{} \widetilde{f}(x) \text{ a.s.} \tag{4.5.100}$$

Proof. The algorithm (4.5.55) can be rewritten as

$$\begin{aligned}\mu_k(x) =&\Big[\mu_{k-1}(x) - \frac{1}{k}(\mu_{k-1}(x) - p(x)) - \frac{1}{k}\bar{e}_k(x)\Big]\\ &\cdot I_{\Big[\left|\mu_{k-1}(x)-\frac{1}{k}(\mu_{k-1}(x)-p(x))-\frac{1}{k}\bar{e}_k(x)\right| \leq M_{\delta_k^{(\mu)}(x)}\Big]},\end{aligned}$$

where

$$\begin{aligned}\bar{e}_k(x) =&p(x) - \omega_k(x) = \Big[p(x) - E\omega_k(x)\Big]\\ &+\Big[E\omega_k(x) - \omega_k(x)\Big] + \Big[\omega_k(x) - \widehat{\omega}_k(x)\Big].\end{aligned} \tag{4.5.101}$$

Since $p(x)$ is the single root of the linear function $-(y-p(x))$, by Theorem 2.3.1, for convergence of $\mu_k(x)$ it suffices to show

$$\lim_{T\to 0}\limsup_{k\to\infty}\frac{1}{T}\Big|\sum_{j=n_k}^{m(n_k,T_k)}\frac{1}{j}\bar{e}_j(x)\Big| = 0 \;\; \forall\, T_k\in[0,T] \tag{4.5.102}$$

along indices $\{n_k\}$ of any convergent subsequence $\mu_{n_k}(x)$, where $m(k,T)\triangleq \max\{m: \sum_{j=k}^{m}\frac{1}{j}\le T\}$. By (4.5.84), (4.5.95), and (4.5.91) it follows that (4.5.102) holds for $\bar{e}_k(x)$.

The proof of (4.5.99) can similarly be carried out, if we rewrite the algorithm (4.5.57) as follows:

$$\beta_k(x) = \Big[\beta_{k-1}(x) - \frac{1}{k}(\beta_{k-1}(x) - p(x)\widetilde{f}(x)) - \frac{1}{k}\tilde{e}_k(x)\Big]$$
$$\cdot I_{\Big[\big|\beta_{k-1}(x)-\frac{1}{k}(\beta_{k-1}(x)-p(x)\widetilde{f}(x))-\frac{1}{k}\tilde{e}_k(x)\big|\le M_{\delta_k^{(\beta)}(x)}\Big]},$$

where

$$\begin{aligned}\tilde{e}_k(x) =& p(x)\widetilde{f}(x) - \omega_k(x)y_k\\ =&(p(x)\widetilde{f}(x) - E\omega_k(x)f(v_k)) + (E\omega_k(x)f(v_k) - \omega_k(x)f(v_k))\\ &-(\omega_k(x)-\omega_k(x))(f(v_k)+\varepsilon_k) - \omega_k(x)\varepsilon_k.\end{aligned} \tag{4.5.103}$$

Each term at the right-hand side of (4.5.103) satisfies the convergence condition of SAAWET by noticing (4.5.85), (4.5.91), (4.5.96), and (4.5.97) as well.

Therefore, the estimate (4.5.59) is strongly consistent. □

Numerical Example

Let the linear subsystem be

$$v_k + c_1 v_{k-1} + c_2 v_{k-2} = u_{k-1}^0 + d_2 u_{k-2}^0 + d_3 u_{k-3}^0,$$

where $c_1 = 0.2$, $c_2 = 0.6$, $d_2 = -0.3$, and $d_3 = 1.2$, and let the nonlinear function be

$$f(x) = x^2 - 0.5x - 1.$$

The input signal $\{u_k^0\}$ is a sequence of iid Gaussian random variables: $u_k^0 \in \mathcal{N}(0,1)$. Both the driven noises $\{\zeta_k\}$ and $\{\varsigma_k\}$ are sequences of mutually independent Gaussian random variables: $\zeta_k \in \mathcal{N}(0, 0.3^2)$, $\varsigma_k \in \mathcal{N}(0,0.3^2)$. The measurement noises η_k and ε_k are ARMA processes:

$$\eta_k - 0.7\eta_{k-1} = \zeta_k + 0.5\zeta_{k-1},$$
$$\varepsilon_k + 0.4\varepsilon_{k-1} = \varsigma_k - 0.6\varsigma_{k-1}.$$

The parameters used in the algorithms are as follows: $b = 4$ and $M_k = 2^k + 10$.

In the figures illustrated below, the solid lines represent the true values of the system, while the dash lines denote the corresponding estimates. Figure 4.5.2 demonstrates the recursive estimates for the coefficients of the linear subsystem, while Figure 4.5.3 gives the performance of the estimate for $\sigma_k^2(e)$. In Figure 4.5.4 the true nonlinear function is denoted by the solid curve, and its estimates at time $k = 10000$ at 31 points equally chosen from the interval $[-3, 3]$ are shown by symbols $+$. The behavior of the estimates at points $\{-2.4, -2, -0.2, 1.8\}$ versus time is demonstrated by Figure 4.5.5.

The simulation results are consistent with the theoretical analysis.

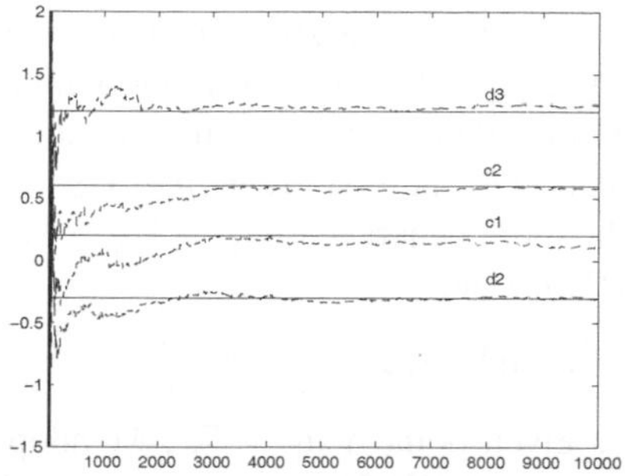

Figure 4.5.2: Estimates for c_1, c_2, d_2, d_3

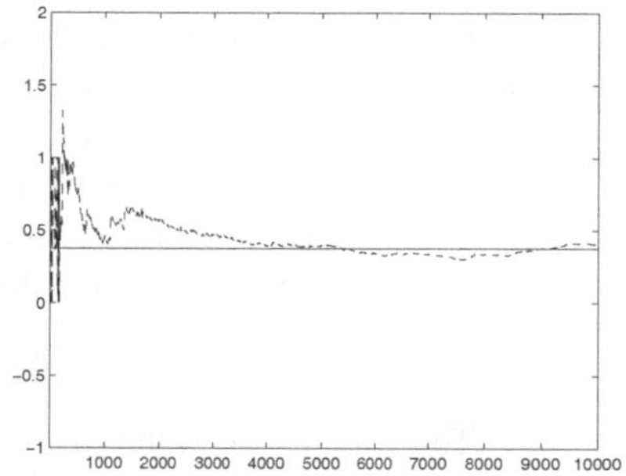

Figure 4.5.3: Estimates for $\sigma_k^2(e)$

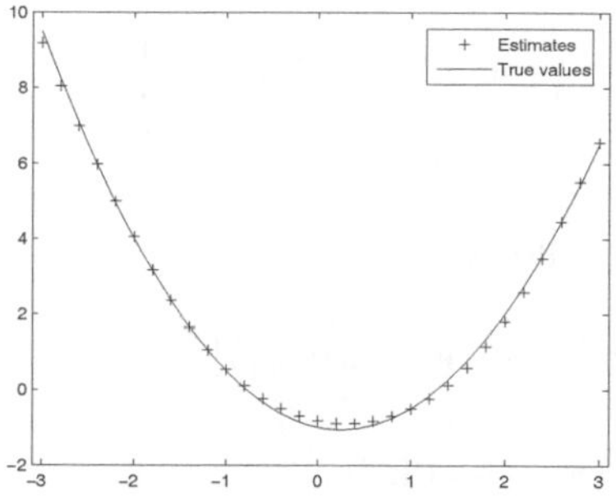

Figure 4.5.4: Estimates for $f(x) = x^2 - 0.5x - 1$

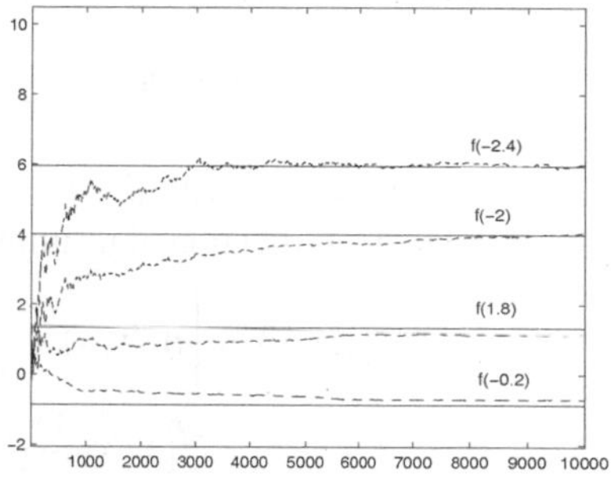

Figure 4.5.5: Estimates for $f(x)$

4.6 Recursive Identification of Nonlinear ARX Systems

In addition to the Wiener system, the Hammerstein system and their variations, the nonlinear ARX (NARX) system is also widely used to model the practical systems appearing in industry, finance, biology, etc. The NARX system was introduced and briefly discussed in Section 1.3.4. Different from the Wiener and Hammerstein systems, the NARX system generally is with no special structure information and the task of identification is to estimate the value of the nonlinear function at any given point.

Let us consider identification of the following single-input single-output (SISO) NARX system

$$y_{k+1} = f(y_k, \cdots, y_{k+1-p}, u_k, \cdots, u_{k+1-q}) + \varepsilon_{k+1}, \tag{4.6.1}$$

where u_k and y_k are the system input and output, respectively, ε_k is the system noise, (p,q) are the known system orders or upper bounds of the true system orders, and $f(\cdot)$ is the unknown nonlinear function.

For the system (4.6.1), the identification task consists in recursively estimating the value of $f(\cdot)$ at an arbitrarily given $\phi^* \in \mathbb{R}^{p+q}$ based on the input-output measurements $\{u_k, y_k\}_{k\geq 0}$. Following the same idea demonstrated in the previous sections, we first consider the recursive identification of the first-order ($p=1, q=1$) NARX system and then extend the results to the more general case ($p \geq 1, q \geq 1$).

Let us consider the first-order NARX system:

$$y_{k+1} = f(y_k, u_k) + \varepsilon_{k+1}. \tag{4.6.2}$$

To estimate the value of $f(\cdot,\cdot)$ at a given point $[y\ u]^T \in \mathbb{R}^2$, we adopt the idea used in previous sections, i.e., incorporating SAAWET with an appropriately defined kernel function serving as local averaging.

Let $\{M_k\}_{k\geq 0}$ be a sequence of positive numbers increasingly diverging to infinity, and let the kernel function be such that

$$w_k(y,u) = \frac{1}{2\pi b_k^2} \exp\left\{ -\frac{1}{2}\left(\frac{y_k - y}{b_k}\right)^2 - \frac{1}{2}\left(\frac{u_k - u}{b_k}\right)^2 \right\}, \tag{4.6.3}$$

where $b_k = 1/k^\delta$ with $\delta \in (0, \frac{1}{4})$.

The identification algorithms are as follows:

$$\begin{aligned} \eta_{k+1}(y,u) = &\left[\eta_k(y,u) - \frac{1}{k+1}(\eta_k(y,u) - w_k(y,u)y_{k+1}) \right] \\ &\cdot I_{\left[\left|\eta_k(y,u) - \frac{1}{k+1}(\eta_k(y,u) - w_k(y,u)y_{k+1})\right| \leq M_{\sigma_k}\right]}, \end{aligned} \tag{4.6.4}$$

$$\sigma_k = \sum_{j=1}^{k-1} I_{\left[\left|\eta_j(y,u) - \frac{1}{j+1}(\eta_j(y,u) - w_j(y,u)y_{j+1})\right| > M_{\sigma_j}\right]}, \quad \sigma_0 = 0, \tag{4.6.5}$$

and

$$\begin{aligned} \theta_{k+1}(y,u) = &\left[\theta_k(y,u) - \frac{1}{k+1}(\theta_k(y,u) - w_k(y,u)) \right] \\ &\cdot I_{\left[\left|\theta_k(y,u) - \frac{1}{k+1}(\theta_k(y,u) - w_k(y,u))\right| \leq M_{\delta_k}\right]}, \end{aligned} \tag{4.6.6}$$

$$\delta_k = \sum_{j=1}^{k-1} I_{\left[\left|\theta_j(y,u) - \frac{1}{j+1}(\theta_j(y,u) - w_j(y,u))\right| > M_{\delta_j}\right]}, \quad \delta_0 = 0. \tag{4.6.7}$$

It is noticed that under the assumptions A1.3.1–A1.3.5, $\eta_k(y,u)$ defined by (4.6.3)–(4.6.5) and $\theta_k(y,u)$ defined by (4.6.6)–(4.6.7) serve as estimates for $f(y,u)f_{\mathrm{IV}}(y,u)$ and $f_{\mathrm{IV}}(y,u)$, respectively, where $f_{\mathrm{IV}}(\cdot,\cdot)$ is the invariant probability density of the chain $\{x_k = [y_k \ u_k]^T\}_{k\geq 0}$ defined by (4.6.2). So, the ratio $\{\eta_k(y,u)/\theta_k(y,u)\}_{k\geq 0}$ naturally serves as the estimate for $f(y,u)$ provided $f_{\mathrm{IV}}(y,u) \neq 0$.

Notice also that $f(y,u)f_{\mathrm{IV}}(y,u)$ and $f_{\mathrm{IV}}(y,u)$ are roots of the functions $x - f(y,u)f_{\mathrm{IV}}(y,u)$ and $x - f_{\mathrm{IV}}(y,u)$, respectively, and are estimated by SAAWET.

Relying on ergodicity and the properties of α-mixing established by Theorems 1.3.6 and 1.4.4, for $\{y_k\}$ defined by (4.6.2), we have the following results. Recall that $P_{\mathrm{IV}}(\cdot)$ is the invariant probability measure of the chain $\{x_k\}_{k\geq 0}$ with density $f_{\mathrm{IV}}(\cdot)$.

Lemma 4.6.1 *Assume that A1.3.1–A1.3.5 hold and the invariant density $f_{\mathrm{IV}}(\cdot)$ is continuous at the given point $[y \ u]^T \in \mathbb{R}^2$. Then*

$$Ew_k(y,u)f(y_k,u_k) \underset{k\to\infty}{\longrightarrow} f(y,u)f_{\mathrm{IV}}(y,u), \tag{4.6.8}$$

$$Ew_k(y,u) \underset{k\to\infty}{\longrightarrow} f_{\mathrm{IV}}(y,u), \tag{4.6.9}$$

$$E\big|w_k(y,u)f(y_k,u_k)\big|^{2+\varepsilon} = O\Big(\frac{1}{b_k^{2(2+\varepsilon)}}\Big), \tag{4.6.10}$$

$$E\big|w_k(y,u)\big|^{2+\varepsilon} = O\Big(\frac{1}{b_k^{2(2+\varepsilon)}}\Big) \tag{4.6.11}$$

for any fixed $\varepsilon \geq 0$, and

$$\sum_{k=0}^{\infty}\frac{1}{k+1}\Big(w_k(y,u)f(y_k,u_k) - Ew_k(y,u)f(y_k,u_k)\Big) < \infty \text{ a.s.} \tag{4.6.12}$$

$$\sum_{k=0}^{\infty}\frac{1}{k+1}\Big(w_k(y,u) - Ew_k(y,u)\Big) < \infty \text{ a.s.} \tag{4.6.13}$$

$$\sum_{k=0}^{\infty}\frac{1}{k+1}w_k(y,u)\varepsilon_{k+1} < \infty \text{ a.s.} \tag{4.6.14}$$

Proof. We first prove (4.6.8). As $P_{\mathrm{IV}}(\cdot)$ is with density $f_{\mathrm{IV}}(\cdot,\cdot)$, we have the following equalities

$$\begin{aligned} &Ew_k(y,u)f(y_k,u_k) \\ &= \iint\limits_{\mathbb{R}^2} \frac{1}{2\pi b_k^2}\exp\left\{-\frac{1}{2}\left(\frac{x_1-y}{b_k}\right)^2 - \frac{1}{2}\left(\frac{x_2-u}{b_k}\right)^2\right\} f(x_1,x_2)P_k(\mathrm{d}x) \\ &= I_{1,k} + I_{2,k}, \end{aligned} \tag{4.6.15}$$

where

$$I_{1,k} \triangleq \iint\limits_{\mathbb{R}^2} \frac{1}{2\pi b_k^2}\exp\left\{-\frac{1}{2}\left(\frac{x_1-y}{b_k}\right)^2 - \frac{1}{2}\left(\frac{x_2-u}{b_k}\right)^2\right\} \cdot f(x_1,x_2)f_{\mathrm{IV}}(x_1,x_2)\mathrm{d}x_1\mathrm{d}x_2,$$

$$I_{2,k} \triangleq \iint_{\mathbb{R}^2} \frac{1}{2\pi b_k^2} \exp\left\{-\frac{1}{2}\left(\frac{x_1-y}{b_k}\right)^2 - \frac{1}{2}\left(\frac{x_2-u}{b_k}\right)^2\right\} \cdot f(x_1,x_2)\Big(P_k(\mathrm{d}x) - P_{\mathrm{IV}}(\mathrm{d}x)\Big).$$

Let $t_1 \triangleq \frac{x_1-y}{b_k}$, $t_2 \triangleq \frac{x_2-u}{b_k}$. Owing to A1.3.4, by Theorem 1.1.8, we see

$$I_{1,k} = \iint_{\mathbb{R}^2} \frac{1}{2\pi} \exp\left\{-\frac{t_1^2+t_2^2}{2}\right\} f(b_k t_1 + y, b_k t_2 + u) \cdot f_{\mathrm{IV}}(b_k t_1 + y, b_k t_2 + u)\,\mathrm{d}t_1\mathrm{d}t_2$$
$$\underset{k\to\infty}{\longrightarrow} f(y,u) f_{\mathrm{IV}}(y,u). \tag{4.6.16}$$

Noting A1.3.4, by Theorem 1.3.6 we have

$$|I_{2,k}| \le \iint_{\mathbb{R}^2} \frac{1}{2\pi b_k^2} \exp\left\{-\frac{1}{2}\left(\frac{x_1-y}{b_k}\right)^2 - \frac{1}{2}\left(\frac{x_2-u}{b_k}\right)^2\right\} |f(x_1,x_2)| \Big|P_k(\mathrm{d}x) - P_{\mathrm{IV}}(\mathrm{d}x)\Big|$$
$$\le \iint_{\mathbb{R}^2} \frac{1}{2\pi b_k^2} \exp\left\{-\frac{1}{2}\left(\frac{x_1-y}{b_k}\right)^2 - \frac{1}{2}\left(\frac{x_2-u}{b_k}\right)^2\right\} \cdot \Big(c_1(|x_1|^l + |x_2|^l) + c_2\Big) \Big|P_k(\mathrm{d}x) - P_{\mathrm{IV}}(\mathrm{d}x)\Big|$$
$$= \iint_{\mathbb{R}^2} \frac{1}{2\pi b_k^2} \exp\left\{-\frac{1}{2}\left(\frac{x_1-y}{b_k}\right)^2 - \frac{1}{2}\left(\frac{x_2-u}{b_k}\right)^2\right\} \cdot \left(c_1\left(\left(\frac{x_1}{b_k}\right)^l + \left(\frac{x_2}{b_k}\right)^l\right) + \frac{c_2}{b_k^l}\right) b_k^l \Big|P_k(\mathrm{d}x) - P_{\mathrm{IV}}(\mathrm{d}x)\Big|$$
$$\le \iint_{\mathbb{R}^2} \frac{1}{b_k^2}\left(c_3 + \frac{c_3}{b_k^l}\right) b_k^l \Big|P_k(\mathrm{d}x) - P_{\mathrm{IV}}(\mathrm{d}x)\Big| \le \frac{c_4}{b_k^2}\rho^k \underset{k\to\infty}{\longrightarrow} 0, \tag{4.6.17}$$

where c_1, c_2, c_3, and c_4 are positive constants. From (4.6.16) and (4.6.17) it follows that (4.6.8) holds. Convergence (4.6.9) can similarly be proved.

To prove (4.6.10), for any fixed $\varepsilon \ge 0$ we note that

$$E|w_k(y,u) f(y_k,u_k)|^{2+\varepsilon}$$
$$= \iint_{\mathbb{R}^2} \exp\left\{-\frac{2+\varepsilon}{2}\left(\frac{x_1-y}{b_k}\right)^2 - \frac{2+\varepsilon}{2}\left(\frac{x_2-u}{b_k}\right)^2\right\} \frac{|f(x_1,x_2)|^{2+\varepsilon}}{(2\pi)^{2+\varepsilon} b_k^{2(2+\varepsilon)}} P_k(\mathrm{d}x)$$
$$\le \iint_{\mathbb{R}^2} \frac{1}{b_k^{2(2+\varepsilon)}} \exp\left\{-\frac{2+\varepsilon}{2}\left(\left(\frac{x_1-y}{b_k}\right)^2 + \left(\frac{x_2-u}{b_k}\right)^2\right)\right\} \cdot \left(c_1\left(\left(\frac{x_1}{b_k}\right)^l + \left(\frac{x_2}{b_k}\right)^l\right) + \frac{c_2}{b_k^l}\right)^{2+\varepsilon} b_k^{(2+\varepsilon)l} P_k(\mathrm{d}x)$$

$$\leq \iint_{\mathbb{R}^2} \frac{1}{b_k^{2(2+\varepsilon)}} \left(c_3 + \frac{c_3}{b_k^{(2+\varepsilon)l}} \right) b_k^{(2+\varepsilon)l} P_k(\mathrm{d}x) = O\left(\frac{1}{b_k^{2(2+\varepsilon)}} \right).$$

Hence (4.6.10) holds, while (4.6.11) can be proved in a similar manner.

Next, we prove (4.6.12) and (4.6.13). Introduce $z_k = \frac{1}{k+1}\big(w_k(y,u)f(y_k,u_k) - Ew_k(y,u)f(y_k,u_k)\big)$, $k \geq 1$ and $\mathscr{F}_n \triangleq \sigma\{z_k,\ 1 \leq k \leq n\}$ and $\mathscr{F}_n \triangleq \{\phi, \Omega\}$, if $n \leq 0$. It is clear that $Ez_k = 0$.

Define $c_k \triangleq \left(E|z_k|^{2+\varepsilon}\right)^{1/(2+\varepsilon)}$. By (4.6.10), we see $c_k = \left(E|z_k|^{2+\varepsilon}\right)^{\frac{1}{2+\varepsilon}} = O\left(1/kb_k^2\right)$. Given that $b_k = \frac{1}{k^\delta}$, $\delta \in (0, \frac{1}{4})$ and by the fact that $\{z_k\}_{k\geq 1}$ is an α-mixing with mixing coefficients tending to zero exponentially fast (Theorem 1.4.4), we have

$$\sum_{k=1}^{\infty} c_k^2 = O\left(\sum_{k=1}^{\infty} \frac{1}{k^2 b_k^4} \right) < \infty,$$

$$\sum_{k=1}^{\infty} (\log k)(\log\log k)^{1+\gamma}(\alpha(k))^{\frac{\varepsilon}{2+\varepsilon}} < \infty.$$

By applying Theorem 1.4.2 we have (4.6.12), while (4.6.13) can be proved in a similar manner.

Finally, we prove (4.6.14). Since ε_{k+1} is independent of $x_k = [y_k\ u_k]^T$, for (4.6.14) we only need to verify

$$\sum_{k=0}^{\infty} \frac{1}{(k+1)^2} w_k^2(y,u) E\varepsilon_{k+1}^2 < \infty \text{ a.s.} \tag{4.6.18}$$

Noticing (4.6.11) and $b_k = \frac{1}{k^\delta}$, $\delta \in \left(0, \frac{1}{4}\right)$, we have

$$\begin{aligned} E\sum_{k=0}^{\infty} \frac{1}{(k+1)^2} w_k^2(y,u) &= \sum_{k=0}^{\infty} \frac{1}{(k+1)^2} E w_k^2(y,u) \\ &= O\left(\sum_{k=1}^{\infty} \frac{1}{k^2 b_k^4} \right) = O\left(\sum_{k=1}^{\infty} \frac{1}{k^{2(1-2\delta)}} \right) < \infty. \end{aligned} \tag{4.6.19}$$

Hence (4.6.18) holds, which in turn implies (4.6.14). □

Theorem 4.6.1 *Assume A1.3.1–A1.3.5 hold and $f_{\mathrm{IV}}(\cdot,\cdot)$ is continuous at the given point $[y,u]^T \in \mathbb{R}^2$. Then*

$$\eta_k(y,u) \underset{k\to\infty}{\longrightarrow} f(y,u) f_{\mathrm{IV}}(y,u) \text{ a.s.} \tag{4.6.20}$$

$$\theta_k(y,u) \underset{k\to\infty}{\longrightarrow} f_{\mathrm{IV}}(y,u) \text{ a.s.}, \tag{4.6.21}$$

and hence

$$\frac{\eta_k(y,u)}{\theta_k(y,u)} \underset{k\to\infty}{\longrightarrow} f(y,u) \text{ a.s.} \tag{4.6.22}$$

Proof. The algorithm (4.6.4) can be rewritten as follows

$$\eta_{k+1}(y,u) = \Big[\eta_k(y,u) - \frac{1}{k+1}\big(\eta_k(y,u) - f(y,u)f_{\rm IV}(y,u) + e_{k+1}\big)\Big] \cdot I_{\big[\big|\eta_k(y,u) - \frac{1}{k+1}(\eta_k(y,u) - f(y,u)f_{\rm IV}(y,u) + e_{k+1})\big| \le M_{\sigma_k}\big]}, \tag{4.6.23}$$

where

$$\begin{aligned} e_{k+1} \triangleq & -w_k(y,u)\varepsilon_{k+1} + \Big(f(y,u)f_{\rm IV}(y,u) - Ew_k(y,u)f(y_k,u_k)\Big) \\ & + \Big(Ew_k(y,u)f(y_k,u_k) - w_k(y,u)f(y_k,u_k)\Big). \end{aligned} \tag{4.6.24}$$

This is SAAWET with the regression function $g(x) = -(x - \pi f(y,u)f_{\rm IV}(y,u))$. By Theorem 2.3.1 for (4.6.20) it suffices to prove

$$\lim_{T\to 0} \limsup_{k\to\infty} \frac{1}{T} \left| \sum_{i=n_k}^{m(n_k,t)} \frac{1}{i+1} e_{i+1} \right| = 0 \quad \text{a.s.} \quad \forall\, t \in (0,T] \tag{4.6.25}$$

along indices n_k of any convergent subsequence $\{\eta_{n_k}(y,u)\}$, where $m(k,t) = \max\{m : \sum_{i=k}^{m} \frac{1}{i+1} \le t\}$. Noticing (4.6.8), (4.6.12), and (4.6.14), we know that (4.6.25) holds. Hence, (4.6.20) is true.

Convergence (4.6.21) can be proved in a similar way, while (4.6.20) and (4.6.21) imply (4.6.25). □

To investigate the convergence rate of the estimates generated by (4.6.13)–(4.6.17), we need the following condition.

A4.6.1 *The unknown function $f(\cdot,\cdot)$ and the densities $f_\varepsilon(\cdot)$ and $f_u(\cdot)$ satisfy the Lipschitz conditions:*

$$|f(x_1,x_2) - f(y_1,y_2)| \le L_f \|(x_1,x_2) - (y_1,y_2)\| \;\forall [x_1,x_2]^T \in \mathbb{R}^2,\; [y_1,y_2]^T \in \mathbb{R}^2,$$
$$|f_\varepsilon(x) - f_\varepsilon(y)| \le L_\varepsilon |x-y|, \;\; |f_u(x) - f_u(y)| \le L_u |x-y| \;\; \forall\, x,\, y \in \mathbb{R},$$

where L_f, L_ε, and L_u are positive constants.

The assertion of Theorem 1.3.6 can be strengthened as follows.

Lemma 4.6.2 *If A1.3.1–A1.3.5 and A4.6.1 hold, then $P_{\rm IV}(\cdot)$ is with a continuous density $f_{\rm IV}(\cdot,\cdot)$. Further, $f_{\rm IV}(\cdot,\cdot)$ is bounded on $\mathbb{R}^2$ and satisfies the Lipschitz condition:*

$$\Big|f_{\rm IV}(s_1+\Delta s_1, s_2+\Delta s_2) - f_{\rm IV}(s_1,s_2)\Big| \le L_{\rm IV}\|(\Delta s_1,\Delta s_2)\| \;\; \forall\, [s_1,s_2]^T \in \mathbb{R}^2,$$

where $L_{\rm IV}$ is a positive constant.

Proof. We first prove that both $f_u(\cdot)$ and $f_\varepsilon(\cdot)$ are bounded on $\mathbb{R}$. We only prove $\sup_{x\in\mathbb{R}} f_u(x) < \infty$, while $\sup_{x\in\mathbb{R}} f_\varepsilon(x) < \infty$ can similarly be shown.

To prove $\sup_{x\in\mathbb{R}} f_u(x) < \infty$, by the continuity of $f_u(\cdot)$, it suffices to show $\lim_{|x|\to\infty} f_u(x) = 0$ or $\lim_{x\to+\infty} f_u(x) = 0$.

If $\limsup_{x\to+\infty} f_u(x) > 0$, then there would exist a positive sequence $\{x_k\}_{k\geq 1}$, $x_k > 0$, $x_k \to \infty$ as $k \to \infty$ such that $0 < \lim_{k\to\infty} f_u(x_k) \triangleq \gamma < \infty$.

Since $\int_{\mathbb{R}} f_u(x)\mathrm{d}x = 1$ and $x_k \to \infty$, the following sequence tends to zero

$$\int_{x_k-\frac{\gamma}{3L_u}}^{x_k+\frac{\gamma}{3L_u}} f_u(x)\mathrm{d}x \underset{k\to\infty}{\longrightarrow} 0.$$

By continuity of $f_u(\cdot)$, there exists $\bar{x}_k \in [x_k - \frac{\gamma}{3L_u}, x_k + \frac{\gamma}{3L_u}]$ such that $\int_{x_k-\frac{\gamma}{3L_u}}^{x_k+\frac{\gamma}{3L_u}} f_u(x)\mathrm{d}x = f_u(\bar{x}_k)\frac{2\gamma}{3L_u}$ and hence $f_u(\bar{x}_k) \underset{k\to\infty}{\longrightarrow} 0$.

Noticing that $f_u(\cdot)$ is Lipschitzian, we have that

$$\begin{aligned} f_u(x_k) \leq & |f_u(x_k) - f_u(\bar{x}_k)| + f_u(\bar{x}_k) \leq L_u|x_k - \bar{x}_k| + f_u(\bar{x}_k) \\ \leq & L_u\frac{2\gamma}{3L_u} + f_u(\bar{x}_k) = \frac{2}{3}\gamma + f_u(\bar{x}_k), \end{aligned}$$

which contradicts with $\lim_{k\to\infty} f_u(x_k) = \gamma$ by noticing $f_u(\bar{x}_k) \to 0$ as $k \to \infty$. So we have $\lim_{x\to+\infty} f_u(x) = 0$, $\lim_{|x|\to\infty} f_u(x) = 0$, and $\sup_{x\in\mathbb{R}} f_u(x) < \infty$.

By (1.3.34), it follows that

$$f_{\mathrm{IV}}(s_1,s_2) = \iint_{\mathbb{R}^2} f_\varepsilon(s_1 - f(x_1,x_2))P_{\mathrm{IV}}(\mathrm{d}x) f_u(s_2) \leq \Big(\sup_{s_1\in\mathbb{R}} f_\varepsilon(s_1)\Big)\Big(\sup_{s_2\in\mathbb{R}} f_u(s_2)\Big) < \infty.$$

Hence $f_{\mathrm{IV}}(\cdot,\cdot)$ is bounded on $\mathbb{R}^2$.

From (1.3.34) and A4.6.1, we have the following inequalities

$$\begin{aligned} &\big|f_{\mathrm{IV}}(s_1+\Delta s_1, s_2+\Delta s_2) - f_{\mathrm{IV}}(s_1,s_2)\big| \\ &\leq \left|\iint_{\mathbb{R}^2} f_\varepsilon(s_1+\Delta s_1 - f(x_1,x_2))P_{\mathrm{IV}}(\mathrm{d}x) f_u(s_2+\Delta s_2) \right.\\ &\quad \left. - \iint_{\mathbb{R}^2} f_\varepsilon(s_1+\Delta s_1 - f(x_1,x_2))P_{\mathrm{IV}}(\mathrm{d}x) f_u(s_2)\right| \\ &\quad + \left|\iint_{\mathbb{R}^2} f_\varepsilon(s_1+\Delta s_1 - f(x_1,x_2))P_{\mathrm{IV}}(\mathrm{d}x) f_u(s_2) - \iint_{\mathbb{R}^2} f_\varepsilon(s_1 - f(x_1,x_2))P_{\mathrm{IV}}(\mathrm{d}x) f_u(s_2)\right| \\ &\leq \Big(\sup_{s_1\in\mathbb{R}} f_\varepsilon(s_1)\Big)\big|f_u(s_2+\Delta s_2) - f_u(s_2)\big| \\ &\quad + \Big(\sup_{s_2\in\mathbb{R}} f_u(s_2)\Big)\iint_{\mathbb{R}^2} \big|f_\varepsilon(s_1+\Delta s_1 - f(x_1,x_2)) - f_\varepsilon(s_1 - f(x_1,x_2))\big| P_{\mathrm{IV}}(\mathrm{d}x) \\ &\leq \Big(\sup_{s_1\in\mathbb{R}} f_\varepsilon(s_1)\Big) L_u|\Delta s_2| + \Big(\sup_{s_2\in\mathbb{R}} f_u(s_2)\Big) L_\varepsilon|\Delta s_1| \leq L_{\mathrm{IV}}\|(\Delta s_1, \Delta s_2)\|, \end{aligned}$$

where L_{IV} is a positive constant. Hence, $f_{\mathrm{IV}}(\cdot,\cdot)$ is Lipschitzian. □

For the convergence rate of the estimates generated by (4.6.3)–(4.6.7), we need the following technical lemma.

Lemma 4.6.3 *If A1.3.1–A1.3.5 and A4.6.1 hold, then*

$$Ew_k(y,u)f(y_k,u_k) - f(y,u)f_{\mathrm{IV}}(y,u) = O\Big(\frac{1}{k^\delta}\Big), \tag{4.6.26}$$

$$Ew_k(y,u) - f_{\mathrm{IV}}(y,u) = O\Big(\frac{1}{k^\delta}\Big) \tag{4.6.27}$$

for any fixed $\varepsilon \geq 0$ *and any* $\delta \in (0, \frac{1}{4})$ *figured in (4.6.3), and*

$$\sum_{k=0}^{\infty} \frac{1}{(k+1)^{1-\delta'}} \Big(w_k(y,u)f(y_k,u_k) - Ew_k(y,u)f(y_k,u_k)\Big) < \infty \text{ a.s.}, \tag{4.6.28}$$

$$\sum_{k=0}^{\infty} \frac{1}{(k+1)^{1-\delta'}} \Big(w_k(y,u) - Ew_k(y,u)\Big) < \infty \text{ a.s.}, \tag{4.6.29}$$

and

$$\sum_{k=0}^{\infty} \frac{1}{(k+1)^{1-\delta'}} w_k(y,u)\varepsilon_{k+1} < \infty \text{ a.s.} \tag{4.6.30}$$

for any $\delta' > 0$ *such that* $0 < \delta' < (\frac{1}{2} - 2\delta)$.

Proof. By the definition of $w_k(y,u)$, we have

$$Ew_k(y,u)f(y_k,u_k) - f(y,u)f_{\mathrm{IV}}(y,u) = I_{1,k} + I_{2,k}, \tag{4.6.31}$$

where

$$I_{1,k} \triangleq \iint\limits_{\mathbb{R}^2} \frac{1}{2\pi b_k^2} \exp\left\{-\frac{1}{2}\left(\frac{x_1-y}{b_k}\right)^2 - \frac{1}{2}\left(\frac{x_2-u}{b_k}\right)^2\right\} f(x_1,x_2)f_{\mathrm{IV}}(x_1,x_2)\mathrm{d}x_1\mathrm{d}x_2$$
$$- f(y,u)f_{\mathrm{IV}}(y,u),$$

$$I_{2,k} \triangleq \iint\limits_{\mathbb{R}^2} \frac{1}{2\pi b_k^2} \exp\left\{-\frac{1}{2}\left(\frac{x_1-y}{b_k}\right)^2 - \frac{1}{2}\left(\frac{x_2-u}{b_k}\right)^2\right\} f(x_1,x_2)\Big(P_k(\mathrm{d}x) - P_{\mathrm{IV}}(\mathrm{d}x)\Big).$$

By setting $\frac{x_1-y}{b_k} = t_1$, $\frac{x_2-u}{b_k} = t_2$, and by noticing that $f_{\mathrm{IV}}(\cdot,\cdot)$ is bounded on $\mathbb{R}^2$ and both $f_{\mathrm{IV}}(\cdot,\cdot)$ and $f(\cdot,\cdot)$ are Lipschitzian, it follows that

$$|I_{1,k}| = \left|\iint\limits_{\mathbb{R}^2} \frac{1}{2\pi} \mathrm{e}^{-\frac{t_1^2+t_1^2}{2}} \Big(f(b_kt_1+y, b_kt_2+u)\, f_{\mathrm{IV}}(b_kt_1+y, b_kt_2+u) \right.$$
$$\left. - f(y,u)f_{\mathrm{IV}}(y,u)\Big)\mathrm{d}t_1\mathrm{d}t_2\right|$$

$$\leq \left| \iint_{\mathbb{R}^2} \frac{1}{2\pi} e^{-\frac{t_1^2+t_1^2}{2}} \Big(f(b_k t_1 + y, b_k t_2 + u) f_{\text{IV}}(b_k t_1 + y, b_k t_2 + u)$$

$$- f(y,u) f_{\text{IV}}(b_k t_1 + y, b_k t_2 + u) \Big) \mathrm{d}t_1 \mathrm{d}t_2 \right|$$

$$+ \left| \iint_{\mathbb{R}^2} \frac{1}{2\pi} e^{-\frac{t_1^2+t_1^2}{2}} \Big(f(y,u) f_{\text{IV}}(b_k t_1 + y, b_k t_2 + u) - f(y,u) f_{\text{IV}}(y,u) \Big) \mathrm{d}t_1 \mathrm{d}t_2 \right|$$

$$\leq \Big(\sup_{(t_1,t_2)\in\mathbb{R}^2} f_{\text{IV}}(t_1,t_2) \Big) L_f \iint_{\mathbb{R}^2} \frac{1}{2\pi} e^{-\frac{t_1^2+t_1^2}{2}} (b_k^2 t_1^2 + b_k^2 t_2^2)^{\frac{1}{2}} \mathrm{d}t_1 \mathrm{d}t_2$$

$$+ |f(y,u)| L_{\text{IV}} \iint_{\mathbb{R}^2} \frac{1}{2\pi} e^{-\frac{t_1^2+t_1^2}{2}} (b_k^2 t_1^2 + b_k^2 t_2^2)^{\frac{1}{2}} \mathrm{d}t_1 \mathrm{d}t_2 = O(b_k) = O\Big(\frac{1}{k^\delta}\Big). \tag{4.6.32}$$

By ergodicity of $\{x_k\}$ and by noticing $f(x_1, x_2)$ is Lipschitzian, it is derived that

$$|I_{2,k}| = O\Big(\iint_{\mathbb{R}^2} \frac{1}{b_k^2} \Big| P_k(\mathrm{d}x) - P_{\text{IV}}(\mathrm{d}x) \Big| \Big) = O\Big(\frac{1}{b_k^2} \rho^k \Big). \tag{4.6.33}$$

Hence (4.6.26) follows from (4.6.31), (4.6.32), and (4.6.33), while (4.6.27) can be proved in a similar manner.

We now sketch the proof of (4.6.28), (4.6.29), and (4.6.30).

Set $z_k \triangleq \frac{1}{(k+1)^{1-\delta'}} \big(w_k(y,u) f(y_k,u_k) - E w_k(y,u) f(y_k,u_k) \big)$, $k \geq 1$ and $\mathscr{F}_n \triangleq \sigma\{z_k,\ 1 \leq k \leq n\}$ with $\mathscr{F}_n \triangleq \{\phi, \Omega\}$ for $n \leq 0$. Notice that the chain $\{x_k = [y_k\ u_k]^T\}_{k\geq 1}$ is an α-mixing with mixing coefficients tending to zero exponentially fast. For any fixed $\varepsilon > 0$, define $c_k \triangleq \big(E|z_k|^{2+\varepsilon} \big)^{\frac{1}{2+\varepsilon}}$. By Lemma 4.6.1 and the assumption that $0 < \delta' < (\frac{1}{2} - 2\delta)$, it follows that

$$\sum_{k-1}^{\infty} c_k^2 = O\Big(\sum_{k-1}^{\infty} \frac{1}{k^{2(1-\delta')} b_k^4} \Big) < \infty. \tag{4.6.34}$$

Given (4.6.34), by Theorem 1.4.2 we derive (4.6.28), while (4.6.29) is verified in a similar manner. Since ε_{k+1} is independent of $x_k = [y_k\ u_k]^T$, for (4.6.30) we only need to verify

$$\sum_{k=0}^{\infty} \frac{1}{(k+1)^{2(1-\delta')}} w_k^2(y,u) E\varepsilon_{k+1}^2 < \infty \ \text{a.s.}, \tag{4.6.35}$$

which can be proved by noticing

$$E \sum_{k=0}^{\infty} \frac{1}{(k+1)^{2(1-\delta')}} w_k^2(y,u) = O\Big(\sum_{k=1}^{\infty} \frac{1}{k^{2(1-\delta')} b_k^4} \Big) < \infty.$$

□

Theorem 4.6.2 *Assume A1.3.1–A1.3.5 and A4.6.1 hold. Then*

$$\left|\eta_k(y,u) - f(y,u)f_{\text{IV}}(y,u)\right| + \left|\theta_k(y,u) - f_{\text{IV}}(y,u)\right| = o\left(\frac{1}{k^{\delta'}}\right) \quad \text{a.s.} \tag{4.6.36}$$

and

$$\left|\frac{\eta_k(y,u)}{\theta_k(y,u)} - f(y,u)\right| = o\left(\frac{1}{k^{\delta'}}\right) \tag{4.6.37}$$

for any $0 < \delta' < \min(\frac{1}{2} - 2\delta, \delta)$.

Proof. By Lemma 4.6.3, we have

$$\sum_{k=0}^{\infty} \frac{1}{(k+1)^{1-\delta'}} \Big(w_k(y,u)f(y_k,u_k) - Ew_k(y,u)f(y_k,u_k)\Big) < \infty \quad \text{a.s.}$$

$$\sum_{k=0}^{\infty} \frac{1}{(k+1)^{1-\delta'}} w_k(y,u)\varepsilon_{k+1} < \infty \quad \text{a.s.}$$

and

$$Ew_k(y,u)f(y_k,u_k) - f(y,u)f_{\text{IV}}(y,u) = O\left(\frac{1}{k^{\delta'}}\right)$$

for any $0 < \delta' < \min(\frac{1}{2} - 2\delta, \delta)$.

Paying attention to (4.6.23) and (4.6.24), by Theorem 2.6.1 we have

$$\left|\eta_k(y,u) - f(y,u)f_{\text{IV}}(y,u)\right| = o\left(\frac{1}{k^{\delta'}}\right) \quad \text{a.s.}$$

Similarly, it is shown that

$$\left|\theta_k(y,u) - f_{\text{IV}}(y,u)\right| = o\left(\frac{1}{k^{\delta'}}\right) \quad \text{a.s.}$$

Hence (4.6.36) holds, while (4.6.37) is implied by (4.6.36). □

We now consider the NARX system with $p \geq 1,\ q \geq 1$, i.e.,

$$y_{k+1} = f(y_k, \cdots, y_{k+1-p}, u_k, \cdots, u_{k+1-q}) + \varepsilon_{k+1}. \tag{4.6.38}$$

The identification algorithms and their strong consistency can be obtained similarly to those for the case $(p = 1,\ q = 1)$. For example, at a fixed $\phi^* = [y^{(1)} \dots y^{(p)} \dots u^{(1)} \dots u^{(q)}]^T \in \mathbb{R}^{p+q}$, the kernel function and the recursive identification algorithms are as follows:

$$w_k(\phi^*) = \frac{1}{(2\pi)^{\frac{p+q}{2}} b_k^{p+q}} \exp\Big\{ -\frac{1}{2}\sum_{i=1}^{p}\Big(\frac{y_{k+1-i} - y^{(i)}}{b_k}\Big)^2 - \frac{1}{2}\sum_{i=1}^{q}\Big(\frac{u_{k+1-i} - u^{(i)}}{b_k}\Big)^2 \Big\} \tag{4.6.39}$$

with $b_k = \frac{1}{k^\delta}$, $\delta \in (0, \frac{1}{2(p+q)})$,

$$\eta_{k+1}(\phi^*) = \left[\eta_k(\phi^*) - \frac{1}{k+1}\left(\eta_k(\phi^*) - w_k(\phi^*)y_{k+1}\right)\right] \cdot I_{\left[\left|\eta_k(\phi^*) - \frac{1}{k+1}(\eta_k(\phi^*) - w_k(\phi^*)y_{k+1})\right| \le M_{\sigma_k}\right]}, \tag{4.6.40}$$

$$\sigma_k = \sum_{j=1}^{k-1} I_{\left[\left|\eta_j(\phi^*) - \frac{1}{j+1}\left(\eta_j(\phi^*) - w_j(\phi^*)y_{j+1}\right)\right| > M_{\sigma_j}\right]}, \quad \sigma_0 = 0, \tag{4.6.41}$$

and

$$\theta_{k+1}(\phi^*) = \left[\theta_k(\phi^*) - \frac{1}{k+1}\left(\theta_k(\phi^*) - w_k(\phi^*)\right)\right] \cdot I_{\left[\left|\theta_k(\phi^*) - \frac{1}{k+1}(\theta_k(\phi^*) - w_k(\phi^*))\right| \le M_{\delta_k}\right]}, \tag{4.6.42}$$

$$\delta_k = \sum_{j=1}^{k-1} I_{\left[\left|\theta_j(\phi^*) - \frac{1}{j+1}\left(\theta_j(\phi^*) - w_j(\phi^*)\right)\right| > M_{\delta_j}\right]}, \quad \delta_0 = 0. \tag{4.6.43}$$

By carrying out a similar discussion as that for Theorem 4.6.2, we have

Theorem 4.6.3 *If A1.3.1–A1.3.3, A1.3.4', and A1.3.5' hold, then the estimates generated by (4.6.39)–(4.6.43) are strongly consistent provided that $f_{\rm IV}(\cdot)$ is continuous at ϕ^*:*

$$\eta_k(\phi^*) \underset{k\to\infty}{\longrightarrow} f(\phi^*) f_{\rm IV}(\phi^*) \text{ a.s.}, \quad \theta_k(\phi^*) \underset{k\to\infty}{\longrightarrow} f_{\rm IV}(\phi^*) \text{ a.s.}, \tag{4.6.44}$$

and hence

$$\frac{\eta_k(\phi^*)}{\theta_k(\phi^*)} \underset{k\to\infty}{\longrightarrow} f(\phi^*) \text{ a.s.} \tag{4.6.45}$$

Further, if $f(\cdot)$, $f_u(\cdot)$, and $f_\varepsilon(\cdot)$ are Lipschitzian, then

(i) the density $f_{\rm IV}(\cdot)$ is continuous and Lipschitzian on $\mathbb{R}^{p+q}$,

(ii) the estimates generated by (4.6.39)–(4.6.43) are with the following convergence rate,

$$\eta_k(\phi^*) - f(\phi^*) f_{\rm IV}(\phi^*) = o\left(\frac{1}{k^{\delta'}}\right) \text{ a.s.}, \tag{4.6.46}$$

$$\theta_k(\phi^*) - f_{\rm IV}(\phi^*) = o\left(\frac{1}{k^{\delta'}}\right) \text{ a.s.}, \tag{4.6.47}$$

and

$$\frac{\eta_k(\phi^*)}{\theta_k(\phi^*)} - f(\phi^*) = o\left(\frac{1}{k^{\delta'}}\right) \text{ a.s.} \tag{4.6.48}$$

for any $0 < \delta' < \min\left(\frac{1}{2} - (p+q)\delta, \delta\right)$.

Remark 4.6.1 *From Theorem 4.6.3, it is clear that if the parameter δ in (4.6.39) is chosen as $\delta_0 = 1/2(p+q+1)$, then the convergence rate is optimized, i.e., δ' in Theorem 4.6.3 may be arbitrarily close to $1/(2(p+q+1))$, where (p,q) are the orders or the upper bounds of orders of the NARX system.*

Numerical Examples

(i) Let the NARX system be

$$y_{k+1} = ay_k + bu_k^3 + \varepsilon_{k+1},$$

where $\{u_k\}$ and $\{\varepsilon_k\}$ are mutually independent and each of them is an iid sequence, $u_k \in \mathcal{N}(0,1)$, $\varepsilon_k \in \mathcal{N}(0,1)$. Let $f(y,u) = ay + bu^3$, $a = 0.5$, and $b = 1$.

Let the parameters used in the algorithms (4.6.3)–(4.6.7) be as follows: $M_k = 2k$, $\delta = \frac{1}{6}$. Figures 4.6.1–4.6.2 show the estimated surfaces $\eta_k(y,u)/\theta_k(y,u)$ and estimation errors $|f(y,u) - \eta_k(y,u)/\theta_k(y,u)|$ at $k = 5000$. In Figure 4.6.1, the solid lines denote the true surfaces while the dashed lines denote the estimated surfaces. As expected, the estimated surface with the optimal factor $\delta = 1/6$ is very close to the true one.

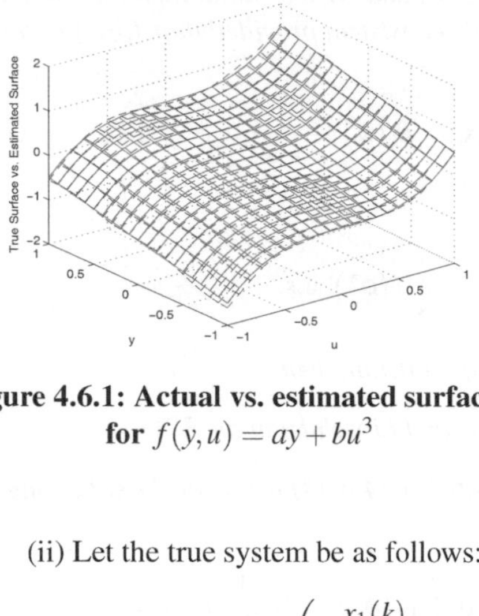

Figure 4.6.1: Actual vs. estimated surfaces for $f(y,u) = ay + bu^3$

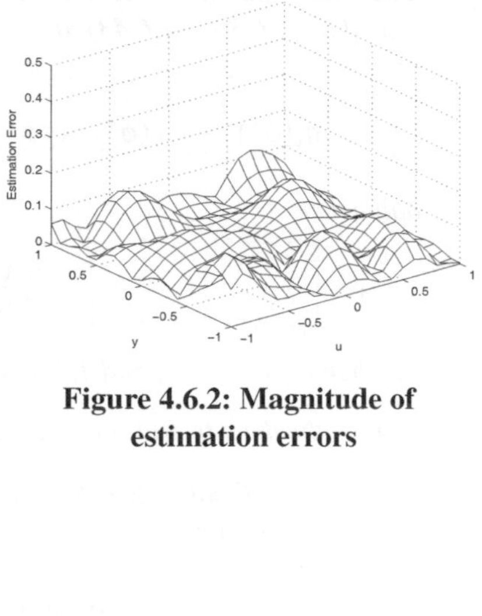

Figure 4.6.2: Magnitude of estimation errors

(ii) Let the true system be as follows:

$$x_1(k+1) = \left(\frac{x_1(k)}{1+x_1^2(k)} + 1\right)\sin x_2(k),$$

$$x_2(k+1) = x_2(k)\cos x_2(k) + x_1(k)\exp\left(-\frac{x_1^2(k)+x_2^2(k)}{8}\right) + \frac{u_k^3}{1+u_k^2+0.5\cos(x_1(k)+x_2(k))},$$

$$y_k = \frac{x_1(k)}{1+0.5\sin x_2(k)} + \frac{x_2(k)}{1+0.5\sin x_1(k)} + \varepsilon_k,$$

where u_k and y_k are the system input and output, respectively, ε_k is the system noise, $\varepsilon_k \in \mathscr{N}(0,\sigma^2)$ with $\sigma = 0.1$, and $x_1(k)$ and $x_2(k)$ are the system states, which are not directly observed.

The following NARX system

$$y_{k+1} = f(y_k, y_{k-1}, y_{k-2}, u_k, u_{k-1}, u_{k-2}) + \varepsilon_{k+1}$$

is used to model (approximate) the true system.

First, $N(= 5000)$ samples $\{u_k, y_k\}_{k=1}^{5000}$ are generated by iid u_k with $u_k \in \mathscr{N}(0,1)$. The function $f(\cdot)$ is recursively estimated based on $\{u_k, y_k\}_{k=1}^{5000}$ by the algorithms (4.6.39)–(4.6.43) with $b_k = 1/k^{\frac{1}{100}}$. Denote the estimate for $f(\cdot)$ by $\widehat{f}_N(\cdot)$.

Then the following input signals

$$u_k = \sin\frac{\pi k}{5} + \sin\frac{2\pi k}{25}, \quad k = N+1, \cdots, N+200$$

are fed to the estimated model to calculate the one-step ahead predicted output

$$\widehat{y}_{k+1} = \widehat{f}_N(y_k, y_{k-1}, y_{k-2}, u_k, u_{k-1}, u_{k-2}),$$

$k = 5001, \cdots, 5200$, which are marked by the dashed line in Figure 4.6.3. The same input signals are also fed to the true system, and the corresponding output is expressed by the solid line in Figure 4.6.3. For a comparison, the dotted line in Figure 4.6.3 plots the predicted outputs generated by the following kernel estimator:

$$\widehat{f}_N = \frac{\sum_{k=1}^{N} K\left(\frac{y_k - y^{(1)}}{r}, \cdots, \frac{y_{k-2} - y^{(3)}}{r}, \frac{u_k - u^{(1)}}{r}, \cdots, \frac{u_{k-2} - u^{(3)}}{r}\right) y_{k+1}}{\sum_{k=1}^{N} K\left(\frac{y_k - y^{(1)}}{r}, \cdots, \frac{y_{k-2} - y^{(3)}}{r}, \frac{u_k - u^{(1)}}{r}, \cdots, \frac{u_{k-2} - u^{(3)}}{r}\right)} \tag{4.6.49}$$

where $K(\cdot)$ is a Gaussian type kernel with bandwidth $r = 0.2$.

In Figure 4.6.4, the solid and dashed lines are the same as those in Figure 4.6.3. But the dotted line indicates the one-step ahead predicted outputs generated by the direct weight optimization (DWO) approach, where $f(\varphi^*)$ is estimated by

$$\widehat{f}_N(\varphi^*) = \sum_{k=1}^{N} \widehat{w}_k y_{k+1}, \tag{4.6.50}$$

where

$$[\widehat{w}_1, \cdots, \widehat{w}_N]^T = \arg\min_{w_1,\cdots,w_N} \frac{1}{4}\left(\sum_{k=1}^{N} |w_k| \|\varphi_k - \varphi^*\|\right)^2 + \sigma^2 \sum_{k=1}^{N} w_k^2 \tag{4.6.51}$$

$$\text{subject to } \sum_{k=1}^{N} w_k = 1 \text{ and } \sum_{k=1}^{N} w_k(\varphi_k - \varphi^*) = 0 \tag{4.6.52}$$

with $\varphi_k = [y_k \; y_{k-1} \; y_{k-2} \; u_k \; u_{k-1} \; u_{k-2}]^T$ and $\varphi^* = [y^{(1)} \; y^{(2)} \; y^{(3)} \; u^{(1)} \; u^{(2)} \; u^{(3)}]^T$.

From Figures 4.6.3 and 4.6.4, it is seen that the performances of the three methods under comparison do not differ from each other essentially. However, since the kernel approach (4.6.49) and the DWO approach (4.6.52) deal with the data set of fixed size, they are unable to update the estimates when new data arrive, i.e., they are nonrecursive. In contrast to this, the SAAWET-based approach proposed in this section is recursive and is shown to be convergent with probability one.

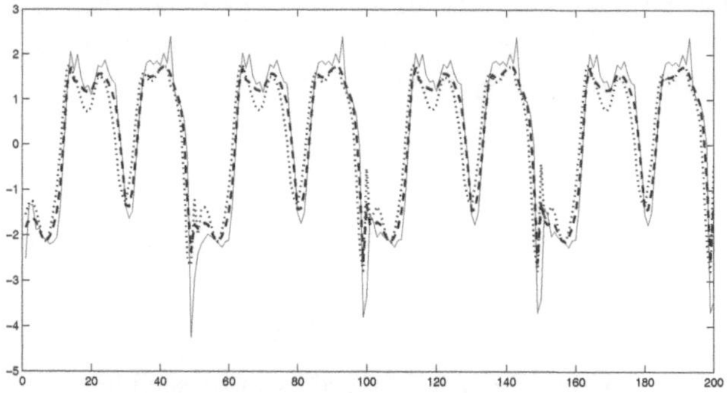

Figure 4.6.3: Predicted and actual outputs (SAAWET algorithm and kernel algorithm)

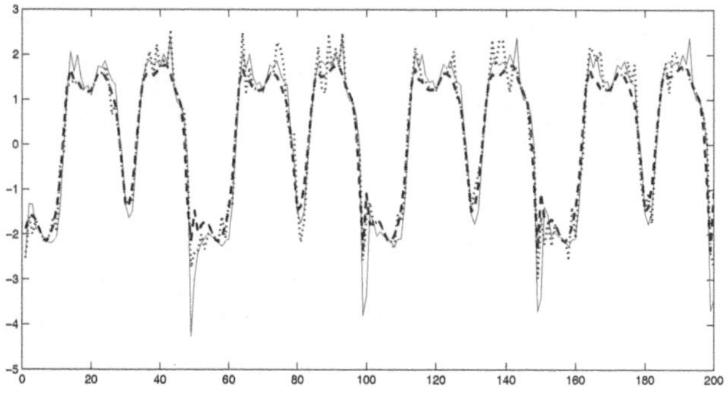

Figure 4.6.4: Predicted and actual outputs (SAAWET algorithm and DWO algorithm)

4.7 Notes and References

For the practical applications of the block-oriented nonlinear systems like the Wiener system, Hammerstein system, and their combinations we refer to [58], [63], and [121], while for their identification we refer to [10], [12], [20], [42], [46], [47], [49], [55], [93], [111], [115], [122], and to [62] and [104] for the EIV situation.

As far as the identification of NARX systems is of concern, the methods can roughly be divided into two categories, the parametric approach and the nonparametric approach. For the parametric approach we refer to [101], while for the nonparametric approach we refer to [7], [8], [38], [39], [52], [97], [108], and [118] and references therein. Particularly, the direct weight optimization (DWO) approach to nonparametric identification of NARX systems is introduced in [97] and then further investigated in [8]; the nonrecursive kernel estimator is discussed in details from a statistic viewpoint in [38] and [39] while from a systems and control viewpoint in [7].

Most of the methods proposed in the above references are designed for a special class of systems and are not directly applicable to other systems. Sections 4.1–4.6 are written on the basis of [29], [83], [84], [85], [86], [118], and [120], from which we can see that the recursive identification of these systems is carried out in a unified framework with SAAWET applied. Figures 4.1.1–4.1.7 are reprinted with permission from [29], Figures 4.2.1–4.2.6 with permission from [86], Figures 4.5.1–4.5.5 with permission from [85], Figures 4.6.1–4.6.2 with permission from [120], and Figures 4.6.3–4.6.4 are reprinted with permission from [118].

We notice that the Wiener, Hammerstein, Wiener–Hammerstein, and NARX systems are typical examples of nonlinear systems among many others. For the identification of other kinds of nonlinear systems and related issues, we refer to [32], [42], [89], [113], etc.

Notes and References

[illegible]

[illegible]

Chapter 5

Other Problems Reducible to Parameter Estimation

CONTENTS

5.1 Principal Component Analysis

In a practical system there may be a large amount of variables involved, but the variables may not be equally important. The principal component analysis (PCA) proposed by Pearson aims at estimating eigenvectors of a symmetric matrix in the decreasing order of importance, i.e., first to select the most important factors and then the less important factors by using linear transformations acting on the variables.

PCA is now widely used in various areas such as data analysis, image compression, pattern recognition, subspace identification, and many others.

Let each component of $x \in \mathbb{R}^n$ represent a variable of the system with huge n. PCA is to find the eigenvectors of $A \triangleq Exx^T$.

We now consider a slightly modified setting. Let a deterministic symmetric matrix A be noisily observed: $A_k = A + N_k$, where A_k is the observation of A at time k

and N_k is the observation noise. On the basis of observations $\{A_k\}$, it is required to recursively estimate the eigenvectors and the corresponding eigenvalues of A in the decreasing order of eigenvalues. We impose no restrictive conditions on A.

Algorithms

Let us first define the recursive algorithm for $\{u_k^{(1)}\}_{k\geq 0}$ estimating the normalized eigenvector or one of the normalized eigenvectors corresponding to the largest eigenvalue of A:

$$\tilde{u}_{k+1}^{(1)} = u_k^{(1)} + a_k A_{k+1} u_k^{(1)}, \quad a_k > 0 \tag{5.1.1}$$

$$u_{k+1}^{(1)} = \tilde{u}_{k+1}^{(1)} / \|\tilde{u}_{k+1}^{(1)}\|, \tag{5.1.2}$$

whenever $\|\tilde{u}_{k+1}^{(1)}\| \neq 0$.

In the case where $\|\tilde{u}_{k+1}^{(1)}\| = 0$, $u_k^{(1)}$ is reset to be some other unit vector making $\|\tilde{u}_{k+1}^{(1)}\| \neq 0$ defined by (5.1.1).

Assuming $u_s^{(i)}$, $i = 1, \cdots, j$, $s = 0, 1, \cdots, k$ have been defined, we define $u_s^{(j+1)}$, $s = 0, 1, \cdots, k+1$ as follows.

For this we first define the $n \times j$-matrix

$$V_s^{(j)} \triangleq [u_s^{(1)}, P_s^{(1)} u_s^{(2)}, \cdots, P_s^{(j-1)} u_s^{(j)}], \quad P_s^{(0)} \triangleq I, \quad s = 1, \cdots, k, \tag{5.1.3}$$

where $P_s^{(i)} \triangleq I - V_s^{(i)} V_s^{(i)+}, i = 1, \cdots, j-1$ is the projection to the subspace orthogonal to the space spanned by columns of $V_s^{(i)}$, where $V_s^{(i)+}$ denotes the pseudo-inverse of $V_s^{(i)}$.

Given an initial unit vector $u_0^{(j+1)}$, recursively define

$$\tilde{u}_{k+1}^{(j+1)} = P_k^{(j)} u_k^{(j+1)} + a_k P_k^{(j)} A_{k+1} P_k^{(j)} u_k^{(j+1)}, \tag{5.1.4}$$

$$u_{k+1}^{(j+1)} = \tilde{u}_{k+1}^{(j+1)} / \|\tilde{u}_{k+1}^{(j+1)}\|, \tag{5.1.5}$$

whenever $\|\tilde{u}_{k+1}^{(j+1)}\| \neq 0$.

Otherwise, reset $u_k^{(j+1)}$ to be a unit vector so that $\|u_k^{(j+1)}\| = 1$ and $\|P_k^{(j)} u_k^{(j+1)}\| = 1$.

Noticing

$$a_k P_k^{(j)} A_{k+1} P_k^{(j)} u_k^{(j+1)} \to 0$$

by A5.1.1 and A5.1.2 to be given later and $\|P_k^{(j)} u_k^{(j+1)}\| = 1$ after a resetting, we find that $\|\tilde{u}_{k+1}^{(j+1)}\| = 0$ may occur at most a finite number of times, and hence resetting $u_k^{(j+1)}$ ceases in a finite number of steps.

From now on we always assume that k is large enough and no resetting occurs.

For eigenvalues of A the recursive estimates $\{\lambda_k^{(j)}\}_{k\geq 1}$, $j = 1, \cdots, n$ with arbitrary initial values $\lambda_0^{(j)}$ are given by the following algorithms

$$\lambda_{k+1}^{(j)} = \lambda_k^{(j)} - a_k(\lambda_k^{(j)} - u_k^{(j)T} A_{k+1} u_k^{(j)}) \tag{5.1.6}$$

Denote by J the set of all unit eigenvectors of A.

Let $V(J) \triangleq \{\lambda^{(1)},\cdots,\lambda^{(n)}\}$ be the set of eigenvalues of A stated in the non-increasing order. Notice that some of the eigenvalues may be identical. The convergence analysis is completed by three steps.

Step 1. We first show that for each $\{u_k^{(j)}\}$ there exists a subset J_j of J such that as k tends to infinity $d(u_k^{(j)},J_j)$, the distance between $u_k^{(j)}$ and J_j converges to zero, and $\lambda_k^{(j)}$ converges to the eigenvalue $\lambda(j)$ associated with J_j.

Step 2. It is shown that the convergence established in Step 1 is ordered in the sense that $\lambda(j) = \lambda^{(j)}$. In other words, J_1 corresponds to the largest eigenvalue of A, and J_2 either coincides with J_1 in the case $\lambda^{(1)}$ is with multiplicity greater than one, or corresponds to the second largest eigenvalue of A and so on.

Step 3. Except the case where all eigenvalues are equal to each other, it is shown there is a unit vector $u^{(j)} \in J_j$ such that $u_k^{(j)} \underset{k\to\infty}{\longrightarrow} u^{(j)}$ and $\|u_k^{(j)} - u^{(j)}\| = O(a_k^{\delta})$ with $\delta > 0$.

For establishing results stated in Step 1, the following assumptions are needed.

A5.1.1 *$a_k > 0$, $a_k \underset{k\to\infty}{\longrightarrow} 0$, $\sum_{k=0}^{\infty} a_k = \infty$, and $\sum a_k^{1+\eta} < \infty$ for any $\eta > 0$. Moreover, there is an $a > 0$ such that*

$$\lim_{k\to\infty} a_{k+1}^{-1} - a_k^{-1} = a > 0.$$

A5.1.2 *$A_k = A + N_k$, $\{N_k, \mathscr{F}_k\}$ is a bounded mds with $\sup_k \|N_{k+1}\| = \zeta < \infty$ a.s., and*

$$\sup_k E(\|N_{k+1}\|^2|\mathscr{F}_k) < \infty \ \text{a.s.},$$

where $\mathscr{F}_k$ is the σ-algebra generated by $\{N_1,\cdots,N_k\}$.

Remark 5.1.1 *It is clear that if $a_k = \frac{c}{k}$ with $c > 0$, then $\lim_{k\to\infty} a_{k+1}^{-1} - a_k^{-1} = \frac{1}{c}$, and A5.1.1 holds. Under A5.1.1, $\frac{a_k}{a_{k+1}} = 1 + O(a_k)$. So, in the sequel, we will not distinguish between $O(a_k)$ and $O(a_{k-s})$ for any finite s.*

Convergence of Estimates

We show the convergence of the algorithms defined by (5.1.1)–(5.1.6).

Theorem 5.1.1 *Assume A5.1.1 and A5.1.2 hold. Then the estimates $\{u_k^{(i)}\}$, $i = 1,\cdots,n$ given by (5.1.1)–(5.1.5) have the following properties:*

(i) There exists a connected subset J_i of J such that $d(u_k^{(i)},J_i) \underset{k\to\infty}{\longrightarrow} 0$.

(ii) There is an eigenvalue $\lambda(i) \in V(J)$ so that

$$d(Au_k^{(i)},\lambda(i)u_k^{(i)}) \to 0, \ \text{and}\ Au = \lambda(i)u \ \text{for any}\ u \in J_i. \tag{5.1.7}$$

(iii) *The recursive algorithm for $u_k^{(i)}$ can be expressed as*

$$\begin{aligned} u_{k+1}^{(i)} =& u_k^{(i)} + a_k(Au_k^{(i)} - (u_k^{(i)T}Au_k^{(i)})u_k^{(i)}) \\ &+ O\big(a_k(a_k + \sum_{s=1}^{i-1} d(u_k^{(s)}, J_s))\big) + a_k\delta_{k+1}(i), \end{aligned} \tag{5.1.8}$$

where $\delta_{k+1}(i)$ is bounded and is a linear combination of mdses being measurable with respect to $\mathscr{F}_{k+1}$, $\mathscr{F}_k$, $\cdots$, and $\mathscr{F}_{k+2-i}$, respectively, and with bounded second conditional moments and $\sum_{k=1}^{\infty} a_k\delta_{k+1}(i) < \infty$ a.s.

(iv) *$\lambda_k^{(j)}$ defined by (5.1.6) converges to $\lambda(j)$ as k tends to infinity, $j = 1, \cdots, n$.*

Proof. For large k, $u_{k+1}^{(1)}$ can be expanded as follows

$$\begin{aligned} u_{k+1}^{(1)} =& (u_k^{(1)} + a_kA_{k+1}u_k^{(1)})(1 + 2a_ku_k^{(1)T}A_{k+1}u_k^{(1)} + a_k^2u_k^{(1)T}A_{k+1}^2u_k^{(1)})^{-\frac{1}{2}} \\ =& (u_k^{(1)} + a_kA_{k+1}u_k^{(1)})\{1 - a_ku_k^{(1)T}A_{k+1}u_k^{(1)} + O(a_k^2)\} \\ =& u_k^{(1)} + a_kA_{k+1}u_k^{(1)} - a_k(u_k^{(1)T}A_{k+1}u_k^{(1)})u_k^{(1)} + O(a_k^2) \\ =& u_k^{(1)} + a_k(Au_k^{(1)} - (u_k^{(1)T}Au_k^{(1)})u_k^{(1)}) + a_k\varepsilon_{k+1}^{(1)} + O(a_k^2), \end{aligned} \tag{5.1.9}$$

where $\varepsilon_{k+1}^{(1)} = N_{k+1}u_k^{(1)} - (u_k^{(1)T}N_{k+1}u_k^{(1)})u_k^{(1)}$.

By A5.1.2, $(\varepsilon_{k+1}^{(1)}, \mathscr{F}_{k+1})$ is an mds. Since $\sum_k a_k^2 < \infty$, $\sup_k E\{\|N_{k+1}\|^2|\mathscr{F}_k\} < \infty$, $\|u_k^{(1)}\| = 1$, by Theorem 1.2.8 we have

$$\sum_k a_k[N_{k+1}u_k^{(1)} - (u_k^{(1)T}N_{k+1}u_k^{(1)})u_k^{(1)}] < \infty \text{ a.s.} \tag{5.1.10}$$

Thus, (5.1.8) with $\delta_{k+1}(1) = \varepsilon_{k+1}^{(1)}$ holds for $i = 1$.

From (5.1.10) it follows that

$$\lim_{T\to 0}\limsup_{k\to\infty}\frac{1}{T}\Big\|\sum_{i=k}^{m(k,T)} a_i[N_{i+1}u_i^{(1)} - (u_i^{(1)T}N_{i+1}u_i^{(1)})u_i^{(1)}]\Big\| = 0, \tag{5.1.11}$$

and hence

$$\lim_{T\to 0}\limsup_{k\to\infty}\frac{1}{T}\Big\|\sum_{i=k}^{m(k,T)} a_i(\varepsilon_{i+1}^{(1)} + O(a_i))\Big\| = 0, \tag{5.1.12}$$

where $m(k,T) \triangleq \max\{m : \sum_{i=k}^{m} a_i \le T\}$.

Define $f(u) \triangleq Au - (u^TAu)u$ on the unit sphere S. It is clear that the root set of $f(\cdot)$ on S is J.

Let $v(u) \triangleq -\frac{1}{2}u^T Au$. Then, for $u \in S$ we have

$$v_u^T(u)f(u) = -u^T A[Au - (u^T Au)u] = -u^T A^2 u + (u^T Au)^2$$
$$\begin{cases} < \|Au\|^2\|u\|^2 - u^T A^2 u = 0, & \text{if } u \notin J \\ = 0, & \text{if } u \in J. \end{cases} \tag{5.1.13}$$

Denote by J_1 the totality of the limiting points of $u_k^{(1)}$. By (5.1.12) applying Theorem 2.3.4 to (5.1.9) leads to

$$d(u_k^{(1)T} Au_k^{(1)}, V(J)) \to 0 \quad \text{and} \quad d(u_k^{(1)}, J_1) \to 0, \tag{5.1.14}$$

where J_1 is a connected subset of J.

Since $V(J)$ is composed of isolated points, by (5.1.14) there is a $\lambda(1) \in V(J)$ such that

$$d(u_k^{(1)T} Au_k^{(1)}, \lambda(1)) \to 0. \tag{5.1.15}$$

We now show that

$$Au = \lambda(1)u \ \ \forall u \in J_1. \tag{5.1.16}$$

Assume the converse: there exist $\widetilde{u} \in J_1$ and $\lambda(1)' \neq \lambda(1)$ such that

$$A\widetilde{u} = \lambda(1)'\widetilde{u} \quad \text{and} \quad \widetilde{u}^T A\widetilde{u} = \lambda(1)'. \tag{5.1.17}$$

Since J_1 is composed of limiting points of $\{u_k^{(1)}\}$, for $\widetilde{u} \in J_1$ there must exist a subsequence $\{u_{n_k}^{(1)}\}$ such that $u_{n_k}^{(1)} \to \widetilde{u}$. By (5.1.17) it follows that

$$d(u_{n_k}^{(1)T} Au_{n_k}^{(1)}, \lambda(1)') \to 0,$$

which contradicts with (5.1.15). Hence, (5.1.16) holds.

Since $d(u_k^{(1)}, J_1) \to 0$ by (5.1.14), from (5.1.16) it follows that $d(Au_k^{(1)}, \lambda(1)u_k^{(1)}) \xrightarrow[k\to\infty]{} 0$.

Thus, we have proved the theorem for $i = 1$.

We need to show the following equalities

$$(V_{k+1}^{(i-1)T} V_{k+1}^{(i-1)})^{-1} = I + O(a_k^2), \tag{5.1.18}$$

$$V_{k+1}^{(i-1)T} u_{k+1}^{(i)} = O\big(a_k(a_k + \sum_{s=1}^{i-1} d(u_k^{(s)}, J_s))\big) + a_k \eta_{k+1}^{(i-1)}, \quad \text{and} \tag{5.1.19}$$

$$V_{k+1}^{(i-1)} V_{k+1}^{(i-1)+} u_{k+1}^{(i)} = O\big(a_k(a_k + \sum_{s=1}^{i-1} d(u_k^{(s)}, J_s))\big) + a_k \gamma_{k+1}(i) \tag{5.1.20}$$

are valid for all $i:\ 2 \le i \le n$, where both $\eta_{k+1}^{(i-1)}$ and $\gamma_{k+1}(i)$ are the linear combinations of mdses measurable with respect to $\mathscr{F}_{k+1}$, $\mathscr{F}_k$, $\cdots$, $\mathscr{F}_{k+3-i}$, respectively, and with bounded second conditional moments. Thus, $\sum_{k=1}^{\infty} a_k \eta_{k+1}^{(i)} < \infty$, and $\sum_{k=1}^{\infty} a_k \gamma_{k+1}(i) < \infty$.

Since $V_{k+1}^{(1)} = u_{k+1}^{(1)}$, from (5.1.2) it is seen that $u_{k+1}^{(1)T} u_{k+1}^{(1)} = 1$, and hence (5.1.18) is valid for $i = 2$.

Let us prove (5.1.19), (5.1.20), and the theorem for $i = 2$.

We have

$$\begin{aligned}
\|\tilde{u}_{k+1}^{(2)}\|^{-1} =&\{[P_k^{(1)} u_k^{(2)} + a_k P_k^{(1)} A_{k+1} P_k^{(1)} u_k^{(2)}]^T \\
&\cdot [P_k^{(1)} u_k^{(2)} + a_k P_k^{(1)} A_{k+1} P_k^{(1)} u_k^{(2)}]\}^{-\frac{1}{2}} \\
=&[u_k^{(2)T} P_k^{(1)} u_k^{(2)} + 2a_k u_k^{(2)T} P_k^{(1)} A_{k+1} P_k^{(1)} u_k^{(2)} \\
&+ a_k^2 u_k^{(2)T} P_k^{(1)} A_{k+1} P_k^{(1)} A_{k+1} P_k^{(1)} u_k^{(2)}]^{-\frac{1}{2}} \\
=&[1 - u_k^{(2)T} V_k^{(1)} V_k^{(1)+} u_k^{(2)} + 2a_k u_k^{(2)T} P_k^{(1)} A_{k+1} P_k^{(1)} u_k^{(2)} \\
&+ a_k^2 u_k^{(2)T} P_k^{(1)} A_{k+1} P_k^{(1)} A_{k+1} P_k^{(1)} u_k^{(2)}]^{-\frac{1}{2}} \\
=&1 + \frac{1}{2} u_k^{(2)T} V_k^{(1)} V_k^{(1)T} u_k^{(2)} - a_k u_k^{(2)T} P_k^{(1)} A_{k+1} P_k^{(1)} u_k^{(2)} + O(a_k^2). \quad (5.1.21)
\end{aligned}$$

By (5.1.9) and noticing $u_k^{(1)T} u_{k+1}^{(2)} = 0$, we have

$$\begin{aligned}
V_k^{(1)T} u_k^{(2)} =&(u_{k-1}^{(1)} + a_{k-1}(Au_{k-1}^{(1)} - (u_{k-1}^{(1)T} A u_{k-1}^{(1)}) u_{k-1}^{(1)}) \\
&+ O(a_{k-1}^2) + a_{k-1}\varepsilon_k^{(1)})^T u_k^{(2)} \\
=&a_{k-1}(Au_{k-1}^{(1)})^T u_k^{(2)} + O(a_{k-1}^2) + a_{k-1}\varepsilon_k^{(1)T} u_k^{(2)}. \quad (5.1.22)
\end{aligned}$$

Since $d(u_k^{(1)}, J_1) \xrightarrow[k\to\infty]{} 0$ and $u_k^{(1)T} u_{k+1}^{(2)} = 0$, continuing (5.1.22) for any $u^{(1)} \in J_1$ we have

$$\begin{aligned}
V_k^{(1)T} u_k^{(2)} =&a_{k-1}(Au_{k-1}^{(1)})^T u_k^{(2)} + O(a_{k-1}^2) + a_{k-1}\varepsilon_k^{(1)T} u_k^{(2)} \\
=&a_{k-1}(Au_{k-1}^{(1)} - Au^{(1)} + Au^{(1)} - \lambda(1) u_{k-1}^{(1)})^T u_k^{(2)} + O(a_{k-1}^2) \\
&+ a_{k-1}\varepsilon_k^{(1)T} u_k^{(2)} \\
=&O\big(a_k d(u_k^{(1)}, J_1)\big) + O(a_k^2) + a_{k-1}\varepsilon_k^{(1)T} u_k^{(2)}, \quad (5.1.23)
\end{aligned}$$

which incorporating with (5.1.21) leads to

$$\|\tilde{u}_{k+1}^{(2)}\|^{-1} = 1 - a_k u_k^{(2)T} P_k^{(1)} A_{k+1} P_k^{(1)} u_k^{(2)} + O(a_k^2)$$

and a rough expression for $u_{k+1}^{(2)}$:

$$\begin{aligned}
u_{k+1}^{(2)} =&(P_k^{(1)} u_k^{(2)} + a_k P_k^{(1)} A_{k+1} P_k^{(1)} u_k^{(2)}) \\
&\cdot [1 - a_k u_k^{(2)T} P_k^{(1)} A_{k+1} P_k^{(1)} u_k^{(2)} + O(a_k^2)] \\
=&P_k^{(1)} u_k^{(2)} + a_k P_k^{(1)} A_{k+1} P_k^{(1)} u_k^{(2)} \\
&- a_k (u_k^{(2)T} P_k^{(1)} A_{k+1} P_k^{(1)} u_k^{(2)}) P_k^{(1)} u_k^{(2)} + O(a_k^2) \\
=&P_k^{(1)} u_k^{(2)} + O(a_k). \quad (5.1.24)
\end{aligned}$$

Substituting this to the right-hand side of (5.1.23) gives

$$\begin{aligned} V_k^{(1)T}u_k^{(2)} =& O\big(a_k d(u_k^{(1)},J_1)\big) + O(a_k^2) + a_{k-1}\varepsilon_k^{(1)T}(P_{k-1}^{(1)}u_{k-1}^{(2)} + O(a_{k-1})) \\ =& O\big(a_k d(u_k^{(1)},J_1)\big) + O(a_k^2) + a_{k-1}\eta_k^{(1)}, \end{aligned} \tag{5.1.25}$$

where $\eta_k^{(1)} \triangleq \varepsilon_k^{(1)T}P_{k-1}^{(1)}u_{k-1}^{(2)}$ is a $\mathscr{F}_k$-measurable mds.

This means that (5.1.19) is valid for $i=2$.

By noticing $u_k^{(1)} = u_{k-1}^{(1)} + O(a_{k-1})$ from (5.1.25) we obtain

$$\begin{aligned} V_k^{(1)}V_k^{(1)T}u_k^{(2)} =& (u_{k-1}^{(1)} + O(a_{k-1}))\Big(O\big(a_k d(u_k^{(1)},J_1)\big) \\ & + O(a_k^2) + a_{k-1}\varepsilon_k^{(1)T}P_{k-1}^{(1)}u_{k-1}^{(2)}\Big) \\ =& O\big(a_k d(u_k^{(1)},J_1)\big) + O(a_k^2) + a_{k-1}\gamma_k(2) \end{aligned} \tag{5.1.26}$$

where $\gamma_k(2) \triangleq u_{k-1}^{(1)}\varepsilon_k^{(1)T}P_{k-1}^{(1)}u_{k-1}^{(2)}$ is an $\mathscr{F}_k$-measurable mds.

By taking (5.1.18) into account, this implies (5.1.20) for $i=2$.

By (5.1.26) we have

$$a_k(A_{k+1}V_k^{(1)}V_k^{(1)T} - V_k^{(1)}V_k^{(1)T}A_{k+1}V_k^{(1)}V_k^{(1)T})u_k^{(2)} = O(a_{k-1}^2), \tag{5.1.27}$$

and for any $u^{(1)}(k) \in J_1$ by noticing $u_{k-1}^{(1)T}u_k^{(2)} = 0$

$$\begin{aligned} & a_kV_k^{(1)}V_k^{(1)T}A_{k+1}u_k^{(2)} \\ =& a_kV_k^{(1)}(Au_k^{(1)})^Tu_k^{(2)} + a_kV_k^{(1)}V_k^{(1)T}N_{k+1}u_k^{(2)} \\ =& a_kV_k^{(1)}(Au_k^{(1)} - Au^{(1)}(k) + \lambda(1)u^{(1)}(k) - \lambda(1)u_{k-1}^{(1)} + \lambda(1)u_{k-1}^{(1)})^Tu_k^{(2)} \\ & + a_kV_k^{(1)}V_k^{(1)T}N_{k+1}u_k^{(2)} \\ =& O\big(a_k d(u_k^{(1)},J_1)\big) + a_kV_k^{(1)}V_k^{(1)T}N_{k+1}u_k^{(2)}. \end{aligned} \tag{5.1.28}$$

We are now in a position to derive (5.1.8) for $i=2$.

Noticing $P_k^{(1)} = I - u_k^{(1)}u_k^{(1)T}$, for $k \geq 1$ we have

$$\begin{aligned} \widetilde{u}_{k+1}^{(2)} =& P_k^{(1)}u_k^{(2)} + a_kP_k^{(1)}A_{k+1}P_k^{(1)}u_k^{(2)} \\ =& u_k^{(2)} - V_k^{(1)}V_k^{(1)T}u_k^{(2)} + a_k(I - V_k^{(1)}V_k^{(1)T})A_{k+1}(I - V_k^{(1)}V_k^{(1)T})u_k^{(2)} \\ =& u_k^{(2)} + a_kA_{k+1}u_k^{(2)} - V_k^{(1)}V_k^{(1)T}u_k^{(2)} - a_k(A_{k+1}V_k^{(1)}V_k^{(1)T} \\ & + V_k^{(1)}V_k^{(1)T}A_{k+1} - V_k^{(1)}V_k^{(1)T}A_{k+1}V_k^{(1)}V_k^{(1)T})u_k^{(2)}. \end{aligned} \tag{5.1.29}$$

By (5.1.26)–(5.1.29) for all sufficiently large k we have

$$\widetilde{u}_{k+1}^{(2)} = u_k^{(2)} + a_kAu_k^{(2)} + O\big(a_k(a_k + d(u_k^{(1)},J_1))\big) + a_k\tilde{\delta}_{k+1}^{(2)}, \tag{5.1.30}$$

where $a_k\tilde{\delta}_{k+1}^{(2)} = a_kN_{k+1}u_k^{(2)} - a_{k-1}\gamma_k(2) - a_kV_k^{(1)}V_k^{(1)T}N_{k+1}u_k^{(2)}$ with $N_{k+1}u_k^{(2)}$, $\gamma_k(2)$, and $V_k^{(1)}V_k^{(1)T}N_{k+1}u_k^{(2)}$ being mdses measurable with respect to $\mathscr{F}_{k+1}$, $\mathscr{F}_k$, and $\mathscr{F}_{k+1}$, respectively.

Therefore, for large k we have

$$\begin{aligned}\|\tilde{u}_{k+1}^{(2)}\|^{-1} =&\Big(\big(u_k^{(2)}+a_kAu_k^{(2)}+O\big(a_k(a_k+d(u_k^{(1)},J_1))\big)+a_k\tilde{\delta}_{k+1}^{(2)}\big)^T\\ &\cdot\big(u_k^{(2)}+a_kAu_k^{(2)}+O\big(a_k(a_k+d(u_k^{(1)},J_1))\big)+a_k\tilde{\delta}_{k+1}^{(2)}\big)\Big)^{-\frac{1}{2}}\\ =&(1+2a_ku_k^{(2)T}Au_k^{(2)}+O\big(a_k(a_k+d(u_k^{(1)},J_1))\big)+a_k\tilde{\delta}_{k+1}^{(2)T}u_k^{(2)}\\ &+a_ku_k^{(2)T}\tilde{\delta}_{k+1}^{(2)})^{-\frac{1}{2}}\\ =&1-a_ku_k^{(2)T}Au_k^{(2)}+O\big(a_k(a_k+d(u_k^{(1)},J_1))\big)-a_k\tilde{\delta}_{k+1}^{(2)T}u_k^{(2)},\end{aligned}\tag{5.1.31}$$

and hence

$$\begin{aligned}u_{k+1}^{(2)} =&(u_k^{(2)}+a_kAu_k^{(2)}+O\big(a_k(a_k+d(u_k^{(1)},J_1))\big)+a_k\tilde{\delta}_{k+1}^{(2)})\\ &\cdot(1-a_ku_k^{(2)T}Au_k^{(2)}+O\big(a_k(a_k+d(u_k^{(1)},J_1))\big)-a_k\tilde{\delta}_{k+1}^{(2)T}u_k^{(2)})\\ =&u_k^{(2)}+a_k(Au_k^{(2)}-(u_k^{(2)T}Au_k^{(2)})u_k^{(2)})+O\big(a_k(a_k+d(u_k^{(1)},J_1))\big)\\ &+a_k\tilde{\delta}_{k+1}^{(2)}-a_k\tilde{\delta}_{k+1}^{(2)T}u_k^{(2)}u_k^{(2)}.\end{aligned}\tag{5.1.32}$$

Here $\tilde{\delta}_{k+1}^{(2)T}u_k^{(2)}u_k^{(2)}$ is not an mds, but replacing $u_k^{(2)}$ with the expression given by (5.1.24) in $a_k\tilde{\delta}_{k+1}^{(2)T}u_k^{(2)}u_k^{(2)}$ of (5.1.32) leads to

$$\begin{aligned}u_{k+1}^{(2)} =&u_k^{(2)}+a_k(Au_k^{(2)}-(u_k^{(2)T}Au_k^{(2)})u_k^{(2)})\\ &+O\big(a_k(a_k+d(u_k^{(1)},J_1))\big)+a_k\delta_{k+1}(2),\end{aligned}\tag{5.1.33}$$

where $a_k\delta_{k+1}(2)=a_k\tilde{\delta}_{k+1}^{(2)}-a_k\tilde{\delta}_{k+1}^{(2)T}(P_{k-1}^{(1)}u_{k-1}^{(2)})(P_{k-1}^{(1)}u_{k-1}^{(2)})$. By the property of $a_k\tilde{\delta}_{k+1}^{(2)}$ mentioned above, $a_k\delta_{k+1}(2)$ is a linear combination of two mdses measurable with respect to $\mathscr{F}_{k+1}$ and $\mathscr{F}_k$. By A5.1.2 $\delta_{k+1}(2)$ is bounded, and hence $\sum_{k=1}^{\infty}a_k\delta_{k+1}(2)<\infty$.

Similar to (5.1.14)–(5.1.16), there exists $\lambda(2)\in V(J)$ such that

$$d(u_k^{(2)T}Au_k^{(2)},\lambda(2))\to 0 \quad\text{and}\quad d(u_k^{(2)},J_2)\to 0,\tag{5.1.34}$$

where $J_2\subset J$ is composed of limiting points of $\{u_k^{(2)}\}$ and

$$Au=\lambda(2)u\ \ \forall u\in J_2.\tag{5.1.35}$$

By (5.1.34) and (5.1.35) it follows that $d(Au_k^{(2)},\lambda(2)u_k^{(2)})\xrightarrow[k\to\infty]{}0$.

Thus, we have shown that (5.1.18)–(5.1.20) are valid for $i=2$, and the theorem is valid for $i=1,2$.

We now inductively prove (5.1.18)–(5.1.20) and the theorem. Assume that

(5.1.18)–(5.1.20) and the theorem itself are valid for $i = 1, 2, \cdots, j \leq n-1$. We show that they are also true for $i = j+1$.

Noticing that the columns of $V_{k+1}^{(j)}$ are orthogonal, we have

$$V_{k+1}^{(j)T} V_{k+1}^{(j)} = [u_{k+1}^{(1)}, \cdots, P_{k+1}^{(j-1)} u_{k+1}^{(j)}]^T [u_{k+1}^{(1)}, \cdots, P_{k+1}^{(j-1)} u_{k+1}^{(j)}]$$

$$= \begin{bmatrix} 1 & 0 & \cdots & 0 \\ 0 & u_{k+1}^{(2)T} P_{k+1}^{(1)} u_{k+1}^{(2)} & \ddots & \vdots \\ \vdots & \ddots & \ddots & 0 \\ 0 & \cdots & 0 & u_{k+1}^{(j)T} P_{k+1}^{(j-1)} u_{k+1}^{(j)} \end{bmatrix}$$

$$= \begin{bmatrix} 1 & 0 & \cdots & 0 \\ 0 & 1 - u_{k+1}^{(2)T} V_{k+1}^{(1)} V_{k+1}^{(1)+} u_{k+1}^{(2)} & \ddots & \vdots \\ \vdots & \ddots & \ddots & 0 \\ 0 & \cdots & 0 & 1 - u_{k+1}^{(j)T} V_{k+1}^{(j-1)} V_{k+1}^{(j-1)+} u_{k+1}^{(j)} \end{bmatrix}.$$

From (5.1.18) and (5.1.19) for any $i = 1, \cdots, j$, we then have

$$(V_{k+1}^{(j)T} V_{k+1}^{(j)})^{-1} = (I + O(a_k^2))^{-1} = I + O(a_k^2) \tag{5.1.36}$$

This proves that (5.1.18) is true for $i = j+1$.

We now show that (5.1.19), (5.1.20) are also true for $i = j+1$.

Multiplying both sides of

$$\widetilde{u}_{k+1}^{(j+1)} = P_k^{(j)} u_k^{(j+1)} + a_k P_k^{(j)} A_{k+1} P_k^{(j)} u_k^{(j+1)}$$

by $V_k^{(j)T}$ from left we see $V_k^{(j)T} \widetilde{u}_{k+1}^{(j+1)} = 0$, and hence

$$V_k^{(j)T} u_{k+1}^{(j+1)} = 0 \tag{5.1.37}$$

for sufficiently large k. From here it follows that

$$\begin{aligned} u_{k+1}^{(j+1)T} V_k^{(j)} &= u_{k+1}^{(j+1)T} [V_k^{(j-1)}, P_k^{(j-1)} u_k^{(j)}] \\ &= u_{k+1}^{(j+1)T} [V_k^{(i-1)}, P_k^{(i-1)} u_k^{(i)}, \cdots, P_k^{(j-1)} u_k^{(j)}] = 0. \end{aligned}$$

Thus, for any $i = 1, \cdots, j$ we have

$$\begin{aligned} 0 =& u_{k+1}^{(j+1)T} P_k^{(i-1)} u_k^{(i)} \\ =& u_{k+1}^{(j+1)T} u_k^{(i)} - u_{k+1}^{(j+1)T} V_k^{(i-1)} V_k^{(i-1)+} u_k^{(i)} \\ =& u_{k+1}^{(j+1)T} u_k^{(i)}, \end{aligned}$$

and hence

$$u_{k+1}^{(j+1)T} [u_k^{(1)}, \cdots, u_k^{(j)}] = 0, \quad k \geq 1. \tag{5.1.38}$$

By the inductive assumptions from (5.1.8) we have

$$\begin{aligned} u_{k+1}^{(i+1)} =& u_k^{(i+1)} + O\big(a_k(a_k + \sum_{s=1}^{i+1} d(u_k^{(s)}, J_s))\big) + a_k \delta_{k+1}(i+1) \\ =& u_k^{(i+1)} + O(a_k), \quad i = 1, \cdots, j-1. \end{aligned} \tag{5.1.39}$$

By the inductive assumption,

$$V_{k+1}^{(i)} V_{k+1}^{(i)+} u_{k+1}^{(i+1)} = O\big(a_k(a_k + \sum_{s=1}^{i} d(u_k^{(s)}, J_s))\big) + a_k \gamma_{k+1}(i+1), \quad i = 1, \cdots, j-1,$$

we have

$$\begin{aligned} V_{k+1}^{(j)} =& [u_{k+1}^{(1)}, P_{k+1}^{(1)} u_{k+1}^{(2)}, \cdots, P_{k+1}^{(j-1)} u_{k+1}^{(j)}] \\ =& [u_{k+1}^{(1)}, u_{k+1}^{(2)} - V_{k+1}^{(1)} V_{k+1}^{(1)+} u_{k+1}^{(2)}, \cdots, u_{k+1}^{(j)} - V_{k+1}^{(j-1)} V_{k+1}^{(j-1)+} u_{k+1}^{(j)}] \\ =& [u_{k+1}^{(1)}, u_{k+1}^{(2)}, \cdots, u_{k+1}^{(j)}] + O\big(a_k(a_k + \sum_{s=1}^{j-1} d(u_k^{(s)}, J_s))\big) \\ & - a_k[0, \gamma_{k+1}(2), \cdots, \gamma_{k+1}(j)] \\ =& [u_{k+1}^{(1)}, u_{k+1}^{(2)}, \cdots, u_{k+1}^{(j)}] + O(a_k) \end{aligned} \tag{5.1.40}$$

and

$$\begin{aligned} & u_{k+1}^{(j+1)T} P_{k+1}^{(i)} u_{k+1}^{(i+1)} = u_{k+1}^{(j+1)T} u_{k+1}^{(i+1)} - u_{k+1}^{(j+1)T} V_{k+1}^{(i)} V_{k+1}^{(i)+} u_{k+1}^{(i+1)} \\ &= u_{k+1}^{(j+1)T} u_{k+1}^{(i+1)} + O\big(a_k(a_k + \sum_{s=1}^{i} d(u_k^{(s)}, J_s))\big) + a_k u_{k+1}^{(j+1)T} \gamma_{k+1}(i+1) \\ &= u_{k+1}^{(j+1)T} \Big(u_k^{(i+1)} + O\big(a_k(a_k + \sum_{s=1}^{i+1} d(u_k^{(s)}, J_s))\big) \\ &\quad + a_k \delta_{k+1}(i+1)\Big) + a_k u_{k+1}^{(j+1)T} \gamma_{k+1}(i+1) \\ &= O\big(a_k(a_k + \sum_{s=1}^{i+1} d(u_k^{(s)}, J_s))\big) + a_k u_{k+1}^{(j+1)T} \big(\delta_{k+1}(i+1) \\ &\quad + \gamma_{k+1}(i+1)\big), \qquad i = 1, \cdots, j-1, \end{aligned} \tag{5.1.41}$$

where for the last two equalities in (5.1.41), (5.1.38) and (5.1.39) are used.

Therefore, by (5.1.41) we obtain

$$\begin{aligned} u_{k+1}^{(j+1)T} V_{k+1}^{(j)} =& u_{k+1}^{(j+1)T} [u_{k+1}^{(1)}, \cdots, P_{k+1}^{(j-1)} u_{k+1}^{(j)}] \\ =& O\big(a_k(a_k + \sum_{s=1}^{j} d(u_k^{(s)}, J_s))\big) \\ & + a_k u_{k+1}^{(j+1)T} [0, \delta_{k+1}(2) + \gamma_{k+1}(2), \cdots, \delta_{k+1}(j) + \gamma_{k+1}(j)]. \end{aligned} \tag{5.1.42}$$

However, this is still not in the form of (5.1.19), because the last term in (5.1.42) is not expressed as linear combinations of mdses. Let us express $u_{k+1}^{(j+1)}$ via vectors of time k.

Noticing that $P_k^{(j)}P_k^{(j)} = P_k^{(j)}$ and $P_k^{(j)} = I - V_k^{(j)}V_k^{(j)+}$, for sufficiently large k we have

$$\begin{aligned}
\|\tilde{u}_{k+1}^{(j+1)}\|^{-1} =&\{[P_k^{(j)}u_k^{(j+1)} + a_kP_k^{(j)}A_{k+1}P_k^{(j)}u_k^{(j+1)}]^T[P_k^{(j)}u_k^{(j+1)} + a_kP_k^{(j)}\\
&\cdot A_{k+1}P_k^{(j)}u_k^{(j+1)}]\}^{-\frac{1}{2}}\\
=&[u_k^{(j+1)T}P_k^{(j)}u_k^{(j+1)} + 2a_ku_k^{(j+1)T}P_k^{(j)}A_{k+1}P_k^{(j)}u_k^{(j+1)} + a_k^2u_k^{(j+1)T}\\
&\cdot P_k^{(j)}A_{k+1}P_k^{(j)}A_{k+1}P_k^{(j)}u_k^{(j+1)}]^{-\frac{1}{2}}\\
=&[1 - u_k^{(j+1)T}V_k^{(j)}V_k^{(j)+}u_k^{(j+1)} + 2a_ku_k^{(j+1)T}P_k^{(j)}A_{k+1}P_k^{(j)}u_k^{(j+1)}\\
&+ a_k^2u_k^{(j+1)T}P_k^{(j)}A_{k+1}P_k^{(j)}A_{k+1}P_k^{(j)}u_k^{(j+1)}]^{-\frac{1}{2}}\\
=&1 + O(a_k^2) + \frac{1}{2}u_k^{(j+1)T}V_k^{(j)}(V_k^{(j)T}V_k^{(j)})^{-1}V_k^{(j)T}u_k^{(j+1)}\\
&- a_ku_k^{(j+1)T}P_k^{(j)}A_{k+1}P_k^{(j)}u_k^{(j+1)},
\end{aligned} \tag{5.1.43}$$

which combining with (5.1.42) leads to

$$\|\tilde{u}_{k+1}^{(j+1)}\|^{-1} = 1 - a_ku_k^{(j+1)T}P_k^{(j)}A_{k+1}P_k^{(j)}u_k^{(j+1)} + O(a_k^2). \tag{5.1.44}$$

Therefore, for large enough k we have

$$\begin{aligned}
u_{k+1}^{(j+1)} &= (P_k^{(j)}u_k^{(j+1)} + a_kP_k^{(j)}A_{k+1}P_k^{(j)}u_k^{(j+1)})\cdot[1 - a_ku_k^{(j+1)T}P_k^{(j)}A_{k+1}P_k^{(j)}\\
&\quad\cdot u_k^{(j+1)} + O(a_k^2)]\\
&= P_k^{(j)}u_k^{(j+1)} + a_kP_k^{(j)}A_{k+1}P_k^{(j)}u_k^{(j+1)} - a_k(u_k^{(j+1)T}P_k^{(j)}A_{k+1}P_k^{(j)}\\
&\quad\cdot u_k^{(j+1)})P_k^{(j)}u_k^{(j+1)} + O(a_k^2)\\
&= P_k^{(j)}u_k^{(j+1)} + O(a_k),
\end{aligned} \tag{5.1.45}$$

and hence by (5.1.36) and (5.1.40)

$$\begin{aligned}
u_{k+1}^{(j+1)} =&P_k^{(j)}\cdot(P_{k-1}^{(j)}u_{k-1}^{(j+1)} + O(a_{k-1})) + O(a_k)\\
=&P_{k-1}^{(j)}u_{k-1}^{(j+1)} - V_k^{(j)}V_k^{(j)+}(P_{k-1}^{(j)}u_{k-1}^{(j+1)}) + O(a_k)\\
=&P_{k-1}^{(j)}u_{k-1}^{(j+1)} - [u_{k-1}^{(1)},\cdots,u_{k-1}^{(j)}][u_{k-1}^{(1)},\cdots,u_{k-1}^{(j)}]^TP_{k-1}^{(j)}u_{k-1}^{(j+1)}\\
&+ O(a_k).
\end{aligned} \tag{5.1.46}$$

Again by (5.1.36) and (5.1.40) we have

$$\begin{aligned}
P_{k-1}^{(j)}u_{k-1}^{(j+1)} &= P_{k-1}^{(j)}\cdot(P_{k-2}^{(j)}u_{k-2}^{(j+1)} + O(a_{k-2}))\\
&= P_{k-2}^{(j)}u_{k-2}^{(j+1)} - V_{k-1}^{(j)}V_{k-1}^{(j)+}P_{k-2}^{(j)}u_{k-2}^{(j+1)} + O(a_{k-2})
\end{aligned}$$

$$
\begin{aligned}
&= P_{k-2}^{(j)}u_{k-2}^{(j+1)} - ([u_{k-2}^{(1)},\cdots,u_{k-2}^{(j)}]+O(a_{k-2}))(I+O(a_{k-2}^2))\\
&\quad\cdot([u_{k-2}^{(1)},\cdots,u_{k-2}^{(j)}]+O(a_{k-2}))^T\cdot(P_{k-2}^{(j)}u_{k-2}^{(j+1)})+O(a_{k-2})\\
&= P_{k-2}^{(j)}u_{k-2}^{(j+1)} - [u_{k-2}^{(1)},\cdots,u_{k-2}^{(j)}][u_{k-2}^{(1)},\cdots,u_{k-2}^{(j)}]^T P_{k-2}^{(j)}u_{k-2}^{(j+1)}+O(a_{k-2})\\
&= (I-[u_{k-2}^{(1)},\cdots,u_{k-2}^{(j)}][u_{k-2}^{(1)},\cdots,u_{k-2}^{(j)}]^T)P_{k-2}^{(j)}u_{k-2}^{(j+1)}+O(a_{k-2}).
\end{aligned}
\tag{5.1.47}
$$

Putting (5.1.47) into (5.1.46) yields

$$
\begin{aligned}
u_{k+1}^{(j+1)} =&(I-[u_{k-1}^{(1)},\cdots,u_{k-1}^{(j)}][u_{k-1}^{(1)},\cdots,u_{k-1}^{(j)}]^T)\\
&\cdot(I-[u_{k-2}^{(1)},\cdots,u_{k-2}^{(j)}][u_{k-2}^{(1)},\cdots,u_{k-2}^{(j)}]^T)P_{k-2}^{(j)}u_{k-2}^{(j+1)}+O(a_k)\\
=&(I-[u_{k-1}^{(1)},\cdots,u_{k-1}^{(j)}][u_{k-1}^{(1)},\cdots,u_{k-1}^{(j)}]^T)(I-[u_{k-2}^{(1)},\cdots,u_{k-2}^{(j)}]\\
&\cdot[u_{k-2}^{(1)},\cdots,u_{k-2}^{(j)}]^T)\cdots(I-[u_{k-j}^{(1)},\cdots,u_{k-j}^{(j)}]\\
&\cdot[u_{k-j}^{(1)},\cdots,u_{k-j}^{(j)}]^T)P_{k-j}^{(j)}u_{k-j}^{(j+1)}+O(a_k).
\end{aligned}
\tag{5.1.48}
$$

By (5.1.39) $u_{k+1}^{(i+1)} = u_k^{(i+1)} + O(a_k) = u_{k-j}^{(i+1)} + O(a_k), i=1,\cdots,j-1$, (5.1.48) can be rewritten as

$$
\begin{aligned}
u_{k+1}^{(j+1)} =&(I-[u_{k-j}^{(1)},\cdots,u_{k-j}^{(j)}][u_{k-1}^{(1)},\cdots,u_{k-j}^{(j)}]^T)^j\\
&\cdot P_{k-j}^{(j)}u_{k-j}^{(j+1)}+O(a_k).
\end{aligned}
\tag{5.1.49}
$$

Putting this into the right-hand side of (5.1.41) yields

$$
\begin{aligned}
&u_{k+1}^{(j+1)T}P_{k+1}^{(i)}u_{k+1}^{(i+1)}\\
&=O\big(a_k(a_k+\sum_{s=1}^{i+1}d(u_k^{(s)},J_s))\big)\\
&\quad+a_k\big((I-[u_{k-j}^{(1)},\cdots,u_{k-j}^{(j)}][u_{k-1}^{(1)},\cdots,u_{k-j}^{(j)}]^T)^jP_{k-j}^{(j)}u_{k-j}^{(j+1)}\\
&\quad+O(a_k)\big)^T(\delta_{k+1}(i+1)+\gamma_{k+1}(i+1))\\
&=O\big(a_k(a_k+\sum_{s=1}^{i+1}d(u_k^{(s)},J_s))\big)+a_k\eta_{k+1}^{(j)}(i+1),\ \ i=1,\cdots,j-1,
\end{aligned}
\tag{5.1.50}
$$

where

$$
\begin{aligned}
\eta_{k+1}^{(j)}(i+1) \triangleq& u_{k-j}^{(j+1)T}P_{k-j}^{(j)}\big((I-[u_{k-j}^{(1)},\cdots,u_{k-j}^{(j)}]\\
&\cdot[u_{k-1}^{(1)},\cdots,u_{k-j}^{(j)}]^T)^j(\delta_{k+1}(i+1)+\gamma_{k+1}(i+1)\big).
\end{aligned}
$$

By the inductive assumptions, from (5.1.8) it is seen that $\delta_{k+1}(j)$ is a linear combination of mdses measurable with respect to $\mathscr{F}_{k+1}$, $\mathscr{F}_k$, $\cdots$, and $\mathscr{F}_{k+2-j}$, respectively, and $\gamma_{k+1}(j)$ is a linear combination of mdses measurable with respect to $\mathscr{F}_{k+1}$,

$\mathscr{F}_k, \cdots$, and $\mathscr{F}_{k+3-j}$, respectively. Therefore, $\eta^{(j)}_{k+1}(i+1)$ is a linear combination of mdses measurable with respect to $\mathscr{F}_{k+1}$, $\mathscr{F}_k, \cdots$, and $\mathscr{F}_{k+2-j}$, respectively.

From (5.1.50) it follows that

$$\begin{aligned}u^{(j+1)T}_{k+1}V^{(j)}_{k+1} =& u^{(j+1)T}_{k+1}[u^{(1)}_{k+1},\cdots,P^{(j-1)}_{k+1}u^{(j)}_{k+1}]\\ =& O\big(a_k(a_k+\sum_{s=1}^{j}d(u^{(s)}_k,J_s))\big)+a_k\eta^{(j)}_{k+1},\end{aligned} \tag{5.1.51}$$

where $\eta^{(j)}_{k+1}\triangleq[\eta^{(1)}_{k+1}(1),\eta^{(j)}_{k+1}(2),\cdots,\eta^{(j)}_{k+1}(j)]$ is a linear combination of mdses measurable with respect to $\mathscr{F}_{k+1}$, $\mathscr{F}_k, \cdots$, and $\mathscr{F}_{k+2-j}$, respectively, where $\eta^{(1)}_{k+1}(1)\triangleq\eta^{(1)}_{k+1}$. Thus, (5.1.19) is proved for $i=j+1$.

By (5.1.36) and (5.1.40) from (5.1.51) it follows that

$$\begin{aligned}V^{(j)}_{k+1}V^{(j)+}_{k+1}u^{(j+1)}_{k+1} =& V^{(j)}_{k+1}(V^{(j)T}_{k+1}V^{(j)}_{k+1})^{-1}V^{(j)T}_{k+1}u^{(j+1)}_{k+1}\\ =& ([u^{(1)}_{k-j},\cdots,u^{(j)}_{k-j}]+O(a_k))(I+O(a_k^2))\\ &\cdot\Big(O\big(a_k(a_k+\sum_{s=1}^{j}d(u^{(s)}_k,J_s))\big)+a_k\eta^{(j)}_{k+1}\Big)^T\\ =& O\big(a_k(a_k+\sum_{s=1}^{j}d(u^{(s)}_k,J_s))\big)+a_k\gamma_{k+1}(j+1),\end{aligned} \tag{5.1.52}$$

where $\gamma_{k+1}(j+1)\triangleq[u^{(1)}_{k-j},\cdots,u^{(j)}_{k-j}]\eta^{(j)T}_{k+1}$ is a linear combination of mdses measurable with respect to $\mathscr{F}_{k+1}$, $\mathscr{F}_k, \cdots$, and $\mathscr{F}_{k+2-j}$, respectively.

This means that (5.1.20) is valid for $i=j+1$.

We are now in a position to show that the theorem is true for $i=j+1$ as well.

By (5.1.40) we then have

$$\begin{aligned}&a_kV^{(j)}_kV^{(j)+}_kA_{k+1}u^{(j+1)}_k\\ =& a_kV^{(j)}_k(V^{(j)T}_kV^{(j)}_k)^{-1}V^{(j)T}_kAu^{(j+1)}_k+a_kV^{(j)}_kV^{(j)+}_kN_{k+1}u^{(j+1)}_k\\ =& a_kV^{(j)}_k(V^{(j)T}_kV^{(j)}_k)^{-1}(A[u^{(1)}_k,\cdots,u^{(j)}_k])^Tu^{(j+1)}_k\\ &+O(a_k^2)+a_kV^{(j)}_kV^{(j)+}_kN_{k+1}u^{(j+1)}_k\end{aligned} \tag{5.1.53}$$

for sufficiently large k.

By inductive assumptions, for any $i\in\{1,\cdots,j\}$ there is a sequence $\{u^{(i)}(k)\}\in J_i$ such that $d(u^{(i)}_k,u^{(i)}(k))\xrightarrow[k\to\infty]{}0$ and $Au^{(i)}(k)=\lambda(i)u^{(i)}(k)$.

By (5.1.38) and (5.1.39) we have

$$\begin{aligned}&a_kV^{(j)}_k(V^{(j)T}_kV^{(j)}_k)^{-1}(Au^{(i)}_k)^Tu^{(j+1)}_k+O(a_k^2)+a_kV^{(j)}_kV^{(j)+}_kN_{k+1}u^{(j+1)}_k\\ =& a_kV^{(j)}_k(V^{(j)T}_kV^{(j)}_k)^{-1}\cdot(Au^{(i)}_k-Au^{(i)}(k)+\lambda(i)u^{(i)}(k)-\lambda(i)u^{(i)}_k(k)\\ &+\lambda(i)u^{(i)}_k)^Tu^{(j+1)}_k+O(a_k^2)+a_kV^{(j)}_kV^{(j)+}_kN_{k+1}u^{(j+1)}_k\end{aligned}$$

$$
\begin{aligned}
&= \lambda(i) a_k V_k^{(j)} (V_k^{(j)T} V_k^{(j)})^{-1} u_k^{(i)T} u_k^{(j+1)} + O(a_k(a_k + d(u_k^{(i)}, J_i))) \\
&\quad + a_k V_k^{(j)} V_k^{(j)+} N_{k+1} u_k^{(j+1)} \\
&= \lambda(i) a_k V_k^{(j)} (V_k^{(j)T} V_k^{(j)})^{-1} (u_{k-1}^{(i)} + O(a_k))^T u_k^{(j+1)} \\
&\quad + O(a_k(a_k + d(u_k^{(i)}, J_i))) + a_k V_k^{(j)} V_k^{(j)+} N_{k+1} u_k^{(j+1)} \\
&= O\big(a_k(a_k + d(u_k^{(i)}, J_i))\big) + a_k V_k^{(j)} V_k^{(j)+} N_{k+1} u_k^{(j+1)} \quad \forall i = 1, \cdots, j. \qquad (5.1.54)
\end{aligned}
$$

Putting the expression given by (5.1.54) into (5.1.53) yields

$$
a_k V_k^{(j)} V_k^{(j)+} A_{k+1} u_k^{(j+1)} = O\big(a_k(a_k + \sum_{s=1}^{j} d(u_k^{(s)}, J_s))\big) + a_k V_k^{(j)} V_k^{(j)+} N_{k+1} u_k^{(j+1)}. \qquad (5.1.55)
$$

Then we have

$$
\begin{aligned}
\widetilde{u}_{k+1}^{(j+1)} &= P_k^{(j)} u_k^{(j+1)} + a_k P_k^{(j)} A_{k+1} P_k^{(j)} u_k^{(j+1)} \\
&= (I - V_k^{(j)} V_k^{(j)+}) u_k^{(j+1)} + a_k (I - V_k^{(j)} V_k^{(j)+}) A_{k+1} (I - V_k^{(j)} V_k^{(j)+}) u_k^{(j+1)} \\
&= u_k^{(j+1)} + a_k A_{k+1} u_k^{(j+1)} - V_k^{(j)} V_k^{(j)+} u_k^{(j+1)} - a_k (A_{k+1} V_k^{(j)} V_k^{(j)+} \\
&\quad + V_k^{(j)} V_k^{(j)+} A_{k+1} - V_k^{(j)} V_k^{(j)+} A_{k+1} V_k^{(j)} V_k^{(j)+}) u_k^{(j+1)}. \qquad (5.1.56)
\end{aligned}
$$

Notice that (5.1.52) implies that

$$
a_k (A_{k+1} V_k^{(j)} V_k^{(j)+} - V_k^{(j)} V_k^{(j)+} A_{k+1} V_k^{(j)} V_k^{(j)+}) u_k^{(j+1)} = O(a_k^2).
$$

From (5.1.56) by (5.1.52) and (5.1.55) it follows that

$$
\widetilde{u}_{k+1}^{(j+1)} = u_k^{(j+1)} + a_k A u_k^{(j+1)} + O\big(a_k(a_k + \sum_{s=1}^{j} d(u_k^{(s)}, J_s))\big) + a_k \tilde{\delta}_{k+1}^{(j+1)}, \qquad (5.1.57)
$$

where $a_k \tilde{\delta}_{k+1}^{(j+1)} = a_k N_{k+1} u_k^{(j+1)} - a_{k-1} \gamma_k (j+1) - a_k V_k^{(j)} V_k^{(j)+} N_{k+1} u_k^{(j+1)}$ is a linear combination of mdses measurable with respect to $\mathscr{F}_{k+1}$, $\mathscr{F}_k$, $\cdots$, and $\mathscr{F}_{k+1-j}$, respectively.

An analysis similar to (5.1.31), (5.1.32), and (5.1.33) leads to (5.1.8) for $i = j+1$:

$$
\begin{aligned}
u_{k+1}^{(j+1)} &= u_k^{(j+1)} + a_k (A u_k^{(j+1)} - (u_k^{(j+1)T} A u_k^{(j+1)}) u_k^{(j+1)}) \\
&\quad + O\big(a_k(a_k + \sum_{s=1}^{j} d(u_k^{(s)}, J_s))\big) + a_k \delta_{k+1}(j+1), \qquad (5.1.58)
\end{aligned}
$$

where $\delta_{k+1}(j+1)$ is a linear combination of mdses measurable with respect to $\mathscr{F}_{k+1}$, $\mathscr{F}_k$, $\cdots$, and $\mathscr{F}_{k+1-j}$, respectively.

Similar to (5.1.14), (5.1.15), and (5.1.16) it is shown that there exists a $J_{j+1} \subset J$ such that

$$
\lim_{k \to \infty} d(u_k^{(j+1)}, J_{j+1}) = 0, \qquad (5.1.59)
$$

where J_{j+1} is composed of limiting points of $\{u_k^{(j+1)}\}$.

Correspondingly, there is a $\lambda(j+1) \in V(J)$ such that

$$d(u_k^{(j+1)T} A u_k^{(j+1)}, \lambda(j+1)) \to 0, \text{ and } Au = \lambda(j+1)u \ \forall u \in J_{j+1}, \qquad (5.1.60)$$

which incorporating with (5.1.59) implies $d(Au_k^{(j+1)}, \lambda(j+1)u_k^{(j+1)}) \xrightarrow[k\to\infty]{} 0$.

Thus, (i), (ii), and (iii) of the theorem have been proved by induction.

Finally, (5.1.6) can be rewritten as

$$\lambda_{k+1}^{(j)} = \lambda_k^{(j)} - a_k(\lambda_k^{(j)} - \lambda(j) + \varepsilon_{k+1}^{(j)}),$$

where

$$\varepsilon_{k+1}^{(j)} \triangleq \lambda(j) - u_k^{(j)T} A u_k^{(j)} - u_k^{(j)T} N_{k+1} u_k^{(j)}.$$

Since $\lambda(j) - u_k^{(j)T} A u_k^{(j)} \to 0$ and $\sum_{k=1}^{\infty} a_k u_k^{(j)T} N_{k+1} u_k^{(j)} < \infty$, the conclusion (iv) follows from Theorem 2.5.1. □

Ordered Convergence

We show that convergence established in Theorem 5.1.1 is actually ordered in the sense that $\lambda(i) = \lambda^{(i)}$, where the eigenvalues $\{\lambda^{(i)}\}$ are ordered: $\lambda^{(1)} \geq \lambda^{(2)} \geq \cdots \geq \lambda^{(n)}$. For this we need the following fact listed as a proposition.

Proposition 5.1.1 *Assume the random sequence $\{X_k, k \geq 0\}$ is generated by the following recursion*

$$X_{k+1} = X_k + a_k \alpha_k X_k + a_k \varepsilon_{k+1} + O(a_k^2), \qquad (5.1.61)$$

where the real sequence $\{a_k\}$ is such that $a_k > 0$, $\sum_k a_k = \infty$, and $\sum_k a_k^2 < \infty$; the real number α_k has a positive limit: $\alpha_k \xrightarrow[k\to\infty]{} \alpha > 0$; and $\varepsilon_k = \sum_{i=1}^{l} \varepsilon_{k-i+1}^{(i)}$, $l \in [1, \infty)$, where $\{\varepsilon_k^{(i)}, \mathcal{F}_k\}$ is an mds for any $i \in [1, l]$. Moreover, $\liminf_k E\{\|\varepsilon_{k+1}^{(1)}\| | \mathscr{F}_k\} > 0$, and $\varepsilon_k^{(i)} \xrightarrow[k\to\infty]{} 0$ for any $i \geq 2$.

Then $\mathbb{P}(X_k \to 0) = 0$.

The proof is omitted and here we only present some intuitive explanations: Notice that $\alpha > 0$ and $x(t) \equiv 0$ is an unstable equilibrium of the differential equation $\dot{x} = \alpha x$. Under some conditions on the noise $\{\varepsilon_k\}$, no sample path of the recursion (5.1.61) can converge to the unstable equilibrium of $\dot{x} = \alpha x$ associated with (5.1.61) with possible exception on a set of probability zero.

Lemma 5.1.1 *Assume A5.1.1 and A5.1.2 hold. Then*

$$V_k^{(i)} V_k^{(i)+} A P_k^{(i)} = V_k^{(i)} V_k^{(i)+} A (I - V_k^{(i)} V_k^{(i)+}) = o(1), \qquad (5.1.62)$$

$$V_{k+1}^{(i)} V_{k+1}^{(i)+} P_k^{(i)} = V_{k+1}^{(i)} V_{k+1}^{(i)+} - V_{k+1}^{(i)} V_{k+1}^{(i)+} V_k^{(i)} V_k^{(i)+} = o(a_k) + a_k \varepsilon_{k+1}^{(i+1)'}, \qquad (5.1.63)$$

where $V_k^{(i)}, i = 1, \cdots, n$ are given by (5.1.1)–(5.1.3), and $\varepsilon_{k+1}^{(i+1)'}$ is a linear combination of mdses measurable with respect to $\mathscr{F}_{k+1}$, $\mathscr{F}_k$, $\cdots$, $\mathscr{F}_{k+2-i}$, and each mds is with one of $u_{k-i}^{(1)}$, $\cdots$, $u_{k-i}^{(i)}$ as its left factor.

Proof. By (5.1.18), (5.1.62) and (5.1.63) are equivalent to the following expressions

$$V_k^{(i)}V_k^{(i)T}A(I-V_k^{(i)}V_k^{(i)T})=o(1), \tag{5.1.64}$$

$$V_{k+1}^{(i)}V_{k+1}^{(i)T}-V_{k+1}^{(i)}V_{k+1}^{(i)T}V_k^{(i)}V_k^{(i)T}=o(a_k)+a_k\varepsilon_{k+1}^{(i+1)'}. \tag{5.1.65}$$

Let us prove (5.1.64) and (5.1.65) by induction.

For $i=1$, by Theorem 5.1.1 $d(u_k^{(1)},J_1)\xrightarrow[k\to\infty]{}0$ *a.s.* and $u_k^{(1)T}A-\lambda^{(j)}u^{(1)T}=(u_k^{(1)T}-u^{(1)T})A$ for some j and any $u^{(1)}\in J_1$. Assume that $u^{(1)}(k)\in J_1$ and $d(u_k^{(1)},J_1)=\|u_k^{(1)}-u^{(1)}(k)\|$. Consequently, by noticing $u^{(1)T}(k)u^{(1)}(k)=1$, we have

$$\begin{aligned}
&u_k^{(1)}u_k^{(1)T}A-u_k^{(1)}u_k^{(1)T}Au_k^{(1)}u_k^{(1)T}\\
=&u_k^{(1)}(u_k^{(1)T}A-\lambda^{(j)}u^{(1)T}(k))+\lambda^{(j)}u_k^{(1)}u^{(1)T}(k)\\
&-u_k^{(1)}(u_k^{(1)T}A-\lambda^{(j)}u^{(1)T}(k))u_k^{(1)}u_k^{(1)T}\\
&-\lambda^{(j)}u_k^{(1)}u^{(1)T}(k)u_k^{(1)}u_k^{(1)T}\\
=&O(\|u_k^{(1)}-u^{(1)}(k)\|)=O(d(u_k^{(1)},J_1))=o(1).
\end{aligned} \tag{5.1.66}$$

Thus, (5.1.64) holds for $i=1$. We now show that (5.1.65) also takes place for $i=1$.

By (5.1.9) it follows that

$$\begin{aligned}
u_{k+1}^{(1)}u_{k+1}^{(1)T}=&[u_k^{(1)}+a_k(A_{k+1}u_k^{(1)}-(u_k^{(1)T}A_{k+1}u_k^{(1)})u_k^{(1)})+O(a_k^2)]\\
&\cdot[u_k^{(1)}+a_k(A_{k+1}u_k^{(1)}-(u_k^{(1)T}A_{k+1}u_k^{(1)})u_k^{(1)})+O(a_k^2)]^T\\
=&u_k^{(1)}u_k^{(1)T}+a_kA_{k+1}u_k^{(1)}u_k^{(1)T}-2a_k(u_k^{(1)T}A_{k+1}u_k^{(1)})u_k^{(1)}u_k^{(1)T}\\
&+a_ku_k^{(1)}u_k^{(1)T}A_{k+1}+O(a_k^2).
\end{aligned}$$

Consequently,

$$\begin{aligned}
&u_{k+1}^{(1)}u_{k+1}^{(1)T}-u_{k+1}^{(1)}u_{k+1}^{(1)T}u_k^{(1)}u_k^{(1)T}\\
=&[u_k^{(1)}u_k^{(1)T}+a_kA_{k+1}u_k^{(1)}u_k^{(1)T}-2a_k(u_k^{(1)T}A_{k+1}u_k^{(1)})u_k^{(1)}u_k^{(1)T}\\
&+a_ku_k^{(1)}u_k^{(1)T}A_{k+1}+O(a_k^2)](I-u_k^{(1)}u_k^{(1)T})\\
=&a_ku_k^{(1)}u_k^{(1)T}A_{k+1}-a_ku_k^{(1)}u_k^{(1)T}A_{k+1}u_k^{(1)}u_k^{(1)T}+O(a_k^2)\\
=&a_k(u_k^{(1)}u_k^{(1)T}A-u_k^{(1)}u_k^{(1)T}Au_k^{(1)}u_k^{(1)T})+O(a_k^2)+a_k\varepsilon_{k+1}^{(2)'},
\end{aligned} \tag{5.1.67}$$

where $\varepsilon_{k+1}^{(2)'}=u_k^{(1)}u_k^{(1)T}N_{k+1}-u_k^{(1)}u_k^{(1)T}N_{k+1}u_k^{(1)}u_k^{(1)T}$ is an $\mathscr{F}_{k+1}$-measurable mds and is with $u_k^{(1)}$ as its left factor.

By (5.1.66) from (5.1.67) it follows that

$$V_{k+1}^{(1)}V_{k+1}^{(1)T}-V_{k+1}^{(1)}V_{k+1}^{(1)T}V_k^{(1)}V_k^{(1)T}=o(a_k)+a_k\varepsilon_{k+1}^{(2)'}. \tag{5.1.68}$$

Thus, (5.1.65) holds for $i = 1$.

Assume that for all sufficiently large k (5.1.64) and (5.1.65) hold for $i = 1, 2, \cdots, j-1,\ j \geq 2$.

We now show that (5.1.64) and (5.1.65) are also true for $i = j \leq n$. Let us first verify (5.1.64) for $i = j$.

From (5.1.20) it is seen that $V_k^{(j)} = [V_k^{(j-1)}, P_k^{(j-1)} u_k^{(j)}] = [V_k^{(j-1)}, u_k^{(j)} + o(a_{k-1}) + a_{k-1}\gamma_k(j)]$, and hence

$$\begin{aligned}
V_k^{(j)} V_k^{(j)T} =& [V_k^{(j-1)}, u_k^{(j)} + o(a_{k-1}) + a_{k-1}\gamma_k(j)] \\
& \cdot [V_k^{(j-1)}, u_k^{(j)} + o(a_{k-1}) + a_{k-1}\gamma_k(j)]^T \\
=& V_k^{(j-1)} V_k^{(j-1)T} + u_k^{(j)} u_k^{(j)T} + o(a_{k-1}) + a_{k-1}\gamma_k(j) u_k^{(j)T} \\
& + a_{k-1} u_k^{(j)} \gamma_k^T(j) \\
=& V_k^{(j-1)} V_k^{(j-1)T} + u_k^{(j)} u_k^{(j)T} + O(a_k).
\end{aligned} \tag{5.1.69}$$

Consequently,

$$\begin{aligned}
& V_k^{(j)} V_k^{(j)T} A (I - V_k^{(j)} V_k^{(j)T}) \qquad (5.1.70) \\
=& V_k^{(j)} V_k^{(j)T} A - V_k^{(j)} V_k^{(j)T} A V_k^{(j)} V_k^{(j)T} \\
=& (V_k^{(j-1)} V_k^{(j-1)T} + u_k^{(j)} u_k^{(j)T}) A - (V_k^{(j-1)} V_k^{(j-1)T} \\
& + u_k^{(j)} u_k^{(j)T}) A (V_k^{(j-1)} V_k^{(j-1)T} + u_k^{(j)} u_k^{(j)T}) + O(a_k) \\
=& V_k^{(j-1)} V_k^{(j-1)T} A - V_k^{(j-1)} V_k^{(j-1)T} A V_k^{(j-1)} V_k^{(j-1)T} \\
& - V_k^{(j-1)} V_k^{(j-1)T} A u_k^{(j)} u_k^{(j)T} - u_k^{(j)} u_k^{(j)T} A V_k^{(j-1)} V_k^{(j-1)T} \\
& + u_k^{(j)} u_k^{(j)T} A - u_k^{(j)} u_k^{(j)T} A u_k^{(j)} u_k^{(j)T} + O(a_k).
\end{aligned} \tag{5.1.71}$$

We want to show that the right-hand side of the above expression is $o(1)$. First, by the inductive assumption, the first two terms at the right-hand side of (5.1.70) give $o(1)$. Second, for its third term by Theorem 1, $d(Au_k^{(j)}, \lambda^{(m)} u_k^{(j)}) \xrightarrow[k\to\infty]{} 0$ for some m and by (5.1.19) it follows that

$$V_k^{(j-1)} V_k^{(j-1)T} A u_k^{(j)} u_k^{(j)T} = \lambda^{(m)} V_k^{(j-1)} V_k^{(j-1)T} u_k^{(j)} u_k^{(j)T} + o(1) = o(1). \tag{5.1.72}$$

Similar to (5.1.72) we can show

$$u_k^{(j)} u_k^{(j)T} A V_k^{(j-1)} V_k^{(j-1)T} = \lambda^{(m)} u_k^{(j)} u_k^{(j)T} V_k^{(j-1)} V_k^{(j-1)T} + o(1) = o(1). \tag{5.1.73}$$

Finally, by Theorem 1 $u_k^{(j)T} A - (u_k^{(j)T} A u_k^{(j)}) u_k^{(j)T} \to 0$, hence

$$u_k^{(j)} u_k^{(j)T} A - u_k^{(j)} u_k^{(j)T} A u_k^{(j)} u_k^{(j)T} = u_k^{(j)} (u_k^{(j)T} A - (u_k^{(j)T} A u_k^{(j)}) u_k^{(j)T}) = o(1). \tag{5.1.74}$$

Thus, we have shown that (5.1.64) is true for $i = j$.

We now prove that (5.1.65) is true for $i = j$.

Using the expression of $V_k^{(j)}V_k^{(j)T}$ given after the second equality in (5.1.69) we derive

$$
\begin{aligned}
&V_{k+1}^{(j)}V_{k+1}^{(j)T} - V_{k+1}^{(j)}V_{k+1}^{(j)T}V_k^{(j)}V_k^{(j)T}\\
=&V_{k+1}^{(j-1)}V_{k+1}^{(j-1)T} + u_{k+1}^{(j)}u_{k+1}^{(j)T} - (V_{k+1}^{(j-1)}V_{k+1}^{(j-1)T} + u_{k+1}^{(j)}u_{k+1}^{(j)T})\\
&\cdot(V_k^{(j-1)}V_k^{(j-1)T} + u_k^{(j)}u_k^{(j)T}) + o(a_k) + a_k\gamma_{k+1}(j)u_{k+1}^{(j)T}\\
&+a_k u_{k+1}^{(j)}\gamma_{k+1}^T(j) - a_k\gamma_{k+1}(j)u_{k+1}^{(j)T}(V_k^{(j-1)}V_k^{(j-1)T} + u_k^{(j)}u_k^{(j)T})\\
&-a_k u_{k+1}^{(j)}\gamma_{k+1}^T(j)\cdot(V_k^{(j-1)}V_k^{(j-1)T} + u_k^{(j)}u_k^{(j)T})\\
&-a_{k-1}(V_{k+1}^{(j-1)}V_{k+1}^{(j-1)T} + u_{k+1}^{(j)}u_{k+1}^{(j)T})\gamma_k(j)u_k^{(j)T}\\
&-a_{k-1}(V_{k+1}^{(j-1)}V_{k+1}^{(j-1)T} + u_{k+1}^{(j)}u_{k+1}^{(j)T})u_k^{(j)}\gamma_k(j)^T\\
=&V_{k+1}^{(j-1)}V_{k+1}^{(j-1)T} - V_{k+1}^{(j-1)}V_{k+1}^{(j-1)T}V_k^{(j-1)}V_k^{(j-1)T}\\
&-u_{k+1}^{(j)}u_{k+1}^{(j)T}V_k^{(j-1)}V_k^{(j-1)T} - V_{k+1}^{(j-1)}V_{k+1}^{(j-1)T}u_k^{(j)}u_k^{(j)T}\\
&+u_{k+1}^{(j)}u_{k+1}^{(j)T} - u_{k+1}^{(j)}u_{k+1}^{(j)T}u_k^{(j)}u_k^{(j)T} + o(a_k) + a_k\gamma_{k+1}(j)u_{k+1}^{(j)T}\\
&+a_k u_{k+1}^{(j)}\gamma_{k+1}(j)^T - a_k\gamma_{k+1}(j)u_{k+1}^{(j)T}(V_k^{(j-1)}V_k^{(j-1)T} + u_k^{(j)}u_k^{(j)T})\\
&-a_k u_{k+1}^{(j)}\gamma_{k+1}(j)^T(V_k^{(j-1)}V_k^{(j-1)T} + u_k^{(j)}u_k^{(j)T}) - a_{k-1}(V_{k+1}^{(j-1)}V_{k+1}^{(j-1)T}\\
&+u_{k+1}^{(j)}u_{k+1}^{(j)T})\gamma_k u_k^{(j)T} - a_{k-1}(V_{k+1}^{(j-1)}V_{k+1}^{(j-1)T}\\
&+u_{k+1}^{(j)}u_{k+1}^{(j)T})u_k^{(j)}\gamma_k(j)^T.
\end{aligned}
\tag{5.1.75}
$$

Noting $u_{k+1}^{(i)} = u_k^{(i)} + o(1) = u_{k-j}^{(i)} + o(1)$ by (5.1.39) and $V_{k+1}^{(j)} = [u_{k-j}^{(1)}, \cdots, u_{k-j}^{(j)}] + o(1)$ by (5.1.40), we can rewrite (5.1.75) as

$$
\begin{aligned}
&V_{k+1}^{(j)}V_{k+1}^{(j)T} - V_{k+1}^{(j)}V_{k+1}^{(j)T}V_k^{(j)}V_k^{(j)T}\\
=&V_{k+1}^{(j-1)}V_{k+1}^{(j-1)T} - V_{k+1}^{(j-1)}V_{k+1}^{(j-1)T}V_k^{(j-1)}V_k^{(j-1)T}\\
&-u_{k+1}^{(j)}u_{k+1}^{(j)T}V_k^{(j-1)}V_k^{(j-1)T} - V_{k+1}^{(j-1)}V_{k+1}^{(j-1)T}u_k^{(j)}u_k^{(j)T} + u_{k+1}^{(j)}u_{k+1}^{(j)T}\\
&-u_{k+1}^{(j)}u_{k+1}^{(j)T}u_k^{(j)}u_k^{(j)T} + o(a_k) + a_k\gamma_{k+1}(j)u_{k-j}^{(j)T} + a_k u_{k-j}^{(j)}\gamma_{k+1}(j)^T\\
&-a_k\gamma_{k+1}(j)u_{k-j}^{(j)T}([u_{k-j}^{(1)}, \cdots, u_{k-j}^{(j-1)}][u_{k-j}^{(1)}, \cdots, u_{k-j}^{(j-1)}]^T + u_{k-j}^{(j)}u_{k-j}^{(j)T})\\
&-a_k u_{k-j}^{(j)}\gamma_{k+1}(j)^T([u_{k-j}^{(1)}, \cdots, u_{k-j}^{(j-1)}][u_{k-j}^{(1)}, \cdots, u_{k-j}^{(j-1)}]^T\\
&+u_{k-j}^{(j)}u_{k-j}^{(j)T}) - a_{k-1}([u_{k-j-1}^{(1)}, \cdots, u_{k-j-1}^{(j-1)}]\\
&\cdot[u_{k-j-1}^{(1)}, \cdots, u_{k-j-1}^{(j-1)}]^T + u_{k-j-1}^{(j)}u_{k-j-1}^{(j)T})\gamma_k(j)u_{k-j-1}^{(j)T} - a_{k-1}\\
&\cdot([u_{k-j-1}^{(1)}, \cdots, u_{k-j-1}^{(j-1)}][u_{k-j-1}^{(1)}, \cdots, u_{k-j-1}^{(j-1)}]^T\\
&+u_{k-j-1}^{(j)}u_{k-j-1}^{(j)T})u_{k-j-1}^{(j)}\gamma_k(j)^T.
\end{aligned}
\tag{5.1.76}
$$

We want to express the right-hand side of (5.1.76) in the form of the right-hand side of (5.1.65) for $i = j$. Let us analyze each term at the right-hand side of (5.1.76).

First, the first two terms are estimated by the inductive assumption:

$$V_{k+1}^{(j-1)}V_{k+1}^{(j-1)T} - V_{k+1}^{(j-1)}V_{k+1}^{(j-1)T}V_{k}^{(j-1)}V_{k}^{(j-1)T} = o(a_k) + a_k\varepsilon_{k+1}^{(j)'}, \qquad (5.1.77)$$

where $\varepsilon_{k+1}^{(j)'}$ is a linear combination of mdses measurable with respect to $\mathscr{F}_{k+1}$, $\mathscr{F}_k$, $\cdots$, $\mathscr{F}_{k+3-j}$, and each mds is with one of the vectors $u_{k-j}^{(1)}, \cdots, u_{k-j}^{(j-1)}$ as its left factor.

Second, by (5.1.37) its third term equals zero.

Replacing $u_{k+1}^{(j)}$ in (5.1.19) with the expression given by the first equality of (5.1.39), we derive

$$\begin{aligned} V_{k+1}^{(j-1)T}u_k^{(j)} &= o(a_k) + a_k\eta_{k+1}^{(j-1)} - a_kV_{k+1}^{(j-1)T}\delta_{k+1}(j) \\ &= o(a_k) + a_k\eta_{k+1}^{(j-1)} - a_k[u_{k-j}^{(1)},\cdots,u_{k-j}^{(j-1)}]^T\delta_{k+1}(j), \end{aligned} \qquad (5.1.78)$$

where for the last equality (5.1.40) is used.

By (5.1.39), (5.1.40), and (5.1.78) it follows that

$$\begin{aligned} &V_{k+1}^{(j-1)}V_{k+1}^{(j-1)T}u_k^{(j)}u_k^{(j)T} \\ &= ([u_{k-j}^{(1)},\cdots,u_{k-j}^{(j-1)}] + o(1))(o(a_k) + a_k\eta_{k+1}^{(j-1)} \\ &\quad - a_k[u_{k-j}^{(1)},\cdots,u_{k-j}^{(j-1)}]^T\delta_{k+1}(j))\cdot(u_{k-j}^{(j)} + o(1))^T \\ &= o(a_k) + a_k[u_{k-j}^{(1)},\cdots,u_{k-j}^{(j-1)}]\eta_{k+1}^{(j-1)}u_{k-j}^{(j)T} - a_k[u_{k-j}^{(1)},\cdots,u_{k-j}^{(j-1)}] \\ &\quad \cdot[u_{k-j}^{(1)},\cdots,u_{k-j}^{(j-1)}]^T\delta_{k+1}(j)u_{k-j}^{(j)T}. \end{aligned} \qquad (5.1.79)$$

This gives the required expression for the fourth term at the right-hand side of (5.1.76).

Finally, by the first equality in (5.1.39) it follows that

$$\begin{aligned} u_{k+1}^{(j)}u_{k+1}^{(j)T} &= (u_k^{(j)} + o(a_k) + a_k\delta_{k+1}(j))\cdot(u_k^{(j)} + o(a_k) + a_k\delta_{k+1}(j))^T \\ &= u_k^{(j)}u_k^{(j)T} + o(a_k) + a_k\delta_{k+1}(j)u_k^{(j)T} + a_ku_k^{(j)}\delta_{k+1}(j)^T, \end{aligned} \qquad (5.1.80)$$

and hence

$$\begin{aligned} &u_{k+1}^{(j)}u_{k+1}^{(j)T} - u_{k+1}^{(j)}u_{k+1}^{(j)T}u_k^{(j)}u_k^{(j)T} \\ &= [u_k^{(j)}u_k^{(j)T} + o(a_k) + a_k\delta_{k+1}(j)u_k^{(j)T} + a_ku_k^{(j)}\delta_{k+1}(j)^T](I - u_k^{(j)}u_k^{(j)T}) \\ &= o(a_k) + a_ku_k^{(j)}\delta_{k+1}(j)^T - a_ku_k^{(j)}\delta_{k+1}(j)^Tu_k^{(j)}u_k^{(j)T} \\ &= o(a_k) + a_ku_{k-j}^{(j)}\delta_{k+1}(j)^T - a_ku_{k-j}^{(j)}\delta_{k+1}(j)^Tu_{k-j}^{(j)}u_{k-j}^{(j)T}, \end{aligned} \qquad (5.1.81)$$

where the last equality is because $u_{k+1}^{(j)} = u_k^{(j)} + o(1) = u_{k-j}^{(j)} + o(1)$ by (5.1.39).

This gives expression for the fifth and sixth terms at the right-hand side of (5.1.76).

Notice that in (5.1.76), (5.1.79), and (5.1.81) $\eta_{k+1}^{(j-1)}$, $\delta_{k+1}(j)$, and $\gamma_{k+1}(j)$ are involved. They are the linear combinations of mdses. To be precise, $\eta_{k+1}(j)$ is a linear combination of mdses measurable with respect to $\mathscr{F}_{k+1}, \mathscr{F}_k, \cdots, \mathscr{F}_{k+3-j}$, $\delta_{k+1}(j)$ is a linear combination of mds' measurable with respect to $\mathscr{F}_{k+1}, \mathscr{F}_k, \cdots, \mathscr{F}_{k+2-j}$, and $\gamma_{k+1}(j)$ is a linear combination of mdses measurable with respect to $\mathscr{F}_{k+1}, \mathscr{F}_k, \cdots, \mathscr{F}_{k+3-j}$.

Thus, putting (5.1.77), (5.1.79), and (5.1.81) into (5.1.76) leads to

$$V_{k+1}^{(j)} V_{k+1}^{(j)T} - V_{k+1}^{(j)} V_{k+1}^{(j)T} V_k^{(j)} V_k^{(j)T} = o(a_k) + a_k \varepsilon_{k+1}^{(j+1)'},$$

where $\varepsilon_{k+1}^{(j+1)'}$ is a linear combination of mdses measurable with respect to $\mathscr{F}_{k+1}$, $\mathscr{F}_k, \cdots, \mathscr{F}_{k+2-j}$.

It is noticed that for each terms containing either $\eta_{k+1}^{(j-1)}$ or $\delta_{k+1}(j)$ there is a left factor from $u_{k-j}^{(1)}, \cdots, u_{k-j}^{(j)}$, while, by definition, the expression of $\gamma_{k+1}(j)$ given immediately after (5.1.52) also includes a left factor $[u_{k-j}^{(1)}, \cdots, u_{k-j}^{(j)}]$.

Thus, we have shown that (5.1.65) is true for $i = j$ and have completed the proof of the lemma. □

To establish the ordered convergence of $\{u_k^{(i)}\}$ we need one more assumption A5.1.3 in addition to A5.1.1 and A5.1.2.

A5.1.3

$$\liminf_n E\left\{\|x_k^T N_{k+1} y_k\| | \mathscr{F}_k\right\} > 0$$

for any $\mathscr{F}_k$-measurable x_k and y_k bounded from above and from zero:

$$\begin{aligned} 0 &< \liminf_{k\to\infty} \|x_k\| \le \limsup_{k\to\infty} \|x_k\| < \infty, \\ 0 &< \liminf_{k\to\infty} \|y_k\| \le \limsup_{k\to\infty} \|y_k\| < \infty. \end{aligned} \tag{5.1.82}$$

Remark 5.1.2 *Condition A5.1.3 has excluded the case $A_{k+1} \equiv A$ from consideration. As a matter of fact, in order to achieve the desired limit some observation noise is necessary; otherwise, it may happen that the algorithm is stuck at an undesired vector. To see this, let $A_{k+1} \equiv A$ and let the initial value $u_1^{(1)} \triangleq u^{(i)}$, where $u^{(i)}$ is a unit eigenvector of A corresponding to an eigenvalue $\lambda^{(i)}$ different from the largest one. Then,*

$$\begin{aligned} \tilde{u}_{k+1}^{(1)} &= u^{(i)} + a_k A u^{(i)} = (1 + a_k \lambda^{(i)}) u^{(i)}, \\ u_{k+1}^{(1)} &= u^{(i)}, \end{aligned}$$

and the algorithm will never converge to the desired $u^{(1)}$.

The following proposition gives sufficient conditions on $\{N_k\}$ in order for A5.1.3 to be satisfied.

Proposition 5.1.2 *Assume that $(N_k, \mathscr{F}_k)$ with $N_k = \{N_{ij}(k)\}$ is an mds, $E(N_{ij}(k+$*

$1)N_{st}(k+1)|\mathscr{F}_k) = 0$ *whenever* $(ij) \neq (st)$, $\liminf_{k\to\infty} E(N_{ij}^2(k+1)|\mathscr{F}_k) \geq \sigma > 0$, *and for some* $\alpha > 2$, $\limsup_{k\to\infty} E(\|N_{k+1}\|^\alpha|\mathscr{F}_k) < \infty$. *Then, A5.1.3 holds.*

Proof. Let x_k and y_k be $\mathscr{F}_k$-measurable and satisfy (5.1.82). By the Hölder inequality with $p = \frac{\alpha-1}{\alpha-2}$ and $q = \alpha - 1$ we have

$$E(|x_k^T N_{k+1} y_k|^2|\mathscr{F}_k) \leq \left(E(|x_k^T N_{k+1} y_k||\mathscr{F}_k)\right)^{\frac{\alpha-2}{\alpha-1}} \left(E(|x_k^T N_{k+1} y_k|^\alpha|\mathscr{F}_k)\right)^{\frac{1}{\alpha-1}}.$$

Since $\limsup_{k\to\infty} E(\|N_{k+1}\|^\alpha|\mathscr{F}_k) < \infty$, by (5.1.82) for the proposition it suffices to show

$$\liminf_{k\to\infty} E(|x_k^T N_{k+1} y_k|^2|\mathscr{F}_k) > 0.$$

Writing $x_k = [x_1(k), \cdots, x_n(k)]^T$ and $y_k = [y_1(k), \cdots, y_n(k)]^T$ and noticing the conditional uncorrelatedness of components of N_k, we have

$$\begin{aligned}
&\liminf_{k\to\infty} E\left(|x_k^T N_{k+1} y_k|^2|\mathscr{F}_k\right) \\
&= \liminf_{k\to\infty} E\Big(\sum_{i,j}^n x_i(k) y_j(k) N_{ij}(k+1) \sum_{s,t}^n x_s(k) y_t(k) N_{st}(k+1)|\mathscr{F}_k\Big) \\
&= \liminf_{k\to\infty} \sum_{ij}^n (x_i(k) y_j(k))^2 E(N_{ij}^2(k+1)|\mathscr{F}_k) \\
&\geq \liminf_{k\to\infty} \sum_{i=1}^n x_i^2(k) \sum_{j=1}^n y_j^2(k) \sigma^2 = \liminf_{k\to\infty} \|x_k\|^2 \|y_k\|^2 \sigma^2 > 0.
\end{aligned}$$

This proves the proposition. □

Prior to describing the result on ordered convergence let us diagonalize the matrix

$$A = [\phi_1, \cdots, \phi_n] \begin{bmatrix} \lambda^{(1)} & 0 & \cdots & 0 \\ 0 & \lambda^{(2)} & \ddots & \vdots \\ \vdots & \ddots & \ddots & 0 \\ 0 & \cdots & 0 & \lambda^{(n)} \end{bmatrix} \begin{bmatrix} \phi_1^T \\ \vdots \\ \phi_n^T \end{bmatrix}$$

where ϕ_i is a unit eigenvector corresponding to the eigenvalue $\lambda^{(i)}$.

For any j if there is an i such that $\lambda^{(i)} > \lambda^{(j)}$, then we define

$$a(j) = \max\{i : \lambda^{(i)} > \lambda^{(j)}\}.$$

Similarly, define

$$b(j) = \min\{i : \lambda^{(i)} < \lambda^{(j)}\},$$

if there is an i such that $\lambda^{(i)} < \lambda^{(j)}$.

Let S^i be the set of unit eigenvectors corresponding to $\lambda^{(i)}$, where the identity among some of $\{S^i\}$ is not excluded. Further, let J be the totality of all unit eigenvectors of A. Then

$$J = S^1 \cup \cdots \cup S^n.$$

Theorem 5.1.2 *Assume A5.1.1–A5.1.3 hold. Then* $\lambda(i) = \lambda^{(i)}$, $J_i = S^i$, *and* $d(u_k^{(i)}, S^i) \xrightarrow[k\to\infty]{} 0$.

Proof. If $\lambda^{(1)} = \lambda^{(2)} = \cdots = \lambda^{(n)}$, then the conclusion of the theorem follows from Theorem 5.1.1. So, we need only to consider the case where $b(1)$ is well defined. Let us first prove the theorem for $i = 1$.

By Theorem 5.1.1 we have $d(u_k^{(1)}, J_1) \xrightarrow[k\to\infty]{} 0$. To prove $J_1 = S^1$, it suffices to show $P(d(u_k^{(1)}, S^m) \to 0) = 0$ for any $m \geq b(1)$.

Assume the converse: there is an $m \geq b(1)$ such that $d(u_k^{(1)}, S^m) \to 0$ *a.s.* This means that $\phi^T u_k^{(1)} \to 0$ for any $\phi \in S^1$.

Multiplying ϕ^T to both sides of (5.1.9) from the left and noticing $A\phi = \lambda^{(1)}\phi$ we derive

$$\begin{aligned}\phi^T u_{k+1}^{(1)} &= \phi^T u_k^{(1)} + a_k \phi^T (A u_k^{(1)} - (u_k^{(1)T} A u_k^{(1)}) u_k^{(1)}) + a_k \phi^T (\varepsilon_{k+1}^{(1)} + O(a_k)) \\ &= \phi^T u_k^{(1)} + a_k [\lambda^{(1)} - (u_k^{(1)T} A u_k^{(1)})] \phi^T u_k^{(1)} + a_k \phi^T (\varepsilon_{k+1}^{(1)} + O(a_k)).\end{aligned} \tag{5.1.83}$$

Set $\Gamma_1 \triangleq \{\omega : \phi^T u_k^{(1)} \to 0\}$.

By definition of $\varepsilon_{n+1}^{(1)}$ given after (5.1.9) it is clear that

$$\begin{aligned}E(\|\phi^T \varepsilon_{k+1}^{(1)}\| | \mathscr{F}_k) =& E(\|\phi^T N_{k+1} u_k^{(1)} - (u_k^{(1)T} N_{k+1} u_k^{(1)}) \phi^T u_k^{(1)}\| | \mathscr{F}_k) \\ \geq& E(\|\phi^T N_{k+1} u_k^{(1)}\| | \mathscr{F}_k) - E(|u_k^{(1)T} N_{k+1} u_k^{(1)}| | \mathscr{F}_k) \|\phi^T u_k^{(1)}\|.\end{aligned}$$

Thus, on Γ_1 by A5.1.3 we have $\liminf_n E\{\|\varepsilon_{n+1}^{(1)T} \phi\| | \mathscr{F}_n\} > 0$.

Further, by A5.1.2 $\{\phi^T \varepsilon_{n+1}^{(1)}\}$ is an mds with $E(\|\phi^T \varepsilon_{n+1}^{(1)}\|^2 | \mathscr{F}_n) < \infty$ and $\lambda^{(1)} - (u_k^{(1)T} A u_k^{(1)}) \to \lambda^{(1)} - \lambda^{(m)} > 0$.

Then, $P(\Gamma_1) = 0$ by Proposition 5.1.1. This means that with probability one $\phi^T u_k^{(1)}$ does not converge to zero. The obtained contradiction shows that $u^{(1)} \in S^1$ a.s.

Inductively assume

$$u^{(i)} \in S^i \text{ a.s. } \forall i = 1, \cdots, j, \; j \geq 1.$$

We now show it holds also for $i = j+1$.

We distinguish two cases: 1) $b(j+1)$ is not defined, and 2) $b(j+1)$ is well defined.

1) Consider the case where $b(j+1)$ is not defined. Since not all eigenvalues are equal, $a(j+1) \leq j$ must be well defined and

$$\lambda^{(1)} \geq \cdots \lambda^{(a(j+1))} > \lambda^{(a(j+1)+1))} = \cdots \lambda^{(j+1)} = \cdots = \lambda^{(n)}.$$

The subspace of unit eigenvectors $\{S^1 \cup \cdots \cup S^{a(j+1)}\}$ corresponding to the eigenvalues $\{\lambda^{(1)}, \cdots, \lambda^{(a(j+1))}\}$ is of dimension $a(j+1)$. By (5.1.18) the unit vectors

$\{u_k^{(1)},\cdots,u_k^{(a(j+1))}\}$ are asymptotically orthogonal and by the inductive assumptions they converge to $\{S^1\cup\cdots\cup S^{a(j+1)}\}$. By Theorem 5.1.1 $u_k^{(j+1)}$ converges to J_{j+1}. If $d(u_k^{(j+1)},\{S^1\cup\cdots\cup S^{a(j+1)}\})\xrightarrow[k\to\infty]{}0$, then the set of $a(j+1)+1$ asymptotically orthogonal unit vectors $\{u_k^{(1)},\cdots,u_k^{(a(j+1))},u_k^{(j+1)}\}$ would converge to a subspace of dimension $a(j+1)$. This is impossible.

Therefore, $J_{j+1}\subset\{S^{a(j+1)+1}\cup\cdots\cup S^n\}$. Since $\lambda^{(a(j+1)+1))}=\cdots\lambda^{(j+1)}=\cdots=\lambda^{(n)}$, then $J_{j+1}=S^{j+1}$ and $d(u_k^{(j+1)},S^{j+1})\xrightarrow[k\to\infty]{}0$. In this case the induction is completed.

2) We now complete the induction by considering the case where $b(j+1)$ is well defined.

By Theorem 5.1.1, $d(u_k^{(j+1)},J_{j+1})\xrightarrow[k\to\infty]{}0$ a.s. Since $u_k^{(j+1)T}u_k^{(i)}\xrightarrow[k\to\infty]{}0\ \forall i=1,\cdots,j$ and by the inductive assumption $d(u_k^{(i)},S^i)\xrightarrow[k\to\infty]{}0\ \forall i=1,\cdots,j$, the converse assumption $d(u_k^{(j+1)},S^{j+1})\not\xrightarrow[k\to\infty]{}0$ is equivalent to $d(u_k^{(j+1)},S^m)\xrightarrow[k\to\infty]{}0$ for some $m\geq b(j+1)$. This in turn is equivalent to $\phi^Tu_k^{(j+1)}\to 0$ for those $\phi\in S^{j+1}$ for which $\phi^Tu_k^{(i)}\to 0\ \forall i=1,\cdots,j$.

By using the expression of $u_{k+1}^{(j+1)}$ given by the second equality of (5.1.45) it follows that

$$\begin{aligned}
P_{k+1}^{(j)}u_{k+1}^{(j+1)}=&P_{k+1}^{(j)}[P_k^{(j)}u_k^{(j+1)}+a_kP_k^{(j)}A_{k+1}P_k^{(j)}u_k^{(j+1)}\\
&-a_k(u_k^{(j+1)T}P_k^{(j)}A_{k+1}P_k^{(j)}\cdot u_k^{(j+1)})P_k^{(j)}u_k^{(j+1)}+O(a_k^2)]\\
=&P_k^{(j)}u_k^{(j+1)}+a_kAP_k^{(j)}u_k^{(j+1)}-a_k(u_k^{(j+1)T}P_k^{(j)}AP_k^{(j)}u_k^{(j+1)})P_k^{(j)}u_k^{(j+1)}\\
&+a_k\varepsilon_{k+1}^{(j+1)''}-a_kV_k^{(j)}V_k^{(j)+}AP_k^{(j)}u_k^{(j+1)}-V_{k+1}^{(j)}V_{k+1}^{(j)+}P_k^{(j)}u_k^{(j+1)}\\
&-a_kV_{k+1}^{(j)}V_{k+1}^{(j)+}P_k^{(j)}A_{k+1}P_k^{(j)}u_k^{(j+1)}+a_k(u_k^{(j+1)T}P_k^{(j)}A_{k+1}P_k^{(j)}u_k^{(j+1)})\\
&\cdot V_{k+1}^{(j)}V_{k+1}^{(j)+}P_k^{(j)}u_k^{(j+1)}+O(a_k^2),
\end{aligned}\tag{5.1.84}$$

where $\varepsilon_{k+1}^{(j+1)''}=P_k^{(j)}N_{k+1}P_k^{(j)}u_k^{(j+1)}-(u_k^{(j+1)T}P_k^{(j)}N_{k+1}P_k^{(j)}u_k^{(j+1)})P_k^{(j)}u_k^{(j+1)}$ is an mds measurable with respect to $\mathscr{F}_{k+1}$.

Multiplying both sides of (5.1.84) by ϕ^T from the left and noticing $A\phi=\lambda^{(j+1)}\phi$ we derive

$$\begin{aligned}
\phi^TP_{k+1}^{(j)}u_{k+1}^{(j+1)}=&\phi^TP_k^{(j)}u_k^{(j+1)}+a_k\phi^TAP_k^{(j)}u_k^{(j+1)}\\
&-a_k(u_k^{(j+1)T}P_k^{(j)}AP_k^{(j)}u_k^{(j+1)})\phi^TP_k^{(j)}u_k^{(j+1)}+a_k\phi^T\varepsilon_{k+1}^{(j+1)''}\\
&-a_k\phi^TV_k^{(j)}V_k^{(j)+}AP_k^{(j)}u_k^{(j+1)}-\phi^TrV_{k+1}^{(j)}V_{k+1}^{(j)+}P_k^{(j)}u_k^{(j+1)}\\
&-a_k\phi^TV_{k+1}^{(j)}V_{k+1}^{(j)+}P_k^{(j)}A_{k+1}P_k^{(j)}u_k^{(j+1)}\\
&+a_k(u_k^{(j+1)T}P_k^{(j)}A_{k+1}P_k^{(j)}u_k^{(j+1)})\phi^TV_{k+1}^{(j)}V_{k+1}^{(j)+}P_k^{(j)}u_k^{(j+1)}+O(a_k^2)
\end{aligned}$$

$$
\begin{aligned}
=&\phi^T P_k^{(j)} u_k^{(j+1)} + a_k[\lambda^{(j+1)} - (u_k^{(j+1)T} P_k^{(j)} A P_k^{(j)} u_k^{(j+1)})]\phi^T P_k^{(j)} u_k^{(j+1)} \\
&- a_k \phi^T V_k^{(j)} V_k^{(j)+} A P_k^{(j)} P_k^{(j)} u_k^{(j+1)} - \phi^T V_{k+1}^{(j)} V_{k+1}^{(j)+} P_k^{(j)} P_k^{(j)} u_k^{(j+1)} \\
&+ a_k \phi^T \varepsilon_{k+1}^{(j+1)''} + O(a_k^2) - a_k \phi^T V_{k+1}^{(j)} V_{k+1}^{(j)+} P_k^{(j)} A_{k+1} P_k^{(j)} u_k^{(j+1)} \\
&+ a_k (u_k^{(j+1)T} P_k^{(j)} A_{k+1} P_k^{(j)} u_k^{(j+1)}) \phi^T V_{k+1}^{(j)} V_{k+1}^{(j)+} P_k^{(j)} u_k^{(j+1)}. \qquad (5.1.85)
\end{aligned}
$$

By (5.1.64) (or its equivalent expression (5.1.62)) it is known that

$$
a_k \phi^T V_k^{(j)} V_k^{(j)+} A P_k^{(j)} P_k^{(j)} u_k^{(j+1)} = o(a_k) \phi^T P_k^{(j)} u_k^{(j+1)},
$$

and by (5.1.65) (or its equivalent form (5.1.63))

$$
\begin{aligned}
&- a_k \phi^T V_{k+1}^{(j)} V_{k+1}^{(j)+} P_k^{(j)} A_{k+1} P_k^{(j)} u_k^{(j+1)} + a_k (u_k^{(j+1)T} P_k^{(j)} A_{k+1} P_k^{(j)} u_k^{(j+1)}) \\
&\cdot \phi^T V_{k+1}^{(j)} V_{k+1}^{(j)+} P_k^{(j)} u_k^{(j+1)} = O(a_k^2),
\end{aligned}
$$

we can rewrite (5.1.85) as

$$
\begin{aligned}
\phi^T P_{k+1}^{(j)} u_{k+1}^{(j+1)} =& \phi^T P_k^{(j)} u_k^{(j+1)} + a_k[\lambda^{(j+1)} - u_k^{(j+1)T} P_k^{(j)} A P_k^{(j)} u_k^{(j+1)} \\
&+ o(1)]\phi^T P_k^{(j)} u_k^{(j+1)} + O(a_k^2) + a_k \phi^T \varepsilon_{k+1}^{(j+1)}, \qquad (5.1.86)
\end{aligned}
$$

where $\varepsilon_{k+1}^{(j+1)} = \varepsilon_{k+1}^{(j+1)''} - \varepsilon_{k+1}^{(j+1)'} P_k^{(j)} u_k^{(j+1)}$ is a linear combination of mdses measurable with respect to $\mathscr{F}_{k+1}, \mathscr{F}_k, \cdots, \mathscr{F}_{k+2-j}$.

Set $\Gamma_{j+1} = \{\omega : \phi^T u_k^{(j+1)} \to 0\}$.

We have

$$
\phi^T P_k^{(j)} u_k^{(j+1)} = \phi^T u_k^{(j+1)} - \phi^T V_k^{(j)} V_k^{(j)+} u_k^{(j+1)} \to 0 \text{ on } \Gamma_{j+1}, \qquad (5.1.87)
$$

because $\phi^T u_k^{(j+1)} \to 0$ on Γ_{j+1} and $V_k^{(j)} V_k^{(j)+} u_k^{(j+1)} \to 0$ by (5.1.20).

Notice that

$$
\begin{aligned}
E(\|\phi^T \varepsilon_{k+1}^{(j+1)}\| | \mathscr{F}_k) \geq& E(\|\phi^T \varepsilon_{k+1}^{(j+1)''}\| | \mathscr{F}_k) - E(\|\phi^T \varepsilon_{k+1}^{(j+1)'}\| | \mathscr{F}_k) \\
\geq& E(\|\phi^T P_k^{(j)} N_{k+1} P_k^{(j)} u_k^{(j+1)}\| | \mathscr{F}_k) - E(\|\phi^T \varepsilon_{k+1}^{(j+1)'}\| | \mathscr{F}_k) \\
&- E(|u_k^{(j+1)T} P_k^{(j)} N_{k+1} P_k^{(j)} u_k^{(j+1)}| | \mathscr{F}_k) |\phi^T P_k^{(j)} u_k^{(j+1)}|. \qquad (5.1.88)
\end{aligned}
$$

By Lemma 5.1.1 each term in $\varepsilon_{k+1}^{(j+1)'}$ is headed by one of the vectors $\{u_{k-j}^{(1)}, \cdots, u_{k-j}^{(j)}\}$. Since $\phi^T u_k^{(i)} \to 0 \ \forall i = 1, \cdots, j$, it follows that $\|\phi^T \varepsilon_{k+1}^{(j+1)'}\| \xrightarrow[k\to\infty]{} 0$, and by Theorem 1.1.8 $E(\|\phi^T \varepsilon_{k+1}^{(j+1)'}\| | \mathscr{F}_k) \xrightarrow[k\to\infty]{} 0$.

Then by (5.1.87) from (5.1.88) it follows that on Γ_{j+1}

$$
\begin{aligned}
&\liminf_{k\to\infty} E(\|\phi^T \varepsilon_{k+1}^{(j+1)}\| | \mathscr{F}_k) \\
&\geq \liminf_{k\to\infty} E(\|\phi^T P_k^{(j)} N_{k+1} P_k^{(j)} u_k^{(j+1)}\| | \mathscr{F}_k). \qquad (5.1.89)
\end{aligned}
$$

Since $\phi^T u_k^{(i)} \to 0 \ \forall i = 1, \cdots, j$, by induction it is directly verified that $\phi^T V_k^{(i)} \xrightarrow[k\to\infty]{} 0$, which implies $\phi^T P_k^{(j)} - \phi^T \xrightarrow[k\to\infty]{} 0$. Further, by (5.1.20) it is seen that $P_k^{(j)} u_k^{(j+1)} - u_k^{(j+1)} \xrightarrow[k\to\infty]{} 0$.

Then by A5.1.3 from (5.1.89) we find that $\liminf_{k\to\infty} E(\|\phi^T \varepsilon_{k+1}^{(j+1)}\| | \mathscr{F}_k) > 0$.

By A5.1.2 $\limsup_n E(\|\phi^T \varepsilon_{n+1}^{(j+1)}\|^2 | \mathscr{F}_n) < \infty$, and by noticing $P_k^{(j)} u_k^{(j+1)} - u_k^{(j+1)} \xrightarrow[k\to\infty]{} 0$, we have $\lambda^{(j+1)} - u_k^{(j+1)T} P_k^{(j)} A P_k^{(j)} u_k^{(j+1)} + o(1) \to \lambda^{(j+1)} - \lambda^{(m)} > 0$. Then, by Proposition 5.1.1 we conclude that $P(\Gamma_{j+1}) = 0$, i.e., $\phi^T u_k^{(j+1)}$ cannot converge to zero. The obtained contradiction shows that $u^{(j+1)} \in S^{j+1}$ a.s. □

Convergence Rates

From the viewpoint of PCA the case $\{\lambda^{(1)} = \cdots = \lambda^{(n)}\}$ is less interesting, because this case means that all components are equally important and PCA may play no roll. Except for this less interesting case, we now show the convergence rate when $u_k^{(i)}$ converges to some $u^{(i)} \in S^i$ as k tends to infinity.

Lemma 5.1.2 *Let $\{x_k\}$ be recursively defined by*

$$x_{k+1} = x_k + O(a_k^{1+\delta}) + a_k w_k \ \ \textit{with} \ \ \delta \in (0, 1/2), \tag{5.1.90}$$

where $\{w_k\}$ is such that $W_n \triangleq \sum_{k=1}^n a_k^{1-\delta} w_k$ converges to a finite limit as n tends to ∞, and $\{a_k\}$ is given by A5.1.1. Then x_k converges to a vector x and $\|x_k - x\| = O(a_k^\delta)$.

Proof. Summing up both sides of (5.1.90) from 1 to n leads to

$$x_{n+1} = x_1 + O(\sum_{k=1}^n a_k^{1+\delta}) + \sum_{k=1}^n a_k w_k. \tag{5.1.91}$$

By A5.1.1 there are a small enough $\varepsilon > 0$ and a sufficiently large N so that $a_{i+1}^{-1} - a_i^{-1} > a - \varepsilon \ \forall i \geq N$. Then for $n \geq N+1$ we have

$$\begin{aligned}
a_n^{-\delta} \sum_{k=n}^\infty a_k^{1+\delta} <& \frac{1}{a-\varepsilon} a_n^{-\delta} \sum_{i=n}^\infty \left(\frac{1}{a_i^{-1}}\right)^{1+\delta} (a_i^{-1} - a_{i-1}^{-1}) \\
\leq& \frac{1}{a-\varepsilon} a_n^{-\delta} \sum_{i=n}^\infty \int_{a_{i-1}^{-1}}^{a_i^{-1}} \left(\frac{1}{a_i^{-1}}\right)^{1+\delta} \mathrm{d}x \\
\leq& \frac{1}{a-\varepsilon} a_n^{-\delta} \int_{a_{n-1}{}^{-1}}^{\infty} \left(\frac{1}{x}\right)^{1+\delta} \mathrm{d}x \\
=& \frac{1}{\delta(a-\varepsilon)} (\frac{a_{n-1}}{a_n})^\delta = O(1).
\end{aligned} \tag{5.1.92}$$

Summing by parts, we have

$$
\begin{aligned}
a_n^{-\delta}\sum_{k=n}^{\infty} a_k w_k =& a_n^{-\delta}\sum_{k=n}^{\infty}(W_k - W_{k-1})a_k^{\delta} \\
=& a_n^{-\delta}\sum_{k=n}^{\infty} W_k(a_k^{\delta} - a_{k+1}^{\delta}) - W_{n-1} \\
=& a_n^{-\delta}\sum_{k=n}^{\infty} W_k a_k^{\delta}(1 - (\frac{a_{k+1}}{a_k})^{\delta}) - W_{n-1} = O(1),
\end{aligned} \tag{5.1.93}
$$

because $a_{i+1}^{-1} - a_i^{-1} \xrightarrow[k\to\infty]{} a > 0$ and from here it follows that $1 - (\frac{a_{k+1}}{a_k})^{\delta} - a\delta a_{k+1} = o(a_k)$.

From (5.1.93) it is seen that $\sum_{k=1}^{\infty} a_k w_k$ converges and the rate of convergence is $\sum_{k=n}^{\infty} a_k w_k = O(a_n^{\delta})$. Then, from (5.1.91) we conclude that x_k converges to a vector denoted by x, and

$$
x - x_n = O(\sum_{k=n}^{\infty} a_k^{1+\delta}) + \sum_{k=n}^{\infty} a_k w_k. \tag{5.1.94}
$$

Thus, by (5.1.92) and (5.1.93) we derive $\|x_k - x\| = O(a_k^{\delta})$. □

Theorem 5.1.3 *Assume A5.1.1–A5.1.3 hold. Except the case* $\{\lambda^{(1)} = \cdots = \lambda^{(n)}\}$, *there are* $u^{(i)} \in S^i,\ i = 1, \cdots, n$ *such that*

$$
\lim_{k\to\infty} \|u_k^{(i)} - u^{(i)}\| = O(a_k^{\delta}) \ \textit{with some}\ \ \delta \in (0, 1/2)\ \ \forall i = 1, \cdots, n.
$$

Proof. Since the case $\{\lambda^{(1)} = \cdots = \lambda^{(n)}\}$ is excluded from consideration, $b(1)$ is well defined.

Multiplying both sides of (5.1.9) from the left by $\phi_{b(1)}^T$ and noticing $A\phi_{b(1)} = \lambda^{b(1)}\phi_{b(1)}$ we derive

$$
\begin{aligned}
\phi_{b(1)}^T u_{k+1}^{(1)} =& \phi_{b(1)}^T [u_k^{(1)} + a_k(Au_k^{(1)} - (u_k^{(1)T} A u_k^{(1)})u_k^{(1)}) + a_k \varepsilon_{k+1}^{(1)} + O(a_k^2)] \\
=& \phi_{b(1)}^T u_k^{(1)} + a_k[\lambda^{b(1)} - (u_k^{(1)T} A u_k^{(1)})](\phi_{b(1)}^T u_k^{(1)}) + a_k \phi_{b(1)}^T \varepsilon_{k+1}^{(1)} + O(a_k^2) \\
=& \phi_{b(1)}^T u_k^{(1)} + a_k[(\lambda^{b(1)} - \lambda^{(1)})(\phi_{b(1)}^T u_k^{(1)}) + (\lambda^{(1)} - u_k^{(1)T} A u_k^{(1)}) \\
& \cdot (\phi_{b(1)}^T u_k^{(1)})] + a_k \phi_{b(1)}^T \varepsilon_{k+1}^{(1)} + O(a_k^2),
\end{aligned} \tag{5.1.95}
$$

where $\lambda^{b(1)} - \lambda^{(1)} < 0$ and $\lambda^{(1)} - u_k^{(1)T} A u_k^{(1)} \xrightarrow[k\to\infty]{} 0$.

Define $\varepsilon_{k+1}^{'} \triangleq \phi_{b(1)}^T \varepsilon_{k+1}^{(1)},\ \varepsilon_{k+1}^{''} = O(a_k)$.

By A5.1.2, $\varepsilon_{k+1}^{'}$ is an mds measurable with respect to $\mathscr{F}_{k+1}$.

Since $\sum_k a_k^{2(1-\delta)} < \infty$ for any $\delta \in (0, \frac{1}{2})$, $\sup_k E\{\|N_{k+1}\|^2|\mathscr{F}_k\} < \infty$, and $\|u_k^{(1)}\| = 1$, by Theorem 1.2.8 we have $\sum_k a_k^{1-\delta}\varepsilon'_{k+1} < \infty$.

Further, $\varepsilon''_{k+1} = O(a_k) = o(a_k^\delta)$, and by Theorem 2.6.1 we find $\phi_{b(1)}^T u_k^{(1)} = o(a_{k-1}^\delta)$.

Similarly, we obtain $\phi_i^T u_k^{(1)} = O(a_k^\delta)$, $i = b(1)+1, \cdots, n$.

Consequently, we derive

$$d(u_k^{(1)}, S^1) = \sqrt{(\phi_{b(1)}^T u_k^{(1)})^2 + \cdots + (\phi_n^T u_k^{(1)})^2} = O(a_k^\delta).$$

Then, there is a subsequence $u^{(1)}(k) \in S^1$ such that $\|u_k^{(1)} - u^{(1)}(k)\| = O(a_k^\delta)$ and $\|Au_k^{(1)} - (u_k^{(1)T} Au_k^{(1)})u_k^{(1)}\| = O(\|u_k^{(1)} - u^{(1)}(k)\|) = O(a_k^\delta)$.

From here and (5.1.8) it follows that

$$u_{k+1}^{(1)} = u_k^{(1)} + O(a_k^{1+\delta}) + a_k\delta_{k+1}(1).$$

Since $\sum_{k=1}^\infty a_k^{1-\delta}\delta_{k+1}(1) < \infty$, by Lemma 5.1.2 the theorem is true for $u_k^{(1)}$.

Inductively assume that

$$\|u_k^{(i)} - u^{(i)}\| = O(a_k^\delta) \quad \forall i = 1, \cdots, j$$

for some $\delta > 0$. We now show that it also holds for $i = j+1$.

By (5.1.38) it is seen that

$$\begin{aligned} u^{(i)T} u_{k+1}^{(j+1)} &= (u^{(i)} - u_k^{(i)} + u_k^{(i)})^T u_{k+1}^{(j+1)} \\ &= O\|u_k^{(i)} - u^{(i)}\| = O(a_k^\delta) \quad \forall i = 1, \cdots, j. \end{aligned} \tag{5.1.96}$$

(1) If $b(j+1)$ is not defined, then $a(j+1) \le j$ must be well defined. In this case

$$d^2(u_k^{(j+1)}, S^{j+1}) = O\Big(\sum_{i=1}^{a(j+1)} (u^{(i)T} u_{k+1}^{(j+1)})^2\Big) = O(a_k^{2\delta}), \quad \delta > 0, \tag{5.1.97}$$

where (5.1.96) and the inductive assumption are used.

Similar to $u_k^{(1)}$, there is a sequence $u^{(j+1)}(k) \in S^{j+1}$ such that $\|u_k^{(j+1)} - u^{(j+1)}(k)\| = O(a_k^\delta)$, and hence $\|Au_k^{(j+1)} - (u_k^{(j+1)T} Au_k^{(j+1)})u_k^{(j+1)}\| = O(\|u_k^{(j+1)} - u^{(j+1)}(k)\|) = O(a_k^\delta)$, and by (5.1.8)

$$u_{k+1}^{(j+1)} = u_k^{(j+1)} + O(a_k^{1+\delta}) + a_k\delta_{k+1}(j+1).$$

Then, again by Lemma 5.1.2 the conclusion of the theorem follows for $j+1$. The induction is completed for this case.

(2) Consider the case where $b(j+1)$ is well defined.

Let ϕ be any unit vector from $S^{b(j+1)+i}$ $\forall i = 0, 1, \cdots, n-b(j+1)$. By the inductive assumption from (5.1.8) it is seen that

$$\begin{aligned} u_{k+1}^{(j+1)} =& u_k^{(j+1)} + a_k(Au_k^{(j+1)} - (u_k^{(j+1)T} Au_k^{(j+1)})u_k^{(j+1)}) \\ &+ O(a_k^{1+\delta}) + a_k \delta_{k+1}(j+1). \end{aligned} \tag{5.1.98}$$

Proceeding as in (5.1.95) we derive

$$\begin{aligned} &\phi_{b(j+1)+i}^T u_{k+1}^{(j+1)} \\ &= \phi_{b(j+1)+i}^T u_k^{(j+1)} + a_k[\lambda^{b(j+1)+i} - (u_k^{(j+1)T} Au_k^{(j+1)})](\phi_{b(j+1)+i}^T u_k^{(j+1)}) \\ &+ a_k \phi_{b(j+1)+i}^T \delta_{k+1}(j+1) + O(a_k^{1+\delta}) \\ &= \phi_{b(j+1)+i}^T u_k^{(j+1)} + a_k[(\lambda^{b(j+1)+i} - \lambda^{(j+1)})(\phi_{b(j+1)+i}^T u_k^{(j+1)}) \\ &+ (\lambda^{(j+1)} - u_k^{(j+1)T} Au_k^{(j+1)}) \\ &\cdot (\phi_{b(j+1)+i}^T u_k^{(j+1)})] + a_k \phi_{b(j+1)+i}^T \delta_{k+1}(j+1) + O(a_k^{1+\delta}), \end{aligned}$$

where $\lambda^{b(j+1)+i} - \lambda^{(j+1)} < 0$ and $\lambda^{(j+1)} - u_k^{(j+1)T} Au_k^{(j+1)} \underset{k\to\infty}{\longrightarrow} 0$.

Again by the convergence rate Theorem 5.1.2 we have

$$|\phi_{b(j+1)+i}^T u_k^{(j+1)}| = O(a_k^{\delta}). \tag{5.1.99}$$

Finally, we conclude that

$$\begin{aligned} d^2(u_k^{(j+1)}, S^{j+1}) &= O\Big(\sum_{i=1}^{a(j+1)} (u_k^{(j+1)T} u^{(i)})^2\Big) + O\Big(\sum_{i=0}^{n-b(j+1)} ((\phi_{b(j+1)+i}^T u_k^{(j+1)}))^2\Big) \\ &= O(a_k^{2\delta}), \quad \delta > 0 \end{aligned}$$

by (5.1.96), (5.1.99), and the inductive assumption.

This is the same estimate as (5.1.97), and from here we find that there is a $u^{(j+1)} \in S^{j+1}$ such that $\|u_k^{(j+1)} - u^{(j+1)}\| = O(a_k^{\delta})$. □

5.2 Consensus of Networked Agents

Multi-agent systems can be regarded as a starting point for studying complex systems arising from diverse fields including physics, biology, economy, society, and others. The network communication in practice is often disturbed by various uncertain factors, such as measurement or channel noise, quantization errors, etc. So, the consensus control of dynamic multi-agent networks under a stochastic environment has naturally attracted attention from researchers.

Networked Agents

Consider a network with N agents, which is often modeled by a *directed graph*

$G=(\mathcal{N},\mathcal{E}_t)$ consisting of the set $\mathcal{N}=\{1,2,\cdots,N\}$ of nodes (agents) and the set $\mathcal{E}_t\subset\mathcal{N}\times\mathcal{N}$ of edges at time t. An edge in G is denoted as an ordered pair (i,j), $i\neq j$. A path from i_1 to i_l in G consists of a sequence of nodes $i_1,i_2,\cdots,i_l, l\geq 2$ such that $(i_j,i_{j+1})\in\mathcal{E}_t$, $1\leq j\leq l-1$. A graph is called *connected* if for any distinct agents i and j there exist a path from i to j and also a path from j to i. A connected graph is called *undirected* if ordered pairs (i,j) and (j,i) are edges in G simultaneously. We say that an agent j is a *neighbor* of i if $(j,i)\in\mathcal{E}_t$. The neighbor set of i including i itself at time t is denoted by $\mathcal{N}_i(t)$.

Let $P=[p_{ij}]$ be an $N\times N$-matrix with nonnegative elements $p_{ij}\geq 0$. The matrix P is called *row-wise stochastic* if $P\mathbf{1}=\mathbf{1}$, and is called *column-wise stochastic* if $P^T\mathbf{1}=\mathbf{1}$, where $\mathbf{1}=[1,\cdots,1]^T$ is N-dimensional. A nonnegative matrix P is called *doubly stochastic* if it is both row-wise stochastic and column-wise stochastic.

By the definition of irreducibility of nonnegative matrices (see Definition B.1 in Appendix B), the nonnegative matrix P is irreducible if for any distinct i and j: $1\leq i\leq N,\ 1\leq j\leq N$ there are $1\leq k_s\leq N,\ s=1,2,\cdots,l,\ k_1=i$, and $k_l=j$ such that $p_{k_sk_{s+1}}>0,\ s=1,\cdots,l-1$.

It is clear that a Markov chain with states in $\mathcal{N}$ can always be associated with a stochastic matrix, and irreducibility of the matrix means that any state j can be reached from any other state i with positive probability.

Denote by $x_i(t)\in\mathbb{R}^n$ and $u_i(t)\in\mathbb{R}^n$ the multidimensional state and consensus control at time t of the ith agent, respectively. Consider the time-varying topology, i.e., $\mathcal{N}_i(t)$ may be time-dependent. Assume the states of agents are governed by the following dynamic equation

$$x_i(t+1)=x_i(t)+u_i(t),\ t=0,1,\cdots,\ i=1,2,\cdots,N. \tag{5.2.1}$$

The noisy observation $y_{ij}(t)$ of the agent j, being a neighbor of the agent i, is received by the agent i at time t:

$$y_{ij}(t)=x_j(t)+\omega_{ij}(t),\ j\in\mathcal{N}_i(t), \tag{5.2.2}$$

where $\omega_{ij}(t)$ is the observation noise.

The problem is to define the control $u_i(t)$ as a function of $\{y_{ij}(s),\ x_i(s),\ s\leq t,\ j\in\mathcal{N}_i(t)\}$ leading the network to consensus in a certain sense.

Let us define the control as follows

$$u_i(t)=a_t\Big(\sum_{j\in\mathcal{N}_i(t)}p_{ij}(t)(y_{ij}(t)-x_i(t))\Big), \tag{5.2.3}$$

where

$$\begin{aligned}&p_{ij}(t)>0\ \forall j\in\mathcal{N}_i(t),\ p_{ij}(t)=0\ \forall j\in\mathcal{N}_i^c,\\ &\sum_{j\in\mathcal{N}_i(t)}p_{ij}(t)=1\ \ \forall i=1,2,\cdots,N \text{ and } \forall t>0.\end{aligned} \tag{5.2.4}$$

If (5.2.4) is satisfied, then $P(t)=[p_{ij}(t)]$ is an $N\times N$ stochastic matrix. It is important to note that the connectivity of the graph $\mathcal{N}_i(t)$ implies irreducibility of $P(t)$ by its definition and *vice versa*. It is also important to note that (see Appendix B) for an irreducible stochastic matrix P there is a unique vector $\alpha = [\alpha_1, \cdots, \alpha_N]^T$ which we call the invariant probability vector, satisfying $\alpha_i > 0\ \forall i = 1, \cdots, N$ such that $\alpha^T \mathbf{1} = 1$ and $\alpha^T P = \alpha^T$.

Let us form the $N \times N^2$ matrix $A^{P(t)}$ from $P(t)$: The ith row of $A^{P(t)}$ consists of N sub-rows of equal length, and all sub-rows are set to be zero except the ith sub-row, which is set to equal $p_i(t) \triangleq [p_{i1}(t), \cdots, p_{iN}(t)]$, $i = 1, \cdots, N$.

Let $x(t) \triangleq (x_1^T(t), \cdots, x_N^T(t))^T$ and

$$\omega(t) \triangleq [\omega_{11}^T(t), \cdots, \omega_{1N}^T(t), \cdots, \omega_{N1}^T(t), \cdots, \omega_{NN}^T(t)]^T. \tag{5.2.5}$$

Then the closed-loop network becomes

$$x(t+1) = x(t) + a_t[(P(t) - I_N) \otimes I_n x(t) + (A^{P(t)} \otimes I_n)\omega(t)], \tag{5.2.6}$$

where "$\otimes$" is the Kronecker product, and where and hereafter I_s always denotes the identity matrix of dimension $s \times s$.

For consensus of the network we need the following conditions:

A5.2.1 $a_t > 0$, $a_t \underset{t\to\infty}{\longrightarrow} 0$, $\sum_{t=0}^{\infty} a_t = \infty$;

A5.2.2 *$\omega_{ij}(t)$ can be expressed as $\omega_{ij}(t) = \mu_{ij}(t) + \nu_{ij}(t)$ with $\sum_{t=1}^{\infty} a_t \mu_{ij}(t) < \infty$ and $\nu_{ij}(t) \underset{t\to\infty}{\longrightarrow} 0, \forall i, j \in \mathcal{N}$;*

A5.2.3 *The graph G is connected.*

Remark 5.2.1 *It is observed that the states of agents in the closed-loop network (5.2.6) are updated by an SA algorithm. Thus, the consensus of the network is equivalent to convergence of the SA algorithm. From Theorem 2.2.1, it is known that condition A5.2.2 is necessary for the pathwise convergence of SA. So, the noise condition required here is the weakest possible. It is clear that the condition $\sum_{t=1}^{\infty} a_t \mu_{ij}(t) < \infty$ is satisfied if $\{\mu_{ij}(t)\}\ t \geq 0$ is an mds with bounded second conditional moments and $a_t = \frac{1}{t}$, $t > 0$. Besides, the graph is not required to be undirected. Notice that a connected graph may be directed or undirected, and the undirected connection is more restrictive in comparison with the directed one.*

Network with Fixed Topology

We first consider the consensus problem for networks with fixed topologies: $\mathcal{N}_i(t) \equiv \mathcal{N}$, $P(t) \equiv P$ for all t. In this case the states of agents in the closed-loop network (5.2.6) are defined by the following SA algorithm:

$$x(t+1) = x(t) + a_t[(P - I_N) \otimes I_n x(t) + (A^P \otimes I_n)\omega(t)]. \tag{5.2.7}$$

As claimed above, for the strong consensus we need only to show that $x(t)$ given by (5.2.7) converges with probability one. This is given by the following theorem.

Theorem 5.2.1 *Assume conditions A5.2.1–A5.2.3 hold. Then the algorithm given by (5.2.7) with an arbitrary P defined as* $p_{ij} > 0$ *for* $j \in \mathcal{N}_i$, $p_{ij} = 0$ *for* $j \in \mathcal{N}_i^c$, *and*

$$\sum_{j \in \mathcal{N}_i} p_{ij} = 1 \ \ \forall i \in \mathcal{N}$$

leads to the weighted average consensus, i.e.,

$$x_i(t) - \sum_{j=1}^{N} \alpha_j x_j(t) \xrightarrow[t\to\infty]{} 0 \ \ \forall i \in \mathcal{N} \ \text{ a.s.}$$

Further, if in addition, $v_{ij}(t) \equiv 0$, *then the algorithm leads to the strong consensus:*

$$x_i(t) \xrightarrow[t\to\infty]{} \gamma \ \ \forall i \in \mathcal{N} \ \text{ a.s.}, \tag{5.2.8}$$

where $\gamma = (\alpha^T \otimes I_n)x_0 + \sum_{t=0}^{\infty} a_t (\alpha^T A^P \otimes I_n)\omega(t)$.

Proof. Let λ_i, $i = 1, \cdots, N$ be the eigenvalues of P with $|\lambda_i| \geq |\lambda_{i+1}|$, $i = 1, \cdots, N-1$. Notice that P is irreducible by A5.2.3 and Remark 5.2.1. Then by Theorem B.2 we have $\lambda_1 = 1$ and $\mathrm{Re}\{\lambda_i\} < 1$, $i = 2, \cdots, N$.

Let T be a nonsingular matrix leading P to the Jordan form $TPT^{-1} = \begin{bmatrix} \Lambda & 0 \\ 0 & 1 \end{bmatrix}$, where Λ is an $(n-1)\times(n-1)$-matrix with all eigenvalues inside the closed unit disk.

Let us denote the first $n-1$ rows of T by T_1. As pointed out in Remark 5.2.1 the invariant probability α for P is unique, so the last row in T must be $c\alpha^T$ with c being a constant. Without loss of generality we may assume $c = 1$. Then,

$$T = \begin{bmatrix} T_1 \\ \alpha^T \end{bmatrix}, \ \ T(P - I_N)T^{-1} = \begin{bmatrix} \Lambda - I_{N-1} & 0 \\ 0 & 0 \end{bmatrix}.$$

In what follows the following matrix identity will repeatedly be used:

$$(A \otimes B)(C \otimes D) = AC \otimes BD. \tag{5.2.9}$$

Set $\Gamma \triangleq I_N - \mathbf{1}\alpha^T$, and notice

$$\Gamma(P - I_N) = P - I_N - \mathbf{1}\alpha^T P + \mathbf{1}\alpha^T = P - I_N = (P - I_N)\Gamma.$$

Multiplying (5.2.7) from the left by $T\Gamma \otimes I_n$, by (5.2.9) we derive that

$$\begin{aligned}
&((T\Gamma) \otimes I_n)x(t+1) \\
&= ((T\Gamma) \otimes I_n)x(t) + a_t[((T(P - I_N)\Gamma) \otimes I_n)x(t) + ((T\Gamma A^P) \otimes I_n)\omega(t)] \\
&= ((T\Gamma) \otimes I_n)x(t) + a_t\{((\begin{bmatrix} \Lambda - I_{N-1} & 0 \\ 0 & 0 \end{bmatrix} T\Gamma) \otimes I_n)x(t) \\
&\quad + ((T\Gamma A^P) \otimes I_n)\omega(t)\}.
\end{aligned} \tag{5.2.10}$$

Noticing $T\Gamma = \begin{bmatrix} T_1 - T_1\mathbf{1}\alpha^T \\ 0 \end{bmatrix}$ and setting $\xi(t) \triangleq ((T_1 - T_1\mathbf{1}\alpha^T) \otimes I_n)x(t)$, we have

$$((T\Gamma \otimes I_n)x(t)) = [\xi(t)^T, 0]^T, \tag{5.2.11}$$

and

$$\xi(t+1) = \xi(t) + a_t[((\Lambda - I_{N-1}) \otimes I_n)\xi(t) + (((T_1 - T_1\mathbf{1}\alpha^T)A^P) \otimes I_n)\omega(t)]. \tag{5.2.12}$$

The recursion (5.2.12) is an SA algorithm for which the observation of the linear function $((\Lambda - I_{N-1}) \otimes I_n)x$ at $x = \xi(t)$ is $((\Lambda - I_{N-1}) \otimes I_n)\xi(t) + (((T_1 - T_1\mathbf{1}\alpha^T)A^P) \otimes I_n)\omega(t)$ with observation noise $(((T_1 - T_1\mathbf{1}\alpha^T)A^P) \otimes I_n)\omega(t)$.

Notice that $(\Lambda - I) \otimes I_n$ is stable. Applying Theorem 2.5.1 to (5.2.12) we find that $\xi(t)$ converges to 0 a.s.

From (5.2.11) it follows that $((T\Gamma) \otimes I_n)x(t)$ converges to 0 a.s. Noticing that by (5.2.9) $(T\Gamma) \otimes I_n = (T \otimes I_n)(\Gamma \otimes I_n)$ and that $T \otimes I_n$ is nonsingular since T is nonsingular, we find that $(\Gamma \otimes I_n)x(t)$ also converges to 0 a.s. as $t \to \infty$.

On the other hand, noticing $\Gamma = I_N - \mathbf{1}\alpha^T$ and using (5.2.9) we have

$$\begin{aligned}(\Gamma \otimes I_n)x(t) &= x(t) - (\mathbf{1}\alpha^T \otimes I_n)x(t) \\ &= x(t) - (\mathbf{1} \otimes I_n)(\alpha^T \otimes I_n)x(t) \\ &= x(t) - (\mathbf{1} \otimes I_n)(\sum_{j=1}^{N} \alpha_j x_j(t)).\end{aligned}$$

Convergence to zero of the left-hand side of this chain of equalities implies that

$$x_i(t) - \sum_{j=1}^{N} \alpha_j x_j(t) \xrightarrow[t\to\infty]{} 0 \quad \forall i \in \mathcal{N} \quad \text{a.s.}, \tag{5.2.13}$$

which is the first assertion of the theorem.

Let us now consider the case $v_{ij} \equiv 0 \ \ \forall i, j \in \mathcal{N}$. In this case γ defined in the theorem is meaningful by A5.2.2. From (5.2.7) again by (5.2.9) it is seen that

$$\begin{aligned}(\alpha^T \otimes I_n)x(t+1) &= (\alpha^T \otimes I_n)x(t) + a_t(\alpha^T \otimes I_n)(A^P \otimes I_n)\omega(t) \\ &= (\alpha^T \otimes I_n)x(t) + a_t(\alpha^T A^P \otimes I_n)\omega(t).\end{aligned} \tag{5.2.14}$$

Summing up both sides of the equality from 0 to ∞ leads to

$$(\alpha^T \otimes I_n)x(t) \xrightarrow[t\to\infty]{} \gamma. \tag{5.2.15}$$

Noticing $(\alpha^T \otimes I_n)x(t) = \sum_{j=1}^{N} \alpha_j x_j(t)$, by (5.2.13) and (5.2.15) we conclude

$$x_i(t) \xrightarrow[t\to\infty]{} \gamma \ \forall i \in \mathcal{N} \ \text{a.s.}$$

□

Network with Time-Varying Topology

In this section, we consider the time-varying networks defined by (5.2.1), (5.2.3), and (5.2.4). We keep A5.2.2 unchanged, but strengthen A5.2.1 to A5.2.1' and change A5.2.3 to A5.2.3' as follows.

A5.2.1' *$a_t > 0$, $a_t \xrightarrow[t\to\infty]{} 0$, $\sum_{t=0}^{\infty} a_t = \infty$ and*

$$\frac{1}{a_t} - \frac{1}{a_{t-1}} \xrightarrow[t\to\infty]{} \alpha \geq 0.$$

A5.2.3' *There are integers k_0 and K such that the graph is jointly connected in the fixed period $[k, k+K]$ $\forall k \geq k_0$, i.e., the graph $(\mathcal{N}, \cup_{j=0}^{K}\mathcal{E}_{k+j})$ $\forall k \geq k_0$ is connected.*

Besides, we need an additional condition on $P(t) = [p_{ij}(t)]$.

A5.2.4 *For all nonzero $p_{ij}(t) \neq 0$ there is a uniform lower bound $p_{ij}(t) > \tau > 0$ $\forall t > 0$, and the matrices $P(t)$ $\forall t > 0$ are doubly stochastic.*

Remark 5.2.2 *The condition of an existing uniform lower bound $\tau > 0$ for $p_{ij}(t)$ is obviously satisfied in the case where the network is with fixed topology or is switched over a finite number of networks with fixed topologies. The requirement for $P(t)$ to be doubly stochastic is equivalent to that the graph is balanced.*

Theorem 5.2.2 *Assume that the algorithm given by (5.2.6) is with an arbitrary $P(t)$ defined as $p_{ij}(t) > 0$ for $j \in \mathcal{N}_i(t)$, $p_{ij}(t) = 0$ for $j \in \mathcal{N}_i^c(t)$, and*

$$\sum_{j\in\mathcal{N}_i(t)} p_{ij}(t) = 1 \quad \forall i \in \mathcal{N}.$$

Then, under the conditions A5.2.1', A5.2.2, A5.2.3', and A5.2.4 the average consensus is achieved:

$$x_i - \frac{1}{N}\sum_{j=1}^{N} x_j(t) \xrightarrow[t\to\infty]{} 0 \quad \forall i \in \mathcal{N} \quad \text{a.s.}$$

Further, if in addition, $v_{ij}(t) \equiv 0$ $\forall i, j \in \mathcal{N}$, then the strong consensus takes place:

$$x_i(t) \xrightarrow[t\to\infty]{} \gamma' \quad \forall i \in \mathcal{N} \quad \text{a.s.},$$

where $\gamma' = \frac{1}{N}\sum_{j=1}^{N} x_j(0) + \frac{1}{N}\sum_{t=0}^{\infty} a_t (e^T A^{P(t)} \otimes I_n)\omega(t)$.

Proof: Since $P(t)$ is doubly stochastic, we have $\mathbf{1}^T P(t) = \mathbf{1}^T$ and $P(t)\mathbf{1} = \mathbf{1}$. Let $T = \begin{bmatrix} T_1 \\ \frac{1}{\sqrt{N}}\mathbf{1}^T \end{bmatrix}$ be an orthogonal matrix: $T_1\mathbf{1} = 0$ and $T_1T_1^T = I_{N-1}$. Then, we have

$$TP(t)T^T = \begin{bmatrix} T_1P(t)T_1^T & 0 \\ 0 & 1 \end{bmatrix},$$

and

$$T\Gamma = \begin{bmatrix} T_1 \\ \frac{1}{\sqrt{N}}\mathbf{1}^T \end{bmatrix}(I_N - \frac{1}{N}\mathbf{1}\mathbf{1}^T) = \begin{bmatrix} T_1 \\ 0 \end{bmatrix},$$

where $\Gamma \triangleq I_N - \frac{1}{N}\mathbf{1}\mathbf{1}^T$.

Multiplying (5.2.6) from the left by $T\Gamma \otimes I_n$, similar to (5.2.10) we have

$$\begin{aligned}
&(\begin{bmatrix} T_1 \\ 0 \end{bmatrix} \otimes I_n)x(t+1) \\
&= (\begin{bmatrix} T_1 \\ 0 \end{bmatrix} \otimes I_n)x(t) + a_t\{(\begin{bmatrix} (T_1P(t)T_1^T - I_{N-1})T_1 \\ 0 \end{bmatrix} \otimes I_n)x(t) \\
&+ (\begin{bmatrix} T_1A^{P(t)} \\ 0 \end{bmatrix} \otimes I_n)\omega(t)\}. \qquad (5.2.16)
\end{aligned}$$

Setting $\xi(t) \triangleq (T_1 \otimes I_n)x(t)$, we have

$$\xi(t+1) = \xi(t) + a_t\{((T_1P(t)T_1^T - I_{N-1}))\xi(t) + (T_1A^P \otimes I_n)\omega(t)\}. \qquad (5.2.17)$$

Since $P(t)$ is a doubly stochastic matrix, $(P(t)+P^T(t))/2$ is a symmetric stochastic matrix. Again, by Theorem B.2 its maximal eigenvalue equals one, and hence for any nonzero $z \in R^N$, $z^TP(t)z = z^T(P(t)+P^T(t))/2z \le z^Tz$.

For any nonzero $u \in \mathbb{R}^{N-1}$,

$$u^T(T_1P(t)T_1^T - I_{N-1})u = (u^TT_1)P(t)(T_1^Tu) - (u^TT_1)(T_1^Tu) \le 0,$$

and similarly

$$u^T(T_1P^T(t)T_1^T - I_{N-1})u = (u^TT_1)P^T(t)(T_1^Tu) - (u^TT_1)(T_1^Tu) \le 0.$$

Set $H_t \triangleq T_1P(t)T_1^T - I_{N-1}$. We have that

$$H_t + H_t^T \le 0. \qquad (5.2.18)$$

Similar to the equivalence between irreducibility of the constructed P and connectivity of the graph associated with the network with fixed topology as pointed out in Remark 5.2.1, by construction of $P(t)$ the joint connectivity of the network as defined in A5.2.3' implies that both $\frac{1}{K+1}\sum_{s=t}^{t+K}P(s)$ and $\frac{1}{K+1}\sum_{s=t}^{t+K}P^T(s)$ are irreducible stochastic matrices. By Theorem B.2 the maximal eigenvalue of $\frac{1}{2(K+1)}(\sum_{s=t}^{t+K}(P(s)+P^T(s)))$ is 1 with single multiplicity and the corresponding eigenvector is $\mathbf{1}$. Therefore, for any $z \in \mathbb{R}^N$ we have

$$\frac{1}{2(K+1)}z^T(\sum_{s=t}^{t+K}(P(s)+P^T(s)))z \le z^Tz, \qquad (5.2.19)$$

where the equality takes place if and only if $z = c\mathbf{1}$.

Since $u^TT_1\mathbf{1} = 0$, T_1^Tu can never be expressed as $c\mathbf{1}$ for any c. Consequently, in (5.2.19) the strict inequality must take place for $z = T_1^Tu$ with any nonzero $u \in \mathbb{R}^{N-1}$:

$$(u^TT_1)\Big(\frac{1}{2(K+1)}\sum_{s=t}^{t+K}(P(s)+P^T(s))\Big)(T_1^Tu) < (u^TT_1)(T_1^Tu).$$

Noticing that $T_1 T_1^T = I_{N-1}$, this implies that for any nonzero $u \in \mathbb{R}^{N-1}$

$$\frac{1}{2(K+1)} \sum_{s=t}^{t+K} u^T \Big(T_1 (P(s) + P^T(s)) T_1^T - 2I_{N-1} \Big) u < 0.$$

This means that the sum of matrices is negative definite:

$$\sum_{s=t}^{t+K} (H_s + H_s^T) < 0.$$

As a matter of fact, with condition $p_{ij}(t) > \tau > 0$ $\forall t > 0$ required in A5.2.4 we have a stronger result: There exists a $\beta > 0$ such that

$$\sum_{s=t}^{t+K} (H_s + H_s^T) \le -\beta I.$$

Applying Theorem 2.5.2 to (5.2.17) we find that $\xi(t)$ converges to 0 a.s.

By the argument completely similar to that used in the proof for Theorem 5.2.1, we find that $(\Gamma \otimes I_n)x(t)$ also converges to 0 as $t \to \infty$. Notice that $\Gamma = I_N - \frac{1}{N}\mathbf{1}\mathbf{1}^T$ in the present case in contrast to $\Gamma = I_N - \mathbf{1}\alpha^T$ in Theorem 5.2.1. By using (5.2.9) we have

$$\begin{aligned}
(\Gamma \otimes I_n)x(t) &= x(t) - (\frac{1}{N}\mathbf{1}\mathbf{1}^T \otimes I_n)x(t) \\
&= x(t) - (\mathbf{1} \otimes I_n)(\frac{1}{N}e^T \otimes I_n)x(t) \\
&= x(t) - (\mathbf{1} \otimes I_n)(\sum_{j=1}^{N} \frac{1}{N} x_j(t)).
\end{aligned} \tag{5.2.20}$$

Since the left-hand side of (5.2.20) tends to zero, it follows that

$$x_i(t) - \frac{1}{N}\sum_{j=1}^{N} x_j(t) \xrightarrow[t\to\infty]{} 0 \quad \forall i \in \mathcal{N} \quad \text{a.s.} \tag{5.2.21}$$

We now consider the case $\nu_{ij}(t) \equiv 0$. It is clear that in the present case the unique invariant probability is $\alpha = \frac{1}{N}\mathbf{1}$. Putting this into (5.2.14) leads to

$$\frac{1}{N}\sum_{j=1}^{N} x_j(t+1) = \frac{1}{N}\sum_{j=1}^{N} x_j(t) + a_t \frac{1}{N}(\mathbf{1}^T A^{P(t)} \otimes I_n)\omega(t).$$

Summing up t in both sides from 0 to ∞ and noticing that $\sum_{t=1}^{\infty} a_t \omega_{ij}(t) < \infty$, we find that

$$\frac{1}{N}\sum_{j=1}^{N} x_j(t) \xrightarrow[t\to\infty]{} \frac{1}{N}\sum_{j=1}^{N} x_j(0) + \frac{1}{N}\sum_{t=1}^{\infty} a_t (\mathbf{1}^T A^{P(t)} \otimes I_n)\omega(t) \triangleq \gamma,$$

which combining with (5.2.21) leads to the strong consensus $x_i(t) \xrightarrow[t\to\infty]{} \gamma$. □

5.3 Adaptive Regulation for Hammerstein and Wiener Systems

Identification of Hammerstein and Wiener systems has been discussed in Chapter 4. However, identifying a system may not be the final goal in practice. It is often to require the system output to follow a given signal, even a given constant. In the latter case, the output of an unknown Hammerstein or Wiener system is required to follow a given constant, and the problem is called adaptive regulation.

Problem Description

Let us consider the SISO nonlinear systems presented in Figures 5.3.1 and 5.3.2.

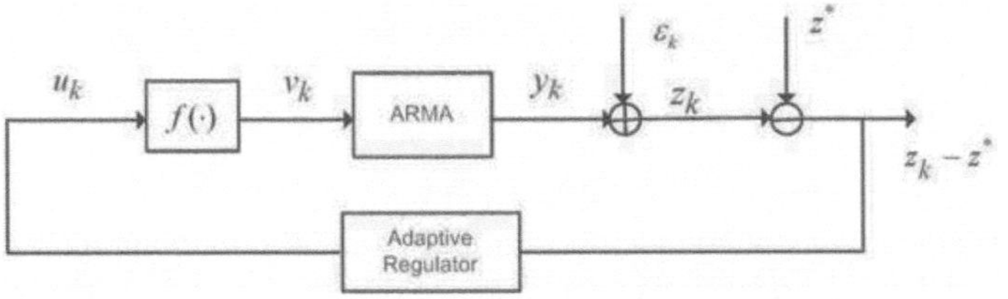

Figure 5.3.1: Adaptive regulation for Hammerstein system

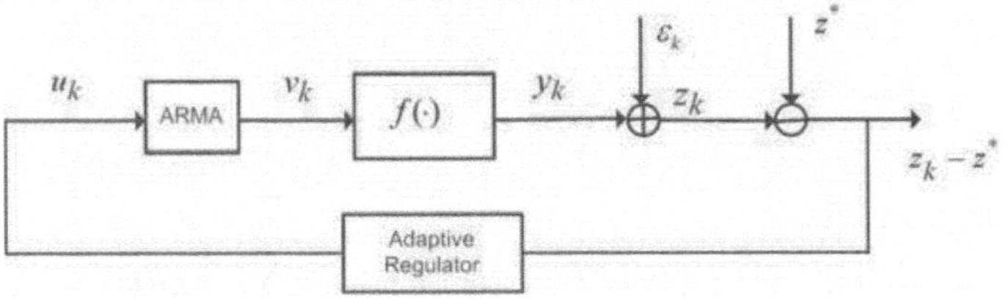

Figure 5.3.2: Adaptive regulation for Wiener system

The linear subsystem is described by

$$C(z)y_k = D(z)v_k \quad \text{and} \quad C(z)v_k = D(z)u_k \tag{5.3.1}$$

for Hammerstein and Wiener system, respectively, where $C(z)$ and $D(z)$ are polynomials in backward-shift operator z:

$$\begin{aligned} C(z) &= 1 + c_1 z + \cdots + c_p z^p, \\ D(z) &= 1 + d_1 z + \cdots + d_q z^q, \quad q < p. \end{aligned} \tag{5.3.2}$$

The polynomials $C(z)$, $D(z)$ and the nonlinear block $f(\cdot)$ are unknown. In Figs 5.3.1 and 5.3.2, $\{\varepsilon_k\}$ is the observation noise, and z_k is the noisy observation of the system output y_k.

Let z^* be a given constant. The problem of adaptive regulation consists in designing the feedback control u_k depending on $(z_0,\dots,z_{k-1})$ in order to minimize

$$\limsup_{n\longrightarrow\infty}\frac{1}{n}\sum_{k=1}^{n}|z_k-z^*|^2 \quad \text{a.s.} \tag{5.3.3}$$

Optimal Regulation

Let us first derive the optimal regulation control when the system is known. Let us denote

$$c\triangleq\sum_{i=0}^{p}c_i,\quad c_0\triangleq 1,\quad d\triangleq\sum_{i=0}^{q}d_i,\ \text{and}\ d_0\triangleq 1. \tag{5.3.4}$$

The following conditions are to be used.

A5.3.1 *All roots of $C(z)$ are outside the closed unit disk, and $c\neq 0$.*

A5.3.2 *$f(\cdot)$ is continuous, and there is a known constant $\mu>0$ such that*

$$|f(u)|=O(|u|^{\mu})\quad \text{as } |u|\longrightarrow\infty. \tag{5.3.5}$$

A5.3.3 *$\{\varepsilon_k\}$ is a sequence of mutually independent random variables with $E\varepsilon_k=0$, $\sup_k E\varepsilon_k^2<\infty$, and*

$$\lim_{n\longrightarrow\infty}\frac{1}{n}\sum_{i=1}^{n}\varepsilon_i^2=R<\infty \ \text{a.s.} \tag{5.3.6}$$

To express the solution of the ARMA system given by (5.3.1) we introduce the following matrix and vector:

$$C\triangleq\begin{bmatrix}-c_1 & 1 & 0 & \cdots & 0\\ \vdots & & \ddots & \ddots & \vdots\\ \vdots & & & \ddots & 0\\ \vdots & & & & 1\\ -c_p & 0 & \cdots & \cdots & 0\end{bmatrix},$$

$$D\triangleq\underbrace{[1\ d_1\ \cdots\ d_q\ 0\ \cdots\ 0]}_{p}{}^T. \tag{5.3.7}$$

Then for the Hammerstein system we have

$$Y_{k+1}=CY_k+Df(u_{k+1})\quad y_k=HY_k,\quad H=\underbrace{[1,0,\dots,0]}_{p}, \tag{5.3.8}$$

and for the Wiener system

$$V_{k+1}=CV_k+Du_{k+1},\quad v_k=HV_k,\quad y_k=f(v_k). \tag{5.3.9}$$

It is clear that

$$Y_{n+j+1} = C^{j+1}Y_n + \sum_{i=n}^{n+j} C^{n+j-i} Df(u_{i+1}), \tag{5.3.10}$$

$$V_{n+j+1} = C^{j+1}V_n + \sum_{i=n}^{n+j} C^{n+j-i} Du_{i+1} \quad \forall n \geq 0 \quad \forall j \geq 0. \tag{5.3.11}$$

By A5.3.1, C is a stable matrix, and hence there are $r > 0$ and $\delta > 0$ such that

$$\|C^k\| \leq re^{-\delta k} \quad \forall k \geq 0. \tag{5.3.12}$$

Lemma 5.3.1 *Assume A5.3.1 and A5.3.3 hold. If there exists a unique solution u^0 to the algebraic equation*

$$\frac{d}{c} f(u^0) = z^*, \tag{5.3.13}$$

then any feedback control sequence $\{u_k\}$ converging to u^0: $u_k \xrightarrow[k\to\infty]{} u^0$ a.s. is the optimal regulation control for the Hammerstein system presented by Fig. 5.3.1. The similar conclusion is also true for the Wiener system presented by Fig. 5.3.2 with the algebraic equation (5.3.13) replaced by the following one:

$$f(\frac{d}{c} u^0) = z^*. \tag{5.3.14}$$

Proof. We now prove the lemma for Hammerstein systems.

Let us denote y_k, z_k, and Y_k by y_k^0, z_k^0, and Y_k^0, respectively, if $u_k \equiv u^0$. In this case, by stability of $C(z)$

$$y_k^0 \xrightarrow[k\to\infty]{} \frac{d}{c} f(u^0) \quad \text{a.s.} \tag{5.3.15}$$

By Theorem 1.2.14 we have

$$\sum_{k=1}^{n} (y_k - z^*)\varepsilon_k = O\Big((\sum_{k=1}^{n} (y_k - z^*)^2)^{\frac{1}{2}+\eta} \Big) \quad \text{a.s.} \quad \forall \eta > 0,$$

and hence

$$\begin{aligned} &\limsup_{n\to\infty} \frac{1}{n} \sum_{k=1}^{n} |z_k - z^*|^2 \\ &= \limsup_{n\to\infty} \frac{1}{n} [\sum_{k=1}^{n} (y_k - z^*)^2 (1 + o(1))] + R \geq R \text{ a.s.} \end{aligned} \tag{5.3.16}$$

On the other hand, by (5.3.13), (5.3.15), and (5.3.6) we have

$$\limsup_{n\to\infty} \frac{1}{n} \sum_{k=1}^{n} |z_k^0 - z^*|^2 = \lim_{n\to\infty} \frac{1}{n} \sum_{k=1}^{n} |o(1) + \varepsilon_k|^2 = R \text{ a.s.,} \tag{5.3.17}$$

and hence u^0 is the optimal control.

By (5.3.12) it follows that

$$\|Y^0_{k+1} - Y_{k+1}\| \xrightarrow[k\to\infty]{} 0 \text{ a.s.} \tag{5.3.18}$$

Hence, $|y_k - y^0_k| \xrightarrow[k\to\infty]{} 0$ a.s. if $u_k \xrightarrow[k\to\infty]{} u^0$ a.s. This by (5.3.17) implies

$$\begin{aligned}
&\limsup_{n\to\infty} \frac{1}{n} \sum_{k=0}^{n} |z_k - z^*|^2 \\
&= \limsup_{n\to\infty} \frac{1}{n} \sum_{k=0}^{n} |(y_k - y^0_k) + (y^0_k - z^*) + \varepsilon_k|^2 = R \text{ a.s.},
\end{aligned}$$

which proves the optimality of $\{u_k\}$ for the Hammerstein system. The conclusion for the Wiener system is proved in a similar way. □

Adaptive Regulation Control

Since $C(z)$, $D(z)$, and $f(\cdot)$ are unknown, the optimal controls given in Lemma 5.3.1 cannot be used. We define the adaptive regulation control in a direct way.

Let

$$a_j = \frac{1}{j}, \quad \text{and} \quad M_j = j^{\frac{1}{1+\mu}}, \tag{5.3.19}$$

where μ is figured in A5.3.2.

Define the adaptive control by SAAWET:

$$u_{j+1} = [u_j + a_j(z_j - z^*)] I_{[|u_j + a_j(z_j - z^*)| \le M_{\sigma_j}]}, \tag{5.3.20}$$

$$\sigma_j = \sum_{i=1}^{j-1} I_{[|u_i + a_i(z_i - z^*)| > M_{\sigma_i}]}, \quad \sigma_0 = 0, \tag{5.3.21}$$

where I_A is the indicator of A.

The control $\{u_j\}$ given by (5.3.20), (5.3.21) is the adaptive regulation control to be applied to both Hammerstein and Wiener systems, where

$$z_j = y_j + \varepsilon_j, \quad C(z)y_j = D(z)v_j, \quad v_j = f(u_j) \tag{5.3.22}$$

for the Hammerstein system, and

$$z_j = y_j + \varepsilon_j, \quad y_j = f(v_j), \quad C(z)v_j = D(z)u_j \tag{5.3.23}$$

for the Wiener system.

Optimality of Adaptive Control for Hammerstein Systems

We now prove optimality of the adaptive regulation control defined by (5.3.20) and (5.3.21) is optimal for Hammerstein systems, for which u_k and z_k are related by (5.3.22).

Lemma 5.3.2 *Assume A5.3.1–A5.3.3 hold. Let* $\lambda \in (0,1)$. *Then for any* j: $j = 0,1,2,\ldots,[n^{\frac{\lambda}{1+\mu}}]+p$, $\sum_{i=n}^{n+j} a_i \xrightarrow[n\to\infty]{} 0$, *and*

$$\sum_{i=n}^{n+j} a_i y_{i-s} \xrightarrow[n\to\infty]{} 0, \quad s = 0,1,\ldots,p, \quad \sum_{i=n}^{n+j} a_i \varepsilon_i \xrightarrow[n\to\infty]{} 0, \tag{5.3.24}$$

where $[x]$ *denotes the integer part of the real number* x, *and* μ *is given in (5.3.5).*

Proof. For $\sum_{i=n}^{n+j} a_i \xrightarrow[n\to\infty]{} 0$ it suffices to note that

$$\sum_{i=n}^{n+[n^{\frac{\lambda}{1+\mu}}]+p} \frac{1}{i} \le \sum_{i=n}^{n+[n^{\frac{\lambda}{1+\mu}}]+p} \int_{i-1}^{i} \frac{dx}{x} = \log \frac{n+[n^{\frac{\lambda}{1+\mu}}]+p}{n-1} \xrightarrow[n\to\infty]{} 0.$$

From (5.3.19), (5.3.20), and (5.3.21) it is known that

$$\sigma_j \le j-1, \text{ and } |u_j| \le M_{\sigma_{j-1}} \le M_{j-2} = (j-2)^{\frac{1}{1+\mu}}. \tag{5.3.25}$$

By (5.3.5) and (5.3.25) it follows that there is $\alpha_1 > 0$ such that

$$|f(u_j)| \le \alpha_1 (j-2)^{\frac{\mu}{1+\mu}} \quad \text{for} \quad j > 2. \tag{5.3.26}$$

By stability of $C(z)$ there is $\alpha_2 > 0$ such that

$$\|Y_j\| \le \alpha_2 j^{\frac{\mu}{1+\mu}}, \text{ and, hence, } \quad |y_j| \le \alpha_2 j^{\frac{\mu}{1+\mu}} \quad \forall j > 2. \tag{5.3.27}$$

Therefore, for $s = 0,1,\ldots,p$ and for any $j: \; j = 0,1,2,\cdots,[n^{\frac{\lambda}{1+\mu}}]+p$

$$\begin{aligned} |\sum_{i=n}^{n+j} a_i y_{i-s}| &\le \alpha_2 \sum_{i=n}^{n+[n^{\frac{\lambda}{1+\mu}}]+p} i^{-\frac{1}{1+\mu}} \le \alpha_2 \sum_{i=n}^{n+[n^{\frac{\lambda}{1+\mu}}]+p} \int_{i-1}^{i} \frac{dx}{x^{\frac{1}{1+\mu}}} \\ &= \frac{\alpha_2(\mu+1)}{\mu}(n-1)^{\frac{\mu}{\mu+1}} [(1+\frac{[n^{\frac{\lambda}{1+\mu}}]+p+1}{n-1})^{\frac{\mu}{1+\mu}} - 1] \xrightarrow[n\to\infty]{} 0 \end{aligned}$$

since $\lambda < 1$.

Finally, the second limit in (5.3.24) follows from the fact that $\sum_{i=1}^{\infty} a_i \varepsilon_i < \infty$ a.s.□

Define

$$m(k,T) \triangleq \max\{m : \sum_{i=k}^{m} a_i \le T\}, \quad T > 0. \tag{5.3.28}$$

Corollary 5.3.1 *Since* $\sum_{i=n}^{n+j} a_i \xrightarrow[n\to\infty]{} 0 \quad \forall j: j=0,1,2,\ldots,[n^{\frac{\lambda}{1+\mu}}]+p$, *for any fixed* $T>0$

$$k+[k^{\frac{\lambda}{1+\mu}}]+p<m(k,T) \tag{5.3.29}$$

for all sufficiently large k.

Lemma 5.3.3 *Assume A5.3.1–A5.3.3 hold. For a fixed sample path let* $\{u_{n_k}\}$ *be a convergent subsequence:* $u_{n_k}\to u^*$. *If* $T>0$ *is small enough, then for all sufficiently large k*

$$u_{j+1}=u_j+a_j(z_j-z^*), \quad j=n_k,\ldots,m(n_k,T), \tag{5.3.30}$$

i.e., there is no truncation for (5.3.20), (5.3.21) when j runs from n_k *to* $m(n_k,T)$.

Proof. If $\lim_{j\to\infty}\sigma_j<\infty$, then $\{u_j\}$ is a bounded sequence and (5.3.20) suffers no truncation starting from some j. This implies (5.3.30).

Assume $\sigma_j \xrightarrow[j\to\infty]{} \infty$.

Let $\alpha>0$ be such that $|u_{n_k}|<\alpha$, and let k be large enough so that $M_{\sigma_k}>2\alpha$. By continuity of $f(\cdot)$ there is a $\gamma>0$ such that

$$|f(u)|<\gamma \qquad \forall u:|u|\le 2\alpha. \tag{5.3.31}$$

Starting from n_k, temporarily consider (5.3.30) without any truncations.

For any $j=1,2,\ldots,[n_k^{\frac{\lambda}{1+\mu}}]+p$ by Lemma 5.3.2 we have

$$|u_{n_k+j}|=|u_{n_k}+\sum_{i=n_k}^{n_k+j-1} a_i(y_i-z^*+\varepsilon_i)|\le\frac{5}{4}\alpha \tag{5.3.32}$$

for all large k, and by (5.3.10)

$$\begin{aligned}\left|y_{n_k+[n_k^{\frac{\lambda}{1+\mu}}]+p}\right| &\le \|Y_{n_k+[n_k^{\frac{\lambda}{1+\mu}}]+p}\|=\|C^{[n_k^{\frac{\lambda}{1+\mu}}]}Y_{n_k+p}\\ &+\sum_{s=n_k+p}^{n_k+[n_k^{\frac{\lambda}{1+\mu}}]+p-1} C^{n_k+[n_k^{\frac{\lambda}{1+\mu}}]+p-1-s}Df(u_{s+1})\|.\end{aligned}$$

From this and by (5.3.12), (5.3.27), and (5.3.31) we have

$$\begin{aligned}\left|y_{n_k+[n_k^{\frac{\lambda}{1+\mu}}]+p}\right| &\le re^{-\delta[n_k^{\frac{\lambda}{1+\mu}}]}\alpha_2(n_k+p)^{\frac{\mu}{1+\mu}}\\ &+r\gamma\|D\|\cdot\sum_{s=n_k+p}^{n_k+[n_k^{\frac{\lambda}{1+\mu}}]+p-1} e^{-\delta(n_k+[n_k^{\frac{\lambda}{1+\mu}}]+p-1-s)},\end{aligned} \tag{5.3.33}$$

where at the right-hand side the first term tends to zero as $k \to \infty$, while the second term is bounded by $\frac{r\gamma\|D\|}{(1-e^{-\delta})}$.

Let $\alpha_3 > 0$ be a constant such that

$$r\alpha_2 e^{-\delta[n_k^{\frac{\lambda}{1+\mu}}]}(n_k+p)^{\frac{\mu}{1+\mu}} \le \alpha_3, \quad k=1,2,\ldots, \tag{5.3.34}$$

and denote

$$\alpha_3 + \frac{r\gamma\|D\|}{1-e^{-\delta}} \triangleq \alpha_4. \tag{5.3.35}$$

Then from (5.3.33), (5.3.34), and (5.3.35) it follows that

$$\left| y_{n_k+[n_k^{\frac{\lambda}{1+\mu}}]+p} \right| \le \alpha_4. \tag{5.3.36}$$

Thus, we have proved for $j=0$

$$\left| u_{n_k+[n_k^{\frac{\lambda}{1+\mu}}]+p+j} \right| \le 2\alpha, \qquad \left| y_{n_k+[n_k^{\frac{\lambda}{1+\mu}}]+p+j} \right| \le \alpha_4. \tag{5.3.37}$$

Assume we have shown (5.3.37) for $j: \ j=0,1,2,\cdots,l$. We now prove that they are also true for $j=l+1$ whenever $n_k+[n_k^{\frac{\lambda}{1+\mu}}]+p+l \le m(n_k,T)$.

Let $T>0$ be small enough so that $(\alpha_4+|z^*|)T < \frac{\alpha}{4}$ and let k be large enough so as

$$\left| \sum_{s=n_k+[n_k^{\frac{\lambda}{1+\mu}}]+p}^{n_k+[n_k^{\frac{\lambda}{1+\mu}}]+p+l} a_s\varepsilon_s \right| < \frac{\alpha}{4} \quad \forall\, l=0,1,2,\cdots.$$

Then by (5.3.30), (5.3.32), and the inductive assumption it follows that

$$\begin{aligned} &\left| u_{n_k+[n_k^{\frac{\lambda}{1+\mu}}]+p+l+1} \right| \\ &= \left| u_{n_k+[n_k^{\frac{\lambda}{1+\mu}}]+p} + \sum_{s=n_k+[n_k^{\frac{\lambda}{1+\mu}}]+p}^{n_k+[n_k^{\frac{\lambda}{1+\mu}}]+p+l} a_s(y_s - z^* + \varepsilon_s) \right| \\ &\le \frac{5}{4}\alpha + (\alpha_4+|z^*|)T + \left| \sum_{s=n_k+[n_k^{\frac{\lambda}{1+\mu}}]+p}^{n_k+[n_k^{\frac{\lambda}{1+\mu}}]+p+l} a_s\varepsilon_s \right| < 2\alpha. \end{aligned} \tag{5.3.38}$$

Similar to (5.3.33) by (5.3.34), (5.3.35), and (5.3.38) we have

$$\begin{aligned}\left|y_{n_k+[n_k^{\frac{\lambda}{1+\mu}}]+p+l+1}\right| &\leq \Bigg\|C^{[n_k^{\frac{\lambda}{1+\mu}}]+l+1}Y_{n_k+p} \\ &+\sum_{s=n_k+p}^{n_k+[n_k^{\frac{\lambda}{1+\mu}}]+p+l} C^{n_k+[n_k^{\frac{\lambda}{1+\mu}}]+p+l-s}Df(u_{s+1})\Bigg\| \\ &\leq re^{-\delta([n_k^{\frac{\lambda}{1+\mu}}]+l+1)}\alpha_2(n_k+p)^{\frac{\mu}{1+\mu}}+\frac{r\gamma\|D\|}{1-e^{-\delta}}\leq \alpha_4.\end{aligned}$$

This completes the induction. Consequently, for a fixed small $T>0$ and all large k we have

$$|u_j|\leq 2\alpha \quad \forall j: j=n_k,\ldots,m(n_k,T).$$

In other words, (5.3.20) has no truncation for all j: $j=n_k,\ldots,m(n_k,T)$ if $T>0$ is small and k is large. □

Theorem 5.3.1 *Assume A5.3.1–A5.3.3 hold. Further assume that there is a unique* u^0 *satisfying (5.3.13) and there is a constant* $c_0>0$ *such that* $\frac{1}{c_0}\int_{-c_0}^{0}f(x)dx > f(u^0) > \frac{1}{c_0}\int_0^{c_0}f(x)dx$ *if* $\frac{d}{c}>0$, *and* $\frac{1}{c_0}\int_{-c_0}^{0}f(x)dx < f(u^0) < \frac{1}{c_0}\int_0^{c_0}f(x)dx$ *if* $\frac{d}{c}<0$. *Then* $\{u_k\}$ *defined by (5.3.20), (5.3.21) is the optimal adaptive regulation control for the Hammerstein system given by Fig. 5.3.1.*

Proof. We rewrite (5.3.20), (5.3.21) as

$$\begin{aligned}u_{j+1}=&[u_j+a_j(\frac{d}{c}f(u_j)-z^*+\varepsilon_j+\delta_j)] \\ &\cdot I_{[|u_j+a_j(\frac{d}{c}f(u_j)-z^*+\varepsilon_j+\delta_j|\leq M_{\sigma_j})]},\end{aligned} \tag{5.3.39}$$

$$\sigma_j=\sum_{i=1}^{j-1}I_{[|u_i+a_i(\frac{d}{c}f(u_i)-z^*+\varepsilon_i+\delta_i)|>M_{\sigma_i}]}, \tag{5.3.40}$$

$$\delta_j=\frac{1}{c}\Big(\sum_{\nu=1}^{q}d_\nu(f(u_{j-\nu})-f(u_j))-\sum_{\nu=1}^{p}c_\nu(y_{j-\nu}-y_j)\Big). \tag{5.3.41}$$

Notice that by (5.3.13)

$$\frac{d}{c}f(u_j)-z^*=\frac{d}{c}[f(u_j)-f(u^0)]. \tag{5.3.42}$$

Therefore, SAAWET defined by (5.3.39), (5.3.40) or by (5.3.20), (5.3.21) is with regression function

$$\frac{d}{c}[f(\cdot)-f(u^0)] \tag{5.3.43}$$

and with observation noise $\varepsilon_j + \delta_j$.

To prove that u_k converges to the root u^0 of the regression function given by (5.3.43) we apply Theorem 2.3.1.

For this it suffices to verify:

(i) There is a continuously differentiable function $v(\cdot)$ such that

$$\sup_{|u-u^0|>0} v_u(u)\frac{d}{c}[f(u)-f(u^0)] < 0,$$
$$v(0) < \inf_{|u|=c_0} v(u) \quad \text{for some } c_0 > 0, \tag{5.3.44}$$

where $v_u(u)$ denotes the derivative of $v(\cdot)$ at u.

(ii)

$$\lim_{T\to 0}\limsup_{k\to\infty}\frac{1}{T}\left|\sum_{i=n_k}^{m(n_k,T)} a_i\varepsilon_i\right| = 0,$$
$$\lim_{T\to 0}\limsup_{k\to\infty}\frac{1}{T}\left|\sum_{i=n_k}^{m(n_k,T)} a_i\delta_i\right| = 0 \tag{5.3.45}$$

for any n_k such that u_{n_k} converges.

Other conditions required in Theorem 2.3.1 are automatically satisfied.

As $v(u)$ we may take

$$v(u) = -\frac{d}{c}\left(\int_0^u f(x)dx - f(u^0)u\right).$$

Then the first inequality in (5.3.44) holds by the uniqueness of u^0 for (5.3.13).

Since $v(0) = 0$, the second part of (5.3.44) turns out to be the following inequalities

$$0 < -\frac{d}{c}\left(\int_0^{c_0} f(x)dx - f(u^0)c_0\right),$$
$$0 < -\frac{d}{c}\left(\int_0^{-c_0} f(x)dx + f(u^0)c_0\right)$$

for some $c_0 > 0$, which are guaranteed by the condition of the theorem. Thus, (i) is satisfied.

We proceed to verify (ii). Since $\sum_{i=1}^{\infty} a_i\varepsilon_i < \infty$ a.s., the first part of (5.3.45) holds for almost all sample paths.

For the second part of (5.3.45) we now prove

$$\lim_{T\to 0}\limsup_{k\to\infty}\frac{1}{T}\left|\sum_{i=n_k}^{m(n_k,T)} a_i\sum_{v=1}^{q} d_v(f(u_{i-v})-f(u_i))\right| = 0, \tag{5.3.46}$$

and

$$\lim_{T\to 0}\limsup_{k\to\infty}\frac{1}{T}\left|\sum_{i=n_k}^{m(n_k,T)} a_i \sum_{v=1}^{p} c_v (y_{i-v}-y_i)\right| = 0. \tag{5.3.47}$$

By (5.3.25) for (5.3.46) it suffices to prove

$$\lim_{T\to 0}\limsup_{k\to\infty}\frac{1}{T}\left|\sum_{i=n_k+q}^{m(n_k,T)} a_i \sum_{v=1}^{q} d_v (f(u_{i-v})-f(u_i))\right| = 0. \tag{5.3.48}$$

By Lemmas 5.3.2 and 5.3.3 we have

$$|u_{j+i}-u_j| \xrightarrow[k\to\infty]{} 0 \ \forall\, i=1,\ldots,q \ \text{ and } \forall\, j: j=n_k,\ldots,m(n_k,T), \tag{5.3.49}$$

which incorporating with continuity of $f(\cdot)$ implies (5.3.48) and hence (5.3.46).

Writing the summation at the left-hand side of (5.3.47) as

$$\sum_{i=n_k}^{n_k+[n_k^{\frac{\lambda}{1+\mu}}]+p} a_i \sum_{v=1}^{p} c_v (y_{i-v}-y_i) + \sum_{i=n_k+[n_k^{\frac{\lambda}{1+\mu}}]+p+1}^{m(n_k,T)} a_i \sum_{v=1}^{p} c_v (y_{i-v}-y_i),$$

we see that by (5.3.24) its first term tends to zero as $k\to\infty$.

Consequently, for (5.3.47) it suffices to show

$$\left|y_{n_k+[n_k^{\frac{\lambda}{1+\mu}}]+j+1} - y_{n_k+[n_k^{\frac{\lambda}{1+\mu}}]+j}\right| \xrightarrow[k\to\infty]{} 0, \tag{5.3.50}$$

for $j=0,1,\ldots,m(n_k,T)-n_k-[n_k^{\frac{\lambda}{1+\mu}}]-1$. By (5.3.10) it follows that

$$\begin{aligned}
\left|y_{n_k+[n_k^{\frac{\lambda}{1+\mu}}]+j+1} - y_{n_k+[n_k^{\frac{\lambda}{1+\mu}}]+j}\right| &\le \left\|C^{[n_k^{\frac{\lambda}{1+\mu}}]+j+1} Y_{n_k} - C^{[n_k^{\frac{\lambda}{1+\mu}}]+j} Y_{n_k}\right\| \\
&+ \left\| \sum_{s=n_k}^{n_k+[n_k^{\frac{\lambda}{1+\mu}}]+j} C^{n_k+[n_k^{\frac{\lambda}{1+\mu}}]+j-s} D f(u_{s+1}) \right. \\
&- \left. \sum_{s=n_k}^{n_k+[n_k^{\frac{\lambda}{1+\mu}}]+j-1} C^{n_k+[n_k^{\frac{\lambda}{1+\mu}}]+j-1-s} D f(u_{s+1}) \right\|.
\end{aligned} \tag{5.3.51}$$

By (5.3.12) and (5.3.27) the first term at the right-hand side of (5.3.51) is bounded by

$$re^{\delta([n_k^{\frac{\lambda}{1+\mu}}]+j)}|re^{-\delta}-1|\alpha_2 n_k^{\frac{\mu}{1+\mu}} \xrightarrow[k\to\infty]{} 0 \qquad \forall j = 0,1,\ldots,$$

while the last term in (5.3.51) is bounded by

$$\left\|C^{[n_k^{\frac{\lambda}{1+\mu}}]+j} D f(u_{n_k+1})\right\| + \left\| \sum_{s=n_k-1}^{n_k+[n_k^{\frac{\lambda}{1+\mu}}]+j-1} C^{n_k+[n_k^{\frac{\lambda}{1+\mu}}]+j-1-s} D \right.$$
$$\left. \cdot \big(f(u_{s+2}) - f(u_{s+1})\big)\right\|. \tag{5.3.52}$$

By (5.3.12) and (5.3.26) the first term in (5.3.52) tends to zero as $k \to \infty$, while its last term is bounded by $\frac{r\|D\|}{1-e^{-\delta}} \max_{n_k \le s \le m(n_k,T)} |f(u_{s+1}) - f(u_s)|$, which tends to zero by (5.3.49).

Thus, we have verified (5.3.50) and hence (5.3.47). This means that both (i) and (ii) hold. By Theorem 2.3.1 we conclude that $u_k \xrightarrow[k\to\infty]{} u^0$ a.s. □

Optimality of Adaptive Control for Wiener Systems

We now consider the system given by Fig. 5.3.2, or by (5.3.23). The adaptive control applied to the system is still defined by (5.3.20), (5.3.21).

Similar to (5.3.24), by (5.3.5), (5.3.25), and stability of $C(z)$ we have

$$\sum_{i=n}^{n+j} a_i |f(v_i)| \xrightarrow[n\to\infty]{} 0 \qquad \forall j: j = 0, 1, \ldots, [n^{\frac{\lambda}{1+\mu}}] + p. \tag{5.3.53}$$

It is directly verified that Lemma 5.3.3 still holds true for the present case.

Theorem 5.3.2 *Assume A5.3.1–A5.3.3 hold. Further assume that there is a unique u^0 satisfying (5.3.16) and there is $c_0 > 0$ such that $\frac{1}{c_0}\int_0^{c_0} f(\frac{d}{c}x)dx < f(\frac{d}{c}u^0) < \frac{1}{c_0}\int_{-c_0}^{0} f(\frac{d}{c}x)dx$. Then $\{u_k\}$ defined by (5.3.20)–(5.3.21) is the optimal adaptive regulation control for the Wiener system given by Fig. 5.3.2.*

Proof. Following the proof of Theorem 5.3.1 we need only to point out the modifications that should be made in the proof.

Corresponding to (5.3.39)–(5.3.40), the algorithm is now rewritten as

$$\begin{aligned} u_{j+1} &= \Big[u_j + a_j\Big(f(\frac{d}{c}u_j) - z^* + \varepsilon_j + \delta_j\Big)\Big] \\ &\quad \cdot I_{[|u_j + a_j(f(\frac{d}{c}u_j) - z^* + \varepsilon_j + \delta_j| \le M_{\sigma_j}]} \end{aligned} \tag{5.3.54}$$

$$\begin{aligned} \sigma_j &= \sum_{i=1}^{j-1} I_{[|u_i + a_i(f(\frac{d}{c}u_i) - z^* + \varepsilon_i + \delta_i)| > M_{\sigma_i}]} \\ \delta_j &= f\Big(\frac{d}{c}u_j + \frac{1}{c}\sum_{s=1}^{q} d_s(u_{j-s} - u_j) \\ &\quad - \frac{1}{c}\sum_{s=1}^{p} c_s(v_{j-s} - v_j)\Big) - f(\frac{d}{c}u_j). \end{aligned} \tag{5.3.55}$$

Noticing $f(\frac{d}{c}u_j) - z^* = f(\frac{d}{c}u_j) - f(\frac{d}{c}u^0)$, we see that instead of (5.3.43) the regression function now is $f(\frac{d}{c}u) - f(\frac{d}{c}u^0)$.

As the Lyapunov function we may take

$$v(u) = -\left[\int_0^u f(\frac{d}{c}x)dx - f(\frac{d}{c}u^0)u\right].$$

Then (5.3.44) is straightforwardly obtained.

It remains to show the second limit in (5.3.45) with δ_j given by (5.3.55).

By (5.3.26) and (5.3.53) it follows that

$$\left|\sum_{i=n_k}^{n_k+[n_k^{\frac{\lambda}{1+\mu}}]+p} a_i(f(v_i) - f(\frac{d}{c}u_i))\right| \xrightarrow[k\to\infty]{} 0.$$

Consequently, we need only to show

$$\lim_{T\to 0}\limsup_{k\to\infty}\frac{1}{T}\left|\sum_{i=n_k+[n_k^{\frac{\lambda}{1+\mu}}]+p+1}^{m(n_k,T)} a_i \cdot\left(f\Big(\frac{d}{c}u_i + \frac{1}{c}\sum_{s=1}^{q} d_s(u_{i-s} - u_i) - \frac{1}{c}\sum_{s=1}^{p} c_s(v_{i-s} - v_i)\Big) - f(\frac{d}{c}u_i)\right)\right| = 0. \tag{5.3.56}$$

By Lemmas 5.3.2, 5.3.3, and (5.3.53) we still have (5.3.49), while from (5.3.50)–(5.3.52) with y_k, Y_k, and $f(u_k)$ replaced by v_k, V_k, and u_k, respectively, we conclude

$$\left|v_{n_k+[n_k^{\frac{\lambda}{1+\mu}}]+j+1} - v_{n_k+[n_k^{\frac{\lambda}{1+\mu}}]+j}\right| \xrightarrow[n\to\infty]{} 0,$$

$$j = 0, 1, \cdots, m(n_k, T) - n_k - [n_k^{\frac{\lambda}{1+\mu}}] - 1. \tag{5.3.57}$$

Applying (5.3.49) and (5.3.57) to (5.3.56) and using continuity of $f(\cdot)$ we find that (5.3.56) is true.

Thus, by Theorem 2.3.1 we conclude $u_k \xrightarrow[k\to\infty]{} u^0$ a.s. □

Numerical Example

Let $C(z)$ and $D(z)$ be defined by

$$C(z) = 1 + 0.75z + 0.6z^2 + 0.45z^3, \quad \text{and} \quad D(z) = 1 - 1.4z - 2.4z^2.$$

Then, c and d defined by (5.3.4) take values 2.8 and -2.8, respectively.

Let the nonlinearity be $f(u) = (u+1)^3$ and let $\{\varepsilon_i\}$ be Gaussian iid, $\varepsilon_i \in \mathcal{N}(0,1)$.

The control object is to lead the system output to track the desired constant z^*, which equals -8 for the Hammerstein system and 8 for the Wiener system. According to Lemma 5.3.1 the optimal control u^0 is equal to 1 and -1 for Hammerstein and Wiener system, respectively. It is directly verified that the inequalities figured in Theorems 5.3.1 and 5.3.2 hold.

The adaptive control is generated by the algorithm defined in (5.3.20) and (5.3.21) with μ in (5.3.19) equal to 3.

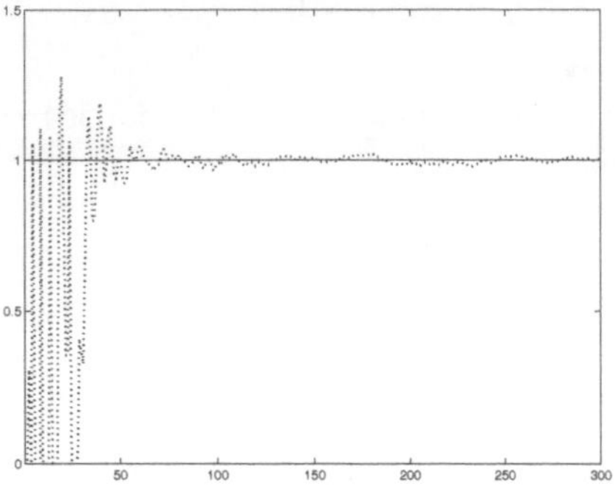

Figure 5.3.3: Adaptive control for Hammerstein system

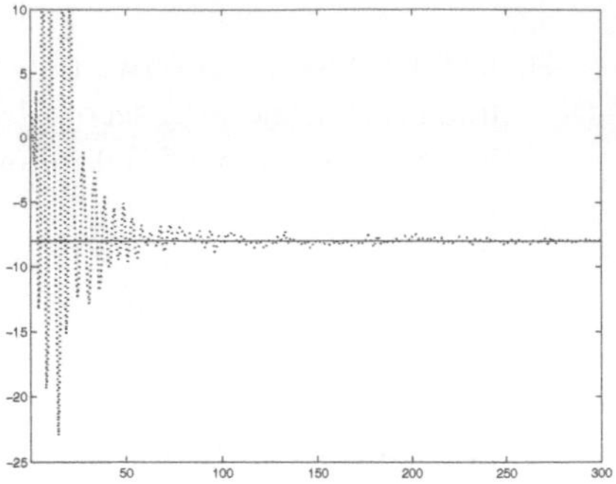

Figure 5.3.4: Output of Hammerstein system

In Figures 5.3.3 and 5.3.5 the dotted lines denote the adaptive controls produced by (5.3.20)–(5.3.21), while the solid lines are the optimal controls, respectively, for Hammerstein and Wiener systems. The figures show that the adaptive controls approach to the optimal ones. In Figures 5.3.4 and 5.3.6 the dotted lines denote the system outputs y_k for Hammerstein and Wiener systems, respectively. From the figures it is seen that y_k approaches the desired value z^*.

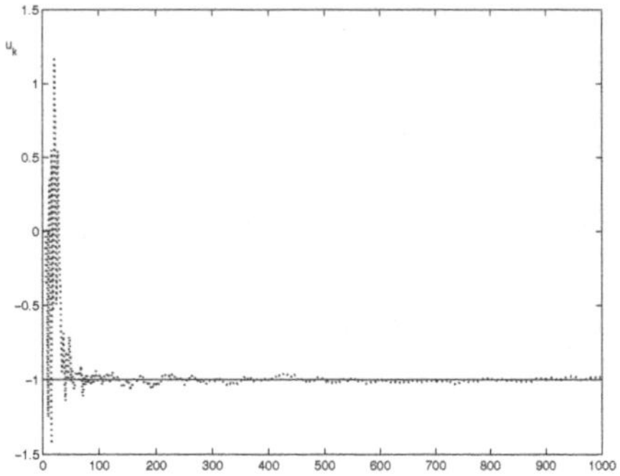

Figure 5.3.5: Adaptive control for Wiener system

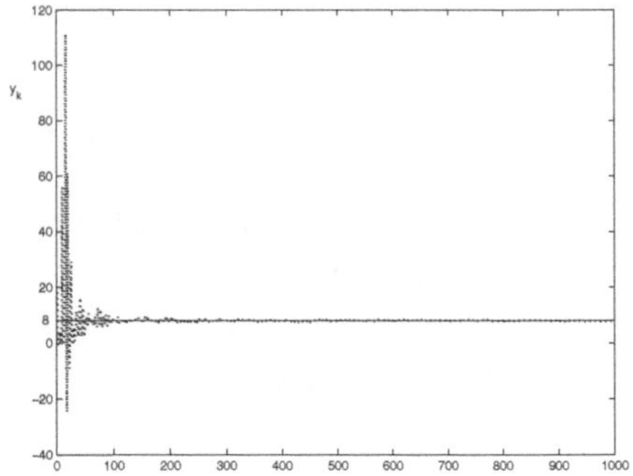

Figure 5.3.6: Output of Wiener system

5.4 Convergence of Distributed Randomized PageRank Algorithms

The PageRank algorithm employed by Google quantifies the importance of each page by the link structure of the web and it has achieved a great success as a commercial searching engine. Let us first recall the PageRank problem. Consider a web with n pages. The web is modeled by a direct graph $\mathscr{G} = (\mathscr{V}, \mathscr{E})$, where $\mathscr{V} = \{1, 2, \cdots, n\}$ is the index set of the pages and $\mathscr{E} \subset \mathscr{V} \times \mathscr{V}$ is the set of links representing the structure of the web. If $(i, j) \in \mathscr{E}$, then page i has an outgoing link to page j. Without losing generality, we assume $n > 2$.

Denote by S_j the set of those pages which have incoming links from page j, and by n_j the number of pages in S_j. Thus we have associated with the graph $\mathscr{G}$ a link matrix

$$A = [a_{ij}]_{n \times n}, \quad a_{ij} = \begin{cases} \frac{1}{n_j}, & j \in \mathscr{L}_i, \\ 0, & \text{otherwise}, \end{cases} \tag{5.4.1}$$

where $\mathscr{L}_i = \{j : (j, i) \in \mathscr{E}\}$. It is clear that $\sum_{i=1}^{n} a_{ij}$ equals either 1 or 0.

The importance of a page i is characterized by its PageRank value $x_i^* \in [0, 1]$, $i \in \mathscr{V}$. Let us assume $\sum_{i=1}^{n} x_i^* = 1$. The basic idea of the PageRank algorithm is that a page which has links from important pages is also important. Mathematically, this suggests to define the PageRank value of page i by

$$x_i^* = \sum_{j \in \mathscr{L}_i} \frac{x_j^*}{n_j}, \tag{5.4.2}$$

or, equivalently, to define $x^* = [x_1^*, \cdots, x_n^*]^T$ from the following linear algebraic e-quation

$$x^* = Ax^*, \quad x_i^* \in [0, 1]. \tag{5.4.3}$$

The normalization condition $\sum_{i=1}^{n} x_i^* = 1$ is possible to be met because if x^* satisfies (5.4.3) then λx^* with $\lambda \in (0, 1]$ also satisfies (5.4.3).

Distributed Randomized PageRank Algorithm (DRPA)

We recall some basic results in the PageRank computation. Notice that in the real world there exist nodes which have no outgoing links to other nodes and thus correspond to zero columns of the link matrix A. To avoid the computational difficulty caused by this, the following assumption A5.4.1 is often made on the matrix A.

A5.4.1 *$A \in \mathbb{R}^{n \times n}$ is a* column-wise stochastic matrix, *i.e., $A = [a_{ij}]$ with $a_{ij} \geq 0$, $i, j = 1, \cdots, n$ and $\sum_{i=1}^{n} a_{ij} = 1$, $j = 1, \cdots, n$.*

Denote by $S \in \mathbb{R}^{n \times n}$ and $\mathbf{1} \in \mathbb{R}^n$ the matrix and vector with all entries being 1. We say that (see Appendix B) a matrix or a vector is positive if all its entries are positive. A probability vector $x = [x_1 \cdots x_n]^T \in \mathbb{R}^n$ is defined as $x_i \geq 0$, $i = 1, \cdots, n$ and $\sum_{i=1}^{n} x_i = 1$.

From (5.4.3) it is clear that the PageRank value of the web is the eigenvector corresponding to eigenvalue 1 of the matrix A. In order for the eigenvalue 1 to have multiplicity 1, the following technique is adopted. Define the matrix $M \in \mathbb{R}^{n\times n}$ by

$$M \triangleq (1-\alpha)A + \alpha\frac{S}{n}, \tag{5.4.4}$$

where $\alpha \in (0,1)$.

Let λ_i, $i = 1,\cdots,n$ be the eigenvalues of M ordered as follows: $|\lambda_i| \geq |\lambda_{i+1}|$, $i = 1,\cdots,n-1$.

The following result directly follows from Theorems B.1 and B.2 given in Appendix B.

Lemma 5.4.1 *If A5.4.1 holds, then*

(i) *M is a positive matrix, whose eigenvalue $\lambda_1 = 1$ is with multiplicity 1, and all the other eigenvalues λ_i, $i = 2,\cdots,n$ of M are strictly in the unit disk;*

(ii) *The dimension of both the left and right eigenvector spaces of M corresponding to eigenvalue 1 is 1;*

(iii) *M can be with a strictly positive right eigenvector corresponding to eigenvalue 1 and $\mathbf{1} = [1\cdots 1]^T$ is a left eigenvector of M corresponding to eigenvalue 1.*

Definition 5.4.1 *The PageRank value x^* of web $\mathscr{G}$ is defined by*

$$x^* = Mx^*,\ x_i^* \in [0,1],\ \sum_{i=1}^{n} x_i^* = 1. \tag{5.4.5}$$

A widely used solution to the PageRank problem (5.4.5) is achieved by the *Power method*, which suggests recursively computing

$$x_{k+1} = Mx_k = (1-\alpha)Ax_k + \frac{\alpha}{n}\mathbf{1} \tag{5.4.6}$$

with $x_0 \in \mathbb{R}^n$ being a probability vector.

Lemma 5.4.2 *By the Power method, x_k generated by (5.4.6) has the following convergence rate:*

$$\|x_k - x^*\| = O(\rho^k), \tag{5.4.7}$$

where $0 < \rho < 1$.

Proof. Recall that λ_2 is the eigenvalue of M with the second largest modulus. Denote by m_2 the multiplicity of λ_2. Then by Lemma 5.4.1 and Theorem B.3 given in Appendix B, we have

$$M^k = vu^T + O(k^{m_2-1}|\lambda_2|^k), \tag{5.4.8}$$

$$x_k = M^k x_0 = vu^T x_0 + O(k^{m_2-1}|\lambda_2|^k), \tag{5.4.9}$$

where u and v are the positive left and right eigenvectors of M corresponding to eigenvalue 1, satisfying $u^T v = 1$.

By Lemma 5.4.1, the vector u must be a multiple of $\mathbf{1} = [1 \cdots 1]^T$. Thus, without losing generality, we may assume $u = \mathbf{1}$, and hence v is a positive probability vector. Since the initial vector x_0 is a probability vector, it follows that $u^T x_0 = 1$.

On the other hand, by Lemma 5.4.1 the dimension of the right eigenvector space of M corresponding to eigenvalue 1 is 1. Therefore, the probability vector v must equal the PageRank value x^*. From (5.4.9) we then conclude

$$x_k = M^k x_0 = x^* + O(k^{m_2-1}|\lambda_2|^k) = x^* + O(\rho^k), \tag{5.4.10}$$

for any $0 < |\lambda_2| < \rho < 1$. This proves (5.4.7). □

Due to the huge dimension of the web, the computation of the PageRank value by use of the traditional algorithms such as the Power method is rather time-consuming. DRPA makes the link matrices $\{A_i\}$, to be defined below, to be sparse and thus simplifies the computation.

A5.4.2 *Choose $\{\theta(k)\}_{k\geq 0}$ to be a sequence of iid random variables with probability*

$$\mathbb{P}\{\theta(k) = i\} = \frac{1}{n}, \; i = 1, \cdots, n. \tag{5.4.11}$$

Consider the web $\mathscr{G} = (\mathscr{V}, \mathscr{E})$. The basic idea of DRPA is as follows: At time k, page i updates its PageRank value by locally communicating with the pages which have incoming links from page i and/or outgoing links to page i, and page i which takes the above action is determined in a random manner. To be precise, DRPA is given by

$$x_{1,k+1} = (1-\alpha_1)A_{\theta(k)}x_{1,k} + \frac{\alpha_1}{n}\mathbf{1}, \tag{5.4.12}$$

where $\{\theta_k\}_{k\geq 0}$ is given by A5.4.2, $x_{1,0}$ is an arbitrary probability vector, $\alpha_1 = \frac{2\alpha}{n-\alpha(n-2)}$, and the link matrix

$$(A_i)_{jl} \triangleq \begin{cases} a_{jl}, & \text{if } j = i \text{ or } l = i \\ 1 - a_{il}, & \text{if } j = l \neq i \\ 0, & \text{otherwise} \end{cases} \tag{5.4.13}$$

for $i = 1, \cdots, n$. It is clear that the matrices $\{A_i\}_{i=1}^n$ are sparse and they are also the column-wise stochastic.

Lemma 5.4.3 *If A5.4.1 and A5.4.2 hold and $\alpha_1 = \frac{2\alpha}{n-\alpha(n-2)}$, then the matrix $M_1 \triangleq (1-\alpha_1)EA_{\theta(k)} + \frac{\alpha_1}{n}S$ is a positive stochastic matrix satisfying*

$$M_1 = \frac{\alpha_1}{\alpha}M + \left(1 - \frac{\alpha_1}{\alpha}\right)I, \text{ and } Ex_{1,k+1} = M_1 Ex_{1,k}. \tag{5.4.14}$$

Proof. Denote the (j,l)th element of $EA_{\theta(k)}$ by $(EA_{\theta(k)})_{jl}$. By the definition of $\{A_i\}_{i=1}^n$ and noticing A5.4.2, we have

$$(EA_{\theta(k)})_{jl} = \frac{1}{n}\sum_{i=1}^{n}(A_i)_{jl} = \begin{cases} \frac{1}{n}\Big[a_{jj} + \sum\limits_{i=1, i\neq j}^{n}(1-a_{ij})\Big] & \text{if } j = l, \\ \frac{2}{n}a_{jl} & \text{if } j \neq l. \end{cases} \tag{5.4.15}$$

By the definition of A, we have $a_{jj} = 0$ and $\sum\limits_{i=1, i\neq j}^{n} a_{ij} = 1$. From (5.4.15) we have

$$EA_{\theta(k)} = \frac{2}{n}A + \frac{n-2}{n}I. \tag{5.4.16}$$

Since A is a column-wise stochastic matrix, from (5.4.16) and the definition of M_1 we know that each element of M_1 is positive and both $EA_{\theta(k)}$ and M_1 are column-wise stochastic matrices.

By noticing that $\alpha_1 = \frac{2\alpha}{n-\alpha(n-2)}$, a direct calculation leads to $\frac{\alpha_1}{\alpha} = \frac{2(1-\alpha_1)}{(1-\alpha)n}$ and $1 - \frac{\alpha_1}{\alpha} = \frac{(1-\alpha_1)(n-2)}{n}$. Then from (5.4.16) and the definitions of M_1 and M (see (5.4.4)), it follows that

$$\begin{aligned} M_1 =& (1-\alpha_1)EA_{\theta(k)} + \frac{\alpha_1}{n}S \\ =& (1-\alpha_1)\left[\frac{2}{n}A + \frac{n-2}{n}I\right] + \frac{\alpha_1}{n}S \\ =& \frac{2(1-\alpha_1)}{n}A + \frac{\alpha_1}{n}S + \frac{(1-\alpha_1)(n-2)}{n}I \\ =& \frac{2(1-\alpha_1)}{(1-\alpha)n}(1-\alpha)A + \frac{\alpha_1}{\alpha}\frac{\alpha}{n}S + \frac{(1-\alpha_1)(n-2)}{n}I \\ =& \frac{\alpha_1}{\alpha}M + \left(1 - \frac{\alpha_1}{\alpha}\right)I. \end{aligned} \tag{5.4.17}$$

Noticing that $\{A_i\}_{i=1}^n$ are column-wise stochastic matrices, from (5.4.12) we know that starting from an arbitrary positive probability vector $x_{1,0}$, $\{x_{1,k},\ k \geq 0\}$ are all probability vectors satisfying

$$x_{1,k+1} = \left((1-\alpha_1)A_{\theta(k)} + \frac{\alpha_1}{n}S\right)x_{1,k}, \tag{5.4.18}$$

which by the mutual independence of $\{\theta(k)\}_{k\geq 0}$ implies

$$Ex_{1,k+1} = M_1 Ex_{1,k}. \tag{5.4.19}$$

Combining (5.4.17) and (5.4.19) gives (5.4.14). □

By Lemma 5.4.3 M_1 is a positive stochastic matrix. Applying Lemma 5.4.2 to (5.4.19) we conclude that $Ex_{1,k}$ tends to the eigenvector ϕ^* corresponding to the biggest eigenvalue of M_1, i.e., $M_1\phi^* = \phi^*$. On the other hand, noticing $M_1 = \frac{\alpha_1}{\alpha}M + \left(1 - \frac{\alpha_1}{\alpha}\right)I$, we derive $\left(\frac{\alpha_1}{\alpha}M + \left(1 - \frac{\alpha_1}{\alpha}\right)I\right)\phi^* = \phi^*$ which implies $M\phi^* = \phi^*$. Noticing that ϕ^* is a probability vector and that M is a positive column-wise stochastic matrix, by Lemma 5.4.1 we find that ϕ^* equals the PageRank value x^* defined by (5.4.5).

We now introduce the following algorithm to estimate x^*.

The estimate $\overline{x}_{1,k}$ for the PageRank value is given by averaging the estimates generated by (5.4.12),

$$\overline{x}_{1,k+1} = \frac{1}{k+1}\sum_{l=0}^{k} x_{1,l}, \tag{5.4.20}$$

which can be written in a recursive way:

$$\overline{x}_{1,k+1} = \overline{x}_{1,k} - \frac{1}{k+1}(\overline{x}_{1,k} - x_{1,k}). \tag{5.4.21}$$

We note that (5.4.21) is an SA algorithm with the linear regression function $f(x) = -(x - x^*)$. The value of $f(\cdot)$ at $\overline{x}_{1,k}$ is $-(\overline{x}_{1,k} - x^*)$. Therefore, $\overline{x}_{1,k} - x_{1,k}$ in (5.4.21) can be treated as $f(\overline{x}_{1,k}) + \varepsilon_{k+1}$, where $\varepsilon_{k+1} = -(x^* - x_{1,k})$ playing the role of observation noise.

Theorem 5.4.1 *If A5.4.1 and A5.4.2 hold and* $\alpha_1 = \frac{2\alpha}{n - \alpha(n-2)}$, *then the estimate* $\overline{x}_{1,k}$ *generated by (5.4.12) and (5.4.21) converges to the true PageRank value almost surely:*

$$\overline{x}_{1,k} - x^* \underset{k\to\infty}{\longrightarrow} 0 \text{ a.s.} \tag{5.4.22}$$

To prove the theorem, we need some technical lemmas. Let us first consider the following *separation of integers*.

Let $\{\alpha_k\}_{k\geq 0}$ be a sequence of strictly increasing nonnegative integers such that $\alpha_0 = 0,\ \alpha_1 = 1,\ \{\alpha_{k+1} - \alpha_k\}$ also increasingly diverges to infinity and

$$0 < \liminf_{k\to\infty} \frac{\alpha_k}{k^a} \leq \limsup_{k\to\infty} \frac{\alpha_k}{k^a} < \infty \text{ for some } a > 1. \tag{5.4.23}$$

Define

$$A_0(0) = \{\alpha_k,\ k = 1, 2, \cdots\}, \tag{5.4.24}$$

$$A_j(i) = \{\alpha_k + i : k = j, j+1, \cdots\},\ A_j = \bigcup_{i=i_1(j)}^{i_2(j)} A_j(i), \tag{5.4.25}$$

where $i_1(j) = \alpha_j - \alpha_{j-1}$ and $i_2(j) = \alpha_{j+1} - \alpha_j - 1$. If $i_2(j) < i_1(j)$, then we set $A_j = \emptyset$. Clearly, α_k may be chosen as $\alpha_k = k^2$.

Lemma 5.4.4 *The sets $A_0(0)$, $A_j(i)$, $i=i_1(j),\cdots,i_2(j)$, $j=1,2,\cdots$ are disjoint and*

$$A_0(0)\bigcup\bigcup_{j=1}^{\infty}\bigcup_{i=i_1(j)}^{i_2(j)}A_j(i)=\{0,1,2,3,\cdots\}. \tag{5.4.26}$$

Proof. By the definition of $A_j(i)$ and by noting $i\in[\alpha_j-\alpha_{j-1},\alpha_{j+1}-\alpha_j-1]$, it is clear that the sets $A_j(i)$ are disjoint. Further, we have

$$\begin{aligned}\bigcup_{j=1}^{\infty}\bigcup_{i=i_1(j)}^{i_2(j)}A_j(i)&=\bigcup_{j=1}^{\infty}\bigcup_{i=\alpha_j-\alpha_{j-1}}^{\alpha_{j+1}-\alpha_j-1}\bigcup_{k=j}^{\infty}\{\alpha_k+i\}\\&=\bigcup_{j=1}^{\infty}\bigcup_{k=j}^{\infty}\bigcup_{i=\alpha_j-\alpha_{j-1}}^{\alpha_{j+1}-\alpha_j-1}\{\alpha_k+i\}\\&=\bigcup_{k=1}^{\infty}\bigcup_{j=1}^{k}\{\alpha_k+\alpha_j-\alpha_{j-1},\alpha_k+\alpha_j-\alpha_{j-1}+1,\cdots,\alpha_k+\alpha_{j+1}-\alpha_j-1\}\\&=\bigcup_{k=1}^{\infty}\{\alpha_k+1,\cdots,\alpha_{k+1}-1\}.\end{aligned} \tag{5.4.27}$$

This proves (5.4.26). □

The following lemma is an extension of the a.s. convergence of the sum of mutually independent random variables.

Lemma 5.4.5 *Let $\{\xi_k\}_{k\geq 1}$ be a sequence of random variables with $E\xi_k=0$ and $\sup_k E\xi_k^2<\infty$. If for any fixed $i,j:\ i=i_1(j),\cdots,i_2(j),\ j=1,2,3,\cdots$ whenever $i_2(j)\geq i_1(j)$, the subsequence $\{\xi_k,\ k\in A_j(i)\}$ is composed of mutually independent random variables with the possible exception of a finite number of ξ_k, then*

$$\sum_{k=1}^{\infty}\frac{1}{k^s}\xi_k<\infty\ \text{a.s.} \tag{5.4.28}$$

for any $s>\frac{3}{2}-\frac{1}{a}$ with a given in (5.4.23).

Proof. From $s>\frac{3}{2}-\frac{1}{a}$ it is clear that $\frac{1}{2}+as-a>\frac{a-1}{2}$. Let $\delta\in\left(\frac{a-1}{2},as-a+\frac{1}{2}\right)$ and let $\delta_1>0$ such that

$$as-a+\frac{1}{2}-\delta-\delta_1>0. \tag{5.4.29}$$

Since $s>\frac{1}{2}$, under the conditions of Lemma 5.4.5 it is clear that $\sum_{k\in A_0(0)}\frac{1}{k^s}\xi_k<\infty$ a.s.. Consequently, the index set $A_0(0)$ can be excluded from the consideration.

For any fixed positive integers j, l, and $i=i_1(j),\cdots,i_2(j)$, define

$$\zeta_l(j,i)=\frac{j^{as-\delta_1-1/2}}{(\alpha_{j-1+l}+i)^s}\xi_{\alpha_{j-1+l}+i},\ \ j\geq 1. \tag{5.4.30}$$

By (5.4.23) we have

$$\frac{j^{as-\delta_1-1/2}}{(\alpha_{j-1+l}+i)^s} < \frac{j^{as-\delta_1-1/2}}{(j-1+l)^{as}} = O\left(\frac{1}{(j-1+l)^{\delta_1+1/2}}\right) = O\left(\frac{1}{l^{\delta_1+1/2}}\right). \quad (5.4.31)$$

From here by (5.4.30) it follows that

$$E\zeta_l(j,i) = 0, \;\; E\zeta_l^2(j,i) < \frac{C}{l^{1+2\delta_1}} \quad (5.4.32)$$

for some $C > 0$.

Then by noticing that $\{\xi_k, k \in A_j(i)\}$ are mutually independent with the possible exception of a finite number of ξ_k, we know that for any fixed i and j, the following series

$$S_{ji}(m) = \sum_{l=1}^{m} \zeta_l(j,i) \quad (5.4.33)$$

converges a.s. to some random variable S_{ji} as $m \to \infty$.

Given $K > 0$, by (5.4.32) and (5.4.33) we have

$$\mathbb{P}\left(|S_{ji}(m)| > Kj^{\delta}\right) \le \frac{1}{K^2 j^{2\delta}} ES_{ji}^2(m) \le \frac{C}{K^2 j^{2\delta}} \sum_{l=1}^{\infty} \frac{1}{l^{1+2\delta_1}} = O\left(\frac{1}{K^2 j^{2\delta}}\right). \quad (5.4.34)$$

Letting $m \to \infty$ in (5.4.34), by Theorem 1.1.8 we have

$$\mathbb{P}\left(|S_{ji}| > Kj^{\delta}\right) = O\left(\frac{1}{K^2 j^{2\delta}}\right). \quad (5.4.35)$$

Define

$$\Omega_j(K) = \{\omega : |S_{ji}| \le Kj^{\delta}, \; i = i_1(j), \cdots, i_2(j)\}, \quad (5.4.36)$$

$$\Omega(K) = \bigcap_{j=1}^{\infty} \Omega_j(K). \quad (5.4.37)$$

In the case $a \ge 2$ it is clear that

$$\begin{aligned} i_2(j) - i_1(j) + 1 &= \alpha_{j+1} - 2\alpha_j + \alpha_{j-1} \\ &= O((j+1)^a - 2j^a + (j-1)^a) = O\left(j^{a-2}\right), \end{aligned} \quad (5.4.38)$$

from which for any $\omega \in \Omega(K)$ it follows that

$$\begin{aligned}
\left|\sum_{k=1}^{\infty}\frac{1}{k^s}\xi_k\right| &= \left|\sum_{j=1}^{\infty}\sum_{i=i_1(j)}^{i_2(j)}\sum_{l=1}^{\infty}\frac{1}{(\alpha_{j+l-1}+i)^s}\xi_{\alpha_{j+l-1}+i}\right| \\
&= \left|\sum_{j=1}^{\infty}\sum_{i=i_1(j)}^{i_2(j)}\frac{1}{j^{as-\delta_1-1/2}}\sum_{l=1}^{\infty}\zeta_l(j,i)\right| \\
&\le \sum_{j=1}^{\infty}\sum_{i=i_1(j)}^{i_2(j)}\frac{|S_{ji}|}{j^{as-\delta_1-1/2}} \\
&= O\left(K\sum_{j=1}^{\infty}\frac{1}{j^{as-\delta_1-\delta-(a-2)-1/2}}\right) < \infty. \qquad (5.4.39)
\end{aligned}$$

By (5.4.35), (5.4.37), and (5.4.38) we have

$$\begin{aligned}
\mathbb{P}(\Omega(K)) &= 1-\mathbb{P}\left(\bigcup_{j=1}^{\infty}\Omega_j^c(k)\right) > 1-\sum_{j=1}^{\infty}\mathbb{P}\left(\Omega_j^c(k)\right) \\
&\ge 1-\sum_{j=1}^{\infty}\sum_{i=i_1(j)}^{i_2(j)}\mathbb{P}\left\{\omega : |S_{ji}| > Kj^{\delta}\right\} \\
&\ge 1-O\left(\sum_{j=1}^{\infty}j^{a-2}\frac{1}{K^2 j^{2\delta}}\right). \qquad (5.4.40)
\end{aligned}$$

By the definition of δ, we have that $2+2\delta-a>1$, and $\mathbb{P}(\Omega(K)) \underset{K\to\infty}{\longrightarrow} 1$. Hence, (5.4.39) and (5.4.40) imply (5.4.28) for $a \ge 2$.

In the case $1<a<2$ we note that for sufficiently large j, $(\alpha_{j+1}-\alpha_j-1)-(\alpha_j-\alpha_{j-1})$ equals either 0 or -1. By noticing $i_1(j)=i_2(j)$, we see that $j=O(i^{\frac{1}{a-1}})$. Similar to (5.4.39) for $\omega \in \Omega(K)$ we derive that

$$\begin{aligned}
\left|\sum_{k=1}^{\infty}\frac{1}{k^s}\xi_k\right| &\le \sum_{j=1,A_j\neq\emptyset}^{\infty}\sum_{i=i_1(j)}^{i_2(j)}\frac{|S_{ji}|}{j^{as-\delta_1-1/2}} \\
&\le K\sum_{j=1,A_j\neq\emptyset}^{\infty}\frac{1}{j^{as-\delta_1-\delta-1/2}} \\
&\le K\sum_{i=1}^{\infty}\frac{1}{i^{\frac{1}{a-1}(as-\delta_1-\delta-1/2)}} < \infty. \qquad (5.4.41)
\end{aligned}$$

Similar to (5.4.40) we have

$$\begin{aligned}\mathbb{P}(\Omega(K)) &= 1-\mathbb{P}\left(\bigcup_{j=1,A_j\neq\emptyset}^{\infty}\Omega_j^c(k)\right) > 1-\sum_{j=1,A_j\neq\emptyset}^{\infty}\mathbb{P}\left(\Omega_j^c(k)\right)\\ &\geq 1-O\left(\sum_{j=1,A_j\neq\emptyset}^{\infty}\frac{1}{K^2 j^{2\delta}}\right)\\ &= 1-O\left(\sum_{i=1}^{\infty}\frac{1}{K^2 i^{\frac{2\delta}{a-1}}}\right)\underset{K\to\infty}{\longrightarrow} 1. \end{aligned} \tag{5.4.42}$$

From (5.4.41) and (5.4.42) we derive (5.4.28) for $1 < a < 2$. This completes the proof. □

Proof of Theorem 5.4.1. Algorithm (5.4.21) is rewritten as

$$\overline{x}_{1,k+1} = \overline{x}_{1,k} + \frac{1}{k+1}(-(\overline{x}_{1,k}-x^*)+e_{1,k+1}+e_{2,k+1}), \tag{5.4.43}$$

where $e_{1,k+1} = -(x^* - Ex_{1,k})$ and $e_{2,k+1} = -(Ex_{1,k} - x_{1,k})$.

Since both $x_{1,k}$ and $\overline{x}_{1,k}$ are probability vectors for all $k \geq 0$, we see that $\|\overline{x}_{1,k}\| \leq 1$ and the algorithm (5.4.43) is in fact an SA evolving in a bounded subspace of $\mathbb{R}^n$. By Theorem 2.3.4, for (5.4.22), we only need to find a Lyapunov function to meet A2.3.3' and to verify the noise condition A2.3.4.

Define $f(x) = -(x - x^*)$ and the Lyapunov function $V(x) \triangleq \|x - x^*\|^2$. It follows that

$$\sup_{\delta<\|x-x^*\|<\Delta,\|x\|\leq 1} \nabla V(x)^T f(x) < 0 \tag{5.4.44}$$

for any $0 < \delta < \Delta$. Hence assumption A2.3.3' holds.

So, for (5.4.22) it suffices to show

$$Ex_{1,k} - x^* \underset{k\to\infty}{\longrightarrow} 0 \tag{5.4.45}$$

and

$$\sum_{k=0}^{\infty}\frac{1}{k+1}(x_{1,k+1} - Ex_{1,k+1}) < \infty \text{ a.s.} \tag{5.4.46}$$

By Lemma 5.4.3, M_1 and M share the same eigenvector x^* corresponding to eigenvalue 1. Then by (5.4.19) and Lemma 5.4.2, we know that (5.4.45) holds. In what follows we show that (5.4.46) takes place.

Define the matrix

$$\Phi(k,j) \triangleq \begin{cases} A_{\theta(k)}A_{\theta(k-1)}\cdots A_{\theta(j)}, & \text{if } j \leq k,\\ I, & \text{if } j = k+1. \end{cases} \tag{5.4.47}$$

Then equation (5.4.12) can be rewritten as

$$x_{1,k+1} =(1-\alpha_1)^{k+1}\Phi(k,0)x_{1,0} + \frac{\alpha_1}{n}\sum_{l=1}^{k+1}(1-\alpha_1)^{k+1-l}\Phi(k,l)\mathbf{1}, \tag{5.4.48}$$

from which it follows that

$$\begin{aligned} & x_{1,k+1} - Ex_{1,k+1} \\ =&(1-\alpha_1)^{k+1}\Big(\Phi(k,0)x_{1,0} - E\Phi(k,0)x_{1,0}\Big) \\ &+ \frac{\alpha_1}{n}\sum_{l=1}^{k+1}(1-\alpha_1)^{k+1-l}\Big(\Phi(k,l)\mathbf{1} - E\Phi(k,l)\mathbf{1}\Big). \end{aligned} \tag{5.4.49}$$

By noticing that $x_{1,0}$ is a probability vector and $\{A_{\theta(k)}\}$ are stochastic matrices, from (5.4.47) it is seen that

$$\sum_{k=1}^{\infty}\frac{1}{k+1}(1-\alpha_1)^{k+1}\Big(\Phi(k,0)x_{1,0} - E\Phi(k,0)x_{1,0}\Big) < \infty \text{ a.s.} \tag{5.4.50}$$

Thus, for (5.4.46) it remains to show that

$$\sum_{k=1}^{\infty}\frac{1}{k+1}\xi_{k+1} < \infty \text{ a.s.,} \tag{5.4.51}$$

where

$$\xi_{k+1} \triangleq \sum_{l=1}^{k+1}(1-\alpha_1)^{k+1-l}(\Phi(k,l)\mathbf{1} - E\Phi(k,l)\mathbf{1}).$$

For any fixed $a > 1$, define $\alpha_k \triangleq [k^a]$, $k \geq 0$. Further, as for Lemma 5.4.5 define $I_0(0) \triangleq \{\alpha_0, \alpha_1, \alpha_2, \cdots\}$, $I_j(i) \triangleq \{\alpha_j + i, \alpha_{j+1} + i, \cdots\}$, and $I_j \triangleq \bigcup_{i=i_1(j)}^{i_2(j)} I_j(i)$, where $i_1(j) \triangleq \alpha_j - \alpha_{j-1}$, $i_2(j) \triangleq \alpha_{j+1} - \alpha_j - 1$.

By Theorem 5.4.5, the sets $\{I_j(i)\}_{j,i}$ are disjoint and

$$I_0(0)\bigcup\Big\{\bigcup_{j=1}^{\infty}\Big[\bigcup_{i=i_1(j)}^{i_2(j)} I_j(i)\Big]\Big\} = \{0, 1, 2, 3, \cdots\}.$$

Take $\tau \in \left(0, 1 - \frac{1}{a}\right)$ and define

$$\overline{\xi}_{k+1} \triangleq \sum_{l=k+1-[k^\tau]}^{k+1}(1-\alpha_1)^{k+1-l}(\Phi(k,l)\mathbf{1} - E\Phi(k,l)\mathbf{1}). \tag{5.4.52}$$

Notice that the random vectors in $\{\xi_k\}_{k\geq 1}$ are not mutually independent. For any fixed $j \geq 1$ and $i \in [\alpha_j - \alpha_{j-1}, \alpha_{j+1} - \alpha_j - 1]$, let us consider the set $\{\overline{\xi}_{k+1} : k+1 \in I_j(i)\}$ and show that $\{\overline{\xi}_{k+1},\ k+1 \in I_j(i)\}$ are mutually independent with the possible exception of a finite number of $\overline{\xi}_{k+1}$.

If $k+1 \in I_j(i)$, then $\overline{\xi}_{k+1} = \overline{\xi}_{[m^a]+i}$ for some integer m. By definition $\overline{\xi}_{[m^a]+i}$ is measurable with respect to $\sigma\{\theta([m^a]+i-1), \cdots, \theta([m^a]+i-[([m^a]+i-1)^\tau])\}$.

Since $\{\theta(k)\}$ is an iid sequence, for the mutual independence of random vectors in $\{\overline{\xi}_{k+1} : k+1 \in I_j(i)\}$ it suffices to show that $\overline{\xi}_{[m^a]+i}$ and $\overline{\xi}_{[(m-1)^a]+i}$ are independent. It is clear that for this it suffices to show $[m^a]+i-[([m^a]+i-1)^\tau] > [(m-1)^a]+i$.

Noticing

$$[m^a]-[(m-1)^a] = am^{a-1} + o(m^{a-1}) \text{ as } m \to \infty$$

and $\tau \in (0, 1-\frac{1}{a})$, we find that as $m \to \infty$

$$\frac{[([m^a]+i-1)^\tau]}{[m^a]-[(m-1)^a]} = O\left(\frac{m^{a\tau}}{m^{a-1}}\right) = O\left(m^{a\tau+1-a}\right) = o(1). \tag{5.4.53}$$

Thus, for any fixed i and j the random vectors in the set $\{\overline{\xi}_{k+1} : k+1 \in I_j(i)\}$ are mutually independent with the possible exception of a finite number of vectors. Then by noticing $\sup\limits_k E\|\overline{\xi}_k\|^2 < \infty$ from Lemma 5.4.5 it follows that

$$\sum_{k=1}^{\infty} \frac{1}{k+1}\overline{\xi}_{k+1} < \infty \text{ a.s.} \tag{5.4.54}$$

Further, we have

$$\begin{aligned} \left\|\sum_{k=1}^{\infty} \frac{1}{k+1}(\xi_{k+1}-\overline{\xi}_{k+1})\right\| &= \left\|\sum_{k=1}^{\infty} \frac{1}{k+1} \sum_{l=1}^{k-[k^\tau]} (1-\alpha_1)^{k+1-l}(\Phi(k,l)\mathbf{1} - E\Phi(k,l)\mathbf{1})\right\| \\ &= O\left(\sum_{k=1}^{\infty} \frac{1}{k+1}(1-\alpha_1)^{[k^\tau]}\right) = O(1), \end{aligned} \tag{5.4.55}$$

which combining with (5.4.54) yields (5.4.51). Thus, (5.4.46) has been proved.

Noticing (5.4.43), (5.4.45), and (5.4.46), by Theorem 2.3.4 we derive the assertion of Theorem 5.4.1. □

Theorem 5.4.2 *Assume A5.4.1 and A5.4.2 hold. For $\overline{x}_{1,k}$ generated by (5.4.12) and (5.4.21) with $\alpha_1 = \frac{2\alpha}{n-\alpha(n-2)}$, the following convergence rate takes place for any $\varepsilon \in (0, \frac{1}{2})$:*

$$\|\overline{x}_{1,k} - x^*\| = o\left(\frac{1}{k^{\frac{1}{2}-\varepsilon}}\right) \text{ a.s.} \tag{5.4.56}$$

Proof. The proof can be carried out similarly to that for Theorem 5.4.1 by using Theorems 2.3.4 and 2.6.1. We only outline the key points.

First by Lemma 5.4.2 we have the exponential rate of convergence $\|Ex_{1,k} - x^*\| = O(\rho^k)$ for some $0 < \rho < 1$.

By Lemma 5.4.5 and carrying out a discussion similar to that for (5.4.46), (5.4.50), and (5.4.51), we can also prove that

$$\sum_{k=0}^{\infty} \frac{1}{(k+1)^s}(x_{1,k+1} - Ex_{1,k+1}) < \infty \text{ a.s.} \tag{5.4.57}$$

with $s > \frac{3}{2} - \frac{1}{a}$ for any fixed $a > 1$, which implies $s > \frac{1}{2}$.

Then by Theorem 2.6.1, we obtain (5.4.56). □

Extension to Simultaneously Updating Multiple Pages

The protocol in (5.4.12) is based on the assumption that only one page updates its PageRank value each time. We now consider the convergence of DRPA for multiple pages simultaneously updating.

Assume that the sequences of Bernoulli random variables $\{\eta_i(k)\}_{k\geq 0}$, $i = 1, \cdots, n$ are mutually independent and each sequence is of iid random variables with probabilities

$$\mathbb{P}\{\eta_i(k) = 1\} = \beta, \tag{5.4.58}$$

$$\mathbb{P}\{\eta_i(k) = 0\} = 1 - \beta, \tag{5.4.59}$$

where $\beta \in (0, 1]$. If $\eta_i(k) = 1$, then page i updates at time k, sending its PageRank value to the pages that page i has outgoing links to and requiring PageRank values from those pages which page i has incoming links from. While if $\eta_i(k) = 0$, no communication is required by page i.

Set $\eta(k) \triangleq (\eta_1(k), \cdots, \eta_n(k))$. The vector $\eta(k)$ reflects updating pages at time k. The corresponding link matrix is given by

$$(A_{p_1,\cdots,p_n})_{ij} \triangleq \begin{cases} a_{ij}, & \text{if } p_i = 1 \text{ or } p_j = 1, \\ 1 - \sum\limits_{h:p_h=1} a_{hj}, & \text{if } p_i = 0 \text{ and } i = j, \\ 0, & \text{if } p_i = p_j = 0 \text{ and } i \neq j, \end{cases} \tag{5.4.60}$$

where $(p_1, \cdots, p_n)$ is a realization of $\eta(k)$. It is clear that $A_{p_1,\cdots,p_n}$ is a sparse matrix.

Similar to (5.4.12) and (5.4.20), the DRPA for the multiple pages updating is given by

$$x_{2,k+1} = (1 - \alpha_2) A_{\eta(k)} x_{2,k} + \frac{\alpha_2}{n}\mathbf{1}, \tag{5.4.61}$$

$$\overline{x}_{2,k+1} = \frac{1}{k+1}\sum_{l=0}^{k} x_{2,l}, \tag{5.4.62}$$

where $x_{2,0}$ is an arbitrary probability vector and $\alpha_2 = \frac{\alpha(1-(1-\beta)^2)}{1-\alpha(1-\beta)^2}$. Clearly, there are 2^n different link matrices.

Define $M_2 \triangleq (1-\alpha_2)EA_{\eta(k)} + \alpha_2\frac{S}{n}$. By a direct calculation with $\alpha_2 = \frac{\alpha[1-(1-\beta)^2]}{1-\alpha(1-\beta)^2}$, we have $M_2 = \frac{\alpha_2}{\alpha}M + \left(1-\frac{\alpha_2}{\alpha}\right)I$. Noticing that $\{A_{\eta(k)}\}_{k\geq 0}$ is iid and carrying out the same discussion as that for Theorems 5.4.1 and 5.4.2, we have the following result.

Theorem 5.4.3 *Assume that A5.4.1 holds. Then the estimates generated by (5.4.61)–(5.4.62) with* $\alpha_2 = \frac{\alpha(1-(1-\beta)^2)}{1-\alpha(1-\beta)^2}$ *are strongly consistent with convergence rate*

$$\|\bar{x}_{2,k} - x^*\| = o\left(\frac{1}{k^{\frac{1}{2}-\varepsilon}}\right) \quad \text{a.s.}$$

for any $\varepsilon \in \left(0, \frac{1}{2}\right)$.

Numerical Example

Let us consider a web with 6 pages, and let the link matrix A be given by

$$A \triangleq \begin{bmatrix} 0 & \frac{1}{2} & 0 & \frac{1}{4} & \frac{1}{3} & \frac{1}{2} \\ \frac{1}{3} & 0 & \frac{1}{3} & \frac{1}{4} & \frac{1}{3} & 0 \\ \frac{1}{3} & 0 & 0 & 0 & 0 & 0 \\ \frac{1}{3} & \frac{1}{2} & \frac{1}{3} & \frac{1}{4} & 0 & \frac{1}{2} \\ 0 & 0 & \frac{1}{3} & \frac{1}{4} & 0 & 0 \\ 0 & 0 & 0 & 0 & \frac{1}{3} & 0 \end{bmatrix}.$$

Choose $\alpha \triangleq 0.15$ in (5.4.4). The absolute values of the estimation errors generated by algorithms (5.4.11)–(5.4.13) and (5.4.20)–(5.4.21) are presented in Figs. 5.4.1–5.4.3, while the estimation errors generated by algorithms (5.4.58)–(5.4.62) with $\beta = 0.1$ are demonstrated in Figs. 5.4.4–5.4.6. The figures show that the estimation errors asymptotically tend to zero as expected.

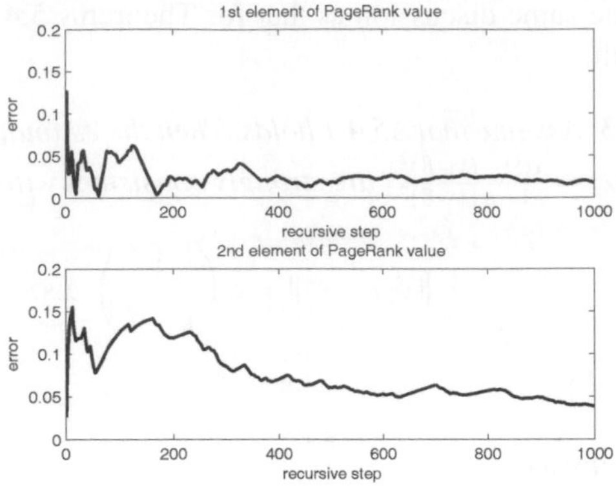

Figure 5.4.1: Estimation error of 1st and 2nd elements of PageRank value

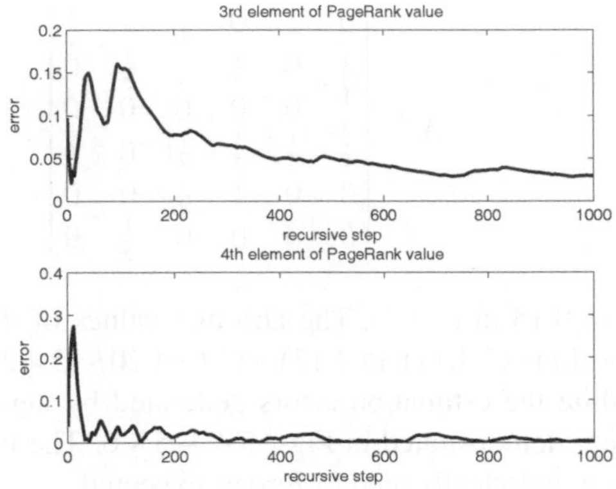

Figure 5.4.2: Estimation error of 3rd and 4th elements of PageRank value

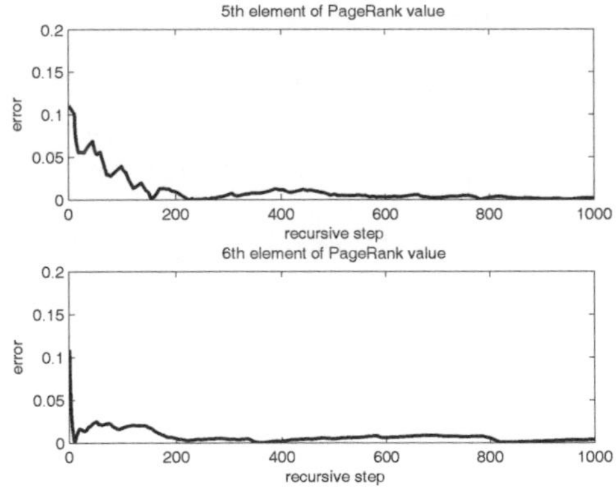

Figure 5.4.3: Estimation error of 5th and 6th elements of PageRank value

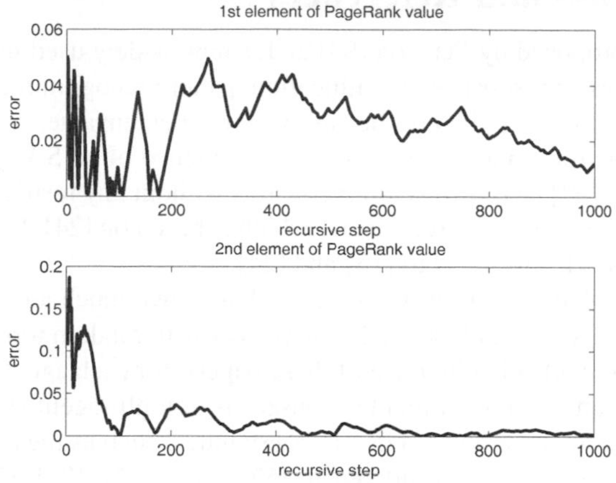

Figure 5.4.4: Estimation error of 1st and 2nd elements of PageRank value

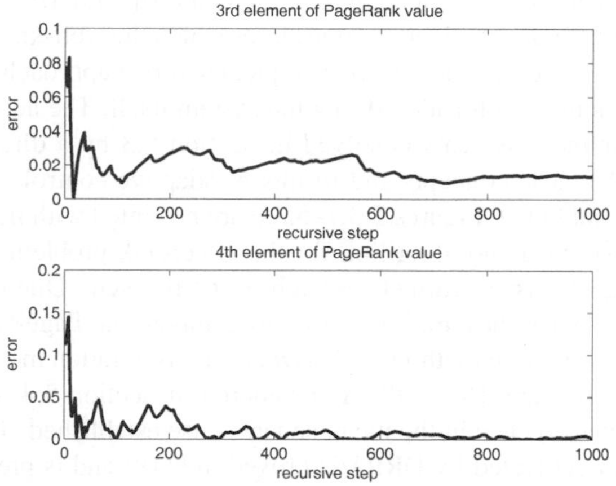

Figure 5.4.5: Estimation error of 3rd and 4th elements of PageRank value

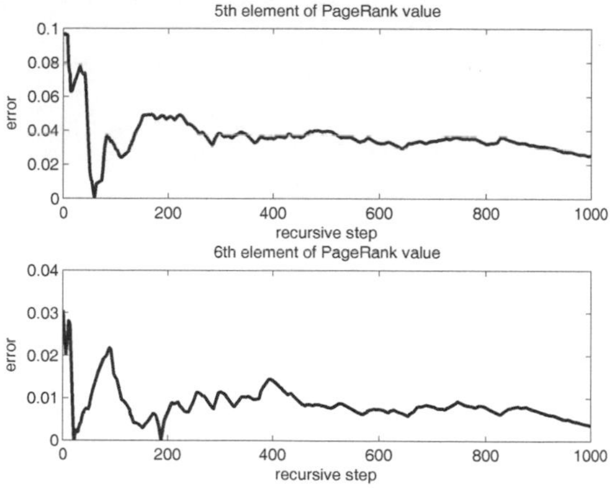

Figure 5.4.6: Estimation error of 5th and 6th elements of PageRank value

5.5 Notes and References

PCA was proposed by Pearson [94] and is now widely used in various areas such as signal processing, subspace identification, pattern recognition, and many others, see, e.g., [31], [35], and [91]. For the case where all eigenvalues of the symmetric matrix are of single multiplicity the ordered convergence of an SA type algorithm was established in [92].The corresponding results without any restriction on multiplicity of eigenvalues are given in Section 5.1 which is based on [24]. For the detailed proof of Proposition 5.1.1 we refer to [13] and [24].

The problem of consensus of agents has drawn much attention from researchers in recent years [60] and [110]. The networks with random topologies are considered in [53], [99], and [117]. In most of these papers, for each agent the states of its neighbors are assumed to be available. Consensus of multi-agent systems under a random environment with various kinds of uncertainties, such as measurement noises, quantization errors, etc., is considered in [57], [65], [73], [72], [75], and [112] among others. Pathwise consensus of multi-agent systems, called the strong consensus ([57] and [73]), is described in Section 5.3 based on [40] with SAAWET applied.

Unlike the indirect adaptive control, for which the well-known self-tuning regulator [5][48] may serve as a typical example, the direct approach proposes the adaptive control without need for identifying the system itself. The adaptive regulation problem for nonlinear systems is solved in Section 5.3 by a direct approach based on [21]. For the basic concepts and results of adaptive control, we refer to [26], [44], [66], [68], and [114]. Figures 5.3.1–5.3.6 are reprinted with permission from [21].

The basic idea and algorithm for the PageRank problem presented in [14] and [71] have given rise to further research of the problem. Due to the huge size of the web, many approaches are introduced to compute the PageRank value, such as the adaptive computation method [64], asynchronous iteration method [34] and [67], and others [6], [16], and [61]. DRPA considered in Section 5.4 was first introduced in [59] with convergence in the mean square sense established. The strong consistency of estimates generated by DRPA is proved in [119] and is presented in Section 5.4. For the convergence analysis, the separation of integer set is based on [55], and some results on the nonnegative matrices are given in Appendix B based on [54] and [100].

Appendix A: Proof of Some Theorems in Chapter 1

Proof of Theorem 1.2.1.

Since $\{\xi_k\}_{k\geq 1}$ is iid, the σ-algebras $\sigma\{\xi_i, 1\leq i\leq k\}$ and $\sigma\{\xi_j, j>k\}$ are mutually independent for each $k\geq 1$. Thus $\mathscr{D}\triangleq\bigcap_{k=1}^{\infty}\sigma\{\xi_j, j>k\}$ is independent of $\sigma\{\xi_j, j\geq k\}\ \forall\, k\geq 1$, and hence is independent of $\mathscr{A}\triangleq\bigcup_{k=1}^{\infty}\sigma\{\xi_i, 1\leq i\leq k\}$.Therefore, $\sigma(\mathscr{A})$ and $\mathscr{D}$ are mutually independent. Noticing $\mathscr{D}\subset\sigma(\mathscr{A})$, we find that the tail-events are independent of themselves, i.e., $\mathbb{P}\{A\}=\mathbb{P}\{A\cap A\}=\mathbb{P}^2\{A\}\ \forall\, A\in\mathscr{D}$, which implies $\mathbb{P}\{A\}=0$ or 1. □

Proof of Theorem 1.2.2.

Denote by F_ξ, F_η, and $F_{\xi,\eta}$ the distributions of ξ, η, and (ξ,η), respectively, and denote the corresponding Lebesgue–Stieljies measures by ν_ξ, ν_η, and $\nu_{\xi,\eta}$. By the independence of ξ and η, it follows that $(\mathbb{R}^{l+m},\mathscr{B}^{l+m},\nu_{\xi,\eta})=(\mathbb{R}^l,\mathscr{B}^l,\nu_\xi)\times(\mathbb{R}^m,\mathscr{B}^m,\nu_\eta)$.

Denote $g_+(x)=E(f(x,\eta))_+$, $g_-(x)=E(f(x,\eta))_-$, and $D=\{x\in\mathbb{R}^l : g_+(x)=g_-(x)=\infty\}$. Consider a set $A=\{\omega : \xi(\omega)\in B\}$, where $B\in\mathscr{B}^l$ is a Borel set. By Theorem 1.1.4, it follows that

$$
\begin{aligned}
\int_A g_\pm(\xi)\mathrm{d}\mathbb{P} &= \int_B g_\pm(x)\mathrm{d}\nu_\xi(x) = \int_B\int_{\mathbb{R}^m}(f(x,y))_\pm \mathrm{d}\nu_\eta(y)\mathrm{d}\nu_\xi(x)\\
&= \int_{B\times\mathbb{R}^m}(f(x,y))_\pm \mathrm{d}\nu_{\xi,\eta} = \int_A (f(\xi,\eta))_\pm \mathrm{d}\mathbb{P}. \qquad (A.1)
\end{aligned}
$$

Since $Ef(\xi,\eta)$ exists, from (A.1) it follows that either $g_+(\xi)<\infty$ a.s. or $g_-(\xi)<\infty$ a.s. and hence $I_D(\xi)=0$ a.s. Thus for $A\in\sigma(\xi)$, by the definition of

$g(\cdot)$, we have

$$\begin{aligned}\int_A g(\xi)\mathrm{d}\mathbb{P} &= \int_A (g_+(\xi) - g_-(\xi)) I_{D^c}(\xi)\mathrm{d}\mathbb{P} \\ &= \int_A g_+(\xi) I_{D^c}(\xi)\mathrm{d}\mathbb{P} - \int_A g_-(\xi) I_{D^c}(\xi)\mathrm{d}\mathbb{P} \\ &= \int_A g_+(\xi)\mathrm{d}\mathbb{P} - \int_A g_-(\xi)\mathrm{d}\mathbb{P} \\ &= \int_A (f(\xi,\eta))_+\mathrm{d}\mathbb{P} - \int_A (f(\xi,\eta))_-\mathrm{d}\mathbb{P} \\ &= \int_A f(\xi,\eta)\mathrm{d}\mathbb{P}. \end{aligned} \tag{A.2}$$

From (A.2) and Theorem 1.1.9, it follows that $g(\xi) = E[f(\xi,\eta)|\sigma(\xi)]$ a.s. □

Proof of Theorem 1.2.5.

We first prove (1.2.6). Define $A \triangleq \left\{\omega : \max_{1\le j\le n} \xi_j \ge \lambda\right\}$, $A_1 \triangleq \{\omega : \xi_1 \ge \lambda\}$, and $A_k \triangleq \{\omega : \xi_j < \lambda,\ 1 \le j < k;\ \xi_k \ge \lambda\}$, $k = 2, \cdots, n$. Then A_k is $\mathscr{F}_k$-measurable, the A_k's are disjoint, and $\bigcup_{k=1}^{n} A_k = A$. Using the submartingale inequality and the fact $\xi_k(\omega) \ge \lambda$ on A_k, we find that

$$\begin{aligned}\int_A \xi_n \mathrm{d}\mathbb{P} &= \sum_{k=1}^{n} \int_{A_k} \xi_n \mathrm{d}\mathbb{P} = \sum_{k=1}^{n} \int_{A_k} E[\xi_n|\mathscr{F}_k]\mathrm{d}\mathbb{P} \\ &\ge \sum_{k=1}^{n} \int_{A_k} \xi_k \mathrm{d}\mathbb{P} \ge \lambda \sum_{k=1}^{n} \mathbb{P}\{A_k\} = \lambda \mathbb{P}\{A\}, \end{aligned} \tag{A.3}$$

which proves (1.2.6).

We now proceed to prove (1.2.7). Denote $\xi_n^* \triangleq \max_{1\le k\le n} \xi_k$.

Noticing that $\{\xi_k\}_{k\ge 1}$ is nonnegative, for any fixed $p > 1$ we have

$$E\xi_n^{*p} = \int_0^\infty x^p \mathrm{d}\mathbb{P}\{\xi_n^* < x\} = \int_0^\infty x^p \mathrm{d}\Big(1 - \mathbb{P}\{\xi_n^* \ge x\}\Big). \tag{A.4}$$

By noticing that x^p is increasing on $[0,\infty)$, it follows that

$$\begin{aligned} x^p \mathbb{P}\{\xi_n^* \ge x\}\Big|_{x=0}^{\infty} &= \lim_{c\to\infty} c^p \mathbb{P}\{\xi_n^* \ge c\} = \lim_{c\to\infty} c^p (1 - \mathbb{P}\{\xi_n^* < c\}) \\ &\le \limsup_{c\to\infty} \int_c^\infty x^p \mathrm{d}\mathbb{P}\{\xi_n^* < x\}. \end{aligned} \tag{A.5}$$

The right-hand side of (A.5) equals zero because of $E\xi_n^{*p} < \infty$. Thus from (A.4), we have

$$E\xi_n^{*p} = -\int_0^\infty x^p \mathrm{d}\mathbb{P}\{\xi_n^* \ge x\} = \int_0^\infty \mathbb{P}\{\xi_n^* \ge x\} p x^{p-1} \mathrm{d}x, \tag{A.6}$$

from which and (1.2.6) we further have

$$E\xi_n^{*p} \le \int_0^\infty px^{p-2} \int\limits_{[\xi_n^* \ge x]} \xi_n \mathrm{d}\mathbb{P}\mathrm{d}x$$
$$= \int_\Omega \xi_n \int_0^{\xi_n^*} px^{p-2}\mathrm{d}x\mathrm{d}\mathbb{P} = \frac{p}{p-1} E\xi_n \xi_n^{*(p-1)}. \tag{A.7}$$

Noticing that $p > 1$ and $\{\xi_k\}_{k\ge 1}$ is nonnegative, by (A.7) and the Hölder inequality, we find that

$$E\xi_n^{*p} \le \frac{p}{p-1} (E\xi_n^p)^{\frac{1}{p}} (E\xi_n^{*p})^{\frac{p-1}{p}}, \tag{A.8}$$

which implies (1.2.7). □

Proof of Theorem 1.2.6.

For $n \ge m$, first by the submartingale property and then by the Jensen inequality, we have

$$(\xi_m - a)_+ \le [E(\xi_n - a|\mathscr{F}_m)]_+ \le E[(\xi_n - a)_+|\mathscr{F}_m].$$

Thus $\{(\xi_k - a)_+, \mathscr{F}_k\}$ is a submartingale.

Note that $\beta(a,b)$ equals $\beta(0,b-a)$ which is the number of up-crossings of the interval $(0,b-a)$ by the submartingale $\{\xi_k - a, \mathscr{F}_k\}$ or $\{(\xi_k - a)_+, \mathscr{F}_k\}$. Thus, for (1.2.10) it suffices to prove that for a nonnegative submartingale $\{\xi_k, \mathscr{F}_k\}_{k=1}^N$,

$$E\beta(0,b) \le \frac{E\xi_N}{b}. \tag{A.9}$$

For $i = 1, \cdots, N$, define

$$\eta_i = \begin{cases} 0, & \text{if } T_{m-1} < i \le T_m \text{ for some odd } m, \\ 1, & \text{if } T_{m-1} < i \le T_m \text{ for some even } m. \end{cases}$$

Define $\xi_0 = 0$. Then for even m, ξ_k crosses $(0,b)$ from time T_{m-1} to T_m, and hence

$$\sum_{k=T_{m-1}+1}^{T_m} \eta_k(\xi_k - \xi_{k-1}) = \sum_{k=T_{m-1}+1}^{T_m} (\xi_k - \xi_{k-1}) = \xi_{T_m} - \xi_{T_{m-1}} = \xi_{T_m} \ge b$$

and

$$\sum_{k=1}^N \eta_k(\xi_k - \xi_{k-1}) \ge b\beta(0,b). \tag{A.10}$$

Further, the set $\{\eta_k = 1\}$ is $\mathscr{F}_{k-1}$-measurable since T_i, $i \ge 1$ are stopping times and

$$\{\eta_k = 1\} = \bigcup_{l \ge 1} \Big(\{T_{2i-1} < k\} \cap \{T_{2i} < k\}^{\mathrm{c}} \Big).$$

Then taking expectation of both sides of (A.10) and by Theorem 1.1.9, we have

$$
\begin{aligned}
bE\beta(0,b) &\le E\sum_{k=1}^{N}\eta_k(\xi_k-\xi_{k-1})\\
&=\sum_{k=1}^{N}\int_{[\eta_k=1]}(\xi_k-\xi_{k-1})\mathrm{d}\mathbb{P}\\
&=\sum_{k=1}^{N}\int_{[\eta_k=1]}E\left[\xi_k-\xi_{k-1}\middle|\mathscr{F}_{k-1}\right]\mathrm{d}\mathbb{P}\\
&\le\sum_{k=1}^{N}\int_{\Omega}E\left[\xi_k-\xi_{k-1}\middle|\mathscr{F}_{k-1}\right]\mathrm{d}\mathbb{P}=E\xi_N.
\end{aligned}
$$

Hence (A.9) and (1.2.10) hold. □

Proof of Theorem 1.2.7.

Set

$$\limsup_{k\to\infty}\xi_k=\xi^*,\quad \liminf_{k\to\infty}\xi_k=\xi_*.$$

Assume the converse: $\mathbb{P}(\xi^*>\xi_*)>0$. Then $\{\xi^*>\xi_*\}=\cup_{x<y}\{\xi^*>y>x>\xi_*\}$, where x and y run over all rational numbers, and there exist rational numbers a and b such that

$$\mathbb{P}(\xi^*>b>a>\xi_*)>0. \tag{A.11}$$

Let $\beta_N(a,b)$ be the number of up-crossings of the interval (a,b) by $\{\xi_k,\mathscr{F}_k\}$, $k=1,\cdots,N$. By Theorem 1.2.6,

$$E\beta_N(a,b)\le\frac{E(\xi_N)_++|a|}{b-a}. \tag{A.12}$$

By Theorem 1.1.6 from (A.12) it follows that

$$E\beta_\infty(a,b)=E\lim_{N\to\infty}\beta_N(a,b)=\lim_{N\to\infty}E\beta_N(a,b)\le\frac{\sup_N E(\xi_N)_++|a|}{b-a}<\infty. \tag{A.13}$$

However, (A.11) implies $\mathbb{P}\big(\beta_\infty(a,b)=\infty\big)>0$, which contradicts with (A.13). Hence

$$\mathbb{P}(\xi^*=\xi_*)=1$$

or ξ_k converges to a limit ξ a.s.

By Theorem 1.1.7, it follows that

$$E(\xi)_+=E\liminf_{k\to\infty}(\xi_k)_+\le\liminf_{k\to\infty}E(\xi_k)_+\le\sup_k E(\xi_k)_+<\infty$$

and

$$E(\xi)_- = E\liminf_{k\to\infty}(\xi_k)_- \le \liminf_{k\to\infty} E(\xi_k)_- \le \sup_k E(\xi_k)_-$$
$$= \sup_k (E(\xi_k)_+ - E\xi_k) \le \sup_k (E(\xi_k)_+ - E\xi_1) < \infty.$$

Hence $E|\xi| < \infty$. □

Proof of Lemma 1.2.2.

Note that $\xi_T I_{[T\le k-1]} = \xi_0 I_{[T=0]} + \xi_1 I_{[T=1]} + \cdots + \xi_{k-1} I_{[T=k-1]}$ is $\mathscr{F}_{k-1}$-measurable. If $\{\xi_k, \mathscr{F}_k\}_{k\ge 0}$ is a martingale, then

$$\begin{aligned} E\left(\xi_{T\wedge k}|\mathscr{F}_{k-1}\right) &= E\left(\xi_T I_{[T\le k-1]} + \xi_k I_{[T>k-1]}|\mathscr{F}_{k-1}\right) \\ &= \xi_T I_{[T\le k-1]} + E\left(\xi_k I_{[T\le k-1]^c}|\mathscr{F}_{k-1}\right) \\ &= \xi_T I_{[T\le k-1]} + I_{[T\le k-1]^c} E\left(\xi_k|\mathscr{F}_{k-1}\right) \\ &= \xi_T I_{[T\le k-1]} + \xi_{k-1} I_{[T>k-1]} = \xi_{T\wedge(k-1)}. \end{aligned}$$

This shows that $\{\xi_{T\wedge k}, \mathscr{F}_k\}_{k\ge 0}$ is a martingale. For supermartingales and submartingales the proof is similar. □

Proof of Theorem 1.2.8.

Since $\sum_{i=1}^{k+1} E\left(\xi_i^2|\mathscr{F}_{i-1}\right)$ is $\mathscr{F}_k$-measurable, for any fixed positive integer M the first exit time

$$T_M \triangleq \begin{cases} \min\left\{k : \sum\limits_{i=1}^{k+1} E\left(\xi_i^2|\mathscr{F}_{i-1}\right) > M\right\} \\ \infty, \text{ if } \sum\limits_{k=1}^{\infty} E\left(\xi_k^2|\mathscr{F}_{k-1}\right) \le M \end{cases}$$

is a stopping time and by Lemma 1.2.2 $\{\eta_{T_M\wedge k}, \mathscr{F}_k\}$ is a martingale.

Noting that $\eta_{k\wedge T_M} = \sum_{i=1}^{k} \xi_i I_{[i\le T_M]}$, we find $E\left(\xi_i I_{[i\le T_M]}\xi_j I_{[j\le T_M]}\right) = E\left(\xi_i\xi_j I_{[j\le T_M]}\right) = E\left[\xi_i I_{[j\le T_M]} E\left(\xi_j|\mathscr{F}_{j-1}\right)\right] = 0$ and

$$\begin{aligned} \left(E|\eta_{k\wedge T_M}|\right)^2 \le E\eta_{k\wedge T_M}^2 &= E\left(\sum_{i=1}^{k} \xi_i^2 I_{[i\le T_M]}\right) = E\left(\sum_{i=1}^{k} E\left(\xi_i^2 I_{[i\le T_M]}|\mathscr{F}_{i-1}\right)\right) \\ &= E\left(\sum_{i=1}^{k} I_{[i\le T_M]} E\left(\xi_i^2|\mathscr{F}_{i-1}\right)\right) = E\left(\sum_{i=1}^{k\wedge T_M} E\left(\xi_i^2|\mathscr{F}_{i-1}\right)\right) \le M, \end{aligned}$$

for any $i < j$, $i, j = 1, \cdots, k$.

By Corollary 1.2.1 $\eta_{k\wedge T_M}$ converges a.s. as $k \to \infty$.

Since $\eta_{k\wedge T_M} = \eta_k$ on $\{T_M = \infty\}$, we find that as $k \to \infty$ η_k converges a.s. on $\{T_M = \infty\}$ and consequently on $\cup_{M=1}^{\infty}[T_M = \infty]$ which equals A by the definition (1.2.12). □

Proof of Theorem 1.2.9.

For any fixed positive integer M, define

$$T_M = \begin{cases} \min\{k : \eta_k > M,\ k \geq 1\} \\ \infty, \ \text{if } \eta_k < M,\ \forall\, k. \end{cases}$$

By Lemma 1.2.2 $\{\eta_{k\wedge T_M}, \mathscr{F}_k\}$ is a martingale. It is clear that

$$\eta_{k\wedge T_M} \begin{cases} \leq M, \ \text{if } k < T_M, \\ = \eta_{T_M-1} + \xi_{T_M} \leq M + \sup_k \xi_k, \ \text{if } k \geq T_M. \end{cases}$$

Consequently,

$$\sup_k E(\eta_{k\wedge T_M})_+ \leq E \sup_k (\eta_{k\wedge T_M})_+ \leq E\left(M + \left(\sup_k \xi_k\right)_+\right) < \infty.$$

By Theorem 1.2.7 $\eta_{k\wedge T_M}$ converges a.s. as $k \to \infty$.

Since $\eta_{k\wedge T_M} = \eta_k$ on $\{T_M = \infty\}$, we find that as $k \to \infty$ η_k converges a.s. on $\{T_M = \infty\}$ and consequently on $\cup_{M=1}^{\infty}[T_M = \infty]$ which equals A_1. This completes the proof. □

Proof of Theorem 1.2.10.

Define

$$\xi_k = \sum_{i=1}^{k} \left[I_{B_i} - E(I_{B_i}|\mathscr{F}_{i-1})\right]. \tag{A.14}$$

Clearly, $\{\xi_k, \mathscr{F}_k\}$ is a martingale and $\{I_{B_k} - E(I_{B_k}|\mathscr{F}_{k-1}), \mathscr{F}_k\}$ is an mds.

Since $|I_{B_k} - E(I_{B_k}|\mathscr{F}_{k-1})| \leq 1$, by Theorem 1.2.9 ξ_k converges on

$$\left\{\omega : \left(\sup_k \xi_k < \infty\right) \bigcup \left(\inf_k \xi_k > -\infty\right)\right\}. \tag{A.15}$$

If $\sum_{k=1}^{\infty} I_{B_k} < \infty$, then from (A.14) it follows that $\sup_k \xi_k < \infty$ a.s., which implies that ξ_k converges a.s. This combining with $\sum_{k=1}^{\infty} I_{B_k} < \infty$ and (A.14) yields $\sum_{k=1}^{\infty} \mathbb{P}(B_k|\mathscr{F}_{k-1}) < \infty$.

Conversely, if $\sum_{k=1}^{\infty} \mathbb{P}(B_k|\mathscr{F}_{k-1}) < \infty$, then from (A.14) and (A.15) it follows that $\inf_k \xi_k > -\infty$ a.s. and ξ_k converges a.s. From the convergence of ξ_k and $\sum_{k=1}^{\infty} \mathbb{P}(B_k|\mathscr{F}_{k-1}) < \infty$, we find that $\sum_{k=1}^{\infty} I_{B_k} < \infty$. □

Proof of Theorem 1.2.11.

Denote by $\mathscr{F}_k$ the σ-algebra generated by $\{B_1, \cdots, B_k\}$.

If $\sum_{k=1}^{\infty} \mathbb{P}(B_k) < \infty$, then

$$E\left(\sum_{k=1}^{\infty} E(I_{B_k}|\mathscr{F}_{k-1})\right) = \sum_{k=1}^{\infty} \mathbb{P}(B_k) < \infty,$$

and $\sum_{k=1}^{\infty} E(I_{B_k}|\mathscr{F}_{k-1}) < \infty$ a.s. which, by Theorem 1.2.10, implies $\sum_{k=1}^{\infty} I_{B_k} < \infty$ a.s. and hence $\mathbb{P}\big(B_k \text{ i.o.}\big) = 0$.

When B_k, $k \geq 1$ are mutually independent and $\sum_{k=1}^{\infty} \mathbb{P}(B_k) = \infty$, then

$$\sum_{k=1}^{\infty} \mathbb{P}(B_k|\mathscr{F}_{k-1}) = \sum_{k=1}^{\infty} \mathbb{P}(B_k) = \infty,$$

which, by Theorem 1.2.10, implies that $\mathbb{P}\big(B_k, \text{i.o.}\big) = 1$. □

Proof of Lemma 1.2.3.

By Theorem 1.2.10 it follows that $A = \big\{\omega : \sum_{k=1}^{\infty} I_{[|y_k|>b_k]} < \infty\big\}$. This means that the events $\{|y_k| > b_k\}$, $k \geq 1$ may occur only a finite number of times on A. Therefore, on A the series $\sum_{k=1}^{\infty} y_k$ converges iff $\sum_{k=1}^{\infty} y_k I_{[|y_k| \leq b_k]}$. □

Proof of Theorem 1.2.12.

Here we adopt the notations used in Lemma 1.2.3. Taking $b_k = c$, by Lemma 1.2.3 we have $S \subset A$ and

$$\left\{\omega : \sum_{k=1}^{\infty} y_k \text{ converges}\right\} \bigcap S = \left\{\omega : \sum_{k=1}^{\infty} y_k I_{[|y_k| \leq c]} \text{ converges}\right\} \bigcap S. \tag{A.16}$$

Define $\xi_k \triangleq y_k I_{[|y_k| \leq c]} - E y_k I_{[|y_k| \leq c]}$, $k \geq 1$. By (1.2.19), it follows that

$$\left\{\omega : \sum_{k=1}^{\infty} y_k \text{ converges}\right\} \bigcap S = \left\{\omega : \sum_{k=1}^{\infty} \xi_k \text{ converges}\right\} \bigcap S. \tag{A.17}$$

Noticing that $E(\xi_k^2|\mathscr{F}_{k-1}) = E\big(y_k^2 I_{[|y_k| \leq c]}|\mathscr{F}_{k-1}\big) - \big(E\big(y_k I_{[|y_k| \leq c]}|\mathscr{F}_{k-1}\big)\big)^2$, from (1.2.20) we see

$$S \subset \left\{\omega : \sum_{k=1}^{\infty} E(\xi_k^2|\mathscr{F}_{k-1}) < \infty\right\}. \tag{A.18}$$

Notice that $\{\xi_k, \mathscr{F}_k\}$ is an mds. From (A.18) and by Theorem 1.2.8, $\sum_{i=1}^{k} \xi_i$ converges on S, or $\big\{\omega : \sum_{k=1}^{\infty} \xi_k \text{ converges}\big\} \bigcap S = S$, which combining with (A.17) implies that $\big\{\omega : \sum_{k=1}^{\infty} y_k \text{ converges}\big\} \bigcap S = S$ or $\sum_{k=1}^{\infty} y_k$ converges on S. □

Proof of Theorem 1.2.13.

By Theorem 1.2.12 it suffices to show that $A \subset S$ where the set S is defined in Theorem 1.2.12 with y_k replaced by ξ_k considered in the present theorem. To this end, we now verify that the three series defined in Theorem 1.2.12 are convergent on A with y_k replaced by ξ_k.

For some constant $c > 0$, we have that

$$\mathbb{P}\big(|\xi_k| \geq c|\mathscr{F}_{k-1}\big) \leq E\left(\frac{|\xi_k|^p}{c^p} I_{[|\xi_k| \geq c]}|\mathscr{F}_{k-1}\right) \leq E\left(\frac{|\xi_k|^p}{c^p}|\mathscr{F}_{k-1}\right), \tag{A.19}$$

and then by (1.2.21)

$$\sum_{k=1}^{\infty} \mathbb{P}\big(|\xi_k| \geq c|\mathscr{F}_{k-1}\big) < \infty. \tag{A.20}$$

We first consider the case $1 \le p \le 2$. Taking into account $E(\xi_k|\mathscr{F}_{k-1}) = 0$, we find

$$\frac{1}{c}\sum_{k=1}^{\infty}\left|E\left(\xi_k I_{[|\xi_k|\le c]}|\mathscr{F}_{k-1}\right)\right| = \frac{1}{c}\sum_{k=1}^{\infty}\left|E\left(\xi_k I_{[|\xi_k|>c]}|\mathscr{F}_{k-1}\right)\right|$$

$$\le \sum_{k=1}^{\infty} E\left(\frac{|\xi_k|}{c} I_{[|\xi_k|>c]}|\mathscr{F}_{k-1}\right) \le \frac{1}{c^p}\sum_{k=1}^{\infty} E\left(|\xi_k|^p|\mathscr{F}_{k-1}\right), \tag{A.21}$$

and then by (1.2.21)

$$\sum_{k=1}^{\infty}\left|E\left(\xi_k I_{[|\xi_k|\le c]}|\mathscr{F}_{k-1}\right)\right| < \infty. \tag{A.22}$$

Noticing $p \ge 1$, by (A.22) we have

$$\frac{1}{c^2}\sum_{k=1}^{\infty} E\left(\xi_k^2 I_{[|\xi_k|\le c]}|\mathscr{F}_{k-1}\right) \le \frac{1}{c^p}\sum_{k=1}^{\infty} E\left(|\xi_k|^p I_{[|\xi_k|\le c]}|\mathscr{F}_{k-1}\right) < \infty. \tag{A.23}$$

Combining (A.22) and (A.23) yields

$$\sum_{k=1}^{\infty}\left\{E\left(\xi_k^2 I_{[|\xi_k|\le c]}|\mathscr{F}_{k-1}\right) - \left(E\left(\xi_k I_{[|\xi_k|\le c]}|\mathscr{F}_{k-1}\right)\right)^2\right\} < \infty. \tag{A.24}$$

Then by Theorem 1.2.12, η_k converges on A for $1 \le p \le 2$.

We now consider the case $0 < p < 1$.

Define

$$\zeta_k \triangleq |\xi_k|^p - E\left(|\xi_k|^p|\mathscr{F}_{k-1}\right), \quad k \ge 1.$$

Since $\{\zeta_k, \mathscr{F}_k\}$ is an mds and on the set A, we have

$$\sum_{k=1}^{\infty} E\left(|\zeta_k||\mathscr{F}_{k-1}\right) \le \sum_{k=1}^{\infty} E\left(|\xi_k|^p + E\left(|\xi_k|^p|\mathscr{F}_{k-1}\right)|\mathscr{F}_{k-1}\right)$$

$$= 2\sum_{k=1}^{\infty} E\left(|\xi_k|^p|\mathscr{F}_{k-1}\right) < \infty.$$

Then applying the result we have just proved for $1 \le p \le 2$ to the mds $\{\zeta_k, \mathscr{F}_k\}$ leads to that $\sum_{k=1}^{\infty}\zeta_k$ converges on A, i.e.,

$$\left\{\omega : \sum_{k=1}^{\infty}\left(|\xi_k|^p - E\left(|\xi_k|^p|\mathscr{F}_{k-1}\right)\right) \text{ converges}\right\}$$

$$\supset \left\{\omega : \sum_{k=1}^{\infty} E\left(|\xi_k|^p|\mathscr{F}_{k-1}\right) \text{ converges}\right\}. \tag{A.25}$$

This is equivalent to

$$\left\{\omega : \sum_{k=1}^{\infty} |\xi_k|^p \text{ converges}\right\} \supset \left\{\omega : \sum_{k=1}^{\infty} E\big(|\xi_k|^p|\mathscr{F}_{k-1}\big) \text{ converges}\right\}. \tag{A.26}$$

Notice that convergence of $\sum_{k=1}^{\infty} |\xi_k|^p$ implies convergence of $\sum_{k=1}^{\infty} |\xi_k|$ since $0 < p < 1$ and $|\xi_k| < 1$ for all k sufficiently large. Consequently, from (A.26) it follows that

$$\left\{\omega : \sum_{k=1}^{\infty} \xi_k \text{ converges}\right\} \supset \left\{\omega : \sum_{k=1}^{\infty} E\big(|\xi_k|^p|\mathscr{F}_{k-1}\big) \text{ converges}\right\}. \tag{A.27}$$

This completes the proof. □

Proof of Lemma 1.2.4.

Set $N_0 = 0$, $N_k = \sum_{i=1}^{k} \frac{1}{b_i} M_i$, and $b_0 = 0$.

By the condition of Lemma 1.2.4, $N_k \underset{k\to\infty}{\longrightarrow} N < \infty$, i.e., for any given $\varepsilon > 0$ there is a positive integer k_ε so that $\|N_k - N\| < \varepsilon$, for $\forall\, k \geq k_\varepsilon$.

Then we have

$$\begin{aligned}
\left\|\frac{1}{b_k}\sum_{i=1}^{k} M_i\right\| &= \left\|\frac{1}{b_k}\sum_{i=1}^{k} b_i(N_i - N_{i-1})\right\| = \left\|N_k + \frac{1}{b_k}\sum_{i=2}^{k}(b_{i-1} - b_i)N_{i-1}\right\| \\
&= \left\|N_k - \frac{b_k - b_1}{b_k}N + \frac{1}{b_k}\sum_{i=2}^{k}(b_{i-1} - b_i)(N_{i-1} - N)\right\| \\
&\leq \|N_k - N\| + \frac{b_1}{b_k}\|N\| + \frac{1}{b_k}\sum_{i=2}^{k_\varepsilon}(b_i - b_{i-1})\,\|N_{i-1} - N\| + \varepsilon.
\end{aligned}$$

This indicates that $\frac{1}{b_k}\sum_{i=1}^{k} M_i$ tends to zero by first letting $k \to \infty$ and then $\varepsilon \to 0$.□

Proof of Theorem 1.2.14.

Without loss of generality, assume $M_0 \neq 0$. Under the conditions of the theorem, for any $\eta > 0$ we have the following chain of inequalities and equalities:

$$\begin{aligned}
&\sum_{k=1}^{\infty} E\left[\left\|\left[s_k(\alpha)\left(\log(s_k^\alpha(\alpha)+e)\right)^{\frac{1}{\alpha}+\eta}\right]^{-1} M_k\xi_{k+1}\right\|^\alpha \Big|\mathscr{F}_k\right] \\
&\leq \sigma\sum_{k=1}^{\infty}\left[s_k^\alpha(\alpha)\left(\log(s_k^\alpha(\alpha)+e)\right)^{1+\alpha\eta}\right]^{-1}\|M_k\|^\alpha \\
&= \sigma\sum_{k=1}^{\infty}\left[s_k^\alpha(\alpha)\left(\log(s_k^\alpha(\alpha)+e)\right)^{1+\alpha\eta}\right]^{-1}\int_{s_{k-1}^\alpha(\alpha)}^{s_k^\alpha(\alpha)} \mathrm{d}x \\
&\leq \sigma\sum_{k=1}^{\infty}\int_{s_{k-1}^\alpha(\alpha)}^{s_k^\alpha(\alpha)} \frac{1}{x(\log(x+e))^{1+\alpha\eta}}\mathrm{d}x \\
&\leq \sigma\int_{s_0^\alpha(\alpha)}^{\infty} \frac{1}{x(\log(x+e))^{1+\alpha\eta}}\mathrm{d}x < \infty.
\end{aligned}$$

By Theorem 1.2.13 it follows that

$$\sum_{k=1}^{\infty} \frac{1}{s_k(\alpha)\left(\log(s_k^{\alpha}(\alpha)+e)\right)^{\frac{1}{\alpha}+\eta}} M_k \xi_{k+1} < \infty \text{ a.s.} \tag{A.28}$$

Notice that $a_k \triangleq s_k(\alpha)\left(\log(s_k^{\alpha}(\alpha)+e)\right)^{\frac{1}{\alpha}+\eta}$ is nondecreasing as $k \to \infty$. If $a_k \to a$ for some $0 < a < \infty$ as $k \to \infty$, then the conclusion of the theorem follows from (A.28). If $a_k \to \infty$ as $k \to \infty$, then by Lemma 1.2.4 the conclusion of the theorem also follows from (A.28). □

In the sequel, we adopt the following notations,

$$E(s(x_n)|x_0 = x) \triangleq \int_{\mathbb{R}^m} s(y) P_n(x, \mathrm{d}y), \quad E_\nu(s) \triangleq \int_{\mathbb{R}^m} s(x)\nu(\mathrm{d}x),$$

where $P_n(x, \cdot)$ is the n-step transition probability of the chain $\{x_k\}_{k\geq 0}$, $s(x)$ is a measurable function on $(\mathbb{R}^m, \mathscr{B}^m)$, and $\nu(\cdot)$ is a measure on $(\mathbb{R}^m, \mathscr{B}^m)$.

Proof of Lemma 1.3.1.

(i) It suffices to prove that for any $x \in \mathbb{R}^m$ and any $A \in \mathscr{B}^m$ with $\nu(A) > 0$, there exists a positive integer $n = n(x, A)$ such that $P_n(x, A) > 0$.

Notice that $\{x_k\}_{k\geq 0}$ satisfies the minorization condition, i.e.,

$$P_{m_0}(x, A) \geq \beta s(x)\nu(A) \ \forall\, x \in \mathbb{R}^m, \ \forall\, A \in \mathscr{B}^m, \tag{A.29}$$

where $s(x)$ satisfies the following condition: $E_\mu s \triangleq \int_{\mathbb{R}^m} s(x)\mu(\mathrm{d}x) > 0$.

Define $S \triangleq \{x : s(x) > 0\}$. Then, $\int_{\mathbb{R}^m} s(x)\mu(\mathrm{d}x) = \int_S s(x)\mu(\mathrm{d}x) > 0$, and we see $\mu(S) > 0$. By the μ-irreducibility of $\{x_k\}_{k\geq 0}$, for the given $x \in \mathbb{R}^m$ there exists a positive integer m' such that

$$P_{m'}(x, S) > 0. \tag{A.30}$$

From (A.29), (A.30), and $\nu(A) > 0$, we have

$$\begin{aligned} P_{m_0+m'}(x, A) &= \int_{\mathbb{R}^m} P_{m_0}(y, A) P_{m'}(x, \mathrm{d}y) \\ &\geq \int_{\mathbb{R}^m} \beta s(y)\nu(A) P_{m'}(x, \mathrm{d}y) \\ &= \beta\nu(A) \int_S s(y) P_{m'}(x, \mathrm{d}y) > 0. \end{aligned} \tag{A.31}$$

Thus the ν-irreducibility of $\{x_k\}_{k\geq 0}$ is proved.

(ii) By the definition of C, from (A.29) we have

$$P_{m_0}(x, A) \geq \beta\gamma\nu(A) \ \forall\, x \in C \text{ and } \forall\, A \in \mathscr{B}^m. \tag{A.32}$$

Thus C is a small set. □

Proof of Lemma 1.3.2.

(i) It can be shown that for a μ-irreducible chain $\{x_k\}_{k\geq 0}$ the totality of small sets is nonempty. Assume that $C' \in \mathscr{B}^m$ with $\mu(C') > 0$ is a small set.

By the μ-irreducibility of $\{x_k\}_{k\geq 0}$, we show that there exists an integer $n > 0$ such that

$$\int_B P_n(x,C')\mu(\mathrm{d}x) > 0. \tag{A.33}$$

If (A.33) were not true, then we would have $\sum_{k=1}^{\infty} P_k(x,C') = 0$ μ-a.e. on the set B. This contradicts with the μ-irreducibility of $\{x_k\}_{k\geq 0}$ and $\mu(C') > 0$.

Define $s(x) \triangleq I_B(x)P_n(x,C')\ \forall\, x \in \mathbb{R}^m$. From (A.33) we have $E_\mu s > 0$. We now show that $s(x)$ is a small function.

Since C' is a small set, there exist an integer $n' > 0$, a constant $\beta > 0$, and a probability measure $\nu(\cdot)$ such that

$$P_{n'}(x,A) \geq \beta I_{C'}(x)\nu(A)\ \ \forall\, x \in \mathbb{R}^m\ \ \forall\, A \in \mathscr{B}^m. \tag{A.34}$$

From the definition of $s(x)$ and (A.34), we have the following chain of equalities and inequalities,

$$\begin{aligned} P_{n+n'}(x,A) &= \int_{\mathbb{R}^m} P_{n'}(y,A)P_n(x,\mathrm{d}y) \\ &\geq \int_{\mathbb{R}^m} \beta I_{C'}(y)\nu(A)P_n(x,\mathrm{d}y) \\ &= \beta P_n(x,C')\nu(A) \\ &\geq \beta s(x)\nu(A)\ \ \forall\, x \in \mathbb{R}^m\ \ \forall\, A \in \mathscr{B}^m. \end{aligned} \tag{A.35}$$

Thus $s(x)$ is a small function.

By the definition of $s(x)$, it is clear that

$$C \triangleq \{x : s(x) \geq \gamma\} \subset \{x : s(x) > 0\} \subset B\ \ \forall\, \gamma > 0.$$

Therefore, there exists a $\gamma > 0$ such that $\mu(C) > 0$. By Lemma 1.3.1 (ii), we conclude that the subset C of B is small.

(ii) Since $P_{n+m_0}(x,A) = \int_{\mathbb{R}^m} P_{m_0}(y,A)P_n(x,\mathrm{d}y)$, by (1.3.16) we have

$$\begin{aligned} P_{n+m_0}(x,A) &\geq \int_{\mathbb{R}^m} \beta s(y)\nu(A)P_n(x,\mathrm{d}y) \\ &= \beta\nu(A)E(s(x_n)|x_0 = x). \end{aligned}$$

So, $E(s(x_n)|x_0 = x)$ is a small function.

(iii) Let the minorization corresponding to the small functions $s(\cdot)$ and $s'(\cdot)$ be denoted by (m_0, β, s, ν) and (m_0', β', s', ν'), respectively. We have

$$P_{m_0}(x,A) \geq \beta s(x)\nu(A) \;\; \forall\, x \in \mathbb{R}^m \;\; \forall\, A \in \mathscr{B}^m, \tag{A.36}$$

$$P_{m_0'}(x,A) \geq \beta' s'(x)\nu'(A) \;\; \forall\, x \in \mathbb{R}^m \;\; \forall\, A \in \mathscr{B}^m. \tag{A.37}$$

Let the positive integers m_1 and m_1' be such that $m_0 + m_1 = m_0' + m_1'$. It can be shown that m_1 and m_1' can be chosen so that $\alpha \triangleq \int_{\mathbb{R}^m}\int_{\mathbb{R}^m} s(y)P_{m_1}(x,\mathrm{d}y)\nu(\mathrm{d}x) > 0$ and $\alpha' \triangleq \int_{\mathbb{R}^m}\int_{\mathbb{R}^m} s(y)P_{m_1'}(x,\mathrm{d}y)\nu'(\mathrm{d}x) > 0$.

We have the following chain of equalities and inequalities,

$$\begin{aligned} P_{m_1+2m_0}(x,A) &= \int_{\mathbb{R}^m} P_{m_1+m_0}(y,A)P_{m_0}(x,\mathrm{d}y) \\ &= \int_{\mathbb{R}^m}\int_{\mathbb{R}^m} P_{m_0}(z,A)P_{m_1}(y,\mathrm{d}z)P_{m_0}(x,\mathrm{d}y) \\ &\geq \int_{\mathbb{R}^m}\int_{\mathbb{R}^m} \beta^2 s(z)\nu(A)P_{m_1}(y,\mathrm{d}z)s(x)\nu(\mathrm{d}y) \\ &= \alpha\beta^2 s(x)\nu(A), \end{aligned} \tag{A.38}$$

and

$$\begin{aligned} P_{m_1+2m_0}(x,A) &= P_{m_1'+m_0'+m_0}(x,A) \\ &= \int_{\mathbb{R}^m} P_{m_0}(y,A)P_{m_0'+m_1'}(x,\mathrm{d}y) \\ &= \int_{\mathbb{R}^m} P_{m_0}(y,A)\int_{\mathbb{R}^m} P_{m_1'}(z,\mathrm{d}y)P_{m_0'}(x,\mathrm{d}z) \\ &\geq \int_{\mathbb{R}^m} \beta s(y)\nu(A)\int_{\mathbb{R}^m} P_{m_1'}(z,\mathrm{d}y)\beta' s'(x)\nu'(\mathrm{d}z) \\ &= \alpha'\beta\beta' s'(x)\nu(A). \end{aligned} \tag{A.39}$$

From (A.38) and (A.39) it follows that

$$P_{m_1+2m_0}(x,A) \geq \frac{1}{2}\Big(\alpha\beta^2 s(x) + \alpha'\beta\beta' s'(x)\Big)\nu(A), \tag{A.40}$$

from which, and by Definition 1.3.4, we know that $\alpha\beta^2 s(x) + \alpha'\beta\beta' s'(x)$ is a small function.

We can further find a constant $\gamma > 0$ such that $\gamma\big(\alpha\beta^2 s(x) + \alpha'\beta\beta' s'(x)\big) > s(x) + s'(x)$. Then by Definition 1.3.4 we conclude that $s(x) + s'(x)$ is also a small function. This finishes the proof. □

Proof of Theorem 1.3.1.

We first prove the theorem under the condition (i).

Assume that $\{x_k\}_{k\geq 0}$ is with period d, $d \geq 1$, i.e., there exist disjoint sets $\{E_1, \cdots, E_d\} \subset \mathscr{B}^m$ such that

$$P(x, E_{i+1}) = 1 \;\; \forall\, x \in E_i, \; i = 1, \cdots, d-1, \tag{A.41}$$

$$P(x, E_1) = 1 \;\; \forall\, x \in E_d, \tag{A.42}$$

and

$$\nu\Big\{\mathbb{R}^m/\bigcup_{i=1}^{d}E_i\Big\}=0 \text{ for some measure } \nu(\cdot). \tag{A.43}$$

We want to prove the assertion of the theorem that d must equal 1.

Step 1. We show

$$\mu(N)=0,\ N\triangleq\Big\{x:\sum_{k=1}^{\infty}P_k\Big(x,\mathbb{R}^m/\bigcup_{i=1}^{d}E_i\Big)>0\Big\}, \tag{A.44}$$

no matter $d=1$ or $d\geq 2$.

Since μ is the maximal irreducible measure, for (A.44) it suffices to prove

$$\mu\Big\{\mathbb{R}^m/\bigcup_{i=1}^{d}E_i\Big\}=0. \tag{A.45}$$

If (A.45) were not true, then by the definition of irreducibility it would follow that

$$\sum_{k=1}^{\infty}P_k\Big(x,\mathbb{R}^m/\bigcup_{i=1}^{d}E_i\Big)>0\ \forall x\in\mathbb{R}^m. \tag{A.46}$$

This implies that for any $x\in\mathbb{R}^m$ there exists some positive integer k' such that

$$P_{k'}\Big(x,\mathbb{R}^m/\bigcup_{i=1}^{d}E_i\Big)>0. \tag{A.47}$$

Taking a fixed point $x_1\in E_1$, by definition of the d-cycle we have that

$$P_k\Big(x_1,\mathbb{R}^m/\bigcup_{i=1}^{d}E_i\Big)=0\ \forall k\geq 1,$$

which, however, contradicts with (A.47). Thus (A.45) and hence (A.44) hold.

Step 2. Since C is a small set, there exist an integer $k_0>0$, a constant $b>0$, and a probability measure $\nu'(\cdot)$ such that

$$P_{k_0}(x,A)\geq b\nu'(A)\ \forall x\in C \text{ and } \forall A\in\mathscr{B}^m. \tag{A.48}$$

We now show that $P_{n+k_0}(x,A)>0$ and $P_{n+k_0+1}(x,A)>0\ \forall x\in C$ and $\forall A\in\mathscr{B}^m$, where n is given in (1.3.18) while k_0 in (A.48).

As a matter of fact, by (1.3.18) we have the following chain of equalities and inequalities,

$$\begin{aligned}P_{n+k_0}(x,A)&=\int_{\mathbb{R}^m}P_{k_0}(y,A)P_n(x,\mathrm{d}y)\\&\geq\int_{C}P_{k_0}(y,A)P_n(x,\mathrm{d}y)\\&>b\nu'(A)P_n(x,C)>0\ \forall x\in C \text{ and } \forall A\in\mathscr{B}^m.\end{aligned} \tag{A.49}$$

Similar to (A.49), we have

$$P_{n+k_0+1}(x,A) \geq b\nu'(A)P_{n+1}(x,C) > 0 \ \forall\, x \in C \text{ and } \forall\, A \in \mathscr{B}^m. \tag{A.50}$$

Step 3. We now show that $d = 1$.

Assume the converse: $d \geq 2$. Under the converse assumption, we show that at least one of (A.49) and (A.50) must be zero. The contradiction will prove that $d = 1$.

For the probability measure $\nu'(\cdot)$, we can further conclude that $\nu'\left(\mathbb{R}^m / \bigcup_{i=1}^d E_i\right) = 0$. This is because $C \subset \left\{x : P_{k_0}\left(x, \mathbb{R}^m / \bigcup_{i=1}^d E_i\right) \geq b\nu'\left(\mathbb{R}^m / \bigcup_{i=1}^d E_i\right)\right\}$ and hence $C \subset \left\{x : \sum_{l=1}^{\infty} P_l\left(x, \mathbb{R}^m / \bigcup_{i=1}^d E_i\right) \geq b\nu'\left(\mathbb{R}^m / \bigcup_{i=1}^d E_i\right)\right\}$.

If $\nu'\left(\mathbb{R}^m / \bigcup_{i=1}^d E_i\right) > 0$, then $C \subset \left\{x : \sum_{l=1}^{\infty} P_l\left(x, \mathbb{R}^m / \bigcup_{i=1}^d E_i\right) > 0\right\}$ and $\mu(C) \leq \mu\left\{x : \sum_{l=1}^{\infty} P_l\left(x, \mathbb{R}^m / \bigcup_{i=1}^d E_i\right) > 0\right\}$. By (A.44) we conclude $\mu(C) = 0$, but this is impossible since C is a small set.

Since $\nu'\left(\bigcup_{i=1}^d E_i\right) > 0$ and $\{E_1, \cdots, E_d\}$ are disjoint, there exists some $i_0 \in \{1, \cdots, d\}$ such that $\nu'\left(E_{i_0+1}\right) > 0$, where $i_0 + 1 \triangleq 1$ if $i_0 = d$.

Denote $\overline{C} \triangleq C \bigcap N^c \bigcap \left(\bigcup_{i=1}^d E_i\right)$. Since $\overline{C}$ is a subset of C, from the definition of small set, $\overline{C}$ is also a small set.

For any fixed $x \in \overline{C} \bigcap E_{i_0}$, we reconsider the $(n + k_0)$th and $(n + k_0 + 1)$th transition probabilities of $\{x_k\}_{k\geq 0}$. Since $x \in \overline{C} \subset N^c$, it follows that $\sum_{k=1}^{\infty} P_k\left(x, \mathbb{R}^m / \bigcup_{i=1}^d E_i\right) = 0$ and

$$\begin{aligned}
P_{n+k_0}(x, E_{i_0+1}) &= \int_{\mathbb{R}^m} P(y, E_{i_0+1}) P_{n+k_0-1}(x, \mathrm{d}y) \\
&= \int_{\bigcup_{i=1}^d E_i} P(y, E_{i_0+1}) P_{n+k_0-1}(x, \mathrm{d}y) \\
&= \int_{E_{i_0}} P(y, E_{i_0+1}) P_{n+k_0-1}(x, \mathrm{d}y) \\
&= P_{n+k_0-1}(x, E_{i_0}) = \cdots = P(x, E_{j_0}) \text{ for some } j_0 \in \{1, \cdots, d\}.
\end{aligned} \tag{A.51}$$

Similar to (A.51), we have

$$P_{n+k_0+1}(x, E_{i_0+1}) = P_{n+k_0}(x, E_{i_0}) = \cdots = P(x, E_{j_0-1}), \tag{A.52}$$

where $j_0 - 1 \triangleq d$ if $j_0 = 1$.

By the converse assumption we have $d \geq 2$. Then at least one of the right-hand sides of (A.51) and (A.52) is zero.

On the other hand, by noticing that $x \in \overline{C} \bigcap E_{i_0} \subset C$ and $\nu'(E_{i_0} + 1) > 0$, from

(A.49), (A.50), and the condition (1.3.18) we have

$$P_{n+k_0}(x, E_{i_0+1}) \geq b\nu'(E_{i_0+1})P_n(x,C) > 0, \tag{A.53}$$

$$P_{n+k_0+1}(x, E_{i_0+1}) \geq b\nu'(E_{i_0+1})P_{n+1}(x,C) > 0, \tag{A.54}$$

which contradict with the assertion that at least one of the probabilities $P_{n+k_0}(x, E_{i_0+1})$ and $P_{n+k_0+1}(x, E_{i_0+1})$ is zero. The contradiction shows $d = 1$ and proves the aperiodicity of the chain $\{x_k\}_{k\geq 0}$.

We now prove the theorem under (ii). For the μ-positive set A, by Lemma 1.3.2 there exists a small set $C \subset A$ with $\mu(C) > 0$. Then by (1.3.19), it follows that

$$P_n(x,C) > 0, \;\; P_{n+1}(x,C) > 0 \;\; \forall\, x \in C,$$

for some positive integer n. Then the aperiodicity of $\{x_k\}_{k\geq 0}$ follows from the result established in case (i). This finishes the proof. □

Proof of Theorem 1.3.2.

(i) By Lemma 1.3.2, $\sum_{k=0}^{l} E(s(x_k)|x_0 = x)$ is a small function. Define $\gamma \triangleq \inf_{x\in C}\sum_{k=0}^{l} E(s(x_k)|x_0 = x) > 0$ and $C' \triangleq \left\{x : \sum_{k=0}^{l} E(s(x_k)|x_0 = x) \geq \gamma\right\}$. It is clear that $C \subset C'$.

By Lemma 1.3.1, we know that C' is a small set, and then from Definition 1.3.4, the subset C of C' is also a small set.

(ii) For the set A, by Lemma 1.3.2 there exists $\overline{B} \subset A$ with $\mu(\overline{B}) > 0$ such that $\overline{B}$ is a small set. Hence $I_{\overline{B}}(x)$ is a small function, which by (1.3.21) satisfies

$$\inf_{x\in C}\sum_{k=0}^{l} P_k I_{\overline{B}}(x) = \inf_{x\in C}\sum_{k=0}^{l} P_k(x,\overline{B}) > 0$$

for some integer $l \geq 0$.

Then by the result established in case (i), C is a small set. □

Proof of Lemma 1.3.3.

Step 1. We first show that $\{x_k\}_{k\geq 0}$ defined by (1.3.32) is μ_2-irreducible.

Notice that $\{\varepsilon_k\}_{k\geq 0}$ and $\{u_k\}_{k\geq 0}$ are mutually independent with densities $f_\varepsilon(\cdot)$ and $f_u(\cdot)$, respectively. For any $x = [\xi_1 \; \xi_2]^T \in \mathbb{R}^2$ and $A = [a_1, b_1) \times [a_2, b_2)$ with $a_1 \leq b_1$, $a_2 \leq b_2$, we have

$$\begin{aligned}
P(x,A) =& \mathbb{P}\{x_1 \in A | x_0 = x\} = \mathbb{P}\{y_1 \in [a_1, b_1), u_1 \in [a_2, b_2) | y_0 = \xi_1, u_0 = \xi_2\} \\
=& \mathbb{P}\{f(y_0, u_0) + \varepsilon_1 \in [a_1, b_1), u_1 \in [a_2, b_2) | y_0 = \xi_1, u_0 = \xi_2\} \\
=& \mathbb{P}\{f(\xi_1, \xi_2) + \varepsilon_1 \in [a_1, b_1), u_1 \in [a_2, b_2)\} \\
=& \int_{a_1 - f(\xi_1,\xi_2)}^{b_1 - f(\xi_1,\xi_2)} f_\varepsilon(s_1)\mathrm{d}s_1 \int_{a_2}^{b_2} f_u(s_2)\mathrm{d}s_2 \\
=& \int_{a_1}^{b_1} f_\varepsilon(s_1 - f(\xi_1,\xi_2))\mathrm{d}s_1 \int_{a_2}^{b_2} f_u(s_2)\mathrm{d}s_2 \\
=& \iint_A f_\varepsilon(s_1 - f(\xi_1,\xi_2)) f_u(s_2)\mathrm{d}s_1\mathrm{d}s_2.
\end{aligned} \tag{A.55}$$

As a matter of fact, (A.55) is valid for any $A \in \mathscr{B}^2$:

$$P(x,A) = \iint_A f_\varepsilon\big(s_1 - f(\xi_1,\xi_2)\big) f_u(s_2) \mathrm{d}s_1 \mathrm{d}s_2 \ \ \forall\, x \in \mathbb{R}^2. \tag{A.56}$$

Since both $f_\varepsilon(\cdot)$ and $f_u(\cdot)$ are positive and continuous on $\mathbb{R}$, it follows from (A.56) that for any $A \in \mathscr{B}^2$ with $\mu_2(A) > 0$ there exists a bounded set $A' \subset A$ with $\mu_2(A') > 0$ such that

$$\begin{aligned} P(x,A) &\geq \iint_{A'} f_\varepsilon\big(s_1 - f(\xi_1,\xi_2)\big) f_u(s_2) \mathrm{d}s_1 \mathrm{d}s_2 \\ &\geq \inf_{(s_1,s_2)\in A'} \big\{ f_\varepsilon\big(s_1 - f(\xi_1,\xi_2)\big) f_u(s_2) \big\} \cdot \mu_2(A') > 0 \ \ \forall\, x \in \mathbb{R}^2. \end{aligned} \tag{A.57}$$

By Definition 1.3.2, the chain $\{x_k\}_{k\geq 0}$ is μ_2-irreducible.

Step 2. Next, we prove that $\mu_2(\cdot)$ is the maximal irreducibility measure of $\{x_k\}_{k\geq 0}$.

Let $\nu(\cdot)$ be also a measure on $(\mathbb{R}^2, \mathscr{B}^2)$ and let $\{x_k\}_{k\geq 0}$ be ν-irreducible. We show that $\nu \ll \mu_2$. Let $A \in \mathscr{B}^2$ and $\mu_2(A) = 0$. From (A.56), $P(x,A) = 0 \ \forall\, x \in \mathbb{R}^2$, and for any $n \geq 2$, we have

$$P_n(x,A) = \iint_{\mathbb{R}^2} P(y,A) P_{n-1}(x, \mathrm{d}y) = 0 \ \ \forall\, n \geq 2. \tag{A.58}$$

From (A.58), we conclude that $\nu(A) = 0$. Otherwise, by the assumption that $\{x_k\}_{k\geq 0}$ is ν-irreducible, from $\nu(A) > 0$ it would follow that

$$\sum_{n=1}^{\infty} P_n(x,A) > 0 \ \ \forall\, x \in \mathbb{R}^2, \tag{A.59}$$

which contradicts with (A.58). Therefore, $\nu \ll \mu_2$.

In (A.58) we have shown that $\sum_{n=1}^{\infty} P_n(x,A) = 0 \ \forall\, x \in \mathbb{R}^2$ if $\mu_2(A) = 0$. Hence the set $\big\{x : \sum_{n=1}^{\infty} P_n(x,A) > 0\big\}$ is null and $\mu_2\big\{x : \sum_{n=1}^{\infty} P_n(x,A) > 0\big\} = 0$. So, by Definition 1.3.2, $\mu_2(\cdot)$ is the maximal irreducibility measure of $\{x_k\}_{k\geq 0}$.

Step 3. We now show that the chain $\{x_k\}_{k\geq 0}$ is aperiodic.

For this by (A.57) and Theorem 1.3.1 it suffices to show that

$$P_2(x,A) > 0 \ \ \forall\, x \in \mathbb{R}^2, \tag{A.60}$$

whenever $\mu_2(A) > 0$.

Similar to (A.55) and (A.56), a direct calculation shows that

$$\begin{aligned} P_2(x,A) = \iint_A \Big(\iint_{\mathbb{R}^2} & f_\varepsilon\big(s_1 - f(f(\xi_1,\xi_2) + t_1, t_2)\big) f_\varepsilon\big(t_1 - f(\xi_1,\xi_2)\big) f_u(t_2) \mathrm{d}t_1 \mathrm{d}t_2 \Big) \\ & \cdot f_u(s_2) \mathrm{d}s_1 \mathrm{d}s_2 \end{aligned} \tag{A.61}$$

for any $x \in \mathbb{R}^2$ and $A \in \mathscr{B}^2$.

Since $f(\cdot,\cdot)$ is continuous on $\mathbb{R}^2$ and $f_\varepsilon(\cdot)$ and $f_u(\cdot)$ are positive and continuous on $\mathbb{R}$, for any $A \in \mathscr{B}^2$ with $\mu_2(A) > 0$ there exists a bounded subset $A' \subset A$ such that $\mu_2(A') > 0$. From (A.61) we have

$$\begin{aligned}
P_2(x,A) \geq & \iint\limits_{A'} \Big(\iint\limits_{A'} f_\varepsilon\big(s_1 - f(f(\xi_1,\xi_2)+t_1,t_2)\big) f_\varepsilon(t_1 - f(\xi_1,\xi_2)) f_u(t_2) \mathrm{d}t_1 \mathrm{d}t_2 \Big) \\
& \cdot f_u(s_2) \mathrm{d}s_1 \mathrm{d}s_2 \\
\geq & \inf_{(s_1,s_2)\in A', (t_1,t_2)\in A'} \Big\{ f_\varepsilon\big(s_1 - f(f(\xi_1,\xi_2)+t_1,t_2)\big) f_\varepsilon(t_1 - f(\xi_1,\xi_2)) \\
& \cdot f_u(t_2) f_u(s_2) \Big\} \mu_2(A') \cdot \mu_2(A') > 0.
\end{aligned}$$

So $P_2(x,A) > 0$ $\forall x \in \mathbb{R}^2$ whenever $\mu_2(A) > 0$. Hence, by Theorem 1.3.1, $\{x_k\}_{k\geq 0}$ is aperiodic.

Step 4. Finally, we prove that any bounded set $S \in \mathscr{B}^2$ with $\mu_2(S) > 0$ is a small set.

Since $f_\varepsilon(\cdot)$ and $f_u(\cdot)$ are positive and continuous on $\mathbb{R}$, from (A.56) it follows that

$$\begin{aligned}
\inf_{x\in S} P(x,B) = & \inf_{x \in S} \iint\limits_{B} f_\varepsilon(s_1 - f(x_1,x_2)) f_u(s_2) \mathrm{d}s_1 \mathrm{d}s_2 \\
\geq & \inf_{(x_1,x_2)\in S, (s_1,s_2)\in B} \Big\{ f_\varepsilon(s_1 - f(x_1,x_2)) f_u(s_2) \Big\} \mu_2(B) > 0
\end{aligned}$$

for any bounded set $S \in \mathscr{B}^2$ with $\mu_2(S) > 0$ and any $B \subset S$ with $\mu_2(B) > 0$. Therefore, by Theorem 1.3.2, the bounded set S with $\mu_2(S) > 0$ is a small set. This finishes the proof. □

Proof of Lemma 1.4.1.

(i) We first prove that if $|\xi| \leq C_1$ a.s. and $|\eta| \leq C_2$ a.s., where C_1 and C_2 are positive constants, then

$$|E\xi\eta - E\xi E\eta| \leq 4C_1C_2\alpha(n). \tag{A.62}$$

Noticing that $\xi \in \mathscr{F}_0^k$, we have that

$$\begin{aligned}
|E\xi\eta - E\xi E\eta| = & |E(\xi(E(\eta|\mathscr{F}_0^k) - E\eta))| \\
\leq & C_1 E|E(\eta|\mathscr{F}_0^k) - E\eta| \\
= & C_1 Eu(E(\eta|\mathscr{F}_0^k) - E\eta),
\end{aligned} \tag{A.63}$$

where $u \triangleq \mathrm{sign}(E(\eta|\mathscr{F}_0^k) - E\eta) \in \mathscr{F}_0^k$.

Noticing that u is measurable with respect to $\mathscr{F}_0^k$, from (A.63) we have

$$|E\xi\eta - E\xi E\eta| \leq C_1|Eu\eta - EuE\eta|. \tag{A.64}$$

Define $v \triangleq \text{sign}(E(u|\mathscr{F}_{k+n}^{\infty}) - Eu)$. Noticing that v is measurable with respect to $\mathscr{F}_{k+n}^{\infty}$, similar to the analysis in (A.63) and (A.64), we have

$$|Eu\eta - EuE\eta| \le C_2|Euv - EuEv|. \tag{A.65}$$

Further define $A_+ \triangleq \{\omega \in \Omega : u = 1\}$, $A_- \triangleq \{\omega \in \Omega : u = -1\}$, $B_+ \triangleq \{\omega \in \Omega : v = 1\}$, and $B_- \triangleq \{\omega \in \Omega : v = -1\}$.

Noticing $A_+ \in \mathscr{F}_0^k$, $A_- \in \mathscr{F}_0^k$, $B_+ \in \mathscr{F}_{k+n}^{\infty}$, $B_- \in \mathscr{F}_{k+n}^{\infty}$, by Definition 1.4.1 we have

$$\begin{aligned}
&|Euv - EuEv| \\
&\quad = |\mathbb{P}(A_+B_+) - \mathbb{P}(A_+)\mathbb{P}(B_+)| + |\mathbb{P}(A_+B_-) - \mathbb{P}(A_+)\mathbb{P}(B_-)| \\
&\quad + |\mathbb{P}(A_-B_+) - \mathbb{P}(A_-)\mathbb{P}(B_+)| + |\mathbb{P}(A_-B_-) - \mathbb{P}(A_-)\mathbb{P}(B_-)| \\
&\quad \le 4\alpha(n).
\end{aligned} \tag{A.66}$$

From (A.64)–(A.66), we know that (A.62) is true.

We now prove the inequality (1.4.8).

For any fixed $a > 0$ and $b > 0$, define $\xi' \triangleq \xi I_{[|\xi| \le a]}$, $\xi'' \triangleq \xi I_{[|\xi| > a]}$, $\eta' \triangleq \eta I_{[|\eta| \le b]}$, and $\eta'' \triangleq \eta I_{[|\eta| > b]}$. For the given constants $p > 1$ and $q > 1$, define $r \triangleq (1 - \frac{1}{p} - \frac{1}{q})^{-1}$, $s \triangleq p(1 - \frac{1}{r})$, and $t \triangleq q(1 - \frac{1}{r})$. Since $\frac{1}{p} + \frac{1}{q} + \frac{1}{r} = 1$, it follows that $1 < s < p$, $1 < t < q$, and $\frac{1}{s} + \frac{1}{t} = 1$.

For the random variables ξ and η, it holds that

$$|E\xi\eta - E\xi E\eta| \le I_1 + I_2 + I_3 + I_4, \tag{A.67}$$

where

$$I_1 \triangleq |E\xi'\eta' - E\xi' E\eta'|, \tag{A.68}$$
$$I_2 \triangleq |E\xi'\eta'' - E\xi' E\eta''|, \tag{A.69}$$
$$I_3 \triangleq |E\xi''\eta' - E\xi'' E\eta'|, \tag{A.70}$$
$$I_4 \triangleq |E\xi''\eta'' - E\xi'' E\eta''|. \tag{A.71}$$

By (A.62) and the definition of ξ' and η', it follows that

$$I_1 \le 4ab\alpha(n). \tag{A.72}$$

We now consider (A.69). By noticing that $|\eta''|/b > 1$ on the set $\{|\eta| > b\}$, we have

$$\begin{aligned}
I_2 =& |E\xi'\eta'' - E\xi' E\eta''| \le E|\xi'\eta''| + E|\xi'|E|\eta''| \\
\le& 2aE|\eta''| = 2abE\left|\frac{\eta''}{b}\right| \le 2abE\left|\frac{\eta}{b}\right|^q.
\end{aligned} \tag{A.73}$$

Similar inequality also holds for (A.70):

$$\begin{aligned} I_3 =&|E\xi''\eta' - E\xi''E\eta'| \le E|\xi''\eta'| + E|\xi''|E|\eta'| \\ \le& 2bE|\xi''| = 2abE\left|\frac{\xi''}{a}\right| \le 2abE\left|\frac{\xi}{a}\right|^p . \end{aligned} \tag{A.74}$$

We now consider (A.71). Noticing $\frac{1}{s}+\frac{1}{t}=1$, by the Hölder inequality and the Lyapunov inequality, we have

$$E|\xi''\eta''| \le (E|\xi''|^s)^{\frac{1}{s}}(E|\eta''|^t)^{\frac{1}{t}}, \tag{A.75}$$

and

$$E|\xi''|E|\eta''| \le (E|\xi''|^s)^{\frac{1}{s}}(E|\eta''|^t)^{\frac{1}{t}}. \tag{A.76}$$

From (A.75) and (A.76), it follows that

$$I_4 \le 2(E|\xi''|^s)^{\frac{1}{s}}(E|\eta''|^t)^{\frac{1}{t}}. \tag{A.77}$$

By noticing $s<p$, $t<q$, $|\xi''|>a$ on the set $\{|\xi|>a\}$ and $|\eta''|>b$ on the set $\{|\eta|>b\}$ we have

$$E|\xi''|^s = a^sE\left|\frac{\xi''}{a}\right|^s \le a^sE\left|\frac{\xi''}{a}\right|^p \le a^sE\left|\frac{\xi}{a}\right|^p, \tag{A.78}$$

and

$$E|\eta''|^t \le b^tE\left|\frac{\eta}{b}\right|^q. \tag{A.79}$$

From (A.77)–(A.79) by noticing $\frac{s}{p}=1-\frac{1}{r}$ and $\frac{t}{q}=1-\frac{1}{r}$, we have

$$\begin{aligned} I_4 \le& 2ab\left(E\left|\frac{\xi}{a}\right|^p\right)^{\frac{1}{s}}\left(E\left|\frac{\eta}{b}\right|^q\right)^{\frac{1}{t}} \\ =& 2ab\left(\frac{(E|\xi|^p)^{\frac{1}{p}}}{a}\right)^{\frac{p}{s}}\left(\frac{(E|\eta|^q)^{\frac{1}{q}}}{b}\right)^{\frac{q}{t}} \\ =& 2ab\left(\frac{(E|\xi|^p)^{\frac{1}{p}}(E|\eta|^q)^{\frac{1}{q}}}{ab}\right)^{\frac{r}{r-1}} . \end{aligned} \tag{A.80}$$

From (A.72)–(A.74) and (A.80), it follows that

$$\begin{aligned} &|E\xi\eta - E\xi E\eta| \\ \le& 2ab\left(2\alpha(n) + E\left|\frac{\xi}{a}\right|^p + E\left|\frac{\eta}{b}\right|^q + \left(\frac{(E|\xi|^p)^{\frac{1}{p}}(E|\eta|^q)^{\frac{1}{q}}}{ab}\right)^{\frac{r}{r-1}}\right). \end{aligned} \tag{A.81}$$

Choosing $a=(E|\xi|^p)^{\frac{1}{p}}(\alpha(n))^{-\frac{1}{p}}$ and $b=(E|\eta|^q)^{\frac{1}{q}}(\alpha(n))^{-\frac{1}{q}}$, from (A.81) we have

$$\begin{aligned}&|E\xi\eta-E\xi E\eta|\\&\le 2(E|\xi|^p)^{\frac{1}{p}}(E|\eta|^q)^{\frac{1}{q}}(\alpha(n))^{-\frac{1}{p}-\frac{1}{q}}\cdot 5\alpha(n)\\&=10(\alpha(n))^{1-\frac{1}{p}-\frac{1}{q}}(E|\xi|^p)^{\frac{1}{p}}(E|\eta|^q)^{\frac{1}{q}}.\end{aligned}$$

Thus (1.4.8) is proved.

(ii) We first assume that both ξ and η are simple, that is,

$$\xi=\sum_{i=1}^{s'}a_iI_{A_i},\quad \eta=\sum_{j=1}^{t'}b_jI_{B_j}, \tag{A.82}$$

where $s'<\infty$, $t'<\infty$, $A_i\cap A_k=\emptyset$ if $i\neq k$, $B_j\cap B_l=\emptyset$ if $j\neq l$, and, without losing generality, $\mathbb{P}(A_i)>0,\ i=1,\cdots,s'$ and $\mathbb{P}(B_j)>0,\ j=1,\cdots,t'$.

By the Hölder inequality, we have

$$\begin{aligned}&|E\xi\eta-E\xi E\eta|\\&=\Big|\sum_{i=1}^{s'}\sum_{j=1}^{t'}a_ib_j\mathbb{P}(A_iB_j)-\sum_{i=1}^{s'}\sum_{j=1}^{t'}a_ib_j\mathbb{P}(A_i)\mathbb{P}(B_j)\Big|\\&=\Big|\sum_{i=1}^{s'}\sum_{j=1}^{t'}a_ib_j(\mathbb{P}(A_iB_j)-\mathbb{P}(A_i)\mathbb{P}(B_j))\Big|\\&=\Big|\sum_{i=1}^{s'}a_i(\mathbb{P}(A_i))^{\frac{1}{p}}\sum_{j=1}^{t'}b_j(\mathbb{P}(B_j|A_i)-\mathbb{P}(B_j))(\mathbb{P}(A_i)^{\frac{1}{q}}\Big|\\&\le\Big(\sum_{i=1}^{s'}|a_i|^p\mathbb{P}(A_i)\Big)^{\frac{1}{p}}\Big(\sum_{i=1}^{s'}\Big|\sum_{j=1}^{t'}b_j(\mathbb{P}(B_j|A_i)-\mathbb{P}(B_j))\Big|^q\mathbb{P}(A_i)\Big)^{\frac{1}{q}}\\&\le\Big(E|\xi|^p\Big)^{\frac{1}{p}}\Big(\sum_{i=1}^{s'}\mathbb{P}(A_i)\Big(\sum_{j=1}^{t'}|b_j|^q|\mathbb{P}(B_j|A_i)-\mathbb{P}(B_j)|\Big)\Big(\sum_{j=1}^{t'}|\mathbb{P}(B_j|A_i)-\mathbb{P}(B_j)|\Big)^{\frac{q}{p}}\Big)^{\frac{1}{q}}\\&\le\max_{1\le i\le s'}\Big(\sum_{j=1}^{t'}|\mathbb{P}(B_j|A_i)-\mathbb{P}(B_j)|\Big)^{\frac{1}{p}}\Big(E|\xi|^p\Big)^{\frac{1}{p}}\\&\quad\cdot\Big(\sum_{i=1}^{s'}\mathbb{P}(A_i)\sum_{j=1}^{t'}|b_j|^q\Big(\mathbb{P}(B_j|A_i)+\mathbb{P}(B_j)\Big)\Big)^{\frac{1}{q}}\\&\le\max_{1\le i\le s'}\Big(\sum_{j=1}^{t'}|\mathbb{P}(B_j|A_i)-\mathbb{P}(B_j)|\Big)^{\frac{1}{p}}(E|\xi|^p)^{\frac{1}{p}}(2E|\eta|^q)^{\frac{1}{q}}.\end{aligned}\tag{A.83}$$

For each $i = 1, \cdots, s'$, define $J_{i,+} \triangleq \{j = 1, \cdots, t' : \mathbb{P}(B_j|A_i) - \mathbb{P}(B_j) \geq 0\}$, $J_{i,-} \triangleq \{j = 1, \cdots, t' : \mathbb{P}(B_j|A_i) - \mathbb{P}(B_j) < 0\}$. Since $\{B_j\}_{j=1}^{t'}$ are disjoint, by the definition of ϕ-mixing we have

$$\begin{aligned}
&\sum_{j=1}^{t'} |\mathbb{P}(B_j|A_i) - \mathbb{P}(B_j)| \\
&= \sum_{j \in J_{i,+}} (\mathbb{P}(B_j|A_i) - \mathbb{P}(B_j)) - \sum_{j \in J_{i,-}} (\mathbb{P}(B_j|A_i) - \mathbb{P}(B_j)) \\
&= \Big(\mathbb{P}\Big(\bigcup_{j \in J_{i,+}} B_j|A_i\Big) - \mathbb{P}\Big(\bigcup_{j \in J_{i,+}} B_j\Big)\Big) - \Big(\mathbb{P}\Big(\bigcup_{j \in J_{i,-}} B_j|A_i\Big) - \mathbb{P}\Big(\bigcup_{j \in J_{i,-}} B_j\Big)\Big) \\
&\leq 2\phi(n).
\end{aligned} \tag{A.84}$$

From (A.83) and (A.84), we know that (1.4.9) holds for simple variables.

We now consider the general case. For ξ, η, and any $N \geq 1$, define

$$\xi_N \triangleq \begin{cases} 0, & \text{if } \xi > N \\ \frac{k}{N}, & \text{if } \frac{k}{N} < \xi \leq \frac{k+1}{N},\ k = 0, \cdots, N-1 \\ -\frac{k}{N}, & \text{if } -\frac{k+1}{N} < \xi \leq -\frac{k}{N},\ k = 0, \cdots, N-1 \\ 0, & \text{if } \xi \leq -N \end{cases}$$

$$\eta_N \triangleq \begin{cases} 0, & \text{if } \eta > N \\ \frac{k}{N}, & \text{if } \frac{k}{N} < \eta \leq \frac{k+1}{N},\ k = 0, \cdots, N-1 \\ -\frac{k}{N}, & \text{if } -\frac{k+1}{N} < \eta \leq -\frac{k}{N},\ k = 0, \cdots, N-1 \\ 0, & \text{if } \eta \leq -N. \end{cases}$$

It is clear that both ξ_N and η_N are simple. Thus the inequality (1.4.9) holds for ξ_N and η_N. It is clear that $|\xi_N| \leq |\xi|$, $|\eta_N| \leq |\eta|$, and $\xi_N \underset{N\to\infty}{\longrightarrow} \xi$ a.s., $\eta_N \underset{N\to\infty}{\longrightarrow} \eta$ a.s. Then by Theorem 1.1.8, it follows that the inequality (1.4.9) holds for ξ and η if $E|\xi|^p < \infty$, $E|\eta|^q < \infty$, and $\frac{1}{p} + \frac{1}{q} = 1$. This completes the proof. □

Proof of Theorem 1.4.1.

The proof can be divided into two parts. First, we derive a maximal inequality for the sum of simple mixingales. Then we establish the almost sure convergence of the sum of mixingales based on the Chebyshev inequality and Theorem 1.1.5.

Set $S_n \triangleq \sum_{k=1}^n \varphi_k$. It is clear that $\varphi_k = \sum_{i=1}^k \big(E[\varphi_k|\mathscr{F}_{k-i+1}] - E[\varphi_k|\mathscr{F}_{k-i}]\big)$ and for any fixed positive sequence $\{a_k\}_{k\geq 0}$

$$\begin{aligned}
S_n &= \sum_{k=1}^n \varphi_k = \sum_{k=1}^n \sum_{i=1}^k \big(E[\varphi_k|\mathscr{F}_{k-i+1}] - E[\varphi_k|\mathscr{F}_{k-i}]\big) \\
&= \sum_{i=1}^n \sum_{k=i}^n \big(E[\varphi_k|\mathscr{F}_{k-i+1}] - E[\varphi_k|\mathscr{F}_{k-i}]\big)
\end{aligned}$$

$$=\sum_{i=1}^{n} a_i^{\frac{1}{2}} a_i^{-\frac{1}{2}} \sum_{k=i}^{n} \left(E[\varphi_k|\mathscr{F}_{k-i+1}]-E[\varphi_k|\mathscr{F}_{k-i}]\right). \tag{A.85}$$

Then by the Hölder inequality (1.1.18), we have

$$S_n^2 \le \sum_{j=1}^{n} a_j \cdot \sum_{i=1}^{n} a_i^{-1} \Big(\sum_{k=i}^{n} \left(E[\varphi_k|\mathscr{F}_{k-i+1}]-E[\varphi_k|\mathscr{F}_{k-i}]\right)\Big)^2, \tag{A.86}$$

and

$$\max_{1\le n\le m} S_n^2 \le \sum_{j=1}^{m} a_j \cdot \sum_{i=1}^{m} a_i^{-1} \max_{1\le n\le m} \Big(\sum_{k=i}^{n} \left(E[\varphi_k|\mathscr{F}_{k-i+1}]-E[\varphi_k|\mathscr{F}_{k-i}]\right)\Big)^2. \tag{A.87}$$

For each i, denote $Y_{i,n} \triangleq \sum_{k=i}^{n} \left(E[\varphi_k|\mathscr{F}_{k-i+1}]-E[\varphi_k|\mathscr{F}_{k-i}]\right)$. Noticing that $\mathscr{F}_k = \{\emptyset,\Omega\}$ if $k \le 0$, we have that $Y_{i,n} = \sum_{k=1}^{n} \left(E[\varphi_k|\mathscr{F}_{k-i+1}]-E[\varphi_k|\mathscr{F}_{k-i}]\right)$. It is clear that $\{E[\varphi_k|\mathscr{F}_{k-i+1}]-E[\varphi_k|\mathscr{F}_{k-i}],\mathscr{F}_{k-i+1}\}_{k=i}^{n}$ is an mds, $\{Y_{i,n}\}_{n\ge i}$ is a martingale, and by the Jensen inequality, $\{|Y_{i,n}|\}_{n\ge i}$ is a nonnegative submartingale. Then by Theorem 1.2.5 for any positive sequence $\{a_k\}_{k\ge 1}$ we have

$$\begin{aligned}
&E\Big[\max_{1\le n\le m} S_n^2\Big] \\
&\le \sum_{j=1}^{m} a_j \cdot \sum_{i=1}^{m} a_i^{-1} E\Big[\max_{1\le n\le m}\Big(\sum_{k=i}^{n}\left(E[\varphi_k|\mathscr{F}_{k-i+1}]-E[\varphi_k|\mathscr{F}_{k-i}]\right)\Big)^2\Big] \\
&\le 4\sum_{j=1}^{m} a_j \cdot \sum_{i=1}^{m} a_i^{-1} E\Big(\sum_{k=i}^{m}\left(E[\varphi_k|\mathscr{F}_{k-i+1}]-E[\varphi_k|\mathscr{F}_{k-i}]\right)\Big)^2 \\
&= 4\sum_{j=1}^{m} a_j \cdot \sum_{i=1}^{m} a_i^{-1} \sum_{k=i}^{m}\Big[E\left(E[\varphi_k|\mathscr{F}_{k-i+1}]\right)^2 - E\left(E[\varphi_k|\mathscr{F}_{k-i}]\right)^2\Big] \\
&= 4\sum_{j=1}^{m} a_j \cdot \sum_{k=1}^{m}\sum_{i=1}^{k} a_i^{-1} \Big[E\left(E[\varphi_k|\mathscr{F}_{k-i+1}]\right)^2 - E\left(E[\varphi_k|\mathscr{F}_{k-i}]\right)^2\Big] \\
&= 4\sum_{j=1}^{m} a_j \cdot \sum_{k=1}^{m}\Big(\frac{1}{a_1}E\varphi_k^2 + \sum_{i=2}^{k} a_i^{-1} E\left(E[\varphi_k|\mathscr{F}_{k-i+1}]\right)^2 - \sum_{i=1}^{k-1} a_i^{-1} E\left(E[\varphi_k|\mathscr{F}_{k-i}]\right)^2\Big) \\
&= 4\sum_{j=1}^{m} a_j \cdot \sum_{k=1}^{m}\Big(\frac{1}{a_1}E\varphi_k^2 + \sum_{i=1}^{k-1}\left(a_{i+1}^{-1}-a_i^{-1}\right) E\left(E[\varphi_k|\mathscr{F}_{k-i}]\right)^2\Big).
\end{aligned} \tag{A.88}$$

Take $a_k = \frac{1}{k\log k(\log\log k)^{1+\gamma}}$ with $\gamma > 0$ given in (1.4.11). Noticing that $\sum_{k=1}^{\infty} a_k < \infty$ and $\frac{\frac{1}{a_{k+1}}-\frac{1}{a_k}}{\log k(\log\log k)^{1+\gamma}} \underset{k\to\infty}{\longrightarrow} 1$, by the definition of simple mixingales, (1.4.10) and (1.4.11), we have

$$E\Big[\max_{1\le n\le m} S_n^2\Big] \le C\sum_{k=1}^{\infty}\Big(c_k^2\psi_0^2 + \sum_{i=1}^{k}\log i(\log\log i)^{1+\gamma} c_k^2 \psi_i^2\Big) < \infty \tag{A.89}$$

where $C > 0$ is a constant.

For any fixed $m \geq 1$, set $\{\widetilde{\varphi}_k \triangleq \varphi_{k+m}, \widetilde{\mathscr{F}}_k \triangleq \mathscr{F}_{k+m}\}_{k\geq 0}$, where $\widetilde{\mathscr{F}}_k \triangleq \{\emptyset, \Omega\}$ for $k < m$. Then $S_{j+m} - S_m = \sum_{k=m+1}^{m+j} \varphi_k = \sum_{k=1}^{j} \widetilde{\varphi}_k$ and $\{\widetilde{\varphi}_k, \widetilde{\mathscr{F}}_k\}_{k\geq 0}$ is also a simple mixingale with

$$E\Big(E[\widetilde{\varphi}_k|\widetilde{\mathscr{F}}_{k-l}]\Big)^2 \leq c_{k+m}^2 \psi_l^2. \tag{A.90}$$

Applying (A.89) to $\{\widetilde{\varphi}_k, \widetilde{\mathscr{F}}_k\}_{k\geq 0}$ and noticing (A.90), for any fixed $\varepsilon > 0$ we have

$$\begin{aligned}
&\mathbb{P}\Big\{\sup_{j\geq 1}|S_{j+m} - S_m| > \varepsilon\Big\} \\
&= \lim_{n\to\infty} \mathbb{P}\Big\{\max_{1\leq j\leq n}|S_{j+m} - S_m| > \varepsilon\Big\} \\
&\leq \limsup_{n\to\infty} \frac{1}{\varepsilon^2} E\Big[\max_{1\leq j\leq n}|S_{j+m} - S_m|^2\Big] \\
&\leq \frac{1}{\varepsilon^2} C\Big(\sum_{j=m+1}^{\infty} c_j^2 \psi_0^2 + \sum_{i=1}^{\infty} \log i (\log\log i)^{1+\gamma} \psi_i^2 \Big(\sum_{k=m+i}^{\infty} c_k^2\Big)\Big),
\end{aligned} \tag{A.91}$$

where the first inequality takes place by the Chebyshev inequality. By (1.4.10), (1.4.11), and Theorem 1.1.8, we know that the right-hand side of (A.91) tends to zero as $m \to \infty$. Then, by Theorem 1.1.5 we conclude that S_n converges almost surely. □

Proof of Theorem 1.4.2.

Define $\mathscr{F}_k \triangleq \sigma\{\varphi_i,\ 0 \leq i \leq k\},\ k \geq 0$. By inequality (1.4.8) for α-mixing, it follows that for any given $\varepsilon > 0$

$$\begin{aligned}
&\mathrm{Cov}\big(\Phi_k(\varphi_k), E(\Phi_k(\varphi_k)|\mathscr{F}_{k-m})\big) \\
&\quad \leq (\alpha(m))^{\frac{\varepsilon}{2(2+\varepsilon)}} \big(E|\Phi_k(\varphi_k)|^{2+\varepsilon}\big)^{\frac{1}{2+\varepsilon}} \big(E|E(\Phi_k(\varphi_k)|\mathscr{F}_{k-m})|^2\big)^{\frac{1}{2}}.
\end{aligned} \tag{A.92}$$

Since $E\Phi_k(\varphi_k) = 0$, we have

$$\begin{aligned}
&\mathrm{Cov}\big(\Phi_k(\varphi_k), E(\Phi_k(\varphi_k)|\mathscr{F}_{k-m})\big) \\
&\quad = E\Big[\Phi_k(\varphi_k) \cdot E(\Phi_k(\varphi_k)|\mathscr{F}_{k-m})\Big] = E\Big[E(\Phi_k(\varphi_k)|\mathscr{F}_{k-m})\Big]^2.
\end{aligned} \tag{A.93}$$

Combining (A.92) and (A.93) yields

$$\Big(E\Big[E(\Phi_k(\varphi_k)|\mathscr{F}_{k-m})\Big]^2\Big)^{\frac{1}{2}} \leq (\alpha(m))^{\frac{\varepsilon}{2(2+\varepsilon)}} \big(E|\Phi_k(\varphi_k)|^{2+\varepsilon}\big)^{\frac{1}{2+\varepsilon}} \triangleq \psi_m c_k, \tag{A.94}$$

where $\psi_m \triangleq (\alpha(m))^{\frac{\varepsilon}{2(2+\varepsilon)}}$ and $c_k \triangleq \big(E|\Phi_k(\varphi_k)|^{2+\varepsilon}\big)^{\frac{1}{2+\varepsilon}}$.

By (1.4.13) and (1.4.14), we have

$$\sum_{k=1}^{\infty} c_k^2 < \infty \tag{A.95}$$

and

$$\sum_{k=1}^{\infty} \log k (\log\log k)^{1+\gamma} \psi_k^2 < \infty. \tag{A.96}$$

Finally, by (A.95), (A.96), and Theorem 1.4.1, we derive (1.4.15). □

Proof of Theorem 1.4.3.

By the Markov property of $\{x_k\}_{k\geq 0}$, we have that

$$\begin{aligned}
&\sup_n E\left[\sup_{B\in\mathscr{F}_{n+k}^{\infty}} |\mathbb{P}(B|\mathscr{F}_0^n) - \mathbb{P}(B)|\right] \\
&= \sup_n E\left[\sup_{B\in\mathscr{F}_{n+k}^{\infty}} |\mathbb{P}(B|x_n) - \mathbb{P}(B)|\right] \\
&= \sup_n \int_{\mathbb{R}^m} \sup_{B\in\mathscr{F}_{n+k}^{\infty}} |\mathbb{P}(B|x_n = x) - \mathbb{P}(B)| P_n(\mathrm{d}x).
\end{aligned} \tag{A.97}$$

For any fixed $s \geq 1$ and any Borel sets $C_0, C_1, \cdots, C_s$ in $\mathbb{R}^m$, define $B \triangleq \{\omega : x_{n+k} \in C_0, x_{n+k+1} \in C_1, \cdots, x_{n+k+s} \in C_s\} \in \mathscr{F}_{n+k}^{\infty}$. We have

$$\begin{aligned}
&|\mathbb{P}\{B|x_n = x\} - \mathbb{P}\{B\}| \\
&= \left| \int_{C_0} P_k(x, \mathrm{d}y_0) \int_{C_1} P(y_0, \mathrm{d}y_1) \cdots \int_{C_s} P(y_{s-1}, \mathrm{d}y_s) \right. \\
&\quad \left. - \int_{C_0} P_{n+k}(\mathrm{d}y_0) \int_{C_1} P(y_0, \mathrm{d}y_1) \cdots \int_{C_s} P(y_{s-1}, \mathrm{d}y_s) \right| \\
&\leq \Big(P_k(x, C_0) - P_{n+k}(C_0)\Big)_+ + \Big(P_k(x, C_0) - P_{n+k}(C_0)\Big)_-.
\end{aligned} \tag{A.98}$$

From (A.98) and by the definition of the total variation norm (1.1.4), it follows that

$$\sup_{B\in\mathscr{F}_{n+k}^{\infty}} |\mathbb{P}(B|x_n = x) - \mathbb{P}(B)| \leq \|P_k(x, \cdot) - P_{n+k}(\cdot)\|_{\mathrm{var}}, \tag{A.99}$$

which combining with (A.97) leads to (1.4.17). □

Proof of Lemma 1.5.1.

Suppose that a_k is the first m-positive term in $\{a_1, \cdots, a_{n+m}\}$ such that

$$a_k + \cdots + a_l > 0, \tag{A.100}$$

and

$$a_k + \cdots + a_p \leq 0 \ \forall\, p : k \leq p \leq l-1 \tag{A.101}$$

for some $l : m \leq l \leq \min(n+m, k+m-1)$.

If there exists some $h: k \leq h \leq l$ such that a_h is not m-positive, then $a_h + \cdots + a_l \leq 0$ and hence

$$a_k + \cdots + a_{h-1} > 0, \tag{A.102}$$

which contradicts with (A.101). Thus, all the terms $a_k, \cdots, a_l$ are m-positive and $a_k + \cdots + a_l > 0$.

Then considering the first m-positive term in $\{a_{l+1}, \cdots, a_{n+m}\}$ and continuing the procedure, we conclude that the successive m-positive terms form disjoint stretches of positive terms. This completes the proof. □

Proof of Lemma 1.5.2.

Let $k = 1, 2, \cdots, n+m$. Define $B^{mk} \triangleq \{\omega : X_k - bY_k \text{ is } m\text{-positive}\}$. Then we have

$$\begin{aligned} B^{mk} &= \Big\{\omega : \sup_{k \leq l \leq \min(n+m, k+m-1)} \Big\{(X_k - bY_k) + \cdots + (X_l - bY_l)\Big\} > 0\Big\} \\ &= \Big\{\omega : \sup_{k \leq l \leq \min(n+m, k+m-1)} \Big\{(X_k + \cdots + X_l) - b(Y_k + \cdots + Y_l)\Big\} > 0\Big\} \\ &= \Big\{\omega : \sup_{k \leq l \leq \min(n+m, k+m-1)} \frac{X_k + \cdots + X_l}{Y_k + \cdots + Y_l} > b\Big\}, \end{aligned} \tag{A.103}$$

and by Lemma 1.5.1,

$$\sum_{k=1}^{n+m} (X_k - bY_k) I_{B^{mk}} \geq 0. \tag{A.104}$$

If $k \leq n$, then $k \leq l \leq k+m-1$. By the definition of B^m given by (1.5.11), we have $B^{mk} = B^m_k$.

For any $Z^n > 0$ and any set C, it follows that

$$\begin{aligned} 0 &\leq \sum_{k=1}^{n+m} \left(\frac{X_k}{Z^n} - b\frac{Y_k}{Z^n}\right) I_{B^{mk}C} \\ &= \sum_{k=1}^{n} \left(\frac{X_k}{Z^n} - b\frac{Y_k}{Z^n}\right) I_{B^{mk}C} + \sum_{k=n+1}^{n+m} \left(\frac{X_k}{Z^n} - b\frac{Y_k}{Z^n}\right) I_{B^{mk}C} \\ &= \sum_{k=1}^{n} \left(\frac{X_k}{Z^n} - b\frac{Y_k}{Z^n}\right) I_{B^m_k C} + \sum_{k=n+1}^{n+m} \left(\frac{X_k}{Z^n} - b\frac{Y_k}{Z^n}\right) I_{B^{mk}C} \\ &\leq \sum_{k=1}^{n} \left(\frac{X_k}{Z^n} - b\frac{Y_k}{Z^n}\right) I_{B^m_k C} + \sum_{k=n+1}^{n+m} \left(\frac{X_k}{Z^n} - b\frac{Y_k}{Z^n}\right)_+ I_C, \end{aligned} \tag{A.105}$$

which yields (1.5.12) by taking mathematical expectation. This completes the proof. □

Proof of Lemma 1.5.3.

By the integral stationarity of $\{X_n\}_{n\geq 1}$ and $\{Y_n\}_{n\geq 1}$, we have

$$\int_{A_k}(X_k - bY_k)\mathrm{d}\mathbb{P} = \int_{A_1}(X_1 - bY_1)\mathrm{d}\mathbb{P}. \tag{A.106}$$

Since C is invariant, it follows that $(B^m C)_k = B_k^m C$. By Lemma 1.5.2, we have

$$\begin{aligned}
0 \leq & \sum_{k=1}^{n}\int_{B_k^m C}\left(\frac{X_k}{n} - b\frac{Y_k}{n}\right)\mathrm{d}\mathbb{P} + \sum_{k=n+1}^{n+m}\int_C\left(\frac{X_k}{n} - b\frac{Y_k}{n}\right)_+\mathrm{d}\mathbb{P}\\
= & \sum_{k=1}^{n}\int_{(B^m C)_k}\left(\frac{X_k}{n} - b\frac{Y_k}{n}\right)\mathrm{d}\mathbb{P} + \sum_{k=n+1}^{n+m}\int_C\left(\frac{X_k}{n} - b\frac{Y_k}{n}\right)_+\mathrm{d}\mathbb{P}\\
= & n\int_{B^m C}\left(\frac{X_1}{n} - b\frac{Y_1}{n}\right)\mathrm{d}\mathbb{P} + m\int_C\left(\frac{X_1}{n} - b\frac{Y_1}{n}\right)_+\mathrm{d}\mathbb{P},
\end{aligned} \tag{A.107}$$

which implies

$$\int_{B^m C}(X_1 - bY_1)\,\mathrm{d}\mathbb{P} + \frac{m}{n}\int_C (X_1 - bY_1)_+\,\mathrm{d}\mathbb{P} \geq 0. \tag{A.108}$$

Letting $n \to \infty$ in (A.108) yields

$$\int_{B^m C}(X_1 - bY_1)\,\mathrm{d}\mathbb{P} \geq 0. \tag{A.109}$$

Recalling the definitions of B^m and $\overline{C}_b$ given by (1.5.11) and (1.5.9), respectively, we find that

$$B^m = \left\{\omega : \sup_{j\leq m}\frac{X^j}{Y^j} > b\right\} \uparrow \overline{B}_b, \quad \text{as } m \to \infty \text{ and } \overline{C}_b \subset \overline{B}_b.$$

By choosing the invariant event $C\overline{C}_b$ to replace C in (A.109), we have

$$\int_{B^m C\overline{C}_b}(X_1 - bY_1)\,\mathrm{d}\mathbb{P} \geq 0. \tag{A.110}$$

Letting $m \to \infty$ in (A.110), by Theorem 1.1.8 and the fact $\overline{C}_b \subset \overline{B}_b$, we have

$$\int_{C\overline{C}_b}(X_1 - bY_1)\,\mathrm{d}\mathbb{P} \geq 0. \tag{A.111}$$

Carrying out the discussion similar to that given above, we find that

$$\int_{C\underline{C}_a}(aY_1 - X_1)\,\mathrm{d}\mathbb{P} \geq 0. \tag{A.112}$$

This completes the proof. □

Proof of Theorem 1.5.1.

Define $\{Y_n = 1\}_{n\geq 1}$ and

$$C = \left\{\omega : \liminf_{n\to\infty} \frac{X^n}{Y^n} < \limsup_{n\to\infty} \frac{X^n}{Y^n}\right\}.$$

It is clear that $Y^n = n$ and

$$C = \bigcup_{a,b\in Q,\ a<b} C_{a,b} = \bigcup_{a,b\in Q,\ a<b} \left\{\omega : \liminf_{n\to\infty} \frac{X^n}{Y^n} < a < b < \limsup_{n\to\infty} \frac{X^n}{Y^n}\right\},$$

where Q is the set of rational numbers. Thus, in order to prove the convergence of $\frac{X^n}{Y^n}$, it suffices to prove $\mathbb{P}\{C_{a,b}\} = 0$.

Notice that $C_{a,b}$ is an invariant set. By Lemma 1.5.3, we have

$$\int_{C_{a,b}} (X_1 - bY_1)\mathrm{d}\mathbb{P} \geq 0, \tag{A.113}$$

$$\int_{C_{a,b}} (aY_1 - X_1)\mathrm{d}\mathbb{P} \geq 0, \tag{A.114}$$

which imply

$$(a-b)\mathbb{P}\{C_{a,b}\} \geq 0, \tag{A.115}$$

and $\mathbb{P}\{C_{a,b}\} = 0$. Thus the almost sure convergence of $\frac{X^n}{n}$ is proved.

The invariance of U is ensured by Definition 1.5.1. We now proceed to prove $U = E(X_1|\mathscr{C})$ a.s.

Noticing that the sets $\{\omega : U = \infty\}$, $\{\omega : U = -\infty\}$, and $\overline{C}_b = \{\omega : U > b\}$ are invariant and $\{\omega : U = \infty\} \subset \overline{C}_b$, by Lemma 1.5.3 for the given constant $b > 0$ we have

$$\int_{\overline{C}_b} (X_1 - bY_1)\mathrm{d}\mathbb{P} \geq 0, \tag{A.116}$$

and

$$\frac{1}{b}\int_{\overline{C}_b} X_1\mathrm{d}\mathbb{P} \geq \mathbb{P}\{\overline{C}_b\} \geq \mathbb{P}\{\omega : U = \infty\}. \tag{A.117}$$

Since $E|X_1| < \infty$, letting $b \to \infty$ in (A.117) leads to $\mathbb{P}\{\omega : U = \infty\} = 0$. By a similar discussion, we conclude $\mathbb{P}\{\omega : U = -\infty\} = 0$. Thus U is a.s. finite.

For any fixed $\varepsilon > 0$, define $C^m = \{\omega : (m-1)\varepsilon \leq U < m\varepsilon\}$. Then $\{C^m\}_{m=-\infty}^{\infty}$ is a sequence of disjoint sets with $\mathbb{P}\{\sum_{m=-\infty}^{\infty} C^m\} = 1$.

Setting $a \triangleq m\varepsilon$, $b \triangleq (m-1)\varepsilon - \frac{1}{n}$, $\underline{C}_a \triangleq \{\omega : U < m\varepsilon\}$, and $\overline{C}_b \triangleq \{\omega : U > (m-$

$1)\varepsilon - \frac{1}{n}\}$, for any invariant set C we have $CC^m \cap \underline{C}_a = CC^m$ and $CC^m \cap \overline{C}_b = CC^m$. Again by Lemma 1.5.3, we have

$$\int_{CC^m\underline{C}_a} X_1 \mathrm{d}\mathbb{P} = \int_{CC^m} X_1 \mathrm{d}\mathbb{P} \le m\varepsilon \mathbb{P}\{CC^m\}, \tag{A.118}$$

$$\int_{CC^m\overline{C}_b} X_1 \mathrm{d}\mathbb{P} = \int_{CC^m} X_1 \mathrm{d}\mathbb{P} \ge \Big((m-1)\varepsilon - \frac{1}{n}\Big)\mathbb{P}\{CC^m\}. \tag{A.119}$$

Tending $n \to \infty$ in (A.119) we derive

$$\int_{CC^m} X_1 \mathrm{d}\mathbb{P} \ge (m-1)\varepsilon \mathbb{P}\{CC^m\}. \tag{A.120}$$

By (A.118), (A.120), and the definition of C^m, we have

$$\int_{CC^m} U \mathrm{d}\mathbb{P} \le m\varepsilon \mathbb{P}\{CC^m\} \le \varepsilon \mathbb{P}\{CC^m\} + \int_{CC^m} X_1 \mathrm{d}\mathbb{P}, \tag{A.121}$$

$$\int_{CC^m} U \mathrm{d}\mathbb{P} \ge (m-1)\varepsilon \mathbb{P}\{CC^m\} \ge \int_{CC^m} X_1 \mathrm{d}\mathbb{P} - \varepsilon \mathbb{P}\{CC^m\}. \tag{A.122}$$

Letting $\varepsilon \to 0$ in (A.121) and (A.122), we obtain

$$\int_{CC^m} U \mathrm{d}\mathbb{P} = \int_{CC^m} X_1 \mathrm{d}\mathbb{P}, \tag{A.123}$$

and

$$\int_{C} U \mathrm{d}\mathbb{P} = \int_{C} X_1 \mathrm{d}\mathbb{P}. \tag{A.124}$$

Since C is an arbitrary invariant set of $\{X_n\}_{n\ge 1}$, by Theorem 1.1.9 we have $U = E(X_1|\mathscr{C})$ a.s. This completes the proof. □

Appendix B: Nonnegative Matrices

We say matrices and vectors are *nonnegative* (*positive*) if all their elements are nonnegative (positive). For two nonnegative matrices M and N, we write $M \geq N$ ($M > N$) if $M - N$ is nonnegative (positive).

It is worth noting that the positiveness of a matrix M should not be confused with the positive definiteness of M though the same notation "$M > 0$" is used.

Denote $\mathbf{1} \triangleq [1 \cdots 1]^T \in \mathbb{R}^n$ and set $M^k \triangleq [m_{ij}^{(k)}]_{i,j=1}^n$ for a square matrix $M = [m_{ij}]_{i,j=1}^n$.

Definition B.1 A nonnegative square matrix $M = [m_{ij}]_{i,j=1}^n \in \mathbb{R}^{n \times n}$ is called *primitive* if there exists some integer $k > 0$ such that M^k is positive while it is called *irreducible* if for any pair (i, j) of the index set, there exists an integer $k = k(i, j)$ such that $m_{ij}^{(k)} > 0$.

From the definition it is clear that a positive matrix is primitive.

Example Let us consider the matrix $P = \begin{bmatrix} 0 & 1 \\ 1 & 0 \end{bmatrix}$. A direct calculation shows that $P^{2k+1} = \begin{bmatrix} 0 & 1 \\ 1 & 0 \end{bmatrix}$ and $P^{2k} = \begin{bmatrix} 1 & 0 \\ 0 & 1 \end{bmatrix}$, $k \geq 0$. Thus the matrix P is irreducible but not primitive.

Lemma B.1 *If the nonnegative matrix $M \in \mathbb{R}^{n \times n}$ is irreducible, then $I + M$ is primitive.*

Proof. We have to show that there exists an integer $k > 0$ such that $(I + M)^k > 0$.

We first notice that

$$(I+M)^s = \sum_{l=0}^{s} C_s^l M^l, \tag{B.1}$$

where C_s^l is the combinatory number, and the (i,j)-element of $(I+M)^s$ is

$$(I+M)^s_{(i,j)} = \sum_{l=0}^{s} C_s^l m_{ij}^{(l)} \tag{B.2}$$

for any $s \geq 1$.

By irreducibility of M, for each (i,j) there exists an integer $k = k(i,j) > 0$ such that $m_{ij}^{(k)} > 0$.

Define $K \triangleq \max_{1 \leq i,j \leq n} k(i,j)$. For K and the fixed index (i,j), it follows that

$$(I+M)^K_{(i,j)} = \sum_{l=0}^{K} C_K^l m_{ij}^{(l)} \geq C_K^{k(i,j)} m_{ij}^{(k(i,j))} > 0. \tag{B.3}$$

Then from (B.1)–(B.3) it follows that $(I+M)$ is primitive. □

Theorem B.1 *(Perron–Frobenius) Suppose $M = [m_{ij}]_{i,j=1}^n \in \mathbb{R}^{n\times n}$ is a nonnegative matrix. If M is primitive, then there exists an eigenvalue λ_1 such that*

(i) λ_1 is real and positive,

(ii) $\lambda_1 > |\lambda|$ for any other eigenvalue $\lambda \neq \lambda_1$,

(iii) λ_1 can be with positive left and right eigenvectors,

(iv) the dimension of both the left and right eigenvector spaces corresponding to λ_1 is one,

(v) the multiplicity of λ_1 is 1.

If M is irreducible, then all the above assertions but (ii) hold. The assertion (ii) should change to

(ii') $\lambda_1 \geq |\lambda|$ for any other eigenvalue $\lambda \neq \lambda_1$.

Proof. (i) We first consider the case where $M = [m_{ij}]_{i,j=1}^n$ is primitive. Let us consider a nonnegative vector $x = [x_1 \cdots x_n]^T \neq 0$ and define

$$\lambda(x) \triangleq \min_{1 \leq j \leq n} \frac{\sum_{i=1}^{n} x_i m_{ij}}{x_j}, \tag{B.4}$$

where $\sum_{i=1}^n x_i m_{ij}/x_j \triangleq \infty$ if $x_j = 0$.

From the definition of $\lambda(x)$, we have

$$x_j\lambda(x) \leq \sum_{i=1}^{n} x_i m_{ij}, \ j = 1, \cdots, n,$$
$$x^T\lambda(x) \leq x^T M,$$

and hence

$$x^T \mathbf{1}\lambda(x) \leq x^T M\mathbf{1}. \tag{B.5}$$

Define $K \triangleq \max_i \sum_{j=1}^{n} m_{ij}$. We have $M\mathbf{1} \leq K\mathbf{1}$, and by noticing (B.5),

$$\lambda(x) \leq \frac{x^T M\mathbf{1}}{x^T\mathbf{1}} \leq \frac{Kx^T\mathbf{1}}{x^T\mathbf{1}} = K. \tag{B.6}$$

So the function $\lambda(x)$ is uniformly bounded for all nonnegative $x \in \mathbb{R}^n$.
Define

$$\lambda_1 \triangleq \sup_{x\geq 0, x\neq 0} \lambda(x) = \sup_{x\geq 0, x\neq 0} \min_{1\leq j\leq n} \frac{\sum_{i=1}^{n} x_i m_{ij}}{x_j}, \tag{B.7}$$

which has no change if x is normalized, i.e., $\lambda_1 = \sup_{x\geq 0, x^T x=1} \min_{1\leq j\leq n} \frac{\sum_{i=1}^{n} x_i m_{ij}}{x_j}$.

Since the set $\{x \in \mathbb{R}^n : x \geq 0, \ x^T x = 1\}$ is compact, there exists a unit vector $\widehat{x} \geq 0$, $\widehat{x}^T\widehat{x} = 1$ such that

$$\lambda_1 = \min_{1\leq j\leq n} \frac{\sum_{i=1}^{n} \widehat{x}_i m_{ij}}{\widehat{x}_j}. \tag{B.8}$$

By primitivity, M contains no zero column, and hence $\lambda(\mathbf{1}) > 0$. By the boundedness of $\lambda(x)$ and the definition of λ_1 given by (B.7), we have that $0 < \lambda(\mathbf{1}) \leq \lambda_1 \leq K < \infty$. Thus λ_1 is positive. We now show that λ_1 is an eigenvalue of M.

From (B.8) it follows that

$$\sum_{i=1}^{n} \widehat{x}_i m_{ij} \geq \lambda_1 \widehat{x}_j, \ j = 1, \cdots, n, \tag{B.9}$$
$$\widehat{x}^T M \geq \lambda_1 \widehat{x}^T, \tag{B.10}$$

and

$$\widehat{y}^T \triangleq \widehat{x}^T M - \lambda_1 \widehat{x}^T \geq 0. \tag{B.11}$$

To prove that λ_1 is an eigenvalue of M, it suffices to show that $\widehat{y}$ defined by (B.11) equals zero. If $\widehat{y}$ is with positive elements, then by primitivity of M there exists an integer $k > 0$ such that $M^k > 0$ and from (B.11)

$$\widehat{y}^T M^k = \widehat{x}^T M^k \cdot M - \lambda_1 \widehat{x}^T M^k > 0, \tag{B.12}$$

or, equivalently,

$$\lambda_1 < \min_{1 \le j \le n} \frac{\sum_{i=1}^{n} (\widehat{x}^T M^k)_i m_{ij}}{(\widehat{x}^T M^k)_j}, \tag{B.13}$$

where $(\widehat{x}^T M^k)_j$ refers to the jth element of $\widehat{x}^T M^k$.

Inequality (B.13) implies that $\lambda_1 < \sup_{x \ge 0, x^T x = 1} \min_{1 \le j \le n} \frac{\sum_{i=1}^{m} x_i m_{ij}}{x_j}$, which contradicts with its definition (B.7). Thus $y = 0$ and

$$\widehat{x}^T M = \lambda_1 \widehat{x}^T, \tag{B.14}$$

so λ_1 is an eigenvalue of M.

(ii) Let λ be any other eigenvalue of M different from λ_1. Then for some vector $x \neq 0$

$$x^T M = \lambda x^T, \tag{B.15}$$

or, equivalently,

$$\sum_{i=1}^{n} x_i m_{ij} = \lambda x_j. \tag{B.16}$$

From (B.16) it follows that

$$|\lambda| \le \frac{\sum_{i=1}^{n} |x_i| m_{ij}}{|x_j|}, \quad j = 1, \cdots, n, \tag{B.17}$$

where $\sum_{i=1}^{n} |x_i| m_{ij} / |x_j| \triangleq \infty$ if $|x_j| = 0$.

From (B.7) and (B.17) we have

$$|\lambda| \le \lambda_1. \tag{B.18}$$

Thus to prove (ii), it remains to show that if $|\lambda| = \lambda_1$, then $\lambda = \lambda_1$.

If $|\lambda| = \lambda_1$, then from (B.17) it follows that

$$\sum_{i=1}^{n} |x_i| m_{ij} \ge |\lambda| |x_j| = \lambda_1 |x_j|, \ j = 1, \cdots, n. \tag{B.19}$$

By primitivity of M, carrying out the discussion similar to that for (B.11)–(B.14), we derive

$$\sum_{i=1}^{n} |x_i| m_{ij} = |\lambda||x_j| = \lambda_1 |x_j|, \; j = 1, \cdots, n, \tag{B.20}$$

and for any integer $k \geq 0$

$$\sum_{i=1}^{n} |x_i| m_{ij}^{(k)} = |\lambda|^k |x_j| = \lambda_1^k |x_j|, \; j = 1, \cdots, n, \tag{B.21}$$

where $m_{ij}^{(k)} > 0, \; i, j = 1, \cdots, n$.

On the other hand, from (B.15) we obtain

$$x^T M^k = \lambda^k x^T, \tag{B.22}$$

$$\sum_{i=1}^{n} x_i m_{ij}^{(k)} = \lambda^k x_j, \; j = 1, \cdots, n, \tag{B.23}$$

and thus

$$\left| \sum_{i=1}^{n} x_i m_{ij}^{(k)} \right| = |\lambda|^k |x_j|, \; j = 1, \cdots, n. \tag{B.24}$$

Combining (B.21) and (B.24), we derive

$$\sum_{i=1}^{n} |x_i| m_{ij}^{(k)} = \left| \sum_{i=1}^{n} x_i m_{ij}^{(k)} \right|, \; j = 1, \cdots, n. \tag{B.25}$$

Since $m_{ij}^{(k)} > 0, \; i, j = 1, \cdots, n$, (B.25) indicates that $x_j, \; j = 1, \cdots, n$ must have the same direction in the complex plane, i.e., $x_j = |x_j| e^{i\varphi}, \; j = 1, \cdots, n$. Then, from (B.23) we obtain

$$\sum_{i=1}^{n} |x_i| m_{ij}^{(k)} = \lambda^k |x_j|, \; j = 1, \cdots, n \tag{B.26}$$

which combining with (B.21) yields that $\lambda = |\lambda| = \lambda_1$. Thus, (ii) is proved.

(iii) We first prove that λ_1 defined by (B.7) can be with positive left eigenvectors. From (B.14), we have

$$\widehat{x}^T M^k = \lambda_1^k \widehat{x}^T \quad \forall \, k \geq 1, \tag{B.27}$$

where $\widehat{x} \geq 0, \; \widehat{x} \neq 0$ is defined in (B.8). By primitivity of M, we know that $\sum_{i=1}^{n} \widehat{x}_i m_{ij}^{(k)} > 0$ for some $k \geq 1, \; j = 1, \cdots, n$. Consequently, from (B.27) it follows that

$$\sum_{i=1}^{n} \widehat{x}_i m_{ij}^{(k)} = \lambda_1^k \widehat{x}_j > 0, \; j = 1, \cdots, n. \tag{B.28}$$

Thus $\widehat{x}$ defined in (B.8) is positive, and by (B.14) $\widehat{x}$ is the left eigenvector corresponding to λ_1.

To prove the remaining assertion in (iii), let us consider the transpose M^T of M. For $x \in \mathbb{R}^n$, $x \geq 0$, $x \neq 0$, define $\overline{\lambda}(x) \triangleq \min\limits_{1\leq i\leq n} \frac{\sum_{j=1}^n x_j m_{ij}}{x_i}$ and $\overline{\lambda}_1 \triangleq \sup\limits_{x\geq 0, x\neq 0} \min\limits_{1\leq i\leq n} \frac{\sum_{j=1}^n x_j m_{ij}}{x_i}$, where $\frac{\sum_{j=1}^n x_j m_{ij}}{x_i} \triangleq \infty$ if $x_i = 0$.

By a similar discussion, we know that $\overline{\lambda}_1$ is a positive eigenvalue of M^T and $\overline{\lambda}_1 > |\lambda|$ for any other eigenvalue $\lambda \neq \overline{\lambda}_1$. Further, $\overline{\lambda}_1$ is with the positive left eigenvectors of M^T. Since M and M^T share the same eigenvalues, and the left eigenvectors of M^T are, in fact, the right eigenvectors of M, we conclude that $\overline{\lambda}_1 = \lambda_1$, which is defined by (B.7), and λ_1 is with the positive right eigenvectors of M.

Hence, (iii) holds.

(iv) Here we only consider the dimension of the left eigenvector space while the assertion for the right eigenvector space can similarly be proved.

Suppose $x = [x_1 \cdots x_n]^T \neq 0$ is a left eigenvector of M corresponding to λ_1, $x^T M = \lambda_1 x^T$. This implies that

$$\sum_{i=1}^{n} |x_i| m_{ij} \geq \lambda_1 |x_j|, \quad j = 1, \cdots, n.$$

Carrying out the discussion similar to that for (B.10)–(B.14), we arrive at

$$\sum_{i=1}^{n} |x_i| m_{ij} = \lambda_1 |x_j|, \quad j = 1, \cdots, n,$$

i.e.,

$$\mathrm{Abs}(x)^T M = \lambda_1 \mathrm{Abs}(x)^T,$$

where $\mathrm{Abs}(x) \triangleq [|x_1| \cdots |x_n|]^T$. Hence, $\mathrm{Abs}(x)$ is also a left eigenvector of M corresponding to λ_1. From the above equality it follows that

$$\mathrm{Abs}(x)^T M^k = \lambda_1^k \mathrm{Abs}(x)^T, \quad k \geq 1.$$

Then by primitivity of M, we have $\mathrm{Abs}(x) > 0$.

Define

$$\eta \triangleq \widehat{x} - cx, \tag{B.29}$$

where x is any left eigenvector of M corresponding to λ_1, $\widehat{x}$ is defined in (B.8), and c is a constant such that $\eta \neq 0$. From the above discussion we see that both η and $\mathrm{Abs}(\eta)$ are the left eigenvectors of M corresponding to λ_1 and $\mathrm{Abs}(\eta) > 0$.

Suppose that the dimension of the left eigenvector space corresponding to λ_1 is bigger than 1. Then we can choose some left eigenvector x and constant c such that $\eta \neq 0$ but some elements of η equal zero. This implies that some elements of $\mathrm{Abs}(\eta)$ equal zero, which contradicts with the just proved assertion that $\mathrm{Abs}(\eta) > 0$.

Hence, the left eigenvector x must be a multiple of $\widehat{x}$ and the dimension of the left eigenvector space corresponding to λ_1 is 1.

(v) We first prove a preliminary result: If a matrix $N = [n_{ij}]_{i,j=1}^{n}$ satisfies $0 \le N \le M$ and α is an eigenvalue of N, then $|\alpha| \le \lambda_1$. Further, if $|\alpha| = \lambda_1$, then $N = M$.

Let $x \neq 0$ be a right eigenvector of N corresponding to α. By noticing $N \le M$, we have

$$|\alpha|\text{Abs}(x) \le N \cdot \text{Abs}(x) \le M \cdot \text{Abs}(x). \tag{B.30}$$

Pre-multiplying $\widehat{x}^T$ defined in (B.8) to both sides of (B.30), we obtain

$$|\alpha|\widehat{x}^T \text{Abs}(x) \le \widehat{x}^T N \cdot \text{Abs}(x) \le \widehat{x}^T M \cdot \text{Abs}(x) = \lambda_1 \widehat{x}^T \text{Abs}(x). \tag{B.31}$$

This implies

$$|\alpha| \le \lambda_1, \tag{B.32}$$

because $\text{Abs}(x) \ge 0$ and $\widehat{x} > 0$ as proved in (iii).

Assume $|\alpha| = \lambda_1$. Then from (B.30) we have

$$\lambda_1 \text{Abs}(x) \le M \cdot \text{Abs}(x). \tag{B.33}$$

Define $y \triangleq M \cdot \text{Abs}(x) - \lambda_1 \text{Abs}(x)$. Carrying out the discussion similar to that for (B.10)–(B.14), we have

$$\lambda_1 \text{Abs}(x) = M \cdot \text{Abs}(x). \tag{B.34}$$

Then from (B.30) we see

$$\lambda_1 \text{Abs}(x) = N \cdot \text{Abs}(x) = M \cdot \text{Abs}(x), \tag{B.35}$$

which indicates that $\text{Abs}(x)$ is a nonnegative right eigenvector of M corresponding to λ_1. Then by (iii) and (iv), $\text{Abs}(x)$ is positive.

Denote by $(\cdot)_i$ the ith element of a vector. We now prove that $N = M$ under the assumption $|\alpha| = \lambda_1$. Assume the converse: $n_{ij} < m_{ij}$ for some i, j. Then by the fact $\text{Abs}(x) > 0$ it follows that $n_{ij}|x_j| < m_{ij}|x_j|$ and hence $(N \cdot \text{Abs}(x))_i < (M \cdot \text{Abs}(x))_i$, which contradicts with (B.35). Thus, if $|\alpha| = \lambda_1$, then $N = M$.

We now prove (v). We first show that each element of $\text{Adj}(\lambda_1 I - M)$ is nonzero.

It is clear that

$$\text{Adj}(\lambda_1 I - M)(\lambda_1 I - M) = \det(\lambda_1 I - M)I = 0, \tag{B.36}$$

which implies that the rows of $\text{Adj}(\lambda_1 I - M)$ are either zeros or the left eigenvectors of M corresponding to λ_1.

We show that all rows of $\text{Adj}(\lambda_1 I - M)$ are nonzero. Let us consider the last row of $\text{Adj}(\lambda_1 I - M)$ and prove its (n,n)-element is nonzero. The other rows of $\text{Adj}(\lambda_1 I - M)$ can be analyzed in a similar manner.

Set $M_{(n-1)\times(n-1)} \triangleq \begin{bmatrix} m_{11} & \cdots & m_{1,n-1} \\ \vdots & \ddots & \vdots \\ m_{n-1,1} & \cdots & m_{n-1,n-1} \end{bmatrix}$. The (n,n)-element of $\mathrm{Adj}(\lambda_1 I - M)$ is $\det\left[\lambda_1 I_{(n-1)\times(n-1)} - M_{(n-1)\times(n-1)}\right]$.

Since

$$0 \le \begin{bmatrix} m_{11} & \cdots & m_{1,n-1} & 0 \\ \vdots & & & \vdots \\ m_{n-1,1} & \cdots & m_{n-1,n-1} & 0 \\ 0 & 0 & 0 & 0 \end{bmatrix} \le M, \text{ and } \begin{bmatrix} m_{11} & \cdots & m_{1,n-1} & 0 \\ \vdots & & & \vdots \\ m_{n-1,1} & \cdots & m_{n-1,n-1} & 0 \\ 0 & 0 & 0 & 0 \end{bmatrix} \ne M,$$

by the preliminary result just proved, the moduli of all eigenvalues of $\begin{bmatrix} M_{(n-1)\times(n-1)} & 0 \\ 0 & 0 \end{bmatrix}$ are smaller than λ_1, and hence $\det(\lambda_1 I - M_{(n-1)\times(n-1)}) \ne 0$.

Thus, the (n,n)-element of $\mathrm{Adj}(\lambda_1 I - M)$ is nonzero. Since the rows of $\mathrm{Adj}(\lambda_1 I - M)$ are left eigenvectors of M corresponding to λ_1, by (iii) and (iv) we know that all elements in the last row of $\mathrm{Adj}(\lambda_1 I - M)$ are positive. Carrying out the similar discussion for the other rows of $\mathrm{Adj}(\lambda_1 I - M)$, we conclude that all rows of $\mathrm{Adj}(\lambda_1 I - M)$ are positive left eigenvectors of M corresponding to λ_1.

Set $f(s) \triangleq \det(sI - M)$. Then from the equality $(sI - M)\mathrm{Adj}(sI - M) = \det(sI - M)I$, we obtain

$$\mathrm{Adj}(sI - M) + (sI - M)\frac{\mathrm{d}}{\mathrm{d}s}\mathrm{Adj}(sI - M) = \frac{\mathrm{d}}{\mathrm{d}s}\det(sI - M)I. \tag{B.37}$$

Substituting s with λ_1 and pre-multiplying both sides of (B.37) by $\widehat{x}^T$ defined in (B.8), we obtain

$$\widehat{x}^T \mathrm{Adj}(\lambda_1 I - M) = \frac{\mathrm{d}}{\mathrm{d}s}\det(sI - M)\Big|_{s=\lambda_1} \widehat{x}^T. \tag{B.38}$$

Since all elements of $\widehat{x}$ and $\mathrm{Adj}(\lambda_1 I - M)$ are positive, from (B.38) it follows that $\frac{\mathrm{d}}{\mathrm{d}s}\det(sI - M)\Big|_{s=\lambda_1} > 0$, which indicates that the multiplicity of λ_1, as a root of the polynomial $\det(sI - M)$, must be one.

We now consider the case where M is irreducible. The proof can be carried out similarly to the previous case. We only sketch it. In order to avoid confusion, let us denote the irreducible matrix by $T = [t_{ij}]_{i,j=1}^n$ instead of M.

(i) The formulas (B.4)–(B.8) only require that the matrix is nonnegative. Hence for the irreducible matrix T we can still define the function

$$\lambda(x) = \min_{1\le j\le n} \frac{\sum_{i=1}^n x_i t_{ij}}{x_j}, \ x \ge 0 \tag{B.39}$$

and the constant

$$\lambda_1 \triangleq \sup_{x\ge 0, x\ne 0} \lambda(x). \tag{B.40}$$

As before, we have

$$0 \le \lambda_1 < \infty. \tag{B.41}$$

It remains to show that λ_1 is an eigenvalue of T.

Along the lines of (B.9)–(B.11), we see

$$\widehat{y}^T \triangleq \widehat{x}^T T - \lambda_1 \widehat{x}^T \ge 0. \tag{B.42}$$

We want to show $\widehat{y} = 0$. Assume the converse: $\widehat{y}$ is with positive elements. Then by Lemma B.1 and noticing that $(I+T)T = T(I+T)$, for some $k > 0$ we have

$$\widehat{y}^T (I+T)^k = \widehat{x}^T (I+T)^k T - \lambda_1 \widehat{x}^T (I+T)^k > 0. \tag{B.43}$$

This implies

$$\lambda_1 < \min_{1 \le j \le n} \frac{\sum_{i=1}^{n} (\widehat{x}^T (I+T)^k)_i t_{ij}}{(\widehat{x}^T (I+T)^k)_j}, \tag{B.44}$$

and hence

$$\lambda_1 < \sup_{x \ge 0, x \ne 0} \min_{1 \le j \le n} \frac{\sum_{i=1}^{n} x_i t_{ij}}{x_j}, \tag{B.45}$$

which contradicts with the definition of λ_1.

Consequently,

$$\widehat{x}^T T = \lambda_1 \widehat{x}^T \tag{B.46}$$

and λ_1 is a positive real eigenvalue of T.

(ii') We notice that for (B.15)–(B.18) only the nonnegativity of the matrix is required, and hence they also hold for the irreducible matrix T. So $\lambda_1 \ge |\lambda|$ for any other eigenvalue λ of T.

(iii) From (B.46) we know that the nonnegative vector $\widehat{x}$ is a left eigenvector of T corresponding to λ_1. It remains to show that $\widehat{x} > 0$.

From (B.46) we obtain

$$\widehat{x}^T T^k = \lambda_1^k \widehat{x}^T, \quad k \ge 1 \tag{B.47}$$

and

$$\sum_{i=1}^{n} (\widehat{x})_i t_{ij}^{(k)} = \lambda_1^k (\widehat{x})_j, \quad k \ge 1, \quad j = 1, \cdots, n. \tag{B.48}$$

Since $\widehat{x} \ne 0$, there exists an index i_0 such that $(\widehat{x})_{i_0} > 0$. For i_0 and any fixed

$j \in \{1,\cdots,n\}$, by irreducibility of T there exists some integer $k_0 = k(i_0, j) > 0$ such that $t_{i_0 j}^{(k_0)} > 0$. Then from (B.48) it follows that

$$\lambda_1^{k_0}(\widehat{x})_j = \sum_{i=1}^{n} (\widehat{x})_i t_{ij}^{(k_0)} \geq (\widehat{x})_{i_0} t_{i_0 j}^{(k_0)} > 0, \tag{B.49}$$

and hence $(\widehat{x})_j > 0,\ j = 1,\cdots,n$. This proves that $\widehat{x}$ is a positive left eigenvector of T corresponding to λ_1. The assertion for the right eigenvectors can similarly be proved.

(iv)–(v) The assertions (iv) and (v) can be shown for the irreducible matrix T by applying the technique adopted in (B.43)–(B.49).

This completes the proof. □

Definition B.2 A nonnegative matrix $M = [m_{ij}]_{i,j=1}^{n}$ is called *column-wise stochastic* if $\sum_{i=1}^{n} m_{ij} = 1,\ j = 1,\cdots,n$, *row-wise stochastic* if $\sum_{j=1}^{n} m_{ij} = 1,\ i = 1,\cdots,n$, and *doubly stochastic* if it is both column-wise and row-wise stochastic.

Theorem B.2 *Suppose the nonnegative M is column-wise stochastic, or row-wise stochastic, or doubly stochastic. If M is primitive, then the biggest modulus of the eigenvalues of M is* 1*; while if M is irreducible, then the biggest modulus of the eigenvalues of M is also* 1 *and any other eigenvalue λ of M satisfies* $\mathrm{Re}\{\lambda\} < 1$.

Proof. Let us first consider the case where M is primitive. Assume M is column-wise stochastic. The results for the row-wise stochastic and doubly stochastic matrices can similarly be proved.

We recall the definitions for $\lambda(x)$, $\overline{\lambda}(x)$, λ_1, and $\overline{\lambda}_1$ given in the proof of Theorem B.1 for the primitive matrix M:

$$\lambda(x) = \min_{1 \leq j \leq n} \frac{\sum_{i=1}^{n} x_i m_{ij}}{x_j}, \quad \lambda_1 = \sup_{x \geq 0, x \neq 0} \lambda(x), \tag{B.50}$$

$$\overline{\lambda}(x) = \min_{1 \leq i \leq n} \frac{\sum_{j=1}^{n} m_{ij} x_j}{x_i}, \quad \overline{\lambda}_1 = \sup_{x \geq 0, x \neq 0} \overline{\lambda}(x). \tag{B.51}$$

From here it is seen that $\lambda_1 = \overline{\lambda}_1$ is the eigenvalue of M with the biggest modulus. We now show that $\lambda_1 = \overline{\lambda}_1 = 1$.

From the definition of $\overline{\lambda}(x)$, we have

$$x_i \overline{\lambda}(x) \leq \sum_{j=1}^{n} m_{ij} x_j, \quad i = 1,\cdots,n, \tag{B.52}$$

$$\overline{\lambda}(x) x \leq Mx, \tag{B.53}$$

$$\overline{\lambda}(x) \mathbf{1}^T x \leq \mathbf{1}^T Mx. \tag{B.54}$$

Since M is column-wise stochastic, it holds that $\mathbf{1}^T M = \mathbf{1}^T$. From (B.54) we have

$$\overline{\lambda}(x) \leq 1 \ \ \forall x \geq 0, \ x \neq 0 \tag{B.55}$$

and

$$\lambda_1 = \overline{\lambda}_1 \leq 1. \tag{B.56}$$

On the other hand, substituting x with $x^0 = \frac{1}{n}\mathbf{1}$ in (B.50), we obtain

$$\lambda(x^0) = \min_{1 \leq j \leq n} \frac{\sum_{i=1}^{n} \frac{1}{n} m_{ij}}{\frac{1}{n}} = \min_{1 \leq j \leq n} \sum_{i=1}^{n} m_{ij} = 1, \tag{B.57}$$

which combining with (B.50), (B.51), and (B.56) indicates that $1 = \lambda(x^0) \leq \lambda_1 = \overline{\lambda}_1 \leq 1$, and hence $\lambda_1 = 1$.

We now consider the irreducible matrix. In order to avoid confusion in notations, denote the irreducible matrix by $T = [t_{ij}]_{i,j=1}^{n}$.

For T, similar to (B.50)–(B.51), we can also define $\lambda(x)$, $\overline{\lambda}(x)$, λ_1, and $\overline{\lambda}_1$. By irreducibility of T, carrying out a discussion similar to (B.39)–(B.46), we see that λ_1 is the eigenvalue of T with the biggest modulus and $\lambda_1 = \overline{\lambda}_1$. Then following the same lines as (B.52)–(B.57), we have that $\lambda_1 = 1$.

We now prove that $\mathrm{Re}\{\lambda\} < 1$ for any other eigenvalue λ of T. By Theorem B.1, the multiplicity of eigenvalue 1 of T is 1. Thus to prove $\mathrm{Re}\{\lambda\} < 1$, we only need to consider two cases: (i) $|\lambda| < 1$ and (ii) $|\lambda| = 1$ but $\lambda \neq 1$.

In case (i) it is clear that $|\mathrm{Re}\{\lambda\}| < 1$ and $\mathrm{Re}\{\lambda\} - 1 < 0$. In case (ii) $\lambda = \mathrm{e}^{i\phi}$ or $\lambda = \cos(\phi) + i\sin(\phi)$ for some $\phi \in (0, 2\pi)$. This means $\mathrm{Re}\{\lambda\} - 1 = \cos(\phi) - 1 < 0$. □

For a primitive matrix M, the following result connects the power matrix M^k with the eigenvalues and eigenvectors of M.

Theorem B.3 *For a primitive matrix $M \in \mathbb{R}^{n \times n}$, it holds that*

$$M^k = \lambda_1^k v u^T + O\left(k^{m_2 - 1} |\lambda_2|^k\right) \tag{B.58}$$

for all large enough k, where u and v are the positive left and right eigenvectors of M corresponding to λ_1, satisfying $u^T v = 1$.

To prove the theorem, we need the following technical lemma.

Lemma B.2 *Assume that M is primitive and u and v are the positive left and right eigenvectors of M corresponding to the eigenvalue λ_1 with the biggest modulus. Without losing generality, we may assume $u^T v = 1$. Then*

$$\frac{\mathrm{Adj}(\lambda_1 I - M)}{\left.\frac{\mathrm{d}}{\mathrm{d}s}\det(sI - M)\right|_{s=\lambda_1}} = v u^T. \tag{B.59}$$

Proof. In the proof of (v) in Theorem B.1 we have shown that $\mathrm{Adj}(\lambda_1 I - M) > 0$. From $(\lambda_1 I - M)\mathrm{Adj}(\lambda_1 I - M) = \det(\lambda_1 I - M)I = 0$ it is clear that each column of $\mathrm{Adj}(\lambda_1 I - M)$ is a positive right eigenvector of M corresponding to λ_1. By (iv) in Theorem B.1, the dimension of both the left and right eigenvector spaces corresponding to λ_1 is 1, so we can write

$$\mathrm{Adj}(\lambda_1 I - M) = yx^T, \tag{B.60}$$

where $y > 0$ is a right eigenvector and $x > 0$ is a left eigenvector of M corresponding to λ_1.

Therefore, we have

$$\mathrm{Adj}(\lambda_1 I - M) = c_1 c_2 v u^T \tag{B.61}$$

for some constants $c_1 > 0$ and $c_2 > 0$.

Substituting s with λ_1 in (B.37) and pre-multiplying both sides by u^T and noticing (B.61), we obtain

$$u^T \frac{\mathrm{d}}{\mathrm{d}s}\det(sI - M)\Big|_{s=\lambda_1} = u^T \mathrm{Adj}(\lambda_1 I - M) = c_1 c_2 u^T v u^T, \tag{B.62}$$

and hence

$$u^T v \frac{\mathrm{d}}{\mathrm{d}s}\det(sI - M)\Big|_{s=\lambda_1} = c_1 c_2 (u^T v)^2. \tag{B.63}$$

By the assumption $u^T v = 1$, from (B.63) we obtain $\frac{\mathrm{d}}{\mathrm{d}s}\det(sI - M)\Big|_{s=\lambda_1} = c_1 c_2$, and hence (B.59) follows from (B.61). This completes the proof. □

Proof of Theorem B.3.

Define the complex-valued function $f(z) \triangleq (I - zM)^{-1} = [f_{ij}(z)]_{i,j=1}^n$, $z \neq \lambda_i^{-1}$, $i = 1, \cdots, t$. Let the multiplicity of the eigenvalue λ_i of M be m_i.

By noticing $f(z) = \mathrm{Adj}(I - zM)/\det(I - zM)$ we write each element of $f(z)$ as follows

$$f_{ij}(z) = \frac{c_{ij}(z)}{(1 - z\lambda_1)(1 - z\lambda_2)^{m_2} \cdots (1 - z\lambda_t)^{m_t}}, \tag{B.64}$$

where $c_{ij}(z)$ is a polynomial of z with degree at most $n-1$.

From (B.64), the (i, j)-element of $f(z)$ can further be expressed as

$$f_{ij}(z) = p_{ij} + \frac{a_{ij}}{1 - z\lambda_1} + \sum_{s=1}^{m_2} \frac{b_{ij}^{(s)}}{(1 - z\lambda_2)^s} + \sum_{l=3}^{t} \sum_{s_l=1}^{m_l} \frac{b_{ij}^{(s_l)}}{(1 - z\lambda_l)^{s_l}}, \tag{B.65}$$

where p_{ij}, a_{ij}, $b_{ij}^{(s)}$, and $b_{ij}^{(s_l)}$ are constants.

Expanding the function $1/(1-x)^s$ to the Taylor series at $x = 0$, we find that

$$\frac{1}{(1-x)^s} = \sum_{k=0}^{\infty} C_{s+k-1}^k x^k.$$

Using this formula and noticing that $f(z)$ is analytic in the domain $\Gamma \triangleq \{z : |z| < \frac{1}{\lambda_1}\}$, we have the following series expansions in Γ

$$(I - zM)^{-1} = \sum_{k=0}^{\infty} M^k z^k, \tag{B.66}$$

$$\frac{a_{ij}}{1 - z\lambda_1} = a_{ij} \sum_{k=0}^{\infty} \lambda_1^k z^k, \tag{B.67}$$

$$\frac{b_{ij}^{(s)}}{(1 - z\lambda_2)^s} = b_{ij}^{(s)} \sum_{k=0}^{\infty} C_{s+k-1}^k \lambda_2^k z^k, \tag{B.68}$$

and for $l = 3, \cdots, t$,

$$\frac{b_{ij}^{(s_l)}}{(1 - z\lambda_l)^{s_l}} = b_{ij}^{(s_l)} \sum_{k=0}^{\infty} C_{s_l+k-1}^k \lambda_l^k z^k, \tag{B.69}$$

where $C_{s+k-1}^k = \frac{(s+k-1)!}{k!(s-1)!}$.

By the Stirling formula $\lim\limits_{n\to\infty} \frac{n!}{\sqrt{2\pi n}\left(\frac{n}{e}\right)^n} = 1$, we have

$$\begin{aligned} C_{s+k-1}^k &= \frac{(s+k-1)!}{k!(s-1)!} = O\Bigg(\frac{\sqrt{2\pi(s+k-1)}\left(\frac{s+k-1}{e}\right)^{s+k-1}}{\sqrt{2\pi k}\left(\frac{k}{e}\right)^k}\Bigg) \\ &= O\Big(\frac{(s+k-1)^{s+k-1}}{k^k}\Big) = O\Big((s+k-1)^{s-1}(1 + \frac{s-1}{k})^k\Big) \\ &= O(k^{s-1}) \end{aligned} \tag{B.70}$$

as $k \to \infty$.

Comparing the coefficients of z^k at both sides of (B.65) and noticing (B.66)–(B.70) and the assumption $|\lambda_2| \geq |\lambda_l|$, $l = 3, \cdots, t$, we obtain

$$M^k = A\lambda_1^k + O(k^{m_2-1}|\lambda_2|^k), \tag{B.71}$$

where $A = [a_{ij}]_{i,j=1}^n$ with a_{ij} defined in (B.65).

Thus to prove the theorem it remains to determine the matrix A.

Since $\lambda_1 > |\lambda_2| \geq 0$, from (B.71) we have

$$\lim_{k\to\infty} \frac{M^k}{\lambda_1^k} = A \geq 0, \tag{B.72}$$

and thus the series $\sum_{k=0}^{\infty} \left(\frac{M}{\lambda_1}\right)^k z^k$ is convergent for all $|z| < 1$.

We now show that

$$\lim_{x\to 1-} (1 - x) \sum_{k=0}^{\infty} \left(\frac{M}{\lambda_1}\right)^k x^k = A. \tag{B.73}$$

We first notice that

$$(1-x)\sum_{k=0}^{\infty} Ax^k = A \ \forall\, 0<x<1. \tag{B.74}$$

Set $A_k \triangleq (M/\lambda_1)^k$. Since $A_k \to A$ as $k \to \infty$, for any fixed $\varepsilon > 0$ there exists an integer $N > 0$ such that $\|A_k - A\| < \varepsilon \ \forall\, k \geq N$. Thus, for any $x \in (0,1)$, we have

$$\begin{aligned}
\left\|(1-x)\sum_{k=0}^{\infty}(A_k - A)x^k\right\| &\leq \left\|(1-x)\sum_{k=0}^{N}(A_k - A)x^k\right\| + \left\|(1-x)\sum_{k=N+1}^{\infty}(A_k - A)x^k\right\| \\
&\leq (1-x)\sum_{k=0}^{N}\|A_k - A\|x^k + \varepsilon(1-x)\sum_{k=N+1}^{\infty} x^k \\
&\leq c_1(1-x)\sum_{k=0}^{N} x^k + \varepsilon(1-x)\sum_{k=N+1}^{\infty} x^k \\
&= c_1(1-x)\frac{1-x^N}{1-x} + \varepsilon(1-x)\frac{x^{N+1}}{1-x} \\
&= c_1(1-x^N) + \varepsilon x^{N+1},
\end{aligned} \tag{B.75}$$

where $c_1 > 0$ is a constant. By first letting $x \to 1-$ and then $\varepsilon \to 0$ we find that

$$\lim_{x\to 1-}(1-x)\sum_{k=0}^{\infty}(A_k - A)x^k = 0. \tag{B.76}$$

Combining (B.74) and (B.76) leads to (B.73).

On the other hand, for $0 < x < 1$ we have

$$\sum_{k=0}^{\infty}\left(\frac{M}{\lambda_1}\right)^k x^k = \left(I - \frac{Mx}{\lambda_1}\right)^{-1} = \frac{\operatorname{Adj}\left(I - \frac{Mx}{\lambda_1}\right)}{\det\left(I - \frac{Mx}{\lambda_1}\right)} = \frac{\lambda_1}{x}\frac{\operatorname{Adj}\left(\frac{\lambda_1}{x}I - M\right)}{\det\left(\frac{\lambda_1}{x}I - M\right)}. \tag{B.77}$$

Substituting $\sum_{k=0}^{\infty}\left(\frac{M}{\lambda_1}\right)^k x^k$ in (B.73) with its expression given by (B.77), we obtain

$$\begin{aligned}
A &= \lim_{x\to 1-}(1-x)\cdot\frac{\lambda_1}{x}\cdot\frac{\operatorname{Adj}\left(\frac{\lambda_1}{x}I - M\right)}{\det\left(\frac{\lambda_1}{x}I - M\right)} \\
&= \lim_{x\to 1-}\frac{(1-x)}{\det\left(\frac{\lambda_1}{x}I - M\right)}\lambda_1 \operatorname{Adj}(\lambda_1 I - M) \\
&= \left(\lim_{x\to 1-}\frac{\det\left(\frac{\lambda_1}{x}I - M\right)}{(1-x)}\right)^{-1}\lambda_1 \operatorname{Adj}(\lambda_1 I - M) \\
&= \left(-\lim_{x\to 1-}\frac{\det(\lambda_1 I - M) - \det\left(\frac{\lambda_1}{x}I - M\right)}{(1-x)}\right)^{-1}\lambda_1 \operatorname{Adj}(\lambda_1 I - M)
\end{aligned}$$

$$
\begin{aligned}
&= -\left(\frac{\mathrm{d}}{\mathrm{d}x}\det(\frac{\lambda_1}{x}I - M)\Big|_{x=1}\right)^{-1}\lambda_1 \mathrm{Adj}\,(\lambda_1 I - M) \\
&= \left(\frac{\mathrm{d}}{\mathrm{d}s}\det(sI - M)\Big|_{s=\lambda_1}\right)^{-1}\frac{1}{\lambda_1}\lambda_1 \mathrm{Adj}\,(\lambda_1 I - M) \\
&= \left(\frac{\mathrm{d}}{\mathrm{d}s}\det(sI - M)\Big|_{s=\lambda_1}\right)^{-1}\mathrm{Adj}\,(\lambda_1 I - M). \qquad \text{(B.78)}
\end{aligned}
$$

By Lemma B.2, from (B.78) it follows that

$$A = vu^T, \qquad \text{(B.79)}$$

where u and v are left and right positive eigenvectors of M satisfying $u^T v = 1$.

Combining (B.71) and (B.79), we arrive at (B.58). □

References

[1] H. Akaike. A new look at the statistical model identification. *IEEE Transaction on Automatic Control*, pages 19(6): 716–723, 1974.

[2] A. Al-Smadi and D. M. Wilkes. Robust and accurate ARX and ARMA model order estimation of non-Gaussian processes. *IEEE Transactions on Signal Processing*, pages 50(3): 759–763, 2002.

[3] H. Z. An, Z. G. Chen, and E. J. Hannan. Autocorrelation, autoregression and autoregressive approximation. *The Annals of Statistics*, pages 10(3): 926–936, 1982.

[4] B. D. O. Anderson and J. B. Moore. *Optimal Filtering*. Prentice Hall, NJ, USA, 1979.

[5] K. J. Åström and B. Wittenmark. On self-tuning regulator. *Automatica*, pages 9(2): 195–199, 1973.

[6] K. Avrachenkov, N. Litvak, D. Nemirovsky, and N. Osipova. Monte Carlo methods in PageRank computation: When one iteration is sufficient. *SIAM Journal of Numerical Analysis*, pages 45(2): 890–904, 2007.

[7] E. W. Bai. Non-parametric nonlinear system identification: an asymptotic minimum mean squared error estimator. *IEEE Transaction on Automatic Control*, pages 55(7): 1615–1626, 2010.

[8] E. W. Bai and Y. Liu. Recursive direct weight optimization in nonlinear system identification: A minimal probability approach. *IEEE Transaction on Automatic Control*, pages 52(7): 1218–1231, 2007.

[9] A. Benveniste, M. Métivier, and P. Priouret. *Adaptive Algorithms and Stochastic Approximation*. Springer-Verlag, New York, USA, 1990.

[10] S.A. Billings and S.Y. Fakhouri. Identification of systems containing linear dynamic and static nonlinear elements. *Automatica*, pages 18(1): 15–26, 1982.

[11] J.R. Blum. Multidimensional stochastic approximation methods. *The Annals of Mathematical Statistics*, pages 25(4): 737–744, 1954.

[12] M. Boutayeb and M. Darouach. Recursive identification method for MISO Wiener-Hammerstein model. *IEEE Transactions on Automatic Control*, pages 40(2): 287–291, 1995.

[13] O. Brandiere. Some pathological traps for stochastic approximation. *SIAM Journal on Control and Optimization*, pages 36(4): 1293–1314, 1998.

[14] S. Brin and L. Page. The anatomy of a large-scale hypertextual web search engine. *Computer Networks and ISDN Systems*, pages 30(1–7): 107–117, 1998.

[15] P. J. Brockwell and R. A. Davis. *Time Series, Theory and Methods (2nd ed.)*. Springer-Verlag, Heidelberg, 2001.

[16] A. Z. Broder, R. Lempel, F. Maghoul, and J. Pedersen. Efficient PageRank approximation via graph aggregation. *Information Retrieval*, pages 9(2): 123–138, 2006.

[17] P. Caines. *Linear Stochastic Systems*. John Wiley and Sons, New York, NY, USA, 1988.

[18] H.-F. Chen. Stochastic approximation and its new applications. *Proceedings of 1994 Hong Kong International Workshop on New Directions of Control and Manufacturing*, pages 2–12, 1994.

[19] H.-F. Chen. *Stochastic Approximation and Its Applications*. Kluwer Academic Publisher, Dordrecht, The Netherlands, 2002.

[20] H.-F. Chen. Pathwise convergence of recursive identification algorithms for Hammerstein systems. *IEEE Transactions on Automatic Control*, pages 49(10): 1641–1649, 2004.

[21] H.-F. Chen. Adaptive regulator for Hammerstein and Wiener systems with noisy observations. *IEEE Transactions on Automatic Control*, pages 52(4): 703–709, 2007.

[22] H.-F. Chen. Recursive identification for multivariate errors-in-variables systems. *Automatica*, pages 43(7): 1234–1242, 2007.

[23] H.-F. Chen. New approach to recursive identification for ARMAX systems. *IEEE Transactions on Automatic Control*, pages 55(4): 879–886, 2010.

[24] H.-F. Chen, H.T. Fang, and L.L. Zhang. Recursive estimation for ordered eigenvectors of symmetric matrix with observation noise. *Journal of Mathematical Analysis and Applications*, pages 382(2): 882–842, 2011.

[25] H.-F. Chen and L. Guo. Convergence rate of least-squares identification and adaptive control for stochastic systems. *International Journal of Control*, pages 44(5): 1459–1476, 1986.

[26] H.-F. Chen and L. Guo. *Identification and Stochastic Adaptive Control*. Birkhäuser, Boston, MA, USA, 1991.

[27] H.-F. Chen and W. X. Zhao. New method of order estimation for ARMA/ARMAX processes. *SIAM Journal of Control and Optimization*, pages 48(6): 4157–4176, 2010.

[28] H.-F. Chen and Y.M. Zhu. Stochastic approximation procedures with randomly varying truncations. *Science in China: Series A*, pages 29: 914–926, 1986.

[29] X.M. Chen and H.-F. Chen. Recursive identification for MIMO Hammerstein systems. *IEEE Transactions on Automatic Control*, pages 56(4): 895–902, 2011.

[30] Y. S. Chow and H. Teicher. *Probability Theory: Independence, Interchangability, Martingales (3rd ed.)*. Springer-Verlag, NY, USA, 1997.

[31] A. Daffertshofer, C.J.C. Lamoth, O.G. Meijer, and P.J. Beek. PCA in studying coordination and variability: A tutorial. *Clinical Biomechanics*, pages 19(4): 415–428, 2004.

[32] M. Dalai, E. Weyer, and M.C. Campi. Parameter identification for nonlinear systems: Guaranteed confidence regions through LSCR. *Automatica*, pages 43(2007): 1418–1425, 2007.

[33] Yu. A. Davydov. Mixing conditions for Markov Chains. *SIAM Theory of Probability and Its Applications*, pages 18(2): 312–328, 1973.

[34] D. V. de Jager and J. T. Bradley. Asynchronous iterative solution for state-based performance metrics. In *Proceedings of ACM SIGMETRICS*, pages 373–374. Accociation for Computing Machinery, USA, 2007.

[35] E. F. (eds.) Deprettere. *SVD and Signal Processing*. Elsevier, Amsterdam The Netherlands, 1988.

[36] J. L. Doob. *Stochastic Processes*. John Wiley & Sons, New York, NY, USA, 1953.

[37] P. Doukhan. *Mixing, Properties and Examples (Lecture Notes in Statistics-85)*. Springer, New York, NY, USA, 1994.

[38] J. Fan and I. Gijbels. *Local Polynomial Modeling and Its Applications*. Chapman & Hall/CRC Press, London, U.K., 1996.

[39] J.Q. Fan and Q.W. Yao. *Nonlinear Time Series: Nonparametric and Parametric Methods*. Springer, New York, USA, 2005.

[40] H.T. Fang, H.-F. Chen, and L. Wen. On control of strong consensus for networked agents with noisy observations. *Journal of Systems Science and Complexity*, pages 25(1): 1–12, 2012.

[41] T. T. Georgiou and A. Lindquist. A convex optimization approach to ARMA modeling. *IEEE Transactions on Automatic Control*, pages 53(5): 1108–1119, 2008.

[42] F. Giri and E.W. Bai. *Block-oriented Nonlinear System Identification.* Springer-Verlag, Berlin, Germany, 2010.

[43] E.G. Gladyshev. On stochastic approximation (in Russian). *SIAM Theory of Probability and Its Applications*, pages 10(2): 275–278, 1965.

[44] G.C. Goodwin and K.S. Sin. *Adaptive Filtering, Prediction, and Control.* Prentice Hall, NJ, USA, 1984.

[45] V. V. Gorodetskii. On the strong mixing property for linear sequences. *SIAM Theory of Probability and Its Applications*, pages 22(2): 411–413, 1978.

[46] W. Greblicki. Nonparametric approach to Wiener system identification. *IEEE Trans Circuits and Systems-I: Fundamental Theory and Applications*, pages 44(6): 538–545, 1997.

[47] W. Greblicki. Stochastic approximation in nonparametric identification of Hammerstein systems. *IEEE Transactions on Automatic Control*, pages 47(11): 1800–1810, 2002.

[48] L. Guo and H.-F. Chen. Åström-Wittenmark self-tuning regulator revisited and ELS-based adaptive trackers. *IEEE Transactions on Automatic Control*, pages 36(7): 802–812, 1991.

[49] A. Hagenblad, L. Ljung, and A. Wills. Maximum likelihood identification of Wiener models. *Automatica*, pages 44(11): 2697–2705, 2008.

[50] E.J. Hannan. The identification of vector mixed autoregressive-moving average systems. *Biometrica*, pages 56(1): 223–225, 1969.

[51] E.J. Hannan. The estimation of ARMA models. *The Annals of Statistics*, pages 3(4): 975–981, 1975.

[52] W. Härdle. *Applied Nonparametric Regression.* Cambridge University Press, Cambridge, U.K., 1990.

[53] Y. Hatano and M. Mesbahi. Agreement over random network. *IEEE Transactions on Automatic Control*, pages 50(11): 1867–1872, 2005.

[54] R. A. Horn and C.R. Johnson. *Topics in Matrix Analysis.* Cambridge University Press, Cambridge, UK, 1991.

[55] X.L. Hu and H.-F. Chen. Identification for Wiener systems with RTF subsystems. *European Journal of Control*, pages 12(6): 581–594, 2006.

[56] D.W. Huang and L. Guo. Estimation of nonstationary ARMAX models based on the Hannan-Rissanen method. *The Annals of Statistics*, pages 18(4): 1729–1756, 1990.

[57] M. Huang and J. H. Manton. Coordination and consensus of networked agents with noisy measurement: Stochastic algorithms and asymptotic behavior. *SIAM Journal on Control Optimization*, pages 48(1): 134–161, 2009.

[58] I. W. Hunter and M. J. Korenberg. The identification of nonlinear biological systems: Wiener and Hammerstein cascade models. *Biological Cybernetics*, pages 55(2–3): 136–144, 1986.

[59] H. Ishii and R. Tempo. Distributed randomized algorithms for the PageRank computation. *IEEE Transactions on Automatic Control*, pages 55(9): 1987–2002, 2010.

[60] A. Jadbabaie, J. Lin, and S. M. Morse. Coordination of groups of mobile autonomous agents using nearest neighbor rules. *IEEE Transactions on Automatic Control*, pages 48(6): 988–1001, 2003.

[61] A. Juditsky and B. Polyak. Robust eigenvector of a stochastic matrix with application to PageRank. In *Proceedings of 51th IEEE Conference on Decision and Control*, pages 3171–3176. IEEE, New York, NY, USA, 2012.

[62] B.E. Jun and D.S. Bernstein. Extended least-correlation estimates for errors-in-variables nonlinear models. *International Journal of Control*, pages 80(2): 256–267, 2007.

[63] A. Kalafatis, N. Arifin, and L. Wang. A new approach to the identification of pH processes based on the Wiener model. *Chemical Engineering and Science*, pages 50(23): 3693–3701, 1995.

[64] S. Kamvar, T. Haveliwala, and G. Golub. Adaptive methods for the computation of PageRank. *Linear Algebra Applications*, pages 386: 51–65, 2004.

[65] A. Kashyap, T. Basar, and R. Srikant. Quantized consensus. *Automatica*, pages 43(7): 1192–1203, 2007.

[66] P.V. Kokotovic. *Foundations of Adaptive Control.* Springer, Berlin, Germany, 1991.

[67] G. Kollias, E. Gallopoulos, and D. B. Szyld. Asynchronous iterative computations with web information retrieval structures: The PageRank case. In *Parallel Computing: Current and Future Issues of High-End Computing (G. R. Joubert, Ed.)*, pages 309–316. John von Neumann Institute for Computing, German, 2006.

[68] P. R. Kumar and P. Varaiya. *Stochastic Systems: Estimation, Identification and Adaptive Control.* Prentice Hall, Englewood Cliffs, NJ, USA, 1986.

[69] H.J. Kushner and D.S. Clark. *Stochastic Approximation for Constrained and Unconstrained Systems.* Springer-Verlag, NY, USA, 1978.

[70] H.J. Kushner and G.G. Yin. *Stochastic Approximation and Recursive Algorithms and Applications.* Springer, New York, USA, 2003.

[71] A. N. Langville and C. D. Meyer. *Google's PageRank and Beyond: The Science of Search Engine Rankings.* Princeton University Press, Princeton, NJ, USA, 2006.

[72] T. Li, M. Fu, L. Xie, and J.F. Zhang. Distributed consensus with limited communication data rate. *IEEE Transaction on Automatic Control*, pages 56(2): 279–292, 2011.

[73] T. Li and J. F. Zhang. Consensus conditions of multi-agent systems with time-varying topologies and stochastic communication noises. *IEEE Transactions on Automatic Control*, pages 55(9): 2043–2056, 2010.

[74] R. Liptser and A.N. Shiryaer. *Statistics of Random Processes I and II.* Springer-Verlag, New York, USA, 1977.

[75] Z.X. Liu and L. Guo. Synchronization of multi-agent systems without connectivity assumptions. *Automatica*, pages 45: 2744–2753, 2009.

[76] L. Ljung. Analysis of recursive stochastic algorithms. *IEEE Transactions on Automatic Control*, pages 22(4): 551–575, 1977.

[77] M. Loève. *Probability theory (4th ed.).* Springer-Verlag, Inc., New York, NY, USA, 1977.

[78] E. Masry and L. Györfi. Strong consistency and rates for recursive probability density estimators of stationary processes. *Journal of Multivariate Analysis*, pages 22(1): 79–93, 1987.

[79] D. L. McLeish. A maximal inequality and dependent strong laws. *The Annals of Probability*, pages 3(5): 829–839, 1975.

[80] C.D. Meyer. Generalized inverse of block matrices. *SIAM Journal on Applied Mathematics*, pages 18(2): 401–406, 1970.

[81] S. P. Meyn and R.L. Tweedie. *Markov Chains and Stochastic Stability (2nd ed.).* Cambridge University Press, Cambridge, UK, 2009.

[82] A. Mokkadem. Mixing properties of ARMA processes. *Stochastic Processes and Their Applications*, pages 29(2): 309–315, 1988.

[83] B.Q. Mu. *Recursive Identification of Block-Oriented Nonlinear Systems.* PhD Thesis, The University of Chinese Academy of Sciences, 2013.

[84] B.Q. Mu and H.-F. Chen. Recursive identification of Wiener-Hammerstein systems. *SIAM Journal of Control Optimization*, pages 50(5): 2621–2658, 2012.

[85] B.Q. Mu and H.-F. Chen. Recursive identification of errors-in-variables Wiener systems. *Automatica*, pages 49(9): 2744–2753, 2013.

[86] B.Q. Mu and H.-F. Chen. Recursive identification of MIMO Wiener systems. *IEEE Transactions on Automatic Control*, pages 58(3): 802–808, 2013.

[87] B.Q. Mu and H.-F. Chen. Hankel matrices for systems identification. *Journal of Mathematical Analysis and Applications*, pages 409(1): 494–508, 2014.

[88] M.B. Nevelson and R.Z. Khasminskii. Stochastic approximation and recursive estimation. *Translation of Mathematical Monographs*, pages 47: 10–16, 1976.

[89] B. Ninness and S.J. Henriksen. Bayesian system identification via Markov Chain Monte Carlo techniques. *Automatica*, pages 46(1): 40–51, 2010.

[90] E. Nummelin. *General Irreducible Markov Chains and Nonnegative Operators*. Cambridge University Press, Cambridge, UK, 1984.

[91] E. Oja. *Subspace Methods of Pattern Recognition*. Research Studies Press and John Wiley, Letchworth, Hertfordshire, UK, 1983.

[92] E. Oja and J. Karhunen. On stochastic approximation of the eigenvectors and eigenvalues of the expectation of a random matrix. *Journal of Mathematical Analysis and Applications*, pages 106(1): 69–84, 1985.

[93] M. Pawlak. On the series expansion approach to the identification of Hammerstein system. *IEEE Transactions on Automatic Control*, pages 36(6): 763–767, 1991.

[94] K. Pearson. On lines and planes of closest fit to systems of points in space. *Philosophical Magazine Series 6*, pages 2(11): 559–572, 1901.

[95] J. Rissanen. Modeling by shortest data description. *Automatica*, pages 14(5): 467–471, 1978.

[96] H. Robbins and S. Monro. A stochastic approximation method. *The Annals of Mathematical Statistics*, pages 22(3): 400–407, 1951.

[97] J. Roll, A. Nazin, and L. Ljung. Nonlinear system identification via direct weight optimation. *Automatica*, pages 41(3): 475–490, 2005.

[98] Yu.A. Rozanov. *Stationary Random Processes*. Holden-Day, CA, USA, 1967.

[99] A. T. Salehi and A. Jadbabaie. A necessary and sufficient condition for consensus over random networks. *IEEE Transactions on Automatic Control*, pages 53(3): 791–795, 2008.

[100] E. Seneta. *Nonnegative Matrices and Markov Chains*. Springer, 1981.

[101] J. Sjoberg, Q. H. Zhang, L. Ljung, and et al. Nonlinear black-box modeling in system identification: a unified overview. *Automatica*, pages 31(12): 1691–1724, 1995.

[102] T. Söderström. Errors-in-variables methods in system identification. *Automatica*, pages 43(6): 939–958, 2007.

[103] J.C. Spall. Multivariate stochastic approximation using a simultaneous perturbation gradient approximation. *IEEE Transactions on Automatic Control*, pages 37(3): 331–341, 1992.

[104] L. Stefanski and R.J. Carroll. Deconvoluting kernel density estimators. *Statistics*, pages 21(2): 169–184, 1990.

[105] P. Stoica. Generalized Yule-Walker equations and testing the order of multivariate time series. *International Journal of Control*, pages 37(5): 1159–1166, 1983.

[106] P. Stoica, T. McKelvey, and J. Mari. MA estimation in polynomial time. *IEEE Transactions on Signal Processing*, pages 48(7): 1999–2012, 2000.

[107] W. F. Stout. *Almost Sure Convergence*. Academic Press, New York, NY, USA, 1974.

[108] K. Takezawa. *Introduction to Nonparametric Regression*. John Wiley & Sons, Hoboken, NJ, USA, 2006.

[109] H. Tong. *Nonlinear Time Series*. Oxford Univ. Press, Oxford, UK, 1990.

[110] J. N. Tsitsiklis, D. P. Bertsekas, and M. Athans. Distributed asynchronous deterministic and stochastic gradient optimization algorithms. *IEEE Transactions on Automatic Control*, pages 31(9): 803–812, 1986.

[111] M. Verhaegen and D. Westwick. Identifying MIMO Hammerstein systems in the context of subspace model identification methods. *International Journal of Control*, pages 63(2): 331–349, 1996.

[112] L. Wang and L. Guo. Robust consensus and soft control of multi-agent systems with noises. *Journal of Systems Science and Complexity*, pages 21(3): 406–415, 2008.

[113] L.Y. Wang, G. Yin, J.F. Zhang, and Y.L. Zhao. *System Identification with Quantized Observations*. Birkhäuser, Basel, Swiss, 2010.

[114] W. Wang and R. Henrikson. Generalized predictive control of nonlinear systems of the Hammerstein form. *International Journal of Modelling, Identification, and Control*, pages 15(4): 253–262, 1994.

[115] A. Wills, T. Schon, L. Ljung, and B. Ninness. Identification of Hammerstein-Wiener models. *Automatica*, pages 1(1): 1–14, 2011.

[116] C. S. Withers. Conditions for linear processes to be strong-mixing. *Z. Wahrscheinlichkeitstheorie verw. Gebiete*, pages 57(4): 477–480, 1981.

[117] C.W. Wu. Synchronization and convergence of linear dynamics in random directed networks. *IEEE Transactions on Automatic Control*, pages 51(7): 1207–1210, 2006.

[118] W. X. Zhao, H. F. Chen, and W. X. Zheng. Recursive identification for nonlinear ARX systems based on stochastic approximation algorithm. *IEEE Transaction on Automatic Control*, pages 55(6): 1287–1299, 2010.

[119] W.X. Zhao, H.-F. Chen, and H.T. Fang. Convergence of distributed randomized PageRank algorithms. *IEEE Transactions on Automatic Control*, pages 58(12): 3255–3259, 2013.

[120] W.X. Zhao, H.-F. Chen, and T. Zhou. New results on recursive identification of NARX systems. *International Journal of Adaptive Control and Signal Processing*, pages 25(10): 855–875, 2011.

[121] Y.C. Zhu. Distillation column identification for control using Wiener model. *Proceedings of American Control Conference*, pages 55: 3462–3466, 1999.

[122] Y.C. Zhu. Identification of Hammerstein models for control using ASYM. *International Journal of Control*, pages 73(18): 1692–1702, 2000.

Index